Biomineralization and Biological Metal Accumulation

Biological and Geological Perspectives

Heinz A. Lowenstam 1983
for the Geology Library

Biomineralization and Biological Metal Accumulation

Biological and Geological Perspectives

Papers presented at the Fourth International Symposium on Biomineralization, Renesse, The Netherlands, June 2-5, 1982

edited by

P. WESTBROEK

and

E. W. DE JONG

Department of Biochemistry, State University of Leiden, The Netherlands

D. REIDEL PUBLISHING COMPANY

Dordrecht, Holland / Boston, U.S.A. / London, England

Library of Congress Cataloging in Publication Data

International Symposium on Biomineralization
(4th : 1982 : Renesse, Netherlands)
Biomineralization and biological metal accumulation.

Includes index.
1. Biomineralization–Congresses. 2. Mineral cycle (Biogeochemistry)–Congresses. I. Westbroek, P. (Pieter), 1937– II. De Jong, E. W. (Elizabeth Wijnanda), 1946– III. Title.
QH512.I566 1982 574.19'214 82-20416
ISBN 90-277-1515-7

Published by D. Reidel Publishing Company,
P.O. Box 17, 3300 AA Dordrecht, Holland.

Sold and distributed in the U.S.A. and Canada
by Kluwer Boston Inc.,
190 Old Derby Street, Hingham, MA 02043, U.S.A.

In all other countries, sold and distributed
by Kluwer Academic Publishers Group,
P.O. Box 322, 3300 AH Dordrecht, Holland.

D. Reidel Publishing Company is a member of the Kluwer Group.

Printed in The Netherlands

TABLE OF CONTENTS

PREFACE

Biominerals are generated by the subtle interaction of biological organization and mineral growth. They belong both to the living and the inanimate world and as such their genesis is among the most intriguing and fundamental subjects in science. However, the conceptual and technical resources that are available in physical chemistry and in the biological sciences is often inadequate for the elucidation of the problems involved, and hence this field is particularly difficult to explore. This may be an important reason why fundamental research on biomineralization mechanisms has traditionally been carried out by a comparatively small group of scientists.

There are signs, however, that the situation is ripe for a change. Various meetings on biomineralization have been organized in the last few years, particularly in the medical sector. It is generally felt that further developments in the therapy of bone and tooth diseases will be largely dependent on an improved understanding of the fundamental underlying mechanisms of biomineralization.

In geology and ecology there is also a growing interest in this field. A large variety of minerals are now known to be formed and accumulated by biological systems. Several of these mechanisms are operating at an immense scale. They contribute enormously to the genesis of many sedimentary rocks and are likely to have a significant impact on the cycling of materials through the outer layers of the Earth. There is, therefore, a growing appreciation of the value that biochemical studies on biomineralization will have for the understanding of important geological processes, such as the genesis of limestones and of sedimentary ores.

A third important and vary new aspect of these studies on biological systems that are capable of metal accumulation, is that they may become exploited technologically, *e.g.* in the retrieval of heavy metals from dilute solutions, or the management of metal fluxes in the environment. There is therefore, even reason to believe that in the future biomineralization research will come to the forefront of the natural sciences as a fascinating interdisciplinary field where a wide spectrum of biological, medical, geological and technological efforts will be integrated.

P. Westbroek and E. W. de Jong (eds.), Biomineralization and Biological Metal Accumulation, ix–xi.

The meeting which is reported in the present volume had this integration as one of its major objectives. From the 2nd to the 4th June 1982, an unusual amalgam of some 70 outstanding scientists with widely different backgrounds settled in the pleasant seclusion of Renesse, a holiday resort on the Dutch coast. Among them were biochemists, microbiologists, bone and tooth physiologists, specialists on calcification in invertebrates and plants, bioinorganic and physical chemists, geochemists, paleontologists, geologists, oceanographers, atmospheric scientists and environmental biotechnologists. Traditional lines of demarcation between these fields seemed to vanish during the meeting as the participants discovered their common interests in heated debates. After the formal gathering a smaller party visited the Ardenne Mountains where Professor Claude Monty (University of Liège, Belgium) convincingly demonstrated the involvement of microorganisms in the genesis of numerous mounts of limestone mud of Upper Devonian age.

We had deliberately announced this meeting as the Fourth International Symposium on Biomineralization, because we wished to give a continuation to the excellent tradition set forth by three earlier symposia in this series: the first being held in Mainz (Federal Republic of Germany) in 1970, the second in Georgetown (South Carolina, U.S.A.) in 1974 and the third in Kashikojima (Japan) in 1977. On these occasions workers on calcium carbonate skeletons in invertebrates and plants were given an opportunity to interact. The Biomineralization Newsletter, skilfully edited by Professor K. Simkiss in Reading, is another outcome of these earlier initiatives. As a consequence, the biosynthesis of calcium carbonate skeletons was also the most central issue in this conference. Although a limited number of contributions on bone, tooth and renal stone formation were presented, a full scale interaction with the medical disciplines was avoided at this stage, since we hope that this will form the basis of a separate subsequent meeting.

Instead we have given special emphasis to the contributions of the geological aspects of biomineralization research. This combined elegantly with a section on the biological accumulation of metals other than calcium, including some systems that lead to other types of mineral formation. The geological implications of some of these mechanisms were clearly highlighted at this meeting and a discussion was held on biotechnological perspectives for environmental management and heavy metal retrieval.

The overall impression was that the interdisciplinary approach advocated in this meeting is a most promising avenue for further research. But it was also realized that we are at the beginning of a long and laborious process of integration. All too often more pertinent questions were raised than could be answered.

We are especially indebted to Professor Lynn Margulis (Boston University, U.S.A.) for her generous and enthusiastic support. Without her intervention and the help of Drs. Ken Nealson and John Stoltz this meeting would not have taken place. Professor C.J.F. Böttcher, Emeritus from the University of Leiden, not only gave us the self-confidence to carry out this experiment by many encouraging discussions, but also used his influence to obtain financial support and contributed in the evaluation of the results. Professors L. Bosch and A. Brouwer (University of Leiden, The Netherlands) gave valuable suggestions and saved us from many mistakes. Professor K. Simkiss reviewed some of the papers in this book, he and Professor R.H. Kretsinger gave some very useful ideas. Professor J.E. Van Hinte (Free University, Amsterdam, The Netherlands) paved the road for indispensable financial support. Margriet Nip and Ineke Boogaard generously helped in the preparations and the organization of this symposium, as did Ton Borman and Paul Van der Wal. The committee of recommendation consisted of Professors L. Bosch, C.J.F. Böttcher, A. Brouwer, J.E. Van Hinte, J. Joosse, J. Lever, J. Reedijk and A.W. Schwartz.

The symposium could take place under the auspices of the Royal Netherlands Academy of Sciences and Arts, thanks to the intervention of the "Commissie voor de Geologische Wetenschappen" (secretary: Dr. H.J.W.G. Schalke). Financial support was obtained from NASA headquarters, Washington (U.S.A.); the Leiden University Fund; The Ministry of Sciences and Education of the Netherlands (this grant was obtained through the intervention of the Royal Netherlands Academy of Sciences and Arts); Shell Research Centre in Sittingbourne (England); Billiton International Metals bv, The Hague (The Netherlands); and the Departments of Biochemistry and Geology, State University of Leiden (The Netherlands).

P. Westbroek

E.W. De Jong

SHINJIRO KOBAYASHI
(1907 - 1981)

DEDICATED TO THE MEMORY OF

SHINJIRO KOBAYASHI

Shinjiro Kobayashi was born October 2, 1907 and died May 27, 1981. He received his doctorate in biology in 1945 from Tokohu University. He served as a director in the institute of the Fuji Pearl Co., Mie Prefecture, between 1946 and 1950 and as a professor in the Faculty of Fisheries in Hokkaido University, Idakodate, between 1951 and 1971. He was a Professor Emeritus of Hokkaido University. Between 1965 and 1967 he was Dean of the Faculty.

He was an organizer for the scientist group on Biomineralization in Japan until his death, and a chairman of the organizing committee for the 3rd International Biomineralization Symposium held at Kashikojima, Mie Prefecture, in 1977.

During a life-span of 73 years he made great contributions to a very wide variety of biological problems, especially the distribution and characteristics of earthworms in Korea, Japan and China and also pearl, shell and fish scale formation. In 1959 he published a book entitled 'Scientific Study on Pearl' (in Japanese) together with N. Watabe. Even now this book is playing a leading part in scientific studies on pearl in Japan.

INTRODUCTION

BIOLOGICAL METAL ACCUMULATION AND BIOMINERALIZATION IN A GEOLOGICAL PERSPECTIVE

P. Westbroek
Department of Biochemistry, State University of Leiden
Wassenaarseweg 64, 2333 AL Leiden, The Netherlands

METAL-CELL INTERACTIONS AND BIOMINERALIZATION

The term "biomineralization" covers a bewildering variety of phenomena. The present state of our knowledge is summarized by Lowenstam and Weiner in this volume. Not only are many minerals produced by organisms, but also is the degree of biological control widely different in the various systems. It is likely that our present survey is still far from complete, particularly if it is considered how little attention has been given to biomineralization in prokaryotes.

Even more complex is the situation with respect to the biological accumulation of metals by adsorption, absorption or other mechanisms. There is a vast literature on this subject, but as yet no comprehensive survey is available. Organisms are even known to be capable of scavenging and accumulating metals that normally do not exceed the ppb range in their environment. Only in a limited number of cases have the mechanisms been studied in more detail.

Figure 1 is a schematic representation of some important interactions between a metal and the cell of a prokaryote. Metals are known to perform many different functions in the cellular organization (Figure 1, F). It has been argued by Faústo da Silva and Williams (1978) that as a rule only those elements are used that are readily available in the environment. Any role that the others might perform can be done as well by a common element. Thus, of only some 25 elements a biological function has been described. For the useful elements there is an optimum concentration at which the function is best performed; at higher concentrations these substances may become toxic. Elements that have no function, such as Cd, Hg, Pb are often deleterious at any concentration. The maintenance of optimum intracellular concentrations of the useful elements and the removal of redundant and toxic materials is a major function of the cellular organization if it is to survive fluctuating environmental conditions.

Transport systems into and out of the cell exist (Figure 1). In general, there will be a requirement for the latter to be more specific than the former. So, arsenate can enter the cell by the same route as phosphate, but in order to avoid phosphate depletion the outwardly

P. Westbroek and E. W. de Jong (eds.), Biomineralization and Biological Metal Accumulation, 1–11.

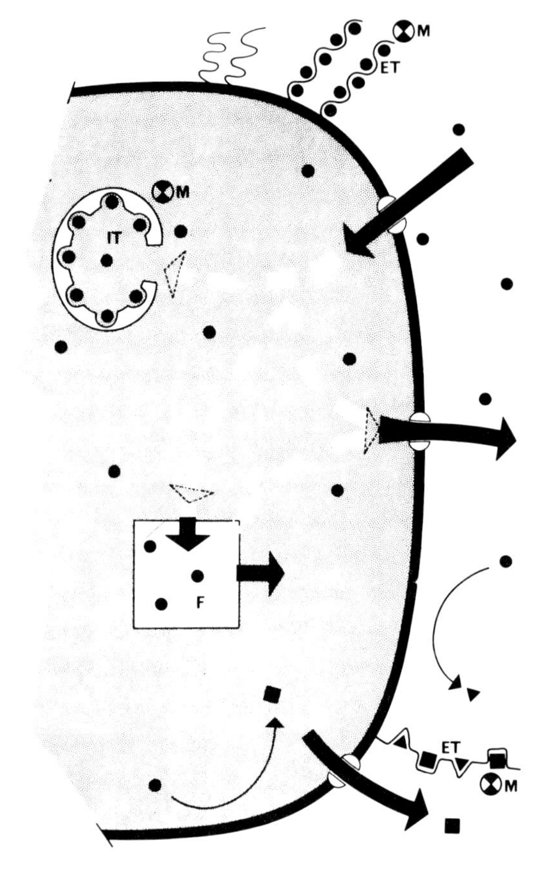

Figure 1. Some interactions between a metal and a simple cell.
F: compartment or site where the metal performs a function; IT: intracellular trap; ET: extracellular trap.
● → ■ : intracellular transformation of metal;
● → ▲: extracellular transformation of metal;
⊗: possible site of metal accumulation; M: possible site of mineralization.
For further explanation see text.

directed pump has to be selective: only arsenate must be removed (Silver, this volume). Another method for the removal of redundant or deleterious metals is their capture by intracellular traps (IT, Figure 1). Metallothioneins are a point in case. These small, cysteine-rich proteins are capable of binding a variety of metals, notably Cu, Zn, Cd and Hg (Vasák and Kägi, this volume); this may not only be functional in the regulation of the intracellular concentrations of the more beneficial elements in this group, but also as instruments of detoxification.

In Figure 1 it is suggested that specific pathways may exist intracellularly by which the metals can be conducted to selected sites without disturbing other cellular activities, in a similar way as the electrons that pass through the wiring of a telephone network. One could think of specific macromolecular tunnels, or tracks along the surface of membranes. There is no convincing experimental evidence available that would support this hypothesis. However, the possibility is attractive: such linear transport systems might only have to be energized on one point along the line, and the supply of ions to the target site could be regulated by breaking and re-connecting the chain.

Before the metal ions can enter the cell through the surrounding cell membrane (by far the most selective barrier they have to traverse), they must travel through a more or less complex macromolecular network: the cell wall or the glycocalyx. This may function as an extracellular trapping device (ET, Figure 1). The traps may function in detoxification by keeping the metals from reaching the cell membrane. Also, it is conceivable that extracellular stores of potentially useful metals are formed in this way. Yet, another hypothetical function of this metal trapping would be the induction of an organized response in the cell, triggered off by a conformational change in the recipient extracellular macromolecules. Various detoxifying mechanisms, such as the production of metallothioneins, Hg^{++} reductase and outwardly directed pumps are known to be induced by the exposure of cells to metals (see *e.g.* Silver, this volume), but it is not known whether the extracellular materials are involved in this process.

Finally, biological systems are known to cause transformations of a variety of metals that they may encounter, in particular redox changes and demethylations. Some of these mechanisms serve for detoxification (*e.g.* the demethylation and subsequent reduction of mercury (see Silver, this volume), but in other cases the transformation appears to be an energy source for the cell (*e.g.* the oxidation of Mn^{2+} to Mn^{4+}, see Nealson, this volume). But the function of these transformations is not always clear (*e.g.* Fe^{2+} and Mn^{2+} oxidation in *Leptothrix*, Hirsch, personal communication). The transformed metals may or may not be bound to extracellular polymers.

In Figure 1 the most likely sites for substantial accumulations of metals are indicated with asterisks: they are the intra- and extracellular macromolecular traps. Indeed, there is an abundant - albeit dispersed - literature on concentrations of many metals on such sites. Under specific conditions such local accumulations may give rise to crystallization of a particular metal compound. The macromolecular traps may then assume the function of heterogenous nucleators of the crystallization and they may also acquire a role in further crystal growth. Such incipient biomineralizations are common features, particularly in the extracellular space of prokaryotes. In eukaryotes specific compartments may be created where the process can take place in isolation. In coccolithophorids and diatoms, *e.g.*, the mineralizing systems are internalized in intracellular vacuoles (De Jong *et al.*; Volcani, this volume); in molluscs and other invertebrates the final stages of skeleton formation are performed in specific extra-epithelial spaces (*e.g.* Nakahara, this volume). This compartmentation provides a high degree of biological control and it leads to the production of the exquisite skeletons that abound in nature. In this situation the original macromolecular traps may also have been internalized and they may have acquired specialized regulatory functions in mineral production (see Weiner and Lowenstam; Krampitz; and Mann *et al.* for comments).

If this hypothetical scenario is correct, the original macromolecular traps form a major common denominator for metal accumulation and biomineralization. A comparative study of these materials in a broad variety of systems seems to be a promising avenue for further research.

Because of their high reproducibility many of these minute biochemical systems are known to have a distinct effect on the environment.

In the past they have profoundly affected the geological record, and the question may be asked if they may even have played a considerable role in the development of the planet Earth as a whole. It is one of the major objectives of this symposium to contribute to the discussion on the geological dimensions that these biochemical systems may have. It is not my intention to discuss these matters in any detail. Rather, I wish to present a panoramic view of a few underlying principles and so contribute to the mutual understanding of interested biochemists and geologists.

TOWARDS A GLOBAL VIEW OF THE PLANET EARTH

In the last two decades our interpretation of the geological record has acquired a planetary perspective. Improvements in dating technology, both radiometric and paleontological, have resulted in a refined chronological framework, especially for the last few hundred million years. From this a world-wide correlation of the geological history is emerging, and developments in the past that have affected the Earth as a whole can be distinguished from more local events.

A major impetus towards the acquisition of a global perspective in the geological sciences was the theory of plate tectonics which came into prominence in the late sixties (see Williams, 1970, for a historical survey). Otherwise unrelated phenomena could now be understood as resulting from a single mechanism. The principle is epitomized in Figure 2. Oceanic crust is constantly being created at the mid-ocean ridges by flows of lava from the earth's interior. This veneer of basaltic rock

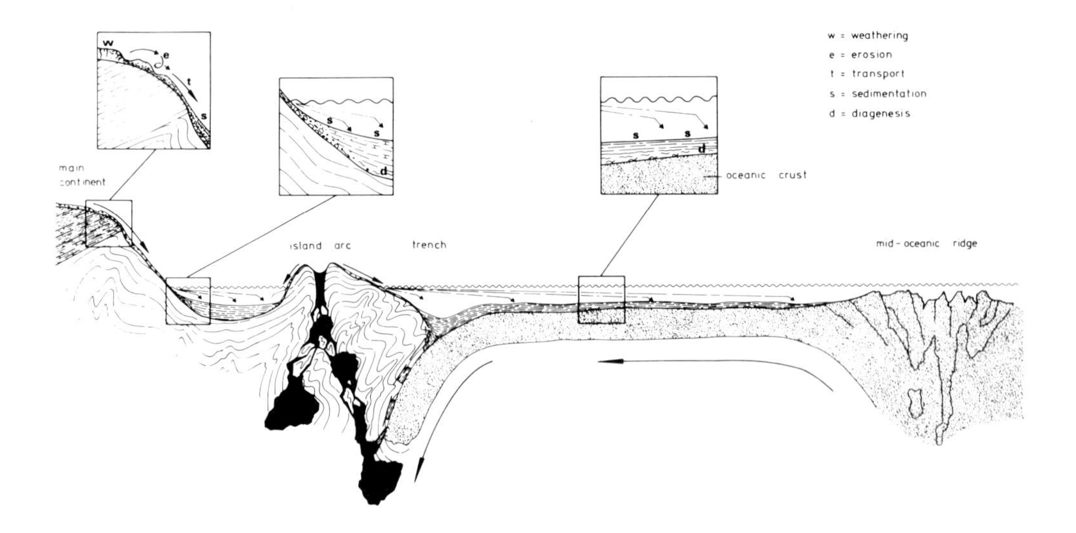

Figure 2. Schematic representation of the geological cycle. For further explanation see text.

rides along on top of the denser mantle with a velocity of about 10 cm per annum. It then plunges at a subduction zone into the deeper recesses of the mantle. The thicker and lighter continents are stable elements in this shifting scene: the oldest continental material has an age of approximately 3.8 billion years, whereas no oceanic crust of more than 200 million years is known. The continents float as scum on the mantle, and large-scale contamination of continental and mantle materials is unlikely. At the subduction zones (trench, Figure 2) the light-weight continent-derived sediments that have accumulated on the ocean floor are scraped off the oceanic basalt, are compressed and re-added to the continental margin; residual lumps of rock and interstitial water that are pulled into the mantle are released to the surface by explosive volcanism.

As a major consequence of plate tectonics, the material of the continental crust is continuously recycled. On the land surface the rocks are rapidly weathered and eroded away; they are then transported by wind, rivers or glaciers to lower regions where they may be sedimented. Through repeated cycles of this kind they end up on the floor of the continental shelves. Finally, they may slide down the continental slopes in spasmodic discharges called turbidity currents and spread out over the ocean floor.

But the atmosphere and the hydrosphere do not only act as transport vehicles for the crustal material; there is also intense chemical interaction that profoundly affects the composition of all three media. Many constituents of rocks are dissolved in water or air and re-assembled elsewhere into new mineral constituents, such as limestone, chert, gypsum and rock salt. Some of these may be sedimented in large amounts on the ocean floor. From there the accumulated deposits are transported back to the continents, whereby the oceanic crust serves as a bandwagon.

Those processes in the cycle that are driven by solar radiation, *viz.* weathering, erosion, sedimentation and early diagnesis (early changes in sediments, such as "hardening") are collectively indicated by the term "exogenic cycle". The "endogenic cycle", on the other hand, is energized by radioactive decay in the earth's interior; it comprises plate tectonics, metamorphism, melting and solidification of rocks, mountain formation, volcanism and other, more or less related phenomena.

It has often been assumed that the entire system in which the continental crust is recycled is in a steady state. The atmosphere, hydrosphere and crust of the earth are considered as reservoirs into which and out of which the constituent materials are transported. As a consequence of the putative steady state the flux of a constituent into a reservoir must be equal to the corresponding outwardly directed one.

It should be understood that this image of a closed steady state global cycle is only a crude approximation of the actual situation. If the cycle were perfectly closed no record of the distant past would be preserved; significant departures from the steady state are known to have occurred in the geological past; and the possibility of additions from the mantle to the crust and *vice versa* is not taken into account in this model (see Garrels and Berner, this volume).

LIFE AS A GEOLOGICAL PROCESS

According to the traditional geological view living systems have had only little effect on the development of the earth. The conditions on the outer side of the planet were considered to be determined largely by physical and chemical parameters; living systems were thought to have no other choice than to adapt to the given environment, or to perish. Today, there is an abundant literature testifying the profound effect that the biota have on the exogenic cycle (see W.E. Krumbein, 1978; Trudinger and Swaine, 1979; Fenchel and Blackburn, 1979; Ehrlich, 1981; Holland and Schidlowski, 1982; Margulis, 1982).

Figure 3 shows the biological cycling of carbon through photosynthesis and aerobic respiration. In photosynthesis, the energy of solar radiation is transformed into a redox potential: organic carbon (CH_2O) and oxygen are produced out of CO_2 and water. The elimination of this redox potential is the driving force for respiration. In principle, it could be a closed cycle, but in the natural environment this is not the case. Less than 1% of the produced CH_2O is trapped in sediments and buried in the crust (Garrels *et al.*, 1976). This reduced material is then carried through the geological cycle, until it becomes exposed at the surface, is weathered, and so undergoes a delayed reaction with atmospheric oxygen.

Plate tectonics is a sluggish process: the mean residence time of organic carbon in the crust is estimated to be in the order of 400×10^6 years. So, in spite of the fact that only a minute fraction of the organic carbon escapes the photosynthesis - respiration cycle, a lithospheric reservoir as big as 10^{21} moles organic carbon exists. It is generally agreed upon that the atmospheric reservoir of oxygen has originated in response to this accumulation of CH_2O in the crust. Examples of

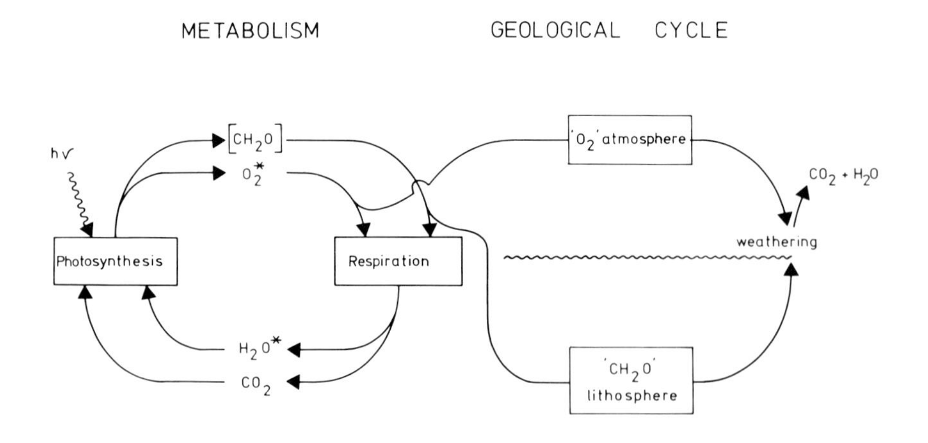

Figure 3. Relation between metabolic and geological cycling of carbon and oxygen. For further explanation see text.

other major sinks of the oxidation power that was (and still is) released are the crustal reservoirs of gypsum ($CaSO_4 \cdot 2H_2O$), ferric oxides and nitrate. But also the reducing power of organic carbon is transferred in part to other substances. So, ferric iron, sulphate and nitrate can be converted into ferrous iron, sulphide and molecular nitrogen, respectively. The redox potential, actively maintained by photosynthesis, results in a multitude of chemical forms on earth and turns the outer domains of the planet into an energized and highly reactive medium that is far away from thermodynamic equilibrium. The reactivity of this system is well illustrated by the fact that the residence time of free oxygen in the atmosphere is of the order of 2000 years.

It is interesting to note that the operation of the endogenic cycle may well be essential for the prolonged maintenance of life on earth, although it is not immediately involved in biological processes by itself. In the first place, it provides the biota with a constant supply of raw materials and nutrients: many of its products are unstable and easily degraded, once they have reached the earth's surface. If the interior of the earth were in a resting state the available supplies would be converted gradually into an unextractable form (see later). Secondly, if organic carbon were not removed from the surface and transported into the crust, the redox potential on which the present extensive biological activity depends could never have built up. From a geological point of view, life is merely an elaboration of the exogenic cycle: it is a complex network of pathways through which many of the products of the endogenic cycle are conducted before they re-enter into the crust.

Recently, the idea has been put forward that the lower part of the atmosphere, the hydrosphere and the upper part of the crust are actively modulated by the biota so as to maintain optimum conditions for life on earth, an idea known as the gaia hypothesis (see Lovelock; Margulis and Stolz, this volume; see, however, Garrels and Berner, this volume). The development of the biota and their environment is perceived as an integrated co-evolutionary process (Dubos, 1979), analogous to a living organism persisting since the origin of life. If this hypothesis is correct, it is likely to have important geological implications. The exogenic domain would - to some extent - be orchestrated by the gaian organization, and individual processes might be functional, in principle, in the system's tendency towards homeostasis. The study of gaian evolution might become an important aspect of geology.

The gaia hypothesis is the most outspoken idea available on the geological involvement of life. It is attractive by its conceptual simplicity and integrating power, but, as it stands, it is still largely based on indirect evidence and on analogy (*e.g.* with cybernetic systems leading to homeostasis in the individual organisms, see, however, Lovelock, this volume).[x]

[x] So far, the proponents have especially concentrated on the Precambrian in their quest for geological arguments. In the future, some of this attention might be focussed on the dramatic last episode of earth history - the Pleistocene; it is known in much greater detail and may lead to interesting clues, especially on biological mechanisms of global temperature control.

By virtue of its global and generalizing perspective the gaia hypothesis easily gives rise to dogmatic and religious overtones. This tendency might become detrimental to the scientific development of the idea. Instead, it should be taken for what it pretends to be: a hypothesis, *i.e.* a new concept (biological regulatory systems operating at a global scale) and a challenge to explore its scientific validity. It is with this intention that the gaia hypothesis was given full consideration during this symposium. Biological metal accumulation and biomineralization are undoubtedly important processes in the global ecology and an attempt was made to prepare the ground for an assessment of the possible role of these mechanisms in global homeostasis.

SOME REMARKS ON BIOLOGICAL METAL ACCUMULATION AND BIOMINERALIZATION AS GEOLOGICAL PROCESSES

Mechanisms of biological metal accumulation and biomineralization have important effects, both on the natural environment and on the geological record. Various examples are treated in this volume. In this context I can only make a few generalizing remarks.

Virtually all the calcium carbonate that is formed both on the land surface and in the oceans is produced by living organisms, and it is likely that most of the $CaCO_3$-reservoir of the crust is derived from biologically produced limestone (Monty, personal communication). The biological involvement in the formation of limestones has a distinct effect on the chemical constitution of these rocks (*e.g.* trace metal content and association with organic materials) and on their structure. Also, the geographical distribution of limestone deposits is profoundly influenced by the biota. Deepsea accumulations of calcium carbonate, *e.g.*, would not be formed at the present scale without the involvement of pelagic calcifying organisms, such as coccolithophorids (see de Jong *et al.*, this volume), foraminifers (see Erez, this volume) and pteropods. Coral reefs form important barriers to marine erosion in the tropical parts of the oceans. Effects of biological calcification on the distribution of calcium and carbonate in the ocean are discussed by Whitfield and Watson (this volume).

The cycle of silicon is profoundly effected by the biota (Volcani; Whitfield and Watson, this volume; see also reviews by Silverman (1979) and Oehler (1979)), and the formation of opaline skeletons by various organisms is an important factor controlling the ecological distribution of silicon. It also has a distinct effect on the chemistry, the structure and the geographical distribution of silica-containing sedimentary rocks.

There is a vast literature on the effects of biological systems, especially microorganisms, on the distribution of heavy metals in the environment. Many studies have concentrated on natural conditions, others on artificial situations, such as activated sludges. There are also many reports on laboratory studies where these processes have been investigated in isolated microbial strains (Nealson; Brierley and Brierley, this volume; Sterritt and Lester, 1980; Kelly *et al.*, 1979; Ernst, 1981).

In spite of the overwhelming evidence that is available of the accumulatory potential in living systems, the biological involvement in the genesis of fossil stratified ores has only been recognized in a few instances (*e.g.* Dexter-Dyer Grosowski, this volume). Recent analogs of most ore-types, where the accumulatory mechanisms can be immediately studied, are hardly known. See Holland and Schidlowski (1982) for discussions.

One example where an important ecological effect of biological accumulation has been reported concerns a Canadian lake where industrial discharges of dissolved uranium appear to be bound by diatoms and humic acids and subsequently sedimented (Degens *et al.*, 1979). As a result, the uranium content of the lake water was 20 ppb, as opposed to a concentration of 200 ppm in the sediment.

OUTLOOK

Figure 4a represents the interrelation between the biosphere and the geological cycle. Raw materials and extracted as nutrients from the environment, utilized, and dispersed in a non-extractable form. Ultimately, it is in the endogenic cycle that the materials are reassembled into minerals that are unstable when they return to the surface, so that they can be re-used as raw materials in the next cycle. In the natural environment these processes must be neatly balanced: the supply of raw materials is the ultimate limiting factor for the proliferation of life. From this point of view evolution may be regarded as the progressive elaboration of the system that mediates this conversion of nutrients

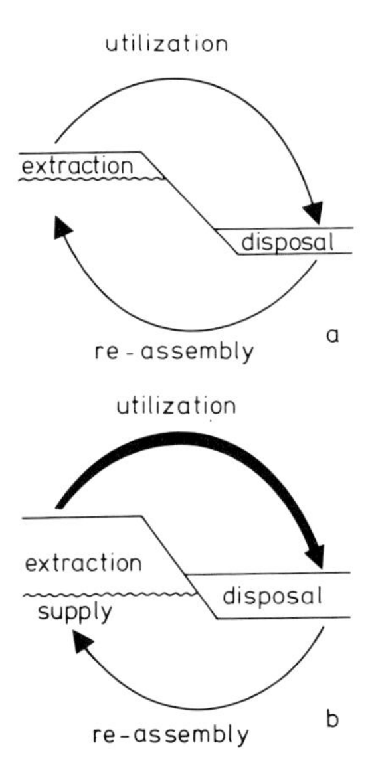

Figure 4. (a) Interrelation between the biosphere and the geological cycle. The supply by the endogenic cycle of raw materials for life is balanced with the disposal of the waste products of the biosphere.
(b) Imbalance in this system due to human culture.
For further explanation see text.

into refuse: the extractive mechanisms have undergone major improvements, especially since the continents became inhabited by plants in the Silurian, and also the network of pathways through which the nutrient materials are recycled has become more and more sophisticated since the origin of life.

The explosive development of human culture has brought about a dangerous imbalance between the supply of fresh resources and the disposal of waste (Figure 4b). Technology has concentrated on the extraction of raw materials in response to growing demands; the sophistication of the system of recycling that is so elaborately developed in the living world has been seriously neglected. The heavy metals from a particularly threatening problem for humanity: supplies are close to being exhausted, while on the dispersal side poisonous wastes are accumulating.

It is obvious, that, in order to obtain a balance between metal wastes and supplies, we are compelled to follow the same strategy as evolution: an elaborate network of pathways has to be created through which the metals are recycled before they can be dispersed. It seems a matter of course that use should be made of the wealth of biological mobilizing and accumulating mechanisms that is available in our surroundings (Brierley and Brierly, this volume).

The study of biological metal accumulation, and in particular of biomineralization, is no longer a matter of only medical and academic importance. In a broader context, comprising geology and technology, a deeper understanding of these subjects may be helpful in grasping the workings of our own planet and may lead to the application of these mechanisms in the adaptation of human wastes to natural supplies.

REFERENCES

Degens, E.T., Van Bonsart, G., How Kin Wong, Khoo, F., and Dickmann, M.D.: 1979. Environmental parameters responsible for the fixation of uranium in recent sediments: test area Bow Lake, Ontario, Canada. Mitt. Geol.-Paläont. Inst. Univ. Hamburg 49, pp. 27-60.

Dubos, R.: 1979. Gaia and creative evolution. Nature 282, pp. 154-155.

Ehrlich, H.L.: 1981. Geomicrobiology, XIII, + 393 pp. Dekker, New York-Basel.

Ernst, W.H.O. (ed.): 1981. Heavy Metals in the Environment. International Conference, Amsterdam.

Faústo da Silva, J.J.R., and Williams, R.J.P.: 1978. The uptake of elements by biological systems. In: New Trends in Bio-inorganic Chemistry (Williams, R.J.P., and Faústo da Silva, J.J.R., eds.), Academic Press, London.

Fenchel, T., and Blackburn, T.H.: 1979. Bacteria and mineral cycling XII, + 225 pp. Academic Press, London.

Garrels, R.M., Lerman, A., and MacKenzie, F.T.: 1976. Controls of atmospheric O_2 and CO_2, past, present and future. Amer. Sci. 64, pp. 306-315.

Holland, H.D., and Schidlowski, M. (eds.): 1982. Mineral Deposits and the Evolution of the Biosphere, X, + 333 pp. Springer, Berlin.

Kelly, D.P., Norris, P.R., and Brierley, C.L.: 1979. Microbiological methods for the extraction and recovery of metals. In: Microbial Technology: Current State, Future Prospects (Bull, A.T., Ellwood, D.C., and Ratledge, C., eds.), Cambridge University Press.

Krumbein, W.E. (ed.): 1978. Environmental biogeochemistry and geomicrobiology. 3 volumes, + 1055 pp. Ann Arbor Science, Mich.

Sterritt, R.M., and Lester, J.N.: 1980. Interactions of heavy metals with bacteria. Science Total Environment 14, pp. 5-17.

Margulis, L.: 1982. Early Life. XVI, p. 160. Science Books, Boston.

Oehler, J.H.: 1979. Deposition and diagnesis of biogenic silica. In: Biogeochemical Cycling of Mineral-forming Elements (Trudinger, P.A., and Swaine, D.J., eds.), Elsevier, Amsterdam, pp. 467-483.

Silverman, M.P.: 1979. Biological and organic chemical decomposition of silicates. In: Biogeochemical cycling of Mineral-forming Elements (Trudinger, P.A., and Swaine, D.J., eds.), Elsevier, Amsterdam, pp. 445-465.

Trudinger, P.A., and Swaine, D.J. (eds.): 1979. Biogeochemical Cycling of Mineral-forming Elements, Elsevier, Amsterdam, VIII, + 612 pp.

Williams, J.T. (ed.): 1970. Continents adrift. Readings from Scientific American. Freeman, San Francisco.

Part I

Global Cycling and Biomineralization

GAIA AS SEEN THROUGH THE ATMOSPHERE

J.E. Lovelock
Coombe Mill Experimental Station
St Giles on the Heath
Launceston, Cornwall, England PL15 9RY

Abstract. Life can flourish only within a narrowly circumscribed range of physical and chemical states and since life began the Earth has kept within this range. This is remarkable for there have been major perturbations such as a progressive increase in solar luminosity, extensive changes in the surface and atmospheric chemical composition and the impact of many planetesimals. The anomalous and chemically unstable composition of the Earth's atmosphere when compared with those of the other terrestrial planets was the first indication of homeostasis by the biota to maintain conditions favourable for their continued survival. This paper will discuss recent evidence in support of the gaia hypothesis and present a simple model of a planetary ecosystem in which homeostasis is a direct and automatic result of the characteristic properties of life.

It is now just over ten years since the first paper on the gaia hypothesis was published, Lovelock and Margulis (9). Many investigators including Redfield (13), Hutchinson (8) and Sillen (15) had previously noted the strong influence of the biota on the composition and properties of the Earth. Like these earlier attempts to unify the biological and geochemical approaches to understanding the Earth, the gaia hypothesis has tended to be ignored rather than criticised by geochemists, almost as if Aristotle still ruled and anything moving towards a circular, even a nonlinear, argument was forbidden. Gaia which uses the circular reasoning of cybernetics was taken to be teleological. Two biologists, Ford Doolittle (4) and Dawkins (3) were prepared to criticise the hypothesis openly on the grounds that it was contrary to the expectations of natural selection but even they did not or were unable to express their views in main stream scientific literature. The purpose of this paper is to try to demystify the gaia hypothesis and to illustrate it by a simple model which is in a form more acceptable for discussion and criticism. The discussion will be drawn mostly from obervations about the composition, properties and regulation of the atmosphere. Lynn Margulis provides in the next paper a parallel account of the gaia hypothesis taken in the context of the biogeochemistry of the sediments.

P. Westbroek and E. W. de Jong (eds.), Biomineralization and Biological Metal Accumulation, 15–25.

WHAT IS GAIA

The gaia hypothesis arose directly from the planetary exploration programme of NASA. There was a need to discover in advance of a landing mission whether or not a planet such as Mars bore life. In 1966 Hitchcock and Lovelock (6) were able to show that information on the atmospheric composition of a planet was sufficient as prima facia evidence of life. The method was based on the high probability that planetary life, through its use of the atmosphere, would drive the chemical composition of this medium far from the near equilibrium steady state of a lifeless planet. This detection method when applied to Mars strongly indicated it to be barren, a conclusion highly unacceptable to exobiologists at that time. The same method applied to the Earth indicated the near certainty of the presence of life. It also suggested that the atmosphere was more than just a biogeochemical mixture. It appeared to be actively maintained in composition at close to an optimum by and for the biota. This way of thinking about the planets was a stunning discouragement for exobiologists whose scientific inspiration came from the search for life, outside the Earth. Some part of the tendency to ignore the joyous counterpart of this approach, the discovery of Gaia perhaps arose from their disappointment.

The evidence drawn from atmospheric compositions which points to life and to a control system on a planet is summarised in Figures 1 and 2, which show the abundances and the fluxes respectively, of the gases of the present atmosphere compared with those of an abiological Earth. From these diagrams it is clear as has been argued in previous papers (9,10,12) that the atmosphere is a highly reactive mixture which would but for life rapidly revert to the stable inert condition of the abio-

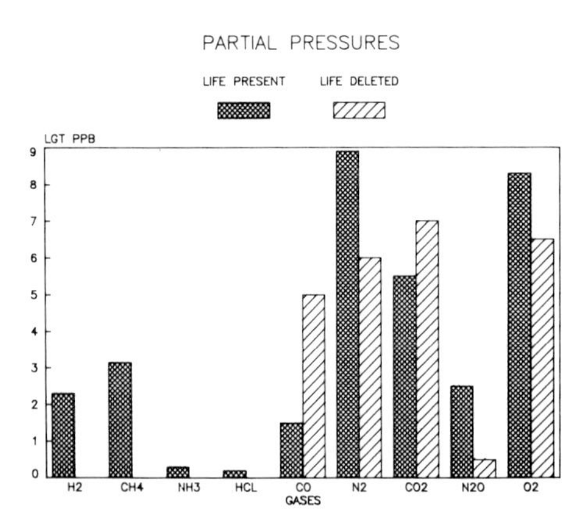

Figure 1. The abundance of gases in the present atmosphereic gas flux compared with that expected of the abiological steady state.

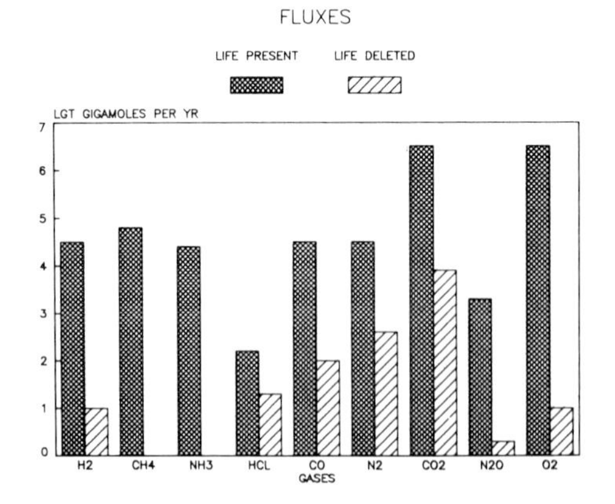

Figure 2. The fluxes of gases (gigamoles per year) through the present atmosphere compared with those expected for the abiological steady state.

logical state. It is the intense disequilibrium of the atmosphere which advertises the presence of life on Earth. It is the maintenance of this reactive and unstable atmosphere at a steady state for times much longer than the residence times of individual gases that suggests the presence of a control system, Gaia.

FEEDBACK AND HOMEOSTASIS

Many geochemists now accept that the Earth's surface features are a result of the coevolution of the biota and the rocks. But they still see the association between life and its environment as passive. Life adapts to environmental change and the evolution of life may change the environment but any feedback, negative or positive between these processes is passive. In sharp contrast the gaia hypothesis sees the Earth as homeostatic, with the biota actively seeking to keep the environment optimal for life. Such a view is usually regarded as teleological or mechanistic, Holland (7), and Gilbert (5).

The first intimation of Gaia comes from the realisation that every evolutionary step of the biota must to a greater or lesser extent alter the environment in which the next generation will evolve. If the change is to a more favourable environment then it will carry more progeny and the environmental change will be reinforced. In the same way a less favourable environment will carry less progeny and hence their unfavourable attributes will become attenuated.

It is not immediately obvious how such a course of events could lead to planetary homeostasis. As Ford Doolittle (4) observed the biota have no capacity for conscious foresight or planning and would not in the pursuit of local selfish interests evolve an altruistic system for planetary improvement and regulation.

The sequential logic of descriptive writing is not designed for the concise explanation of control systems with their inherent circularity recursiveness and non-linearity. Even the formalism of mathematics loses its elegance when an attempt is made to describe a simple non-linear control system such as, for example, an electrical water heater controlled by a bimetallic strip thermostat. I have chosen therefore to present a simple model of an imaginary planet whose temperature is regulated at a biological optimum over a wide range of solar radiation levels as a working example of a Gaian mechanism.

Before describing this model it is useful first to consider the terms active, passive and feedback in the context of their origin, namely systems engineering. Figure 3 illustrates graphically the change of some intrinsic property of a system, such as temperature, with time when there is a constant flux of a related quantity such as heat. The diagonal line across the diagram represents the rate of rise of temperature of an inert body during the constant input of heat. Line (B) illustrates passive negative feedback such as might occur on a watery

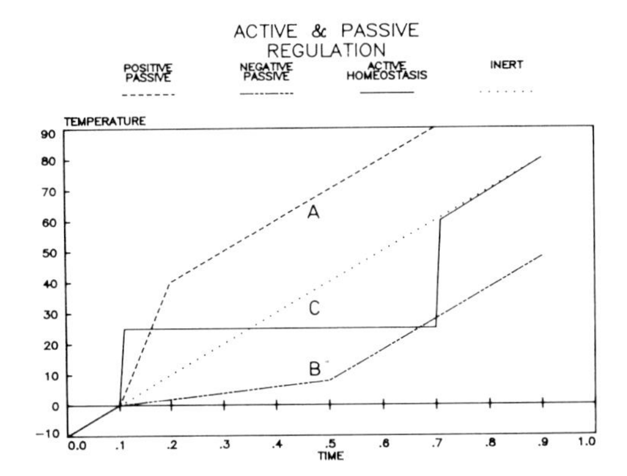

Figure 3. Diagram to illustrate passive negative and positive feedback processes and to compare them with active homeostasis. Energy is supplied at a constant rate to an object of unknown thermal mass. The diagonal line across the diagram illustrates the rate of temperature rise of an inert object. Line (B) illustrates passive negative feedback on the energy supply over a limited range. Line (A) passive positive feedback over the same range and line (C) active regulation at a chosen set temperature.

planet as a result of increasing cloudiness. Line (A) illustrates passive positive feedback such as could take place when an ice covered planet reached the melting point of water and its albedo changed from near 1.0 to a much lower value associated with open oceans and crustal rock. Line (C) is for an active feedback system with the goal of maintaining a set temperature.

With the passive negative feedback some constancy is achieved but at a value arbitrarily set by the properties of water and which cannot be changed. With the active system, constancy is possible at any chosen value. Positive feedback is used constructively so as rapidly to reach the chosen level and negative feedback used to keep the constancy. Furthermore the chosen set point can itself evolve as part of a more intricate system of evolutionary change.

Waler B. Cannon coined the therm "homeostasis" for those coordinated physiological processes which maintain most of the steady states of a living organism. Homeostasis is very much an active process; one in which any departure from the chosen state is sensed and the difference between preference and reality amplified and used to oppose the perturbation and so restore the status quo. The stability of such a system, the quality of its homeostasis, is measured by its capacity to withstand perturbations.

Now let us see how active regulation of a planetary scale might be achieved by the biota without the need for them to have foresight or receive divine assistance.

DAISY WORLD

The dominant plant life of daisy world are black and white daisies. They are grazed by grey cattle but both producers and consumers flourish when the climate permits. Both species of daisy are identical in every respect other than the colour of their flowers and their growth varies with temperature in the same way. Because they absorb more radiation the local temperature of a stand of black daisies will always be higher than that of a stand of white daisies. As a result the rates of growth of the two species will be different at any given intensity of sunlight. To model this planetary ecosystem let us assume that the two species of daisy have a growth rate (Beta) which varies with temperature parabolically as follows:

$$\text{Beta} = AT - BT^2$$

Where (A) and (B) are constants chosen so that growth is zero at below 5 and above 35 degrees Celsius and a maximum at 20 degrees. These limits are those that determine the growth of most contemporary vegetation. Under cool conditions the growth of the black daisies, which are locally warmer than are the white, will be favoured. Under hot conditions the white daisies will have the advantage. The rate of spread of one species into the zone of the other is given by a relationship described and experimentally confirmed, Carter and Prince (2), as follows:

$$dy/dt = \text{Beta}\ xy - \text{Gamma}\ y$$

Where (x) is the number of susceptible sites for growth and (y) is the number of infective sites. Beta and Gamma are the growth and death rates respectively.

Daisy world is a cloudless planet with no greenhouse gases. Figure 4 illustrates the response of the mean planetary temperature with increasing solar luminosity. The dotted line illustrates the temperature of a barren planet and the solid line when the daisies are present. The model illustrates how the powerful capacity of life to grow until a niche is full acts as an amplifier and natural selection a sensor in a control system which is able effectively and precisely to regulate planetary temperature at close to the optimum for the specified life form. No foresight or planning is required by the daisies only their opportunistic local growth when conditions favour them.

This type of model is not limited to the very artifical conditions of daisy world. A more general version would take into account the possibility that once life appears on a planet geochemical evolution will be limited by the circumscribed set of physical and chemical constraints which characterize the biota. Any external or internal chemical or phy-

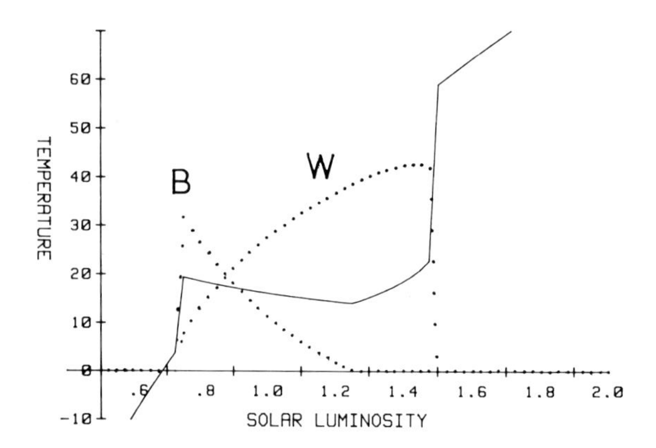

Figure 4. The active regulation of planetary temperature during the increase of solar luminosity by the growth and natural selection of two species of daisy. White daisies (W) albedo 0.7 black daisies (B) albedo 0.3 and the albedo of the bare planetary surface taken as 0.5. The solid line is the mean planetary temperature. The vertical axis also denotes the proportion of planetary area covered by each specie.

sical change away from this set of conditions will lead not only to adaptation but also to the selection of those organisms whose growth alters the environment in so as to oppose the unfavourable development.

PAST, PRESENT AND FUTURE

The gaia hypothesis postulates the existence of active systems for the chemostasis and thermostasis of the planetary environment. It predicts that the environment has been and will be stable and constant in spite of perturbations, whether sudden or from some continuous and progressive change. Strong support for the hypothesis will therefore come from the discovery of instances where there was clear evidence of the rapid and effective restoration of homeostasis after known perturbations.

There are several examples of major perturbations during the course of the history of life on Earth. The first and possibly the greatest was the origin of life itself. Whatever the environment was before life it should have changed profoundly soon after the planet was colonised. We know that conditions were favourable at the start of life so it is worth asking the question; how was it that these conditions remained favourable in spite of the inevitable changes in surface, ocean and atmospheric composition imposed by life? Thus the use of CO_2 as a carbon source by early photosynthetic life could have gravely disturbed the precarious radiation balance of the Earth when warmed by a cooler sun. It is interesting therefore to speculate on the composition of the Archean atmosphere after life began. There is evidence to suggest that the rate of carbon burial in the archean was not greatly different from now. Much of the carbon entering present day anoxic sediments is returned to the atmosphere as methane. If the Archean production of methane was comparable with that which is now produced and bearing in mind that none of it would be oxidised by consumers in the anoxic Archean oceans, the flux to the atmosphere could have been substantial. In addition to methane, sulphur gases and possibly even nitrous oxide may also have been flowing into the archean atmosphere. It would need a general circulation model to predict such an atmosphere in detail. But it can be speculated that the troposphere would have a net reducing tendency with abundant polyatomic organic gases present. The radiative properties of such an atmosphere would solve the infra red problem of less CO_2 and also serve if necessary to filter out ultra violet radiation at wavelengths less than 300 nanometres. The well established glaciation coincident with the first appearance of oxygen in the atmosphere at 2.2 giga years ago is consistent with the oxidative destruction of these organic gases.

The emergence of oxygen as an atmospheric gas may have been the largest perturbation the Earth has yet experienced. Significantly it was internally and biologically driven. The process of photosynthesis whereby oxygen and carbon are segregated and some of the carbon becomes buried, ineluctably drove the planetary surface ever more oxidising during the Archean. It is true that some of this tendency was offset by the return of reducing materials by tectonic processes but until the critical problems posed by the presence of gaseous oxygen began to exert their effect on selection oxidation proceeded unchecked.

Among the minor but startling problems posed by the appearance of oxygen would have been the speciation of the element uranium. In the reduced form uranium is safely locked as water insoluble material dispersed a great dilution. In an oxidising environment uranium is water soluble and readily concentrated by microorganisms. This task was once successfully completed by microorganisms resident at a region which is now Gabon in Africa about 2.2 giga years ago. As a result a nuclear reactor commenced operation and ran for several million years. At that time uranium was substantially richer in the fissionable isotope U235 than now. It is fortunate that the Earth did not become oxidising in the early archean for then the uranium was enriched close to bomb quality. Spectacular nuclear fireworks might have been more than the infant Gaia could have withstood.

By the time the metazoan biota were well established the presence of charcoal in the sediments provides a fossil record of ancient fires. The range of atmospheric oxygen over which fires can take place yet not be so devastating as to threaten all standing vegetation is 15 to 25% by volume. It is therefore tolerably certain that atmospheric oxygen has never ranged beyond these bounds in the last several hundred millions of years. This is a truly remarkable feat of regulation for in the previous 90% of the Earth's history the pE has risen by at least ten units but is now held precisely constant. The mechanism by which oxygen is regulated is not yet known although it has been proposed by Watson *et al.* (17) that the control of the proportion of carbon buried in the anoxic sediments and hence the oxygen abundance is achieved through the regulation of the venting of methane to the atmosphere. Interestingly fires themselves exert a positive feedback on oxygen since carbon as charcoal is resistant to digestion by microorganisms and hence more is buried.

Throughout the existence of life on Earth there have been frequent collisions with planetisimals several kilometres in diameter, the most recent, 65 Myrs ago. The impact energy of these collisions is vast enough to have caused major, albeit temporary environmental changes, and was proposed as the cause of species extinctions in the fossil record, Alvarez *et al.* (1). Figure 5 illustrates the impact craters so far discovered on the Canadian shield. Most of the events recorded by the craters represent an energy yield at the Earth's surface about 10^8 times larger than the detonation of the present global stocks of nuclear weapons. Although the consequences of these impacts are not yet known in detail they do act as impulse tests of the "black box" system. If and when a detailed description of the sequence of events at one of these collisions is uncovered it will chart the course of the perturbation and the rapidity and effectiveness of the return to an optimum environment which follows. It could provide important evidence about the existence and the nature of Gaia.

A progressive change in the environment which spans the past, present and the future is that which relates to the climate of the Earth. One of the more certain conclusions of astrophysics is that stars in-

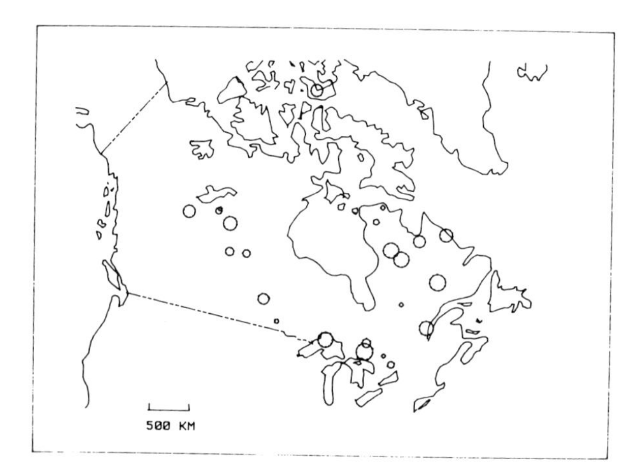

Figure 5. Map showing the impact craters so far discovered on the Canadian Shield.

crease their radiation flux as they age. There is a consenus among astronomers that the sun was very probably about 25% less luminous at the Earth's origin than it is now. We know from the geological record that fluid water has always been present and from the origin of life that the climate cannot have been very different from now 3.5 giga years ago. In this context glaciations represent only minor departures from climatic constancy. Walker *et al.* (16) proposed that a progressive decrease in atmospheric CO_2 from about 10% abundance at the start of life to the present 0.03% could through a decreasing greenhouse effect compensate for the progressive increase of solar luminosty. Although by itself carbon dioxide does not provide a very effective greenhouse, on the Earth its influence is amplified by presence of abundant water vapour.

The mechanism by which CO_2 is varied inversely with the solar luminosity so as to maintain a constant temperature is the weathering of exposed calcium silicate rock. This is the only major sink for CO_2 from the atmosphere and the rate of weathering has a positive temperature coefficient. Walker's proposal provides a plausible abiological mechanism for climatic and CO_2 regulation although with the present information on the fluxes of CO_2 it appears to be only partially able to account for the constancy of the climate throughout the Earth's history.

I do not disagree with the general basis of this interesting abiological control mechanism which would make Gaia redundant but wondered instead how much better it would work if life was included as a part of it. The real world is not abiological and the weathering of calcium silicate is very much a biological concern. At all levels from prokaryotic microorganisms to large trees and soil moving animals the biota participates in the process of rock digestion. The partial pressure of CO_2 in the soil is 10 to 40 times greater than it is in the air. CO_2 is actively pumped from the air by the biota to those regions of the soil where it can react with calcium silicate particles. The rate of CO_2 fixation by plants is a strong function of both temperature and light intensity.

The need to have the biota participate in such a system is best illustrated by considering the consequences of its absence. If all life were deleted the soil CO_2 concentration would rapidly fall to below the present atmospheric level and weathering would be substantially reduced in rate. The input of CO_2 to the atmosphere from volcanoes is on average constant and consequently the atmospheric CO_2 concentration would rise until the current rate of weathering was re-established. The new equilibrium level would probably be above the current levels of the soil, about 1%. This is because diffusion from the air is very slow compared with the active penetration of the soil by plant roots. The ambient temperature under these conditions was calculated by Lovelock and Watson (10) to be about 20 degrees Celsius higher than now. The higher temperature might increase the abiological weathering rate but only slightly if it is limited by the rate of diffusion of CO_2 to the calcium silicate rock. Furthermore as recently modelled by Shukla and Minz (14) the lack of

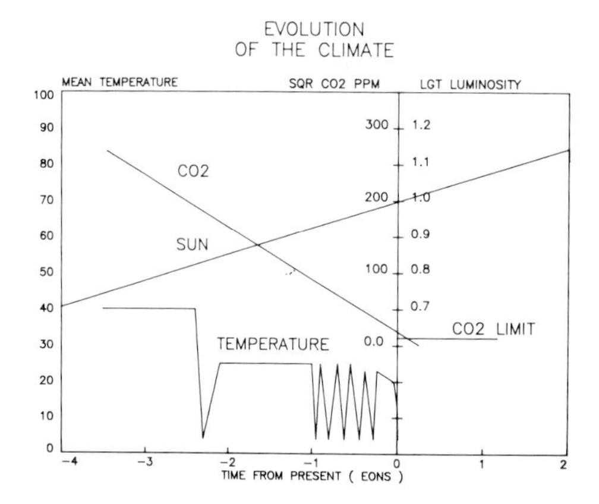

Figure 6. Evolution of the climate showing the variation of solar luminosity from its present value taken as 1.0. Also illustrated on the same time scales are the proposed decline in carbon monoxide concentration, expressed as the square root of its concentration in ppmv and the approximate range of mean surface temperatures in degrees Celsius.

land life would so disturb the planetary water vapour transport that large areas would become desert where the weathering rate would be much reduced. It is also significant that CO_2 and carbohydrate species transport to and from the oceans is very dependent upon the presence of life. The deletion of life from the oceans would lead to a further increase in atmospheric CO_2.

Atmospheric CO_2 abundance and climate is a current environmental concern as a result of the geochemically minor perturbation attributable to fossil fuel combustion. Figure 6 from Lovelock and Whitfield (11) summarizes in simple diagrammatic form the rise of solar luminosity during the Earth's history and the fall of CO_2 abundance needed to compensate for this rise in heat flux.

The most interesting feature of this diagram is its suggestion that a new perturbation of major magnitude is, on a geological timescale imminent. If the climate is to stay constant at near the optimum for the biota then the CO_2 must be further reduced. However, a reduction below 100 ppm could not be suffered by most of the contemporary photosynthesisers. The diagram illustrates that such a level will be reached in only 100 Myrs. If zero CO_2 were tolerable even this is approached in 200 Myrs.

It is unlikely that our descendants will be there to witness this interesting period when it comes. The past history of the Earth suggests that a near optimal planetary environment will be sustained by some other means.

REFERENCES

1. Alvarez, I.W., Alvarez, W., Asaro, F., and Michell, H.V.: 1980, Science 208, p. 1095.
2. Carter, R.N., and Prince, S.D.: 1981, Nature 293, p. 644.
3. Dawkins, R.: 1982, The Extended Phenotype, Freeman, New York, p. 114.
4. Ford Doolittle, W.: 1981, Coevolution 29, p. 58.
5. Gilbert
6. Hitchcock, D.R., and Lovelock, J.E.: 1966, Icarus 7, p. 149.
7. Holland, H.D.: 1978, The Chemistry of the Atmosphere and Oceans, Wiley Interscience, Chichester, U.K.
8. Hutchinson, G.E.: 1954, Biochemistry of the Terrestrial Atmosphere, in the Solar System (ed. Kuiper), Chapter 8, University of Chicago Press.
9. Lovelock, J.E., and Margulis, L.: 1974, Tellus 25, p. 1.
10. Lovelock, J.E., and Watson, A.J.: 1982, J. Planet. Sci., in press.
11. Lovelock, J.E., and Whitfield, M.: 1982, Nature 296, p. 561.
12. Margulis, L., and Lovelock, J.E.: 1974, Icarus 21, p. 471.
13. Redfield, A.C.: 1958, Am. Sci. 46, p. 205.
14. Shukla, J., and Minz, Y.: 1982, Science 215, p. 1498.
15. Sillen, L.G.: 1966, Tellus 18, p. 198.
16. Walker, J.C.J., Hays, P.B., and Kasting, J.F., J. Geophys. Res., in press.
17. Watson, A.J., Lovelock, J.E., and Margulis, L.: 1978, Biosystems 10, p. 293.

MICROBIAL SYSTEMATICS AND A GAIAN VIEW OF THE SEDIMENTS

Lynn Margulis and John Stolz
Department of Biology, Boston University, Boston, MA 02215 USA

ABSTRACT

The "Gaia hypothesis" which asserts that certain properties of the atmosphere are actively maintained by the biota for the biota was developed by J.E. Lovelock to explain the dynamically stable anomalous atmospheric composition, modulated alkalinities and temperatures at the Earth's surface throughout its 3.5 billion-year-old history. Much, if not all, surface sedimentary deposition since the early Archean Aeon has been modulated by organisms. Detailed *in situ* characterization of microbial sedimentary communities such as that currently underway at Laguna Figueroa, Baja California, Mexico are essential for verification of the Gaia hypothesis for the atmosphere and surface sediments. A total inventory of biotic atmospheric gas emission and removal processess as well as of biomineralization potential under realistically varying environmental conditions is a *sine qua non* for the development of the Gaia concept. The current confused state of microbial and "botanical" systematics has precluded even the initiation of an effort that correlates biomineralization and gas exchange potential with taxon. Furthermore, geologists and paleontologists inadvertently create misunderstandings by the use of obsolete taxonomic schemes. This paper, which lists taxa along with their major biogeochemical potentials, is only a tiny first step in the clarification of the current confusion.

1. INTERACTION OF THE BIOTA WITH THE SEDIMENTS

The Gaia hypothesis contends that the composition of the lower atmosphere, its oxidation-reduction state and temperature are modulated by the biota (Lovelock, 1972, 1979). In this paper we extend the gaian concept to include the surface sediments.

In order to understand the production of atmospheric gases by the biota the sedimentary sources of many of the gases must be known. These sources may be highly structured microbial communities such as the one described from Laguna Figueroa, Baja California, Mexico (Horodyski and

P. Westbroek and E. W. de Jong (eds.), Biomineralization and Biological Metal Accumulation, 27–53.

Vonder Haar, 1975; Margulis et al., 1980, 1982). Whereas to the geologist these sediments represent a suite of evaporitic minerals (gypsum, anhydrite, aragonite and halite), the biologist sees complex populations of primary producers, heterotrophs and decomposers. To the atmospheric chemist the sediments represent a major source of reduced gases and the paleontologist uses stratigraphic and structural information to reconstruct paleoenvironments as a clue to past life. Because the gaian concept involves interaction of the biota with the atmosphere and sediments the establishment of this hypothesis requires cooperation between geologists, biologists, atmospheric chemists and paleontologists. Since several different disciplines are involved inevitably inconsistencies and contradictions arise in the use of terminology and nomenclature. The geologist, for example, satisfied that biogenic oxygen was available in a paleoenvironment needs not necessarily to distinguish between cyanobacteria, green algae and plants; these distinctions, however, are crucial to the evolutionist. In the interest of introducing a consistent up-to-date terminology to the interacting disciplines required to establish the gaia hypothesis we present the five kingdom system of classification and a short discussion of current problems of microbial systematics.

From the surface, the evaporite flat at Laguna Figueroa looks desolate, inhospitable and uninhabited. However, a few millimenters or less just beneath the evaporite mineralized crust a diverse microbial community thrives which strongly interacts with the sediment. The photosynthetic community in the upper half centimeter of surface is not only responsible for atmospheric carbon dioxide fixation into organic matter but microbial removal of CO_2 enhances aragonite precipitation (Horodyski and Vonder Haar, 1975). Several heterotrophic bacteria which deposit manganese oxides have been isolated from the community as well as an amoeboflagellate, Paratetramitus jugosus, which eats bacteria that deposit manganese oxides on their spores and apparently resolubilizes the deposits (Margulis et al., 1982). The sulfate-reducing bacteria in the black anaerobic muds produce quantities of hydrogen sulfide which may react with soluble metals to produce insoluble metal sulfides (Renfro, 1974). The cohesive fabric of living organisms and sediment particles, flooded during periods of high tides and fresh water influx from a microbial mat. Microbial mats can be conceived of as living precursors to several types of laminated sedimentary rocks such as carbonate stromatolites (Golubic, 1976), microfossiliferous chert stromatolites (Awramik and Barghoorn, 1977) and perhaps other shallow water laminated cherts (Margulis et al., 1980).

Recent analysis of sediment samples by transmission electron microscopy reveals that a great diversity and abundance of organisms comprise the laminae along with mineral particles. The distinct mat layers clearly visible in the field and in hand samples reflect the highly ordered ecological structure of the vertically laminated microbial community comprising the mat. This ordered structure can be seen, for example in Figure 1. The top 2mm of mat which in hand samples appears green, were dominated by oxygenic filamentous photosynthetic bacteria. The 2mm layer beneath, which in hand samples appears purplish, is

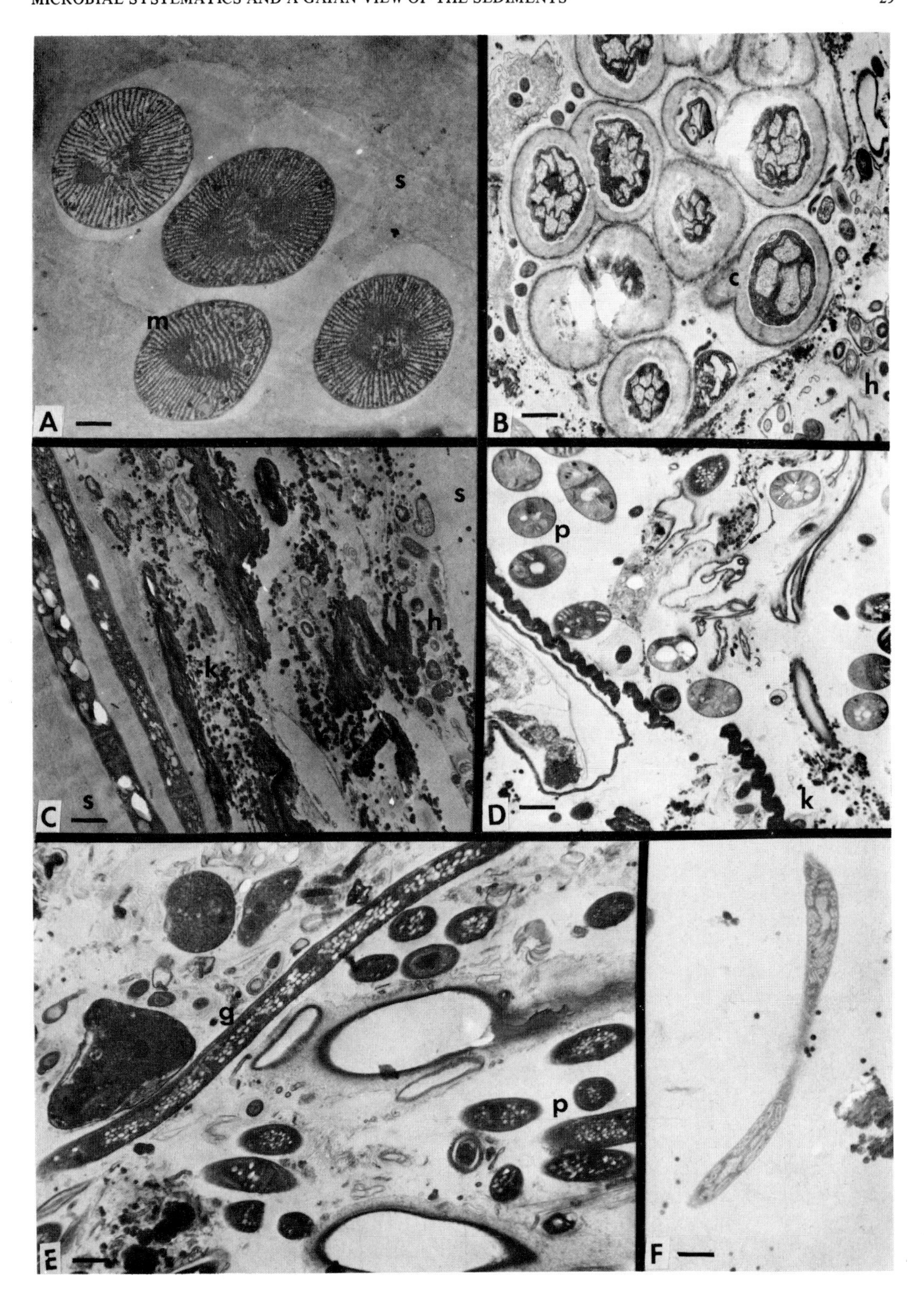

s
m
A
c
h
B
s
h
k
s
C
p
k
D
g
p
E
F

Figure 1. The microbial community from the flat laminated *Microcoleus* mat at North Pond, Laguna Figueroa, Baja California, Mexico. Transmission electron micrographs (for methods see Stolz, 1982). A-D are from the top 2mm of mat. A. Filaments of *Microcoleus* sp. in transverse section. B. Coccoid cyanobacteria and heterotrophic bacteria. C. Bacteria, mineral particles and protokerogenous material between sheaths of *Microcoleus*. D. Photosynthetic bacteria, mineral particles and protokerogenous material. E and F are from the anaerobic photosynthetic layer 4mm below the surface of the mat. E. Purple and green photosynthetic bacteria. F. *Bdellovibrio*-like bacteria-within-a-bacterial host. Key: c coccoid cyanobactera, g green anaerobic photosynthetic bacteria, h heterotrophic bacteria, k protokerogenous material, m *Microcoleus* sp., p purple anaerobic photosynthetic bacteria, s sheath. Bar 1 micrometer.

comprised mainly of anoxygenic photosynthetic bacteria. In addition to these photosynthetic organisms, other bacteria abound, including a diverse assortment of small spirilli and heterotrophic rods which reside between the *Microcoleus* sheaths and sediment particles. This ordered community may even contain a *Bdellovibrio*-like bacterial-bacterial host association (Stolz, 1982). This geological sample is very much alive.

Although this is the first series of studies which utilize *in situ* transmission electron microscopy (Stolz, 1982; Margulis et al., 1982) most other sedimentary samples must contain comparable diverse and abundant microbial populations. The methods used to study the microbial components and their interaction with the sedimentary environment should be reasonably uniform. Furthermore, the identification and classification of microorganisms should be consistent with the rest of biological taxonomic practice. The data obtained from these studies must be accessible to atmospheric scientists, sedimentologists, general ecologists and others who are not necessarily specialists in microbiology.

2. CLASSIFICATION AND SYSTEMATICS

The science of systematics, as applied to fungi, animals, and plants, is well developed. It deals with relationships between organisms as approached through their form and function, their genetic systems and metabolism and their fossil record in an attempt to discern their evolutionary history. The analogous science, microbial systematics is fraught with problems. Attempts to develop a systematics of eukaryotic microorganisms, the protists, has recently been made (Ragan and Chapman, 1978); however, comparable attempts for the prokaryotes are lacking or sources of great contention. Because they are limited in size and morphology, have powerful mechanisms of broad dispersal and genetic transfer, have not left an obvious fossil record, and are nearly unknown in their natural habitats microorganisms are far less amenable than large organisms to systematic analysis. However, this does not mean that they are any less the products of neoDarwinian evolution. Although the development of microbial systematics is difficult it is an intellectually sound and compelling scientific goal. Evolutionary understanding of microbes would be especially helpful to the analysis of ancient ecosystems dominated by bacteria. Knowledge of the order of acquisition of major gas producing metabolic pathways is directly relevant to understanding the history of the Earth's atmosphere and its change through time (Walker, 1980; Margulis and Lovelock, 1980).

Taxonomy is the branch of biology that deals with identifying, naming and classifying organisms into a formal hierarchical system. Fungal, animal and plant taxonomists are also systematists in today's neo-Darwinian world; they try to conform their classification schemes with current concepts of evolution. Hence the larger more inclusive taxa, higher taxa such as kingdoms, phyla and classes diverged from each other in the more remote past relative to lower taxa such as genera and species. In some cases taxa such as classes, orders and families are comparable across kingdom boundaries (plant/animal) and in some they are not (Van Valen, 1973). Given a single phylogeny there are many possible taxonomic schemes. Yet in devising taxonomies evolutionists are unable to base their classifications directly on phylogeny for two major reasons: the first is that the evolutionary history of many organisms is simply not known, a second more germane reason here is that often concise taxonomic criteria are preferable to those requiring complex qualifications, thus for any taxon evolutionary accuracy may be sacrificed to the ease of definition.

Bacteriologists of course have their taxonomists (Sokal and Sneath, 1968; Mandel, 1969) but their practices differ from those of other systematists in that they utilize metabolic and physiological criteria for relatedness rather than the morphological criteria of zoologists or botanists. Bacterial taxonomists tend to ignore evolutionary history. Comparison of the various higher taxa such as Van Valen has made for families of insects and families of mammals (Van Valen, 1973) has not been attempted for microorganisms because there is no concensus even on what constitute higher taxa. There are several reasons that evolution has been ignored by microbiologists, especially bacteriologists. Some think there is no data from which valid evolutionary inferences can be made. Bacteria change so rapidly, they feel, that methods from which their evolutionary trends can be inferred do not exist. Many experimental scientists view microbial evolution as they do all historical reconstructions: not as science but rather as articles of faith. However, recent application of biochemical techniques to the solution of evolutionary problems, especially those involving monomer sequences in nucleic acids and proteins have led to more optimistic attitudes (Dayhoff, 1976; Schwartz and Dayhoff, 1978). The time has come for the establishment of microbial systematics; both microbial phylogenies and a taxonomy consistent with that characteristic of all the rest of biology can be at least attempted.

We assume here that through studies of comparative metabolism and sequences of monomers in macromolecules, coupled with inferences from micropaleontology including the effects of microbes on atmosphere and sediments, an acceptable microbial systematics will be developed. Primarily as a basis for initiation of discussion, a tentative classification scheme for microbes is presented here. There are several principles upon which it has been based. The first is that the living world is divided into two major cell organization styles, prokaryotic and eukaryotic, and that these correspond to superkingdoms. The second is that the most comprehensive and useful classification available is the five kingdom scheme of R.H. Whittaker (1969; as modified by Margulis and Schwartz, 1982). Whittaker's plan (far more than the traditional dichotomous one: plant vs. animal which places algae and bacteria with the plants and protozoa with the animals) is thought to be systematic, that is, it reflects the overall evolutionary relationships of organisms of the microbial world.

For convenience the five kingdoms and their definitional criteria are presented in Table 1. Table 2 (prokaryotes) and Table 3 (eukaryotes) list phyla of microorganisms and their immediate descendants that belong to none of the remaining three kingdoms (Fungi, Animals, Plants). The detailed evolutionary odyssey upon which this classification rests is the subject of (Margulis, 1981). Because they are most important and least recognized by those who study sediment and atmospheric interaction the remainder of this paper deals with only the first two kingdoms: Monera and Protoctista. Those protoctists known to be involved in biogeochemical processes are listed while the Monera are considered in greater detail.

3. MICROBIAL EVOLUTION AND TAXONOMY

Phylogenies are diagrams that attempt to reflect the course of evolutionary history; fantasy must play a significant role in their construction. Several assumptions always underlie the formation of phylogenies and often these are unstated. Those which underlie the phylogeny (Fig. 2), an earlier version of which may be found in Sieburth, 1979 include:

1. All prokaryotes evolved from a minimal replicating unit that would have been recognizable as prokaryotic cells (Morowitz, 1967; Margulis, 1980).
2. No complex multigenic trait of selective advantage (e.g., no seme, Hanson, 1977) was lost from any population of organisms without leaving any trace of evidence for its former presence.
3. Semes in microbes tend to be entire metabolic pathways in which the end products are of selective advantage or the underlying metabolism leading to distinct morphological entities of selective advantage (such as spores, calcareous tests or life cycle stages).
4. Semes are of primary value in phylogeny construction. Many other traits are only of secondary value. These include sequences of monomers in entire macromolecules, characteristic properties of monomers, presence of pigments or metabolites.

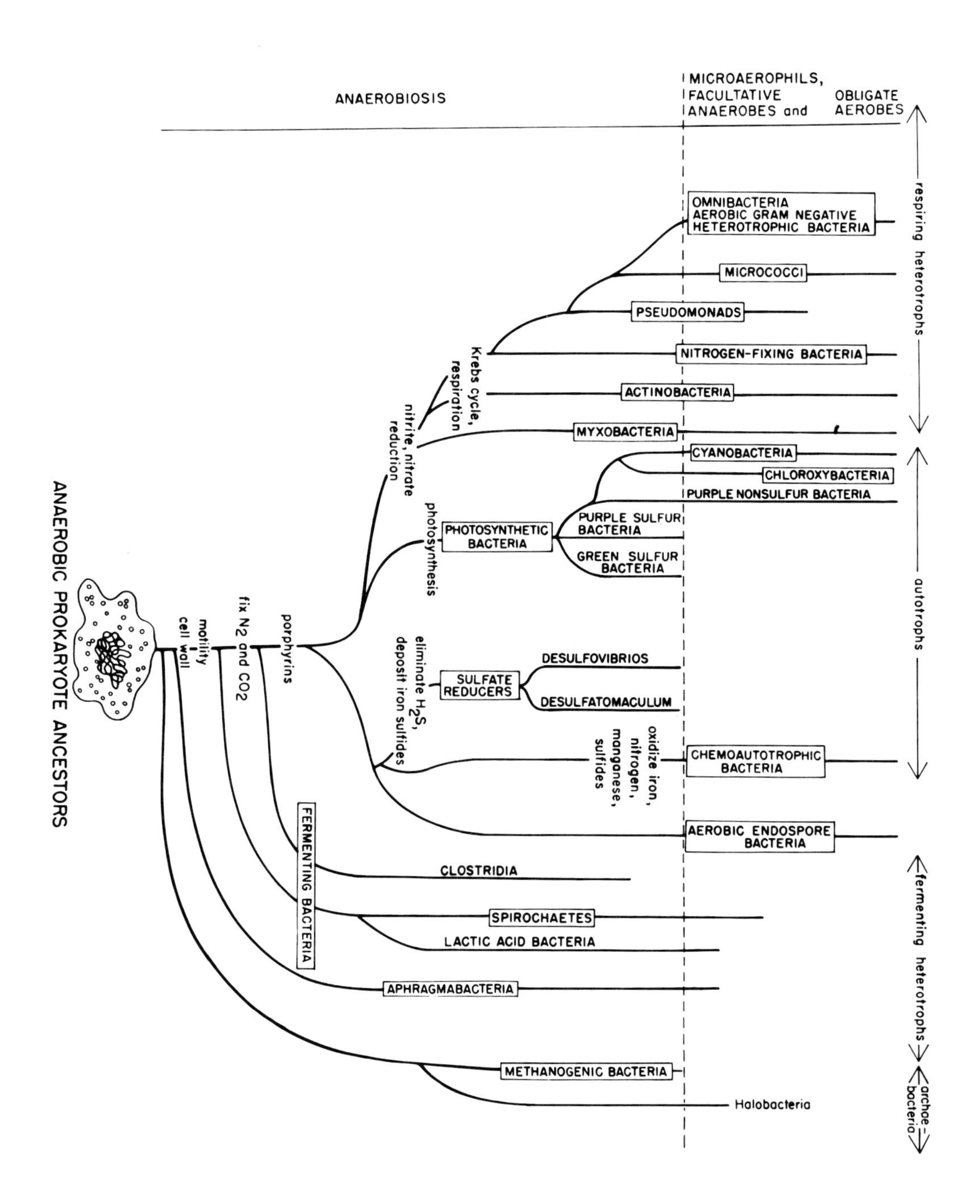

Figure 2. Phylogeny of Monera Kingdom

5. The amount of gene transfer for many semes has been negligible and thus microbial phylogenies can be reliably constructed and tested. (Semes for which there is good evidence for gene transfer: resistance factors, mangangese oxidation, are not used in phylogeny construction.)

6. The acquisition of many semes can be logically ordered (e.g., fermentation preceded oxygen requiring chemoautotrophy, the evolution of parasites followed the appearance of their hosts, and so on). Where logical ordering is equivocal because criteria for ordering are not available it shall be so stated. For example, which originated first, carotenoid (terpenoid) or chlorphyll (tetrapyrole) biosynthetic pathways?

7. A distinction is made between taxonomy and phylogeny. The phylogenetic diagram attempts to order the taxa found at the tips on the basis of divergence of taxa from common ancestors at the branch axes. The taxonomic treatment, in the form of phyla list (Table 2) sacrifices phylogeny when required for the same statement of clear criteria for each phylum.

The distinction between taxonomy and phylogeny can be seen in the following example. Few would doubt that a strong case has been made by Woese and his colleagues (Balch et al., 1979) that the methanogenic bacteria share common ancestry with members of the genus Halobacter. The phylogenetic argument is based on the homology between the 16S and 5S ribosomal RNA sequences, the presence of ether rather than ester linkages in the major lipids and the absence of the standard bacterial peptidoglycan cell wall (Tornabene et al., 1979). However, the methanogens are not high salt requiring and they are obligate anaerobes whereas the halophils are high salt requiring obligate aerobes. (This is one of the many arguments for the convergent origin of salt tolerance and aerobic respiratory pathways.) The halobacters are thus to be placed as a derived genus on an archeabacterial lineage (Fig. 2). However, taxonomically the genus Halobacter fit the criteria of phylum M-10. As Gram negative chemoheterotrophic obligately aerobic respiring flagellated rods they are placed with the pseudomonads. Thus the taxonomy presented here is hoped to be related to the phylogeny but in deference to convenient and logical application of stated criteria, the taxonomic system is given precedence.

4. MAJOR GROUPS OF MICROBES

The protoctist phyla are listed in Table 3, and diagrammed on the phylogency of Fig. 3. Several phyla such as dinoflagellates (Pr-2), testate amoebae (Pr-3), chrysophytes (Pr-4), haptophytes (Pr-5), diatoms (Pr-11), radiolarians (a class of Pr-16), and foraminiferans (Pr-17) are extremely important organisms in sediment forming processes. On Fig. 3 these have been indicated with an asterisk. The protoctists unlike most Moneran genera have a preservation potential high enough such that their fossils can be recognized in sediments (Haq and Boersma, 1978). Several other protoctist groups probably escape ecological and paleontological recognition because their degradation products are not distinctive (Lowenstam and Margulis, 1979). For example, the extent to which deep

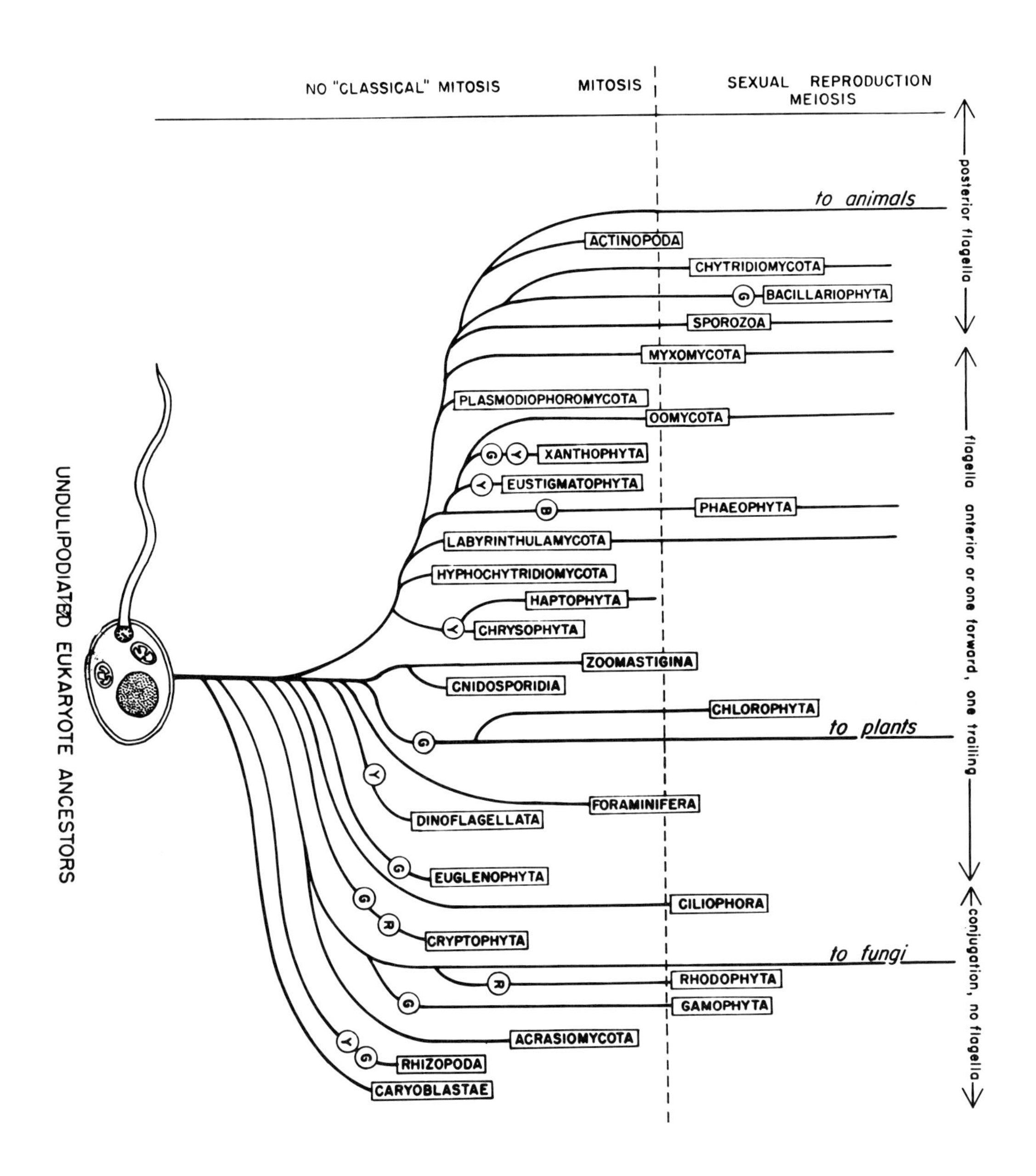

Figure 3. Phylogeny of Protoctista Kingdom

sea silica is diatomaceous and silicoflagellate in origin is not entirely known. Silica spicules such as those in Fig. 4 have not been recognized in sedimentary materials as the spines of fresh water heliozoans. Mineralized tests are common in chrysophytes (Pr-4) and chlorphytes (Pr-16), among others. Mastigonemes (stiff tinsel or hair-like decorations on undulipodia)[1] are very common in the following protoctist phyla: Pr-4, 5, 9, 10, 16, and 27. In most species their composition and micropreservation potential is unknown. Examples such as these serve to alert geochemists studying particulates and their elemental distribution. Many tiny particles such as these are of microbial origin. Their presence and preservation, furthermore, is often related to microbial metabolism and ecology. Potentially these particles could serve as indicators of the environment of deposition.

4.1 Protoctist Phyla

Some thirty-two animal phyla are recognized by zoologists. These animals share the criteria for the kingdom (Table 1) but they differ from each other in major ways such as developmental pattern, body plan and life cycle. They are thought to represent the earliest divergences of major persisting metazoan lineages. Some of these phyla were already established in the late Proterozoic when the Ediacaran fauna appeared nearly 700 million years ago. Others first appear in the lower Paleozoic Era. Some nine plant and five fungal phyla have been recognized. The major trends in evolutionary differentiation in these relatively stationary organisms has involved production of exotic and distinctive metabolites (Swain, 1974). Thus in plants and fungi (relative to animals) only a limited number of fundamentally different body plans and life cycle patterns characterize the phyla. Over two dozen fundamentally different cell or body patterns have been recognized for the protoctists (Table 3), primarily based on electron microscopy. There is no consensus, however, among professional biologists concerning how these eukaryotic microorganisms are related to each other nor how they have evolved. Since some groups (e.g., dinoflagellates, Pr-3; crysophytes, Pr-4; cryptomonads, Pr-6; chlorophytes, Pr-15) have traditionally been classified both by zoologists and botanists using inconsistent criteria the problem of protoctist taxonomy has been exacerbated. The listing here (Table 2) represents our judgment (Whittaker and Margulis, 1978; Schwartz and Margulis, 1982). This listing has not been ratified by any learned societies. A purpose of this paper is to bring to the attention the problems facing systematists of the microorganisms regardless of their origins as bacteriologists, physiologists, mycologists, protozoologists, taxonomists or geologists.

4.2 Bacterial Phyla: A Modest Proposal

There are some 10,000 species of monerans in the literature. To accommodate them into major recognizable, definable groups some sixteen phyla have been erected here. Each of the sixteen is discussed and illustrated by micrographs and analytical drawings in Margulis and Schwartz, 1982. Of course, this is a naive somewhat arbitrary attempt to conform

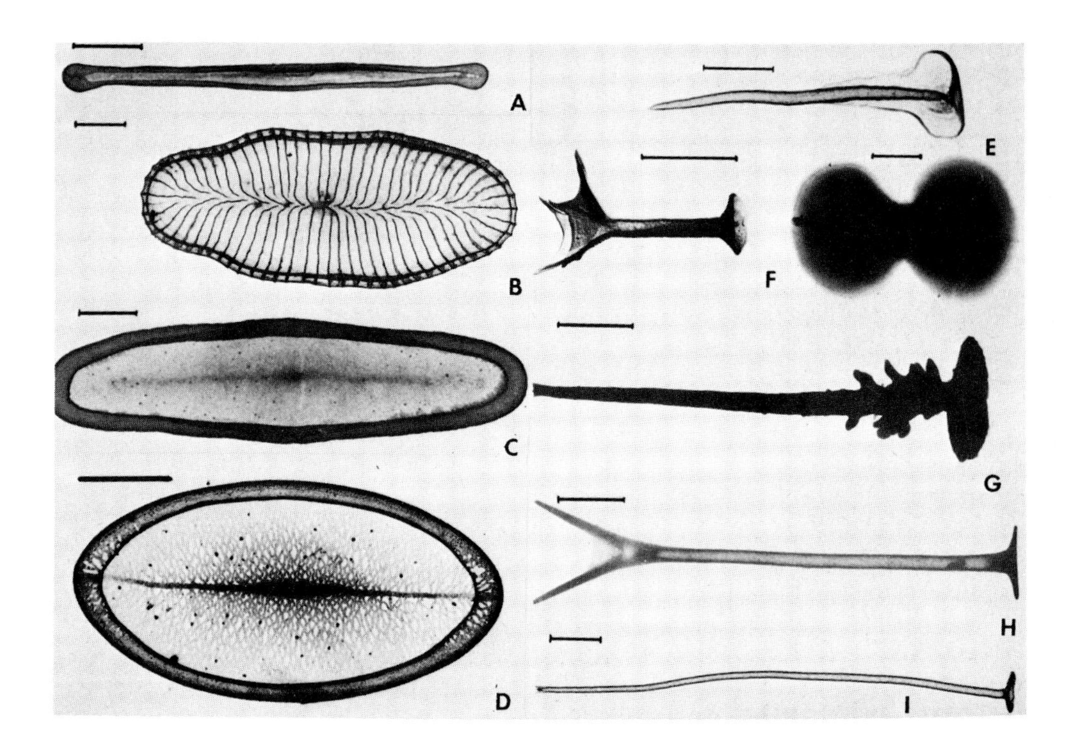

Figure 4. Silica scale and spines of heliozoa (Pr-16) (Courtesy C.D. Bardele, Zool. Inst., Tübingen) Silica scales and spines from centrohelidian heliozoan (Phylum Actinopoda, Kingdom Protoctista). Living cells extracted with sulfur chromic acid, washed 3 times with distilled water and photographed with the electron microscope after placement on formvar coated grids. No staining or shadowing was employed. These structures which are probably as resistant as diatom tests are probably often overlooked by micropaleontologists because they tend to be 3-5 mμ, smaller than the mesh size of many sieves in routine use. They, however, are species specific, can be used for species identification and are formed intracellularly in smooth membrane surface vesicles that show no direct connection with the Golgi. The way in which the membranes and cell metabolism specify the morphology of the spicules is entirely unknown. Photographs and work kindly supplied by Dr. Christian F. Bardele, Institut für Biologie III, Universität Tübingen, Germany. A. Radiophrys neapolitans B. Radiophrys elegans C. Radiophrys marina D. Radiophrys ambigua E. Acanthocystis crinoides F. Acanthocystis sp.*
G. Acanthocystis aculeata H. Acanthocystis turfacea I. Acanthocystis myriospina. Bar 1 micrometer.
*Both unidentified but thought to be species of Acanthocystis.

microbial taxa to that of the rest of biology. However, many microbiologists agree that some consistent taxonomic scheme is required if merely to organize thinking and for teaching. The sixteen phyla system presented here (Table 2) seems logical and generally consistent with current practices as exemplified by Bergey's Manual (Buchanan and Gibbons, 1974) and the new Handbook of Prokaryotes (Starr et al., 1981) as well as with the needs of taxonomists and ecologists who are not trained in the calabalistics of the microbiology guild. Each phylum is listed with a brief description of its characteristics and some representative genera. Those known to be agents in microbial geochemical processes are asterisked. Your criticisms and comments are solicited.

5. BACTERIAL EVOLUTION: MACROMOLECULAR SEQUENCE APPROACHES

One of the most fruitful of the approaches to bacterial evolution has been the inference of bacterial lineages from information on the monomer sequences in widely distributed proteins and nucleic acids, which has led to partial phylogenies such as those in Figs. 5 and 6.

Partial phylogenies based on single macromolecules are intrinsically limited in at least two respects. A bacterium is far more than a single protein or nucleic acid molecule. Cells contain several thousands of different proteins and tens of nucleic acid molecules and such partial phylogenies can only represent one or a few macromolecules at a time. Furthermore, some single genes are subject to episomal or phage transfer and hence are less reliable than other far more conserved genes such as those that code for interacting gene products. For example, genes coding for histones, cytochromes and thylakoid membrane proteins, all of which are restrained by their structural role in the cell change at slower rates than those coding for less constrained products such as globins. In addition, although the rates of substitution of amino acids in proteins and therefore of protein evolution have been considered to be constant with time this assumption is patently false (Dickerson, 1980). Some functional regions, active sites of proteins, are very strongly conserved leading to low rates of mutation, and in others the rates of

---

Figure 5. Composite bacterial phylogeny based on macromolecular sequences (after R. Schwartz and M.O. Dayhoff). Composite evolutionary tree of Dayhoff and Schwartz (1978) with taxonomic assignments based on this paper. M-x, Moneran Kingdom, phylum as numbered in Table 2. Pr-x, Protoctista Kingdom, phylum as numbered in Table 3. Tree based on ferredoxin, cytochromes and 5S RNA sequences. Branch lengths approximately proportional to accepted point mutations per 100 residues for the proteins. Heavy lines represent a tree calculated from matrix of evolutionary distances combining two or three separately derived trees. Light lines represent branches scaled from a single tree. Based on concept of the symbiotic origin of organelles from prokaryotes the time of organelle acquisition is indicated by the stippled and crosshatched symbols.

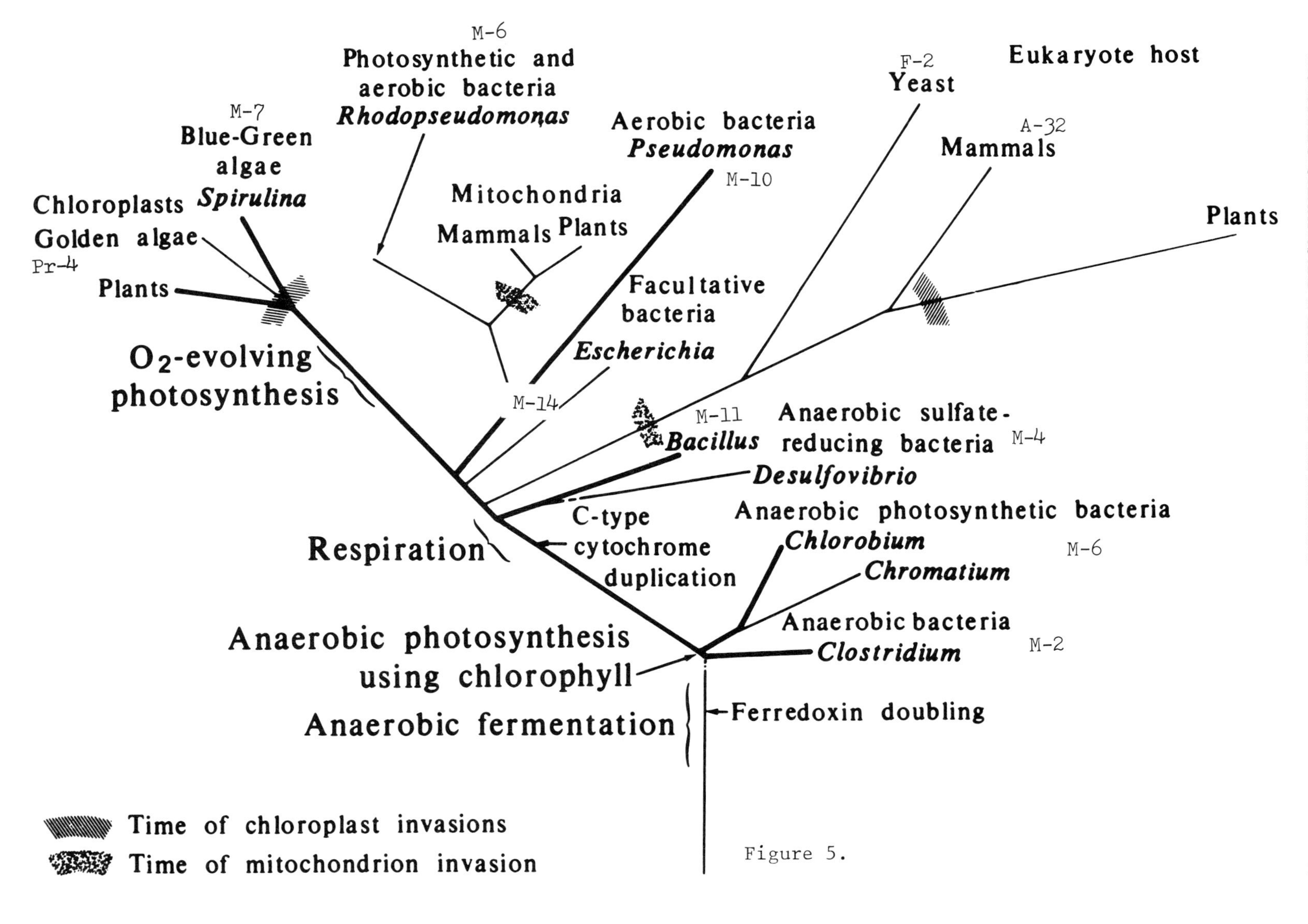
M-6
Photosynthetic and aerobic bacteria
Rhodopseudomonas
F-2
Yeast
Eukaryote host
M-7
Blue-Green algae
Spirulina
Aerobic bacteria
Pseudomonas
M-10
A-32
Mammals
Chloroplasts
Golden algae
Pr-4
Plants
Mitochondria
Mammals
Plants
Plants
Facultative bacteria
Escherichia
O2-evolving photosynthesis
M-14
M-11
Bacillus
Anaerobic sulfate-reducing bacteria
M-4
Desulfovibrio
C-type cytochrome duplication
Anaerobic photosynthetic bacteria
Chlorobium
M-6
Chromatium
Respiration
Anaerobic bacteria
Clostridium
M-2
Anaerobic photosynthesis using chlorophyll
Anaerobic fermentation
Ferredoxin doubling
Time of chloroplast invasions
Time of mitochondrion invasion

Figure 5.

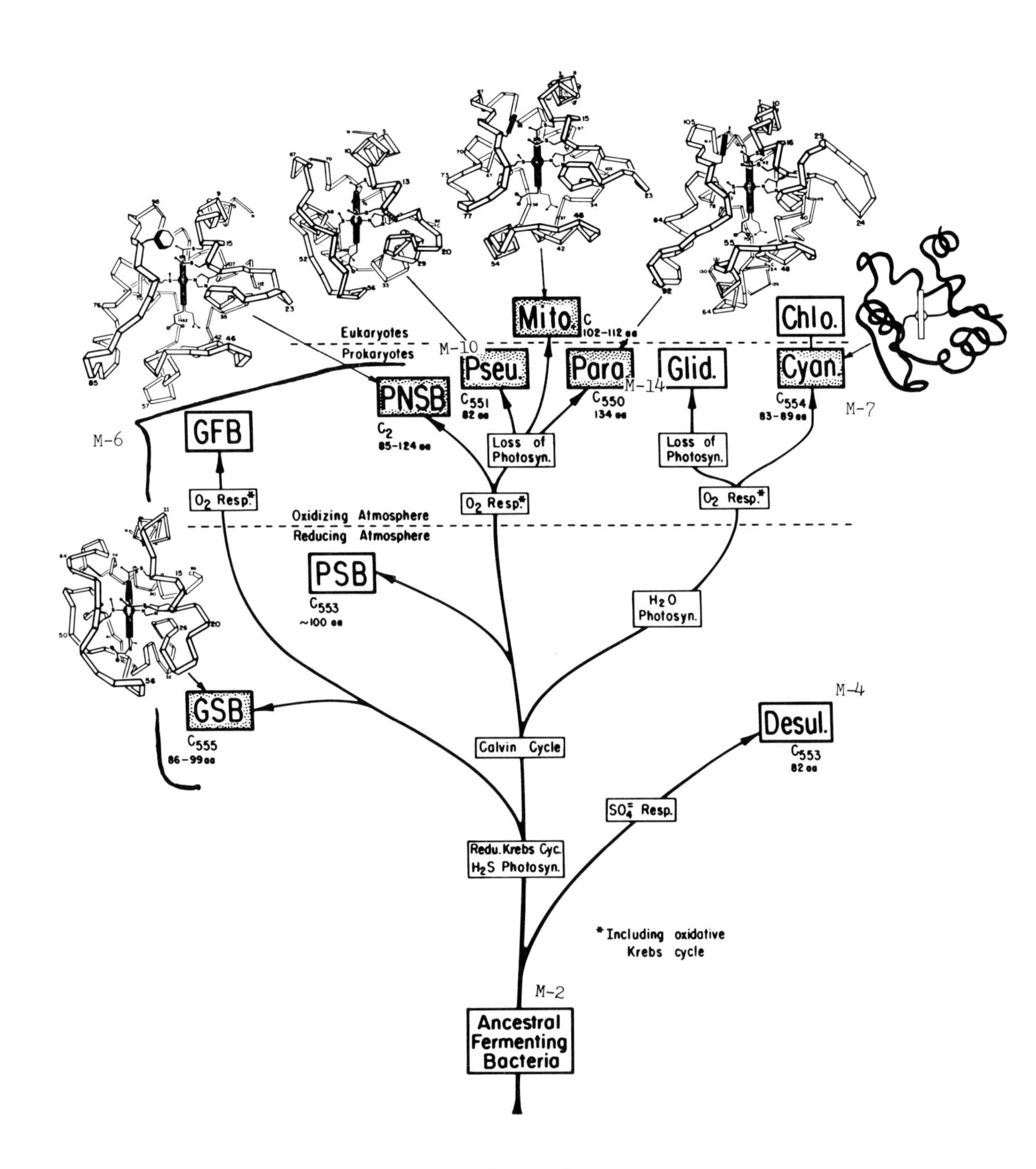
Mito.
C
102–112 aa
Chlo.
Eukaryotes
Prokaryotes
M-10
Pseu.
Para.
M-14
Glid.
Cyan.
PNSB
C_{551}
82 aa
C_{550}
134 aa
C_{554}
83–89 aa
M-7
GFB
C_2
85–124 aa
M-6
Loss of Photosyn.
Loss of Photosyn.
O_2 Resp.*
O_2 Resp.*
O_2 Resp.*
Oxidizing Atmosphere
Reducing Atmosphere
PSB
C_{553}
~100 aa
H_2O Photosyn.
M-4
GSB
C_{555}
86–99 aa
Desul.
C_{553}
82 aa
Calvin Cycle
$SO_4^=$ Resp.
Redu. Krebs Cyc.
H_2S Photosyn.
*Including oxidative Krebs cycle
M-2
Ancestral Fermenting Bacteria

Figure 6.

Figure 6. Bacterial phylogeny based on cytochrome c sequence (after R.E. Dickerson). The stippled boxes indicate a complete structure of the protein by x-ray chrystallographic methods is available for a member of the group. For the others only amino acid sequence data is available. The ancestral fermenting bacteria lack cytochrome c. Abbreviations: M-x, kingdom and phylum (see Table 2); GSB, green sulfur bacteria; GFB, green flexibacteria (Chloroflexus); PSB, purple sulfur bacteria; PNSB, purple non-sulfur bacteria; pseu., pseudomonads; Mito., mitochondria (eukaryotic organelle); Glid., gliding bacteria (Beggiatoa, sulfur oxidizing bacteria); Chlor., chloroplasts (eukaryotic organelle); Cyan., cyanobacteria; Desul., desulfovibrio; Para., Paracoccus.

---

accepted point mutations are far higher. Furthermore, no one ever knows for certain the extent to which the bacterium chosen for analysis is representative of the genus or higher taxon into which it is placed. In fact, the bacterium chosen is representative of only one thing for sure: ease of growth under laboratory conditions.

Traits far less directly related to the genome than sequences of long chain molecules are used for both taxonomy and phylogeny by many authors. Such traits, which are not semes, include the ability to metabolize certain substrates, gas production, catalase production, extracellular slime or hydrolyzing enzyme production, and many others. These traits may be the result of single or very few mutations (such as reistance to a certain antibiotic) or, in some cases (for example, the presence of nitrogenase and correlated nitrogen fixation pathways) they may actually be semes. Hence many traits that are in current use as phylogenetic indicators are of very limited reliability as such. Other traits such as nucleotide base ratios (%GC/GC + AT) are useful for excluding close relationships between bacterial species but are virtually useless for establishing phylogenetic proximity.

6. BACTERIAL FOSSIL RECORD

The direct preservation of bacteria in the fossil record is a rare phenomenon. Nevertheless, objects interpreted as prokaryotes in division have been reported in sediments greater than 3.4 billion years (Knoll and Barghoorn, 1977). Although relative to the Proterozoic Aeon (2.5 to 0.6 x 10^9 years ago) Archean sediments contain less direct morphological evidence for diversified microbial communities they do contain some (Knoll and Barghoorn, 1977; Dunlop et al., 1979). Furthermore, from the extent, distribution, reconstructed environment of deposition and rock textures the existence of Archean microbial ecosystems dominated by photosynthetic prokaryotes can be inferred (Reimer et al., 1979; Knoll, 1979; Margulis et al., 1979). It is probable that the largest insight into microbially dominateprePhanerozoicsedimentary environments will emerge concurrently with an understanding of metalliferous ore deposits (Renfro, 1974; Goodwin et al., 1976; Holland and Schidlowski, 1982) and the nature of microbial metabolic interactions with metals, for example, iron (Nealson, 1982).

Table 1. Summary of Five Kingdoms

Kingdoms	Examples of organisms	Genetic organization	Approximate time of appearance (millions of years ago)	Types of semes produced by action of environmental selection to pressures	Major significant selective factors in environment
SUPERKINGDOM PROKARYOTA (Chromonemal organization) KINGDOM MONERA					
prokaryotic cells, nutrition chemosynthetic, photoautotrophic or absorptive. Anaerobic facultative or aerobic metabolism. Reproduction asexual, nonmitotic; recombination unidirectional or viral mediated. Nonmotile or motile by flagella composed of flagellin or by gliding. Solitary unicellular, colonial or mycelial.	All prokaryotes; eubacteria cyanobacteria, mycelial bacteria, gliding bacteria	No chromosomes, prokaryote chromoneme; mesozygotes only; sex unidirectional	Early Archean > 3400	uv photoprotection, photosynthesis motility and aerobiosis	solar radiation, increasing atmospheric oxygen concentration
SUPERKINGDOM EUKARYOTA (Chromosomal organization) KINGDOM PROTOCTISTA					
Eukaryotic cells, nutrition ingestive, absorptive or, if photoautotropic, in photosynthetic plastids. Premitotic and eumitotic cell division. In eumitotic forms meiosis and fertilization present but life cycle and ploidy levels vary from group to group. Solitary unicellular colonial unicellular or multicellular. Undulipodia organelles of motility composed of microtubules in the (9+2) pattern ("flagella" or cilia)	Nucleated algae: green, yellow-green, red and brown, and golden-yellow; protozoans, "flagellated" fungi, slime molds and slime net molds	chromosome organization varies; ploidy levels, meiotic sexual systems gametes vary (premitotic and eumitotic); zygotes formed in many groups	Proterozoic < 1000	Mendelian genetic systems, obligate recombination each generation; phagocytosis, pinocytosis, intracellular motility, vascularization, histogenesis	depletion of

KINGDOM FUNGI					
Haploid or dikaryotic, mycelial or secondarily unicellular, chitinous walls, absorptive heterotrophic nutrition, lack undulipodia, body plan branched coenocytic filament which may be divided perforate septa. Zygote meiosis, propagate by haploid spores	Amastigomycota: conjugation fungi, sac fungi (molds), club fungi (mushrooms), yeasts	eumitotic: no centrioles, poorly developed mitotic spindle; haploid and dikaryotic; zygote formation followed by meiosis and haploid spore formation	Phanerozoic	tissue development, advanced mycelial development for heterotrophic specializations; absorptive nutrition	transitions from aquatic to terrestrial environments; nature of nutrient source
KINGDOM ANIMALIA					
Develop from diploid blastula, gametic meiosis. Nutrition heterotrophic, digestive, phagocytosis, extensive cellular and tissue differentiation	Metazoa; animals developing from blastulas	eumitotic: centrioles, well-developed mitotic spindle; diploids, meiosis precedes gametogenesis	Proterozoic	embryogenesis, histogenesis, for heterotrophic specializations; complex behaviors	transitions from aquatic to terrestrial and aerial environments
KINGDOM PLANTAE					
Photoautotrophic oxygen eliminating nutrition. Alternation of diploid and haploid generation; advanced tissue differentiation, develop from embryos, plasmads, multilammellar bodies	Metaphyta: byophytes, tacheophytes	eumitotic: no centrioles, well-developed mitotic spindle, alternation between haplo- and diplophases	Phanerozoic	embryogenesis, histogenesis for autotrophic specializations: complex secondary metabolites	transitions from aquatic to terrestrial environments

Table 2. Phyla of the Monera Kingdom[1]
Kingdom I. Monera (Bacteria) (G. moneres = single)

Phylum and description	Meaning of name[2]	Representative genera	Examples of biogeochemical products or activities[3]
M-1 Aphragmabacteria. Bacteria without cell walls.	G. a., without; G. phragma, wall G. bacteria, little sticks	Mycoplasma Thermoplasma	
M-2 Fermenting bacteria. Heterotrophs limited to anaerobic fermentation, unable to synthesize porphyrins.	E.	Bacteroides Clostridia Streptococcus	hydrogen, organic acids
M-3 Spirochaetae. Helical motile heterotrophs with periplasmic flagella (in which the flagella are located within the cell).	G. spiro, coil; G. chaetae, hair	Diplocalyx Cristispira Treponema Spirochaeta Leptospira	Hydrogen
M-4 Thiopneustes. Anaerobic sulfate reducing bacteria.	G. thio, sulfur G. pneustes, breathers	Desulfovibrio Desulfotomaculum Desulfuromonas	Hydrogen sulfide, sulfur
M-5 Methanocreatrices. Archaebacteria. Methanogenic bacteria.	G. methano, methane L. creatrices, makers	Methanobacterium Methanosarcina Methanococcus	methane
M-6 Photosynthobacteria. Anaerobic photosynthetic bacteria. Chlorophyll-containing phototrophs in which the hydrogen donor for photosynthesis is H_2, H_2S or organic compounds.	G. photo, light G. syntho, put together G. bacteria, little sticks	Rhodomicrobium Chlorobium Rhodospirillum Chromatium	hydrogen, sulfur, sulfate
M-7 Cyanobacteria. Cyanophyta, blue-green algae, blue-green bacteria. Chlorophyll a and phycobiliprotein-containing photoautotrophs in which the hydrogen donor for photosynthesis is H_2O.	G. cyano, blue G. bacteria, little sticks	Anabaena Synechococcus Nostoc Lyngbya	oxygen, calcified community structures (stromatolites)
M-8 Chloroxybacteria. Prochlorophyta. Chlorophyll a and chlorophyll b-containing photoautotrophs in which the hydrogen donor for photosynthesis is H_2O	G. chloros, green G. oxy, oxygen	Prochloron	oxygen
M-9 Nitrogen-fixing bacteria. Nitrogenase containing eubacteria with oxygen requiring respiration.	E.	Azobacter Beijerinckia Rhizobium	atmospheric N_2 removal
M-10 Pseudomonads. Nonfermentative, obligatory oxygen respiring eubacteria.	G. pseudo, false G. monas, single	Pseudomonas Hydrogenomonas	*organic transformations
M-11 Aeroendospora. Aerobic endospore forming bacteria.	G. aero, air G. endo, inside	Bacillus	*organic transformations
M-12 Micrococci. Gram positive respiring bacteria.	G. micro, small L. cocci, berries	Micrococcus Sarcina Gaffkya	*organic transformations

M-13 Chemoautotrophobacteria. Aerobic bacteria deriving energy from the direct oxidation of reduced inorganic compounds such as H_2S and NH_3.	G. *chemo*, chemical G. *auto*, self G. *trophic*, feeding	*Beggiatoa* *Nitrosomonas*	nitrite, nitrate, nitrous oxide, sulfate
M-14 Omnibacteria Enterobacteria. Facultative aerobic Gram negative heterotrophic bacteria.	L. *omni*, all G. bacteria, little sticks	*Caulobacter* *Escherichia* *Serratia* *Erwinia*	nitrogen, nitrate, nitrous oxides, *organic transformations
M-15 Actinobacteria. Actinomycota. Mycelial actinospore forming facultatively or obligately aerobic bacteria.	G. actino, ray G. bacteria, little sticks	*Streptomyces* *Nocardia*	*organic transformations
M-16 Myxobacteria. Heterotrophic gliding and fruiting structure forming bacteria.	G. myxo, mucus G. bacteria, little sticks	*Stigmatella* *Chondrococcus*	*organic transformations

[1]These groups are described and illustrated in Schwartz and Margulis, 1982.

[2]E., English; G., Greek; L., Latin. Complete foreign words italicized; anglicized versions not italicized.

[3]Not all members of phylum necessarily capable of these transformations.

*Organic transformations (e.g., cellulose and sugar breakdown, organic oxidation, etc. usually produce CO_2).

Table 3. Phyla of the Protoctista Kingdom[1]
Kingdom II. Protoctista (Algae, slime molds, slime nets, protozoa, protists, "flagellated" fungi, water molds) (G. pro = before, early; ktist = being)

Phylum and Description		Meaning of name[2]	Representative genera	Examples of biogeochemical products or activities[3]
PR-1	Caryoblastea. Pelobiontida. Amoebae which lack mitosis.	G. caryo, nut G. blastea, bud	Pelomyxa	
PR-2	Dinoflagellata. Mesokaryota. Testate mastigotes with girdle and transverse undulipodia.	G. dino, a whirling L. flagella, whips	Gonyaulax Gymnodinium Peridinium	organic tests and cysts (hystrichospheres), oxygen
PR-3	Rhizopoda. Sarcodina. Shelled and naked mitotic but nonmeiotic amoebae.	G. rhiso, root G. poda, foot	Mayorella Difflugia Arcella	organic carbonate and sandy clastic tests
PR-4	Chrysophyta. Golden yellow algae.	G. chrysos, gold G. phyta, plants	Ochromonas Echinochrysis Sarcinochrysis	silica tests, oxygen
PR-5	Haptophyta. Coccolithophorids. Haptonemid bearing nonmeiotic mastigotes with chlorophyll b containing plastids.	G. hapto, join G. phyta, plants	Prymesium Hymenomonas	calcium carbonate tests (coccoliths) oxygen
PR-6	Cryptophyta. Cryptomonads. Gullet bearing bimastigotes with no or varying color plastids, nonmeiotic	G. crypto, hidden G. phyta, plants	Cyathomonas Cryptomonas Cyanomonas	oxygen
PR-7	Euglenophyta. Euglenids. Nonmeiotic mastigotes with chlorophyll b containing plastids.	G. eu, primitive G. gleno, eyeball G. phyta, plants	Euglena Peranema Astasia	oxygen
PR-8	Zoomastigina. Animal flagellates. Heterotrophic mastigotes with a variety of asexual and sexual patterns.	G. zoo, animal G. mastigina, whip	Joenia, Bodo, Trypanosoma Trichonympha	
PR-9	Xanthophyta. Bimastigotes with yellow green plastids, nonmeiotic	G. xanthos, yellow G. phyta, plants	Ophiocytium Tribonema Botryococcus	oxyten
PR-10	Eustigmatophyta. Mastigotes with stigmas (eyespots) and chlorophyll a and b containing plastids.	G. eu, true G. stigmata, spot G. phyta, plants	Vischeria Pleurochloris	oxygen
PR-11	Bacillariophyta. Diatoms. Silica testate diploid eukaryotic microorganisms with gametic meiosis and chlorophyll a and c containing plastids.	L. bacillario, little sticks G. phyta, plants	Thallassiosira Melosira Navicula Planktoniella	silica tests oxygen

PR-12	Rhodophyta. Red seaweeds, chlorophyll and phycobiliprotein containing plastids. Haploids.	G. rhodos, red G. *phyta*, plants	*Polysiphonia* *Porphyra* *Nemalion*	calcified thalli volatile sulfur compounds, oxygen
PR-13	Phaeophyta. Brown seaweeds, chlorphyll *c* and fucoxanthin containing plastids. Diplohaploids.	G. phaeo, brown G. *phyta*, plants	*Fucus* *Ascophyllum* *Dictyota*	oxygen
PR-13	Gamophyta. Conjugating green algae, desmids. Haploid, lack undulipodia.	G. gamo, union G. *phyta*, plants	*Mougeotia* *Spirogyra* *Zygnema*	organic cysts oxygen
PR-15	Chlorophyta. Grass green algae, haploid, bear undulipodia.	G. *chloro*, green G. *phyta*, plants	*Acetabularia* *Volvox* *Chlamydomonas* *Chara*	Acritarchs, calcified tests and cysts, oxygen
PR-16	Actinopoda. Radiolarians, helioxoans, acantharians. Heterotrophs which bear complex microtubule based ray structures (actinopods), eukaryotic microorganisms	G. actino, ray G. poda, foot	*Acanthocystis* *Echinosphaerium* *Thalassicola*	strontium sulfate spines. silica spines
PR-17	Foraminifera Haplodiploid testate heterotrophic eukaryotes.	L. *foramina*, little holes L. fera, bearing	*Rotaliella* *Globigerina* *Nodosaria*	calcium carbonate tests
PR-18	Cliophora. Ciliates. Macronuclear, micronuclear heterotrophs bearing numerous short undulipodia.	L. *cilio*, eyelash G. *phora*, bearing	*Gastrostyla* *Paramecium* *Tetrahymena*	organic, sandy clastic tests
PR-19	Apicomplexa. Sporozoan parasites bearing the apical complex.	L. api, top L. complex, entwine	*Eimeria* *Telosporidea*	organic cysts
PR-20	Cnidosporidia. Myxosporidian and microsporidian parasites bearing filaments for attachment to or injection into hosts.	G. cnido, filament G. sporidia, spores	*Glugea* *Nosema* *Myxobolus*	organic cysts
PR-21	Labyrinthulamycota. Slime nets. Heterotrophic colonial microorganisms motile within an extracellular matrix only.	G. labyrinthula, labyrinth G. mycota, fungus	*Labyrinthula* *Labyrinthorhiza*	organic slimes
PR-22	Acrasiomycota. Cellular slime molds, asexual colonial heterotrophs.	G. *a*, with G. crasio, mixing G. mycota, fungus	*Dictyostelium* *Acrasia*	organic slime
PR-23	Myxomycota. Plasmodial slime molds, sexual undulipodiated colonial heterotrophs.	G. myxo, mucus G. mycota, fungus	*Echinostelium* *Physarium*	organic slime
PR-24	Plasmodiophoromycota. Plasmodiophorans, plasmodial parasites with undulipodiated stages.	L. plasmodio, living fluid G. phoro, bearing G. mycota, fungus	*Plasmodiophora* *Spongospora* Polymixa	

PR-25	Hyphochytridiomycota. Hyphochytrids. Anteriorly mastigonate zoospore forming mycelial asexual heterotrophs.	G. hypho, web G. chytridio, little earthen pot G. mycota, fungus	Hyphochytrium Rhizidiomyces Anisolpidium
PR-26	Chytridiomycota. Chytrid water molds, posteriorly undulipodiated chytrid-thallus producing heterotrophs, some with sexual stages. Zoosporous asexual reproduction.	G. chytridio, little earthen pot G. mycota, fungus	Blastocladiella Chytria Monoblephara
PR-27	Oomycota. Oomycetous water molds, anteriorly mastigonemate zoospore forming mycelial heterotrophs. Conjugating sexual reproduction, zoosporous asexual reproduction.	G. öo, egg G. mycota, fungus	Saprolegnia Albugo Achlya

[1]These groups are described and illustrated in Schwartz and Margulis, 1982.

[2]E., English; G., Greek; L., Latin. Complete foreign words italicized; anglicized versions not italicized.

[3]Not all members of phylum necessarily capable of these transformations.

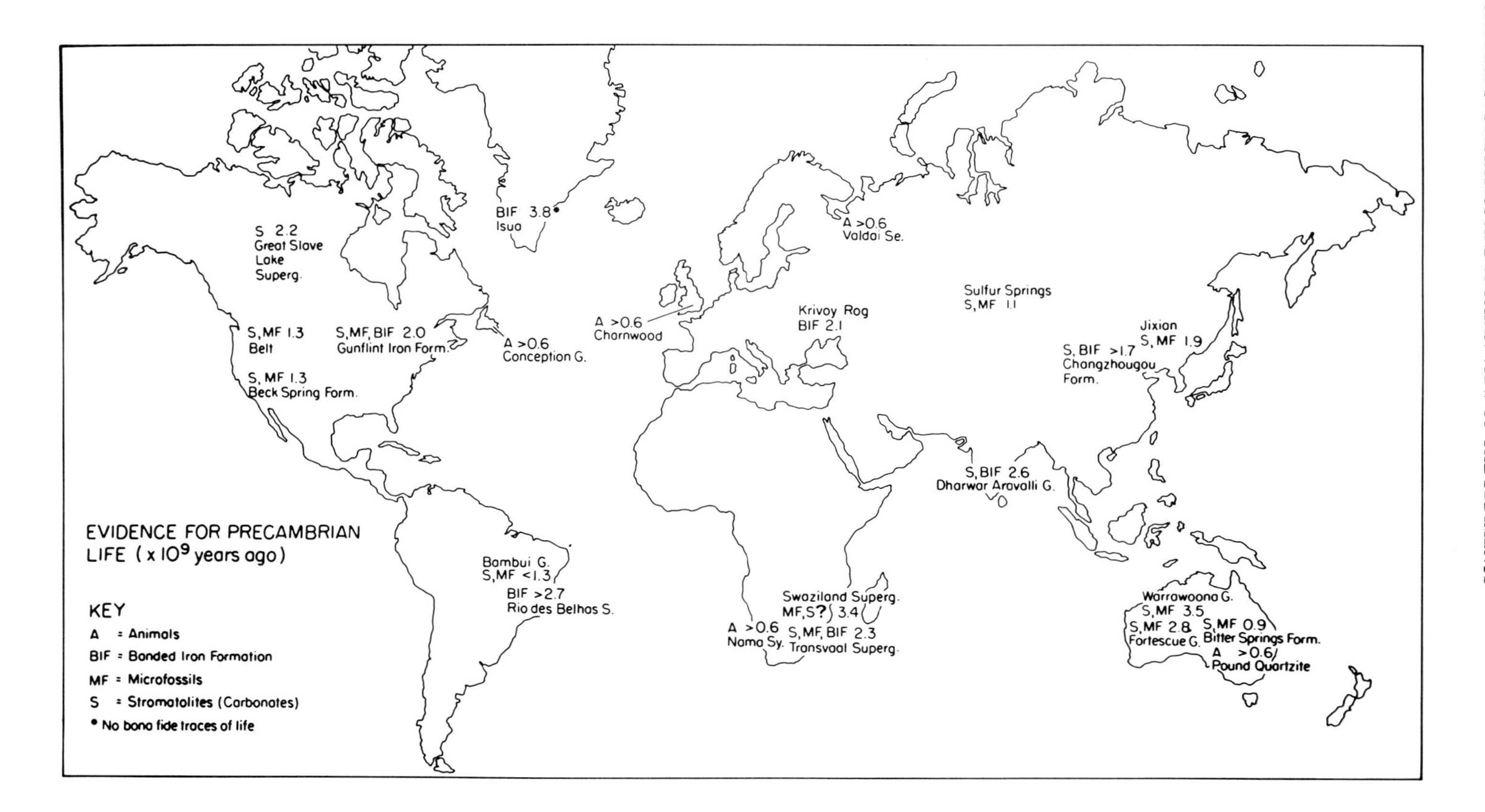

Figure 7. Map summarizing evidence for prePhanerozoic life. (information from S. M. Awramik)

Distinction may be made between a mineral particle that is precipitated as a byproduct of metabolism (i.e., a change in pH) and one that is template mediated and under genetic control by the organism. An example of template mediated precipitation is the magnetite crystals of magnetotactic bacteria (Blakemore and Frankel, 1981). The magnetite crystals have a characteristic morphology and size range (Towe and Moench, 1981). From the discovery of such biogenic magnetite crystals in the fossil record the time of origin of the biochemical pathway for magnetite deposition can be inferred.

The final columns in Tables 2 and 3 very briefly and superficially indicate some of the microbial sources of major biogeochemical and atmospheric impact. There is increasing evidence that the "Superior" type of banded iron formations are related to both microbial oxygenic photosynthesis and microbial metal oxidations (Holland and Schidlowski, 1982). Thus the development of understanding of trends in prePhanerozoic evolution is contingent upon the interaction of microbiologists working with organisms of biogeochemical significance with geologists such as the contributors to the volume on time and strata-bound ore deposits (Klemm and Schneider, 1977).

The three major methods of reconstructing the fossil history of microorganisms are through the study of stromatolites (Walter, 1976) of microfossils and of "chemical fossils" or kerogens. The major locations for study of these clues are indicated in the accompanying map (Fig. 7) which was constructed with the aid of S.M. Awramik.

The need for coordination of prePhanerozoic paleontologists to formally interact with microbiologists toward the development of an appropriate microbial systematics is acute.

7. NOTE

[1]The term undulipodium (Corliss, 1979) is used for cilium or eukaryotic "flagellum" to distinguish it from the entirely different prokaryotic flagellum (Margulis, 1980).

8. ACKNOWLEDGMENTS

Support for this work came from NASA NGR-004-025 and from the Guggenheim Foundation. We are grateful to W. Ford Doolittle for disagreements which led to several concepts developed in this paper; to S.M. Awramik, E.S. Barghoorn, M. Gorczyca, A. Knoll, W.E. Krumbein, K.H. Nealson, and L.K. Read for useful discussion; and to D. Ashendorf, M. Gorczyca, and G. Kline with aid in preparation of the manuscript.

9. REFERENCES

Awramik, S.W., and Barghoorn, E.S.: 1977, "The Gunflint microbiota," Precambrian Research 5, pp. 121-143.

Balch, W.E., Fox, G.E., Magrum, L.J., Woese, C.R., and Wolfe, R.S.: 1979, "Methanogens: Reevaluation of a unique biological group," Microbiological Reviews 43, pp. 260-296.

Blakemore, R.P., and Frankel, R.B.: 1981, "Magnetic navigation in bacteria," Scientific American 245, pp. 58-65.

Buchanan, R.E., and Gibbons, N.E.: 1974, Manual of Determinative Bacteriology, p. 1268. Williams & Wilkins, Boston.

Corliss, J.O.: 1979, The Ciliate Protozoa, 2nd ed., 455 pp., Pergamon, Oxford.

Dayhoff, M.V.: 1976, "The origin and evolution of protein superfamilies," Feder. Proceed. 35, pp. 2132-38.

Dickerson, R.: 1980, "Cytochrome C and serendipity." In: Evolution of Protein Structure and Function (Ed. D.S. Sigman and M.A.B. Brazier). UCLA Forum in Medical Sciences, Academic Press, New York

Dunlop, J.S.R., Muir, M.D., Milne, V.A., and Groves, D.I.: 1978, "A new microfossil assemblage from the Archean of Western Australia," Nature 277, pp. 676-678.

Golubic, S.: 1976, "Organisms that build stromatolites." In: Stromatolites (Ed. M.R. Walter). Elsevier Publishing Co., Amsterdam.

Goodwin, A.M., Monster, J., and Thode, H.G.: 1976, "Carbon and sulfur isotope abundances in Archean iron formations and early Precambrian life," Economic Geology 71, pp. 870-891.

Hanson, E.: 1977, Origin and Early Evolution of Animals, 670 pp. Wesleyan University Press, Middletown, Conn.

Holland, H.D., Schidlowski, M. (eds.): 1982, Mineral Deposits and the Evolution of the Biosphere, Springer Verlag, Berlin, Heidelberg, New York.

Horodyski, R.J., and Vonder Haar, S.J.: 1975, "Recent calacareous stromatolites from Laguna Mormona (Baja California) Mexico," Journal of Sedimentary Petrology 45, pp. 894-906.

Klemm, D.D., and Schneider, H.J.: 1977, Time and Strata-Bound Ore Deposits, pp. 144, Springer-Verlag, Berlin, Heidelberg, New York.

Knoll, A.: 1979, "Archean autotrophy: Some alternatives and limits," Origin of Life 9, pp. 313-327.

Knoll, A., and Barghoorn, E.S.: 1977, "Evidence for cell division in Archean rocks," Science 198, pp. 396-398.

Lowenstam, H.A., and Margulis, L.: 1980, "Evolutionary prerequisites for Phanerozoic skeletons," BioSystems 12, pp. 27-41.

Lovelock, J.E.: 1972, "Gaia as seen through the atmosphere," Atmosphere and Environment 6, pp. 579-580.

Lovelock, J.E.: 1979, Gaia, Oxford University Press, Oxford and New York.

Mandel, M.: 1969, "New approaches to bacterial taxonomy: Perspectives and prospects," Ann. Rev. Microbiol. 23, pp. 239-279.

Margulis, L.: 1980, "Flagella, cilia and undulipodia," BioSystems 12, pp. 105-108.

Margulis, L.: 1981, Symbiosis in Cell Evolution, W.H. Freeman and Co., San Francisco.

Margulis, L., Barghoorn, E.S., Banerjee, S., Giovannoni, S., Francis, S., Ashendorf, D., Chase, D., and Stolz, J.: 1980, "The microbial community in the layered sediments at Laguna Figueroa, Baja California, Mexico: Does it have Precambrian analogues?" Precambrian Res. 11, pp. 93-123.

Margulis, L, and Lovelock, J.E.: 1982, "Atmospheres and evolution." In: Proceedings, NASA Conference on Life in the Universe (Ed. J. Billingham and M. Stoll).

Margulis, L., and Schwartz, K.V.: 1982, Five Kingdoms: Guide to the Phyla of Life on Earth, W.H. Freeman and Co., San Francisco.

Margulis, L., Grosovsky, B.D.D., Stolz, J.F., Gong-Collins, E.J., Lenk, S., Read, D., and López-Cortés, A.: 1982, "Distinctive microbial structures and the prePhanerozoic fossil record," Precambrian Res. (in press).

Morowitz, H.: 1967, "Biological self-replicating systems." In: Progress in Theoretical Biology 1 (Ed. F.M. Shell), pp. 35-38, Academic Press, New York.

Nealson, K.H., and Tebo, B.: 1979, "Structural features of manganese precipitating bacteria." In: Limits of Life: Proceedings of the College Park Colloquium on Chemical Evolution (October 1979) (Ed. C. Ponnamperuma and L. Margulis), pp. 173-182, Reidel Publishing Company, Dordrecht, Holland.

Nealson, K.H.: 1982, p. 51-65, in Mineral Deposits and the Evolution of the Biosphere, Springer Verlag, Berlin, Heidelberg, New York.

Ragan, M.A., and Chapman, D.J.: 1978, A Biochemical Phylogeny of Protists, pp. 317, Academic Press, New York.

Reimer, T.O., Barghoorn, E.S., and Margulis, L.: 1979, "Primary productivity in an early Archean ecosystem," Precambrian Res. 9, pp. 93-104.

Renfro, A.R.: 1974, "Genesis of evaporite-associated stratiform metalliferous deposits--a sabkha process," Economic Geology 69, pp. 33-45.

Schwartz, R., and Dayhoff, M.O.: 1978, "Origins of prokaryotes, eukaryotes, mitochondria and chloropasts," Science 199, pp. 395-403.

Sokal, R.R., and Sneath, P.H.A.: 1963, Principles of Numerical Taxonomy, pp. 359, W.H. Freeman, San Francisco.

Sieburth, J.M.: 1979, Sea Microbes, pp. 491, Oxford University Press, New York.

Starr, M.P., Stolp, H., Trüper, H.G., Balows, A., and Schlegel, H.G.: 1981, The Prokaryotes, A Handbook on the Habitats, Isolation and Identification of Bacteria, Vol. 1 and 2, Springer Verlag, Berlin.

Stolz, J.F.: 1982, "Fine structure of the stratified microbial community at Laguna Figueroa, Baja California, Mexico. I. Methods of in situ study of the laminated sediments," Precambrian Res. (in press).

Swain, T.: 1974, "Biochemical evolution in plants." In: Comprehensive Biochemistry 29 (Ed. A.M. Florkin and E.J. Stotz), pp. 125-302, Elsevier Publishing Co., Amsterdam.

Tornabene, T.G., Langworthy, T.A., Holzer, G., and Oró, J.: 1979, "Squalenes, phytanes and other isoprenoids as major neutral lipids of methanogenic and thermoacidophilic 'Archaebacteria,'" J. Mol. Evol. 13, pp. 73-83.

Towe, K.M., and Moench, T.T.: 1981, "Electron-optical characterization of bacterial magnetite," Earth and Planetary Science Letters 52, pp. 213-220.

Van Valen, L.: 1973, "Are categories in different phyla comparable?" Taxon 22, pp. 333-373.

Walker, J.C.G.: 1980, "Atmospheric constraints on the evolution of metabolism." In: Limits to Life (Ed. C. Ponnamperuma and L. Margulis), pp. 121-132, Reidel Publishing Company, Dordrecht, Holland.

Wallace, D.C., and Morowitz, H.J.: 1973, "Genome size and evolution," Chromosoma (Berl.) 40, pp. 121-126.

Walter, M.R. (ed.): 1976, Stromatolites, pp. 790, Elsevier North Holland, Amsterdam.

FINE STRUCTURE OF THE STRATIFIED MICROBIAL COMMUNITY AT LAGUNA FIGUEROA, BAJA CALIFORNIA, MEXICO

John F. Stolz
Department of Biology, Boston University, Boston, MA 02215 USA

Laguna Figueroa is a lagoonal complex with an evaporite flat and salt marsh separated from the ocean by a barrier dune (Vonder Haar and Gorsline, 1977). The evaporite flat is composed of evaporitic minerals (anhydrite, aragonite, gypsum and halite) and sand clastics (quartz). Beneath a thin surface layer, laminated sediments with a stratified microbial community are found. These laminated sediments are a cohesive fabric of minerals, microbes and gelatinous material and cover an extensive area along the evaporite flat/salt marsh interface (Horodyski et al., 1977). Watered by seawater percolating from beneath the barrier dune these intermittently submerged sediments form characteristic desiccation polygons (Horodyski and Vonder Haar, 1975).

The hypersaline environment greatly restricts the presence of eukaryotic organisms. Two eukaryotes which are components of the community and have been studied in culture are *Dunaliella salinas*, a marine chlorophyte and *Paratetramitus jugosus*, an amoeboflagellate (Margulis et al., 1980; 1982). The diversity of prokaryotes is astounding but not unexpected as a variety of microenvironments exist in the sediment. A range of light intensities, oxygen concentrations, pH and eH can be found along the vertical profile of the laminated sediment. The laminations are due to the trapping and binding of sediment by the topmost cyanobacterial layer and the subsequent colorful stratification is caused by the species of bacteria which dominate each layer.

Light and transmission electron microscopy reveal the microbiota in the laminated sediments. The methods can be found in Stolz (1982). The top 2mm is dominated by oxygenic cyanobacteria. A variety of heterotrophic bacteria, aerotolerant green photosynthetic bacteria (*Chloroflexis* sp.) and mineral particles are packed between sheathed bundles of *Microcoleus*. Other species of cyanobacteria have been seen *in situ* (i.e., *Spirulina*, *Entophysalis*). The next 2mm of sediment comprises the anaerobic photosynthetic layer and is typically reddish-purple in color. Purple and green photosynthetic bacteria predominate. The remaining layers, anaerobic and sulfide rich, contain a plethora of heterotrophic bacteria. Some exhibit unique ultrastructural character-

P. Westbroek and E. W. de Jong (eds.), Biomineralization and Biological Metal Accumulation, 55–56.

istics. Indistinguishable by light microscopy, these very small bacteria (0.5-1 micrometer in diameter) can be distinguished ultrastructurally with the aid of transmission electron microscopy.

The microbiota of the evaporite flat at Laguna Figueroa is actively involved in the production of the laminated sediments. Analysis of these sediments by light and transmission electron microscopy clearly illustrates the ordered stratification of specific microbial species into distinct layers and the interaction of these layers as a microbial ecosystem.

REFERENCES

Horodyski, R.J., and Vonder Haar, S.: 1975, "Recent calcareous stromatolites from Laguna Mormona (Baja California) Mexico," Journal of Sedimentary Petrology 45, pp. 894-906.
Horodyski, R.J., Bloesser, B., and Vonder Haar, S.J.: 1977, "Laminated algal mats from a coastal lagoon, Laguna Mormona, Baja California, Mexico," Journal of Sedimentary Petrology 47, pp. 680-696.
Margulis, L., Barghoorn, E.S., Ashendorf, D., Banerjee, S., Chase, D., Francis, S., Giovannoni, S., and Stolz, J.: 1980, "The microbial community in the layered sediments at Laguna Figueroa, Baja California, Mexico: Does it have Precambrian analogues?" Precambrian Research 11, pp. 93-123.
Margulis, L., Grosovsky, B.D.D., Stolz, J.F., Gong-Collins, E.J., Lenk, S., Read, D., and Lopez-Cortez, A.: 1982, "Distinctive microbial structures and the prePhanerozoic fossil record," Precambrian Research (in press).
Stolz, J.F.: 1982, "Fine structure of the stratified microbial community at Laguna Figueroa, Baja California, Mexico, I. Methods of _in situ_ study of the laminated sediments," Precambrian Research (in press).
Vonder Haar, S.P., and Gorsline, D.S.: 1977, "Hypersaline lagoon deposits and processes in Baja California, Mexico," Geoscience and Man 18, pp. 165-177.

THE INFLUENCE OF BIOMINERALISATION ON THE COMPOSITION OF SEAWATER

M. Whitfield and A.J. Watson
Marine Biological Association of the United Kingdom,
The Laboratory, Citadel Hill, Plymouth PL1 2PB, England

ABSTRACT

Elements whose concentration profiles in the deep ocean are characterised by surface depletion and deep water enrichment are incorporated to a significant extent in biologically produced particles. A large proportion of this particulate material results from biomineralisation reactions in the surface layers of the ocean. Five predominant mineral phases have been identified (calcite, aragonite, celestite, barite and opaline silica). These minerals are progressively dissolved from the settling particulate matter and consequently influence the vertical distribution not only of their major constituent elements (Ca, Ba, Sr, C, Si) but also of a range of trace elements (Ge, Se, Zn, Sc, Cr, Be, Cu, Ni). The extent to which the biomineralisation process contributes to the final removal of these trace elements from the ocean is uncertain although the settling mineral fragments might provide sites for the precipitation of insoluble hydrous metal oxides which are known to scavenge a wide range of trace elements from sea water.

A. INTRODUCTION

The composition of sea water is controlled by the relative rates of input and removal of the various elements. Most elements are lost from solution in the ocean either by incorporation into the settling particulate matter that eventually forms the sediment or by reaction with the hot basalts which are exposed to sea water at the active spreading centres within the mid-oceanic ridge complex. The available evidence suggests that for most elements (with the notable exception of magnesium) removal *via* sedimentation is the dominant process. The bulk of the particulate matter formed within the ocean is of biological origin and the bulk of this material (75% by weight) results from the biologically induced precipitation of mineral phases (biomineralisation). The production and subsequent decomposition of these particles can influence the composition of sea water in two ways - (i) by altering the distribution of the various elements within the oceans and hence

P. Westbroek and E. W. de Jong (eds.), Biomineralization and Biological Metal Accumulation, 57–72.

affecting the homogeneity of sea water composition and (ii) by controlling the net rate of removal of particular elements from the ocean and hence controlling its overall composition. In this short paper we intend to explore both possibilities and to see to what extent biomineralisation might be involved in controlling the composition of sea water. Although a wide range of mineral phases can be produced by biological processes (Lowenstam 1974, 1981) for the sake of brevity we will only consider those phases that have been shown to exert a significant influence on the composition of sea water.

B. THE OBSERVED DISTRIBUTION OF THE ELEMENTS WITHIN THE OCEANS

With the development of clean-handling and sampling techniques, oceanographically consistent profiles for a wide range of trace elements are now becoming available. These profiles fall into three categories (Figure 1) dictated by the relationships between the rates of the various addition and removal processes and the mean oceanic residence times of the elements.

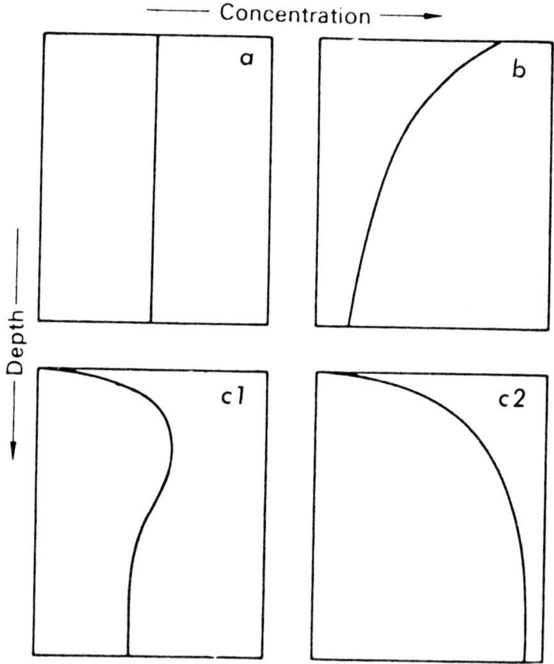

Figure 1. Profile types representative of the depth distribution of the elements in the oceans
A. Uniform distribution of well mixed elements.
B. Component added to the surface of the oceans (via rivers, atmosphere or coastal waters) and removed by scavenging at depth.
C1. Component removed from the surface by biological production of organic soft parts which are largely regenerated within the thermocline.
C2. Component removed from surface by biomineralisation with the mineral detritus being redissolved at depths.

Elements with very long residence times (> $5x10^5$ yr, Whitfield 1979) are uniformly distributed throughout the water column (A, Figure 1) and their concentration profiles are only slightly disturbed by small changes in the total salt content (or salinity). Elements showing this behaviour include the alkali metals, the halogens (with the exception of I) and Mg, S, B, U, V and Mo.

Since, for most elements, the major input is at the surface and the major sink is in the deep ocean one would expect elements that are not so effectively mixed in the oceans to show profiles characterised by high levels at the surface with the concentration decreasing with depth (B, Figure 1). Such behaviour is observed for relatively few elements (Mn, Pb, Co, Th and Cr(III), see Table 1). For these elements the most important removal mechanism appears to be scavenging by adsorption on the surface of other particles - most of which are of biological origin.

The most commonly observed profile is characterised by surface depletion with the concentration increasing, although not always uniformly, with depth (C1 and C2, Figure 1). Elements exhibiting this type of behaviour (Table 1) are involved in biological cycles initiated by the formation of particulate matter in the sun-lit euphotic zone, primarily by photosynthesis. The elements are initially incorporated into phytoplankton cells, either incidentally or by design. These cells, being slightly denser than the surrounding water, will tend to sink but the majority of them will be intercepted by grazing organisms. These organisms will in turn provide a food source for a sequence of predators and the fecal pellets and detrital fragments produced will fall into deeper water where respiration and dissolution will conspire to decompose the particles and return the entrapped elements to the seawater. The surface water consequently becomes depleted and the deep water enriched with the elements recycled in this way. Furthermore a marked fractionation is produced between the Atlantic and Pacific Oceans. Deep water is generated primarily in the North Atlantic and in the vicinity of the Antarctic continent and the general circulation flows southward through the Atlantic, into the circumpolar Antarctic current and hence northward into the Indian and Pacific Oceans. The deep water in the Pacific Ocean is consequently older than that in the Atlantic and it has been able to accumulate a greater burden of the elements released by the dissolution of particulate matter. Hence a second feature, characteristic of the biologically recycled elements, is the occurrence of greater deep water to surface water concentration ratios in the Pacific than in the Atlantic.

Elements may be incorporated into the biogenous particles in at least two ways - (i) by bio-synthesis to produce specific organic compounds containing the elements (organic 'soft parts', type C1, Figure 1) or (ii) by biomineralisation to produce mineral phases (mineral 'hard parts') where the elements are incorporated as major constituents or indicentally as trace components in the mineral structure (type C2, Figure 1).

Table 1. Characteristics of elements showing heterogeneous distributions in the oceans

Element	Concentration ($nmol\ kg^{-1}$)	Residence time (yrs)	Profile type[a]	Notes	Reference
As	15-25	$6.3x10^4$	C1	15-40% deep water enrichment	Braman *et al.* (1977) Andreae (1977, 1979)
Ba	32-150	$1.1-1.4x10^4$	C2	No simple relationship to Si. 30% hydrothermal input	Chen *et al.* (1977) Edmond *et al.* (1979)
Be	$4-30x10^{-3}$	500	C2	Sediment remobilisation apparent	Measures & Edmond (1982)
C	$2-2.5x10^3$	$7.6x10^5$	C2	$CaCO_3$ removal	
Ca	$10.3x10^3$	$0.9-1.04x10^6$	C2	1.5% deep water enrichment in Pacific	
Cd	$1x10^{-3}-1.1$	$4.3x10^2$	C1	Strongly linked to P shallow regeneration	Boyle *et al.* (1976) Bruland *et al.* (1979)
Co	$17-119x10^{-3}$	$1.7x10^4$	B		Knauer *et al.* (1982)
Cr	2-5	$1.0x10^4$	C2(CrVI) B(CrIII)	Cr(VI) shows only slight surface depletion	Cranston & Murray (1978) Campbell & Yeatts (1981)
Cu	0.5-6	$2.6x10^3$	C1/C2	Deep water scavenging plus sediment remobilisation	Boyle & Edmond (1975) Boyle *et al.* (1977) Bruland (1980)
Ge	7-120		C2	Close correlation with Si	Andreae & Froelich (1981)
I	$0.2-0.45x10^3$	$2.9x10^5$	C1(IO_3^-) B(I^-)		Wong & Brewer (1977) Elderfield & Truesdale (1981)
Mn	0.2-3	$9.8x10^2$	B	Very effective scavenging plus some sediment remobilisation	Klinkhammer & Bender (1980) Landing & Bruland(1980) Martin & Knauer (1980)

Table 1 (contd.)

Element	Concentration (nmol kg^{-1})	Residence time (yrs)	Profile type[a]	Notes	Reference
N	$<0.1-45\times10^{3}$	5.8×10^{3}	C1	Thermocline maximum	
Ni	3-12	1.9×10^{5}	C1/C2	Shows both shallow and deep water regeneration	Sclater *et al.* (1976) Bruland (1980) Boyle *et al.* (1981)
P	$<0.1-3.5\times10^{3}$	1.0×10^{5}	C1	Thermocline maximum	
Pb	$5-175\times10^{-3}$	3.4×10^{2}	B	Atmospheric input plus effective scavenging	Schaule & Patterson (1981)
Sc	$8-20\times10^{-3}$	5.0×10^{3}	C2	Only slight deep water enrichment	Brewer *et al.* (1972)
Se	0.5-2.3	3.4×10^{4}	C2	Se IV associated with hard parts, Se VI associated with both hard and soft parts	Measures & Burton (1980a) Measures *et al.* (1980)
Si	$<1-180\times10^{3}$	3.5×10^{4}	C2		
Zn	0.05-9	8.3×10^{3}	C2	Correlation with Si	Bruland *et al.* (1978) Bruland (1980)

[a] See Figure 1

C. INCORPORATION OF THE ELEMENTS INTO ORGANIC 'SOFT PARTS'

Elements incorporated into organic 'soft parts' will be released rapidly into the water column as the organic material is degraded by microbial activity. These elements consequently often show a maximum in the thermocline (type C1, Figure 1) where the microbial respiration is most rapid. The deep water to surface ratios of these elements are consequently not so strongly affected by the general oceanic circulation pattern so that values observed in the Pacific for N and P are only twice as high as those observed in the Atlantic. In addition to N and P, Cd and to a lesser extent I and As also exhibit this behaviour. Cd is a (b)-type metal showing a strong affinity for S-containing ligands (Turner et al., 1981). It forms only moderately strong chloro-complexes in sea water so that it can be effectively bound by S-containing proteinaceous material. For cadmium the link with organic soft parts is so effective that good linear correlations have been observed between the concentration of Cd and P throughout the world oceans. The correlations with I and As are not so strong although both may be expected to behave like the major nutrient elements. Iodine is known to form a wide range of stable organic compounds and arsenic in the pentavalent form ($HAsO_4^{2-}$) is closely related to hydrogen phosphate (HPO_4^{2-}) which is actively taken up by growing phytoplankton. Ni may also show some shallow water regeneration but the concentration profiles of this element are complicated by its propensity for being incorporated in both mineral hard parts and organic soft parts.

D. INCORPORATION OF THE ELEMENTS INTO MINERAL 'HARD PARTS'

The majority of the elements that show increasing concentration with depth appear to be associated with mineral 'hard parts' encased in the fecal and detrital matter raining down from the surface layer. The solubility of these mineral phases increases as a result of the progressive rise in pressure resulting from the increasing overburden of water (Millero 1982). Consequently elements contained in the mineral phases will be gradually released, producing profiles that increase monotonically with depth (C2, Figure 1). The fractionation of these elements between the Atlantic and the Pacific Oceans is consequently very pronounced. The deep water to surface water concentration ratios for barium and silicon, for example, are seven and five times higher, respectively, in the Pacific than in the Atlantic. There are reliable indications (see references to Table 1) to suggest that the concentration profiles of thirteen elements in sea water (C, Ca, Sr, Ba, Si, Ge, Se, Zn, Sc, Cr, Ni, Cu, Be) are strongly influenced by biomineralisation at the surface and subsequent mineral dissolution in the deep ocean. We will now consider the mineral cycles involved (Tables 2 and 3).

Table 2. Source and degree of saturation of biogenous mineral phases

Phase	Source	Saturation index[a] at 1 atm pressure	Reference
Calcite	Foraminifera, Coccolithophorids	5.5	Broecker *et al.* (1979)
Aragonite	Pteropods	3.7	Broecker *et al.* (1979)
Celestite	Acantharid Radiolaria	0.09	Hanor (1969)
Barite	Radiolaria (?)	0.5	Hanor (1969)
Opaline Silica[b]	Diatoms, Radiolaria	1.7 (Quartz) 0.09 (Silica gel)	Nichols (1976)

[a] Saturation index = (activity product in seawater)/solubility product. For values less than unity the water column is undersaturated with respect to the mineral phase. The solubility of all mineral phases increases with increasing pressure and (with the exception of the $CaCO_3$ phases) with increasing temperature.

[b] The solubility of opaline silica is closer to that of silica gel than quartz.

Table 3. Fluxes associated with the formation and dissolution of the major mineral phases (Lerman & Lal 1977, Dehairs *et al.* 1980)[a]

Pathway	C(inorg.)	C(org.)	Sr	Ba	Si
Surface to Deep Ocean	1.5[a]	3.6[a]	1.3[b]	5.0[c]	4.8[d]
Water Column Dissolution	0.53 (35%)[e]	2.8 (78%)	1.17 (90%)	3.3 (66%)	2.4 (50%)
Sediment Surface to Water Column	0.79 (53%)	0.70 (21%)	0.11 (8.5%)	1.1 (22%)	2.3 (48%)
Deep Sediment to Water Column	0.02	0.06	0.03	-	-
Final Burial	0.15	0.04	0.21	0.60	0.10

[a] These figures are close to those obtained in careful recalculations by Berner (1977) and Eppley and Peterson (1979) respectively
[b] Units 10^{14} g yr^{-1}
[c] Units 10^{12} g yr^{-1}. Units for remaining columns 10^{15}g yr^{-1}
[d] Comparable figures are given by De Master (1981)
[e] Values in brackets indicate the percentage of the particulate flux redissolved at each stage

E. CYCLES OF THE MAJOR BIOGENOUS MINERALS

a. Calcium Carbonate

All surface waters are supersaturated with respect to calcite and aragonite (Table 1). Since $CaCO_3$ is more soluble in cold water than in warm water, the extent of supersaturation depends on latitude and varies from 2.5 to 5.5 fold for calcite and from 1.7 to 3.7 fold for aragonite. Calcite is produced primarily by coccolithophorids and foraminifera and aragonite is produced primarily by pteropods. The solubility of calcite increases with depth because of a combination of increasing pressure and falling temperature and pH. The depth at which the water becomes just saturated with respect to $CaCO_3$ is known as the lysocline. Typical lysocline depths of the N. Atlantic are 3.2 km (aragonite) and 4.6 km (calcite) and for the N. Pacific 0.6 km (aragonite) and 0.8 km (calcite) (Takahashi et al. 1981), the older water of the Pacific being more acidic than that of the Atlantic. Although sedimentary $CaCO_3$ will tend to dissolve below this depth there is a region where the particles are settling more rapidly than they are dissolving so that $CaCO_3$ is still found in the sediment. The depth at which $CaCO_3$ eventually disappears from the sediment is known as the carbonate compensation depth. Calcitic oozes are accumulating over less than half of the ocean floor (47.1%, Berger 1976) whereas aragonitic oozes cover less than 1% of the ocean floor (0.6%, Berger 1976) despite the fact that both mineral phases are produced in approximately equal quantities in the euphotic zone (Berner 1977). Apart from the depth restrictions imposed by the solubility horizon, calcareous sediments are distributed throughout the world oceans. The percentage of the total calcium carbonate flux that redissolves ranges from 69% in the Atlantic to 83% in the Pacific (Berner 1977). This is sufficient to produce a 1.5% enrichment of Ca and a 15% enrichment of HCO_3^- in the deep sea. The preservation of $CaCO_3$, notably in the Atlantic, is sufficient to provide the major flux of both calcium and carbon from the oceans. A number of authors have studied the removal of trace elements by co-precipitation with inorganically precipitated $CaCO_3$ (see for example Kitano et al. 1971, 1974, 1980, Lorens 1981) and by association with calcareous tests (Boyle 1981, Dehairs et al. 1981). The low solubility of $CaCO_3$ would appear to make it an ideal vehicle for the removal of trace elements from the ocean. Strontium and barium are possible candidates for removal in this way because of their chemical similarity to calcium and we will consider their cycles separately.

b. Strontium Sulphate (Celestite)

The whole water column is undersaturated with respect to celestite (Table 2) and the degree of undersaturation increases with depth. Celestite is incorporated into the spicules of Acantharid Radiolara and the biogenous precipitation of this mineral in surface waters followed by redissolution in the deep ocean produces a slight (1.5%) enrichment of the deeper waters with strontium (Brass &

Turekian 1972). According to Brass (1980) the flux of acantharid spicules from the surface layer carries with it 30 μg Sr cm^{-2} yr^{-1}. This is equivalent to a whole ocean flux of 1.1 x 10^{14} g Sr yr^{-1}. The redissolution at depth is complete so that celestite is not preserved in deep-sea sediments and it is necessary to look elsewhere for a removal process to balance the river input of strontium. The Ca:Sr ratio in calcite is close to that observed in mean world river water (540:1). Using this ratio and a value of 5 x 10^{15} g Ca yr^{-1} for the transport of Ca via $CaCO_3$ deposition (Table 3) we obtain a flux of 2 x 10^{13} g Sr yr^{-1} associated with the calcareous material. The bulk of the strontium is therefore removed via $CaCO_3$ deposition (Broecker 1974). The rapid recycling of strontium via celestite only serves to redistribute the element within the oceans (Table 3). Although a range of trace elements are known to be associated with acantharid spicules (Brass 1980) it is unlikely that the flux of these fragments makes any significant contribution to their oceanic distribution.

c. Barium Sulphate (Barite)

Thermodynamic calculations suggest that the oceans are undersaturated throughout with respect to barite (Table 2) although not to the same degree as celestite. For example the barium concentration in equilibrium with barite in sea water would be 255 nmol kg^{-1} at 25°C and 1 atm and 175 nmol kg^{-1} at 1°C and 1 atmosphere (Church & Wolgemuth 1972). At 1°C and 500 atm the corresponding value is 358 nmol kg^{-1}. These values may be compared with the typical range of values from 32 nmol kg^{-1} at the surface to 150 nmol kg^{-1} in the deep ocean (Table 1). Despite this degree of undersaturation this mineral is precipitated not as structural units in the exoskeletons of organisms but as virtually pure, isolated barite crystals within fecal and detrital particles and possibly within live planktonic organisms (Dehairs et al. 1980). Flux calculations indicate that crystalline barium accounts for nearly two thirds of the particulate barium produced in the euphotic zone (0.8 μg cm^{-2} yr^{-1}). The balance being made up by barium incorporated into $CaCO_3$ (0.24 μg cm^{-2} yr^{-1}), SiO_2 (0.31 μg cm^{-2} yr^{-1}) and fast settling fecal material (0.03 - 0.5 μg cm^{-2} yr^{-1}). The bulk (80-90%) of the barium associated with the $CaCO_3$ and SiO_2 phases is redissolved in the deep ocean together with half of the barite. The flux of barite crystals to the sediment may make up nearly half of the total barium flux out of the water column. Unlike the more soluble celestite, barite is a widely distributed minor component of shallow to intermediate depth sediments (Elderfield 1976). It rarely constitutes more than 1% of the carbonate-free sediment although higher concentrations (up to 10% by weight of the carbonate-free sediment) can be found along the east Pacific rise. However the barite found in these richer deposits contains 5 mole % of celestite and its distribution pattern is similar to that of hydrous iron oxides suggesting that it might be precipitated from Ba-rich waters of hydrothermal origin (Elderfield 1976). Although it is possible that such Sr-rich barites might be precipitated directly from seawater (Hanor 1969), the wide variation in the Sr:Ba ratios in the authigenic barite crystals found in the water column, and the presence of occasional Sr-free crystals, would suggest biologically mediated formation rather than simple chemical precipitation (Dehairs _et al._ 1980). Barite is formed by the benthic protozoan Xenophyophora (Tendal 1972) and high Ba levels are found in some radiolaria (Martin and Knauer 1973). It is also possible that barite precipitation might be induced in decaying organic matter following sulphate enrichment in zones where sulphide oxidation is occurring (Chow and Goldberg 1960). Barite remains unique in that it is the only biogenous mineral that has been shown to exert a significant influence on the distribution and removal of an element within the oceans without itself performing any specific structural function within the organism.

d. Opaline Silica

Although the oceans are undersaturated throughout with respect to opaline silica (Table 2) this mineral is deposited by diatoms, radiolaria, siliceous sponges and silicoflagellates. The redissolution of silica is very efficient (Table 3) and close to 50% of the silica produced in the euphotic zone is redissolved in the deep ocean with a further 48% being returned to the water column by redissolution of sedimentary material. Because of their higher solubility, siliceous oozes are less common than calcareous oozes and are found over less than 15% of the ocean floor (Berger 1976). Some siliceous tests are more corrosion resistant than others either because of their dense structure or because the inorganic material is covered with a persistent protective coating of organic matter. The major deposits are diatomaceous oozes (77%) with the balance (23%) being made up by radiolarian oozes. The high dissolution rates of opaline silica assure that siliceous sediments only accumulate under regions of high productivity. More than one third of the silica accumulating in the sediment is consequently found beneath the Antarctic polar front and at the northern extremities of the Atlantic and Pacific oceans (De Master 1981) and the general distribution of siliceous oozes is more patchy than for calcareous oozes with no clear trend from the Atlantic to the Pacific (Berger 1976). Despite the high solubility of opaline silica, this mineral is able to account for the removal of up to 70% ($3.2 - 4.4 \times 10^{14}$ g SiO_2 yr^{-1}) of the Si entering the oceans _via_ the world's rivers (4.2×10^{14} g SiO_2 yr^{-1}) and from hydrothermal emanations (1.9×10^{14} g SiO_2 yr^{-1}) (De Master 1981). Consequently the biogenous silica cycle dominates both the oceanic distribution and the overall concentration of Si within the oceans.

Detailed studies of the oceanic profiles of several trace elements (notably Ge and Zn, but, to a lesser degree, Se, Sc, Cr, Be, Cu and Ni) suggest that the distribution of these elements within the ocean may be related in some way to that of silicon. Particularly clear correlations are observed between the profiles of Si and Ge (Andreae and Froelich 1981) as might be expected from the close chemical similarity of these elements. The Ge:Si ratio described in the water column (0.7×10^{-6}) is in fact close to that found in opaline silica suggesting a causal relationship. Unexpectedly strong correlations are also observed between the distributions of Si and Zn in both the Atlantic and the Pacific (Bruland et al. 1978, Bruland 1980). For example in the north Pacific Bruland (1980) was able to show a significant linear regression between Zn and Si concentrations such that

$$[Zn]/\text{nmol kg}^{-1} = (0.0535 \pm 0.0008)\ [Si]/\mu\text{mol kg}^{-1} + (0.02 \pm 0.09)$$

with a regression coefficient of 0.996. Different correlation parameters are required for the Atlantic and the Pacific because the deep water enrichment of Zn in the Pacific is not as pronounced as that of Si suggesting that some auxiliary removal mechanism might be at work for Zn. In addition a good correlation is observed between Ca and Si in the Pacific (Boyle 1981) suggesting that the link between Zn and Si distribution might not be causal.

The deep ocean profiles of the other elements showing enrichment patterns similar to those of Si are rather more complex. Sc, Se (VI) and Cr (VI) show only slight deep water enrichment and the relationship of the profiles of these elements to those of Si is at present more intuitive than quantitative suggesting simply that similar processes are at work in the recycling of these elements. Ni and to some extent Cu show evidence of both shallow and deep water enrichment suggesting that they are taken up in both organic soft parts and in mineral hard parts. In a number of instances (Be, Cu and frequently Se) the deep ocean profiles show excessive deep water enrichment resulting from the release of elements from particulate matter during early diagenesis in the sediment. Such enrichments are also observed for Mn which is generally very efficiently scavenged from the water column.

Only in the case of Ge can clear causal relationships be established to link elemental cycling with the formation and dissolution of opaline silica. For the other elements the evidence is circumstantial and much more work is required to unravel the various processes that may influence their deep ocean concentration profiles.

e. Hydrous Metal Oxides - Deep Water Scavenging

We have already indicated (Section B) that the oceanic profiles of a number of elements (Mn, Pb, Co, Th and Cr (III)) are dominated by progressive removal from solution by surface adsorption on scavenging particles as they settle through the water column. As the vast bulk of the particulate matter formed within the oceans is of biological origin it is interesting to consider the role that biomineralisation might play in this scavenging process. Martin (1970) observed that the concentrations of Fe, Mn, Pb, Cu, Zn and Sr were significantly higher in the exoskeletons of zooplankton collected below 200 m than in those collected at the surface. The first three elements are certainly known to be susceptible to removal by scavenging but any mechanism that is to remain active throughout the water column must be based on detrital material that can persist to great depths in the oceans. A number of scavenging models have been proposed invoking adsorption on to silica and silicate minerals (Schindler 1975), organic coatings (Balistieri et al. 1981) and hydrous metal oxides (Li 1981). In all instances fecal material and the associated biogenous minerals provide the most likely template for the adsorption process. The most wide-ranging model is that of Li (1981) who favours the removal of trace elements by adsorption on to hydrous iron and manganese oxides. Careful analysis of the surface composition of deep sea particulate matter indicates that siliceous and calcareous tests frequently provide a site for the deposition of such phases (W.R. Simpson, personal communication). The effectiveness with which the scavenging process can influence the deep ocean concentration profile of an element will depend on the relationship between the scavenging rate constant (k_s yr^{-1}) and the transit time of particles in the water column (Brewer and Hao 1979). The profiles of Cd ($k_s = 10^{-5.25}$ yr^{-1}) and Ni ($k_s = 10^{-4.2}$ yr^{-1}) show little sign of deep water scavenging whereas the profiles of Th ($k_s = 10^{0.4}$ yr^{-1}) and Pb ($k_s = 10^{-2.7}$ yr^{-1}) are wholly controlled by this process. The profiles of Cu ($k_s = 10^{-3.05}$ yr^{-1}) show clear signs of deep water scavenging but they are further complicated by the release of copper from the sediment during early diagenesis. By providing a site for the deposition of iron and manganese oxyhydroxides within the water column the major biogenous mineral phases might incidentally, be providing an important route for the removal of a range of trace elements from sea water.

F. CONCLUSIONS

As the evidence accumulates it is becoming clear that the cycling of the majority of trace elements within the ocean is dominated by surface depletion and deep-sea enrichment driven by biological processes. Biomineralisation makes a significant contribution to this internal cycling and at least five mineral phases are involved (calcite, aragonite, celestite, barite and opaline silica). Only in the case of the $CaCO_3$ phases is the surface water saturated with respect to the mineral phase (Table 2). Even in this instance inorganic precipitation is a rare occurrence - restricted to warm, shallow, productive waters. The precipitation of all five mineral phases is therefore quite definitely under biological control. The mineral phases are in general produced to perform specific structural functions although the circumstances surrounding the production of barite crystals are still not understood. In addition to the primary constituents of these mineral phases, a number of trace elements (Table 1) show profiles that exhibit comparable deep-water regeneration. At this stage, however, it is not clear to what extent transport to the sediment via biogenous mineral phases provides a route for the final removal of these trace elements from solution. Certainly a considerable amount of dissolution occurs in the upper layers of the sediment (Table 3) and less than 1% of the silica deposited, for example, is actually removed from the system. In the case of celestite the redissolution is complete and Sr removal is achieved by coprecipitation with the much less soluble calcite. It is also possible that the open, filigree structures of the siliceous and calcareous tests lend themselves to a more subtle form of trace element removal. These slowly settling, large surface area structures are ideal sites for the precipitation of highly insoluble iron and manganese hydrous oxides which are able to scavenge out a wide range of trace metals from solution (Li 1981). The elements associated with these oxides will only be remobilised within the sediments if reducing conditions are established. Even under these conditions other minerals might be precipitated (e.g. $MnCO_3$, $Mn_3(PO_4)_2$, FeS, CuS) as a consequence of the microbial oxidation of organic matter and the microbially induced reduction of inorganic compounds. It has been suggested, for example (Piper and Codispoti 1975), that the concentration of phosphate in the oceans might be controlled by the formation of phosphorites (carbonate fluorapatites) on the sea floor beneath regions of high productivity (Atlas and Pytkowicz 1977, Baturin and Bezrukov 1979). Since the N:P ratios in the deep oceans are closely similar to the average value found in marine organisms such a mechanism would also control the nitrogen content of the oceans and hence the biological productivity of the surface layers (Morris 1974, Piper and Codispoti 1975). Although sea water appears to be slightly supersaturated with respect to this mineral (Atlas and Pytkowicz 1977) there is little evidence for phosphorite formation at the present time. However, there is clearly ample scope for discussing the significance of sedimentary biomineralisation.

It is also likely that significant removal fluxes are associated with interactions between sea water and hot basalts along the mid-oceanic ridge complex. For Mg, for example, such removal processes constitute the major output from the oceans (Edmond et al. 1979). The incidental removal of elements from the oceans by scavenging and by hydrothermal reactions would appear to be more general than the removal of elements by the biota in selectively precipitated mineral phases. Certainly the clear correlations observed between fundamental chemical parameters and the partitioning of the element between sea water and the particulate phase (Turner et al. 1980; Turner & Whitfield 1982; Li 1981) would suggest an overall geochemical rather than biological control of seawater composition in the long term (millions of years). To understand the influence of the biota on sea water composition in the short to medium term (years to millenia) it will be necessary to construct dynamic models of the ocean system that take into account the influence of the biogenous particle flux on the vertical partitioning of the elements. It will also be necessary to accumulate much more data on the distribution of the elements within the oceans. In some instances our knowledge is restricted to a single deep-water profile (e.g. for Be, Co, Ge). By developing our understanding in this way we will be able to decide whether the feedback loops resulting from the interlinking of so many elemental cycles by biosynthesis and biomineralisation are able to exert any beneficial influence on the critical environment at the ocean surface.

REFERENCES

Andreae, M.O.: 1979, Limnol. Oceanogr. 24, p.440.
Atlas, E. and Pytkowicz, R.M.: 1977, Limnol. Oceanogr. 22, p.290.
Balistieri, L., Brewer, P.G. and Murray, J.W.: 1981, Deep-Sea Res. 28A, p.101.
Baturin, G.N. and Bezrukov, P.L.: 1979, Mar-Geol., 31, p.317.
Berger, W.H.: 1976, In "Chemical Oceanography" (eds. J.P. Riley and G. Skirrow) Vol. 5, pp.266-388. Academic Press, New York.
Berner, R.A.: 1977, In "The fate of fossil fuel CO_2 in the Oceans" (eds. N.R. Andersen & A. Malahoff), pp.243-260. Plenum Press, New York.
Boyle, E.A.: 1981, Earth planet. Sci. Lett., 53, p.11.
Boyle, E.A. & Edmond, J.M.: 1975, Nature, 253, p.107.
Boyle, E.A., Huested, S.S. and Jones, S.P.: 1981, J. Geophys. Res. 86, p.8018.
Boyle, E.A., Sclater, F. and Edmond, J.M.: 1976, Nature, 263, p.42.
Boyle, E.A., Sclater, F. and Edmond, J.M.: 1977. Anal. Chim. Acta, 91, p.189.
Braman, R.S., Johnson, D.L., Foreback, C.C., Ammons, J.M. and Bricker, J.L.: 1977, Anal. Chem. 49, p.621.
Brass, G.W.: 1980, Limnol. Oceanogr. 25, 146.
Brass, G.W. and Turekian, K.K.: 1972, Earth Planet. Sci. Lett., 16, p.117.
Brewer, P.G. and Hao, W.M.: 1979, In "Chemical Modelling in Aqueous Systems" (ed. E.A. Jenne) ACS Symposium Series 93, pp.261-274. An. Chem. Soc. Washington.
Brewer, P.G., Spencer, D.W. and Robertson, D.E.: 1972, Earth. Planet. Sci. Lett., 16, p.111.
Broecker, W.S.: 1974, Chemical Oceanography. Harcourt, Brace and Jovanovich, New York.
Broecker, W.S., Takahashi, T., Simpson, H.J. and Peng, T.-H.: 1979, Science, 206, p.409.
Bruland, K.W.: 1980, Earth Planet. Sci. Lett., 47, p.176.
Bruland, K.W., Knauer, G.A. and Martin, J.H.: 1978, Nature, 271, p.241.
Bruland, K.W., Franks, R.P., Knauer, G.A. and Martin, J.H.: 1979, Anal. Chim. Acta, 105, p.233.
Campbell, J.A. and Yeats, P.A.: 1981, Earth Planet. Sci. Lett. 53, p.427.
Chan, L.H., Drummond, D., Edmond, J.M. and Grant, B.: 1977, Deep-Sea Res., 24, p.613.
Chow, T.J. and Goldberg, E.D.: 1960, Geochim. Cosmochim. Acta, 20, p.192.
Church, T.M. and Wolgemuth, K.: 1972, Earth Planet. Sci. Lett., 15, p.35.
Cranston, R.E. and Murray, J.W.: 1978, Anal. Chim. Acta, 99, 275.
Dehairs, F., Chessellet, R. and Jedwab, J.: 1980, Earth Planet. Sci. Lett., 49, 528.
DeMaster, D.J.: 1981, Geochim. Cosmochim. Acta, 45, 1715.
Edmond, J.M., Measures, C., McDuff, R.E., Chan, L.H., Collier, R., Grant, B., Gordon, L.I. and Corliss, J.B.: 1979, Earth Planet. Sci. Lett., 46, p.1.

Elderfield, H.: 1976, In "Chemical Oceanography" (eds. J.P. Riley & G. Skirrow) Vol. 5, pp.137-216, Academic Press, New York.
Elderfield, H. and Truesdale, V.W.: 1981, Earth Planet. Sci. Lett., 50, p.105.
Eppley, R.W. and Peterson, B.J.: 1979, Nature, 282, p.677.
Froelich, P.N. and Andreae, M.O.: 1981, Science, 213, p.205.
Hanor, J.S.: 1969, Geochim. Cosmochim. Acta, 33, p.894.
Kitano, Y., Kanamori, N. and Oomori, T.: 1971, Geochem. J., 4, p.183.
Kitano, Y., Kanamori, N., Tokuyama, A. and Oomori, T.: 1974, Proceedings of Symposium on Hydrogeochemistry and Biogeochemistry, Tokyo, September 7-9, 1970. Vol. I - Hydrogeochemistry, pp.484-499. Washington, D.C., The Clarke Company.
Kitano, Y., Okumura, M. and Idogaki, M.: 1980, Geochem. J., 14, p.167.
Klinkhammer, G.P. and Bender, M.L.: 1980, Earth Planet. Sci. Lett., 46, p.361.
Knauer, G.A., Martin, J.H. and Gordon, R.M.: 1982, Nature, 297, p.49.
Landing, K. and Bruland, K.W.: 1980, Earth Planet. Sci. Lett., 49, p.45.
Lerman, A. and Lal, D.: 1977, Am. J. Sci., 277, p.238.
Li, Y-H.: 1981, Geochim. Cosmochim. Acta, 45, p.1659.
Lorens, R.B.: 1981, Geochim. Cosmochim. Acta, 45, p.553.
Lowenstam, H.A.: 1974, In "The Sea" (ed. E.D. Goldberg) Vol. 5, pp.715-796, Wiley, New York.
Lowenstam, H.A.: 1981, Science, 211, p.1126.
Martin, J.H.: 1970, Limnol. Oceanogr., 15, p.756.
Martin, J. and Knauer, G.A.: 1973, Geochim. Cosmochim. Acta, 37, p.1639.
Martin, J.H. and Knauer, G.A.: 1980, Earth Planet. Sci. Lett., 51, p.266.
Measures, C.I. and Burton, J.D.: 1980, Earth Planet. Sci. Lett., 46, p.385.
Measures, C.I. and Edmond, J.M.: 1952, Nature, 297, p.51.
Measures, C.I., McDuff, R.E. and Edmond, J.M.: 1980, Earth Planet. Sci. Lett., 49, p.102.
Millero, F.J.: 1982, Geochim. Cosmochim. Acta, 46, p.11.
Morris, I.: 1974, Sci. Progr. Oxford, 61, p.91.
Nicholls, G.D.: 1976, In "Chemical Oceanography" (eds. J.P. Riley & G. Skirrow), Vol. 5, pp.81-101, Academic Press, London.
Piper, D.Z. and Codispoti, L.A.: 1975, Science, 188, p.15.
Schindler, P.W.: 1975, Thalassia Jugoslavia, 11, p.101.
Schaule, B.K. and Patterson, C.C.: 1981, Earth Planet. Sci. Lett., 54, p.97.
Sclater, F.R., Boyle, E. and Edmond, J.M.: 1976, Earth Planet. Sci. Lett., 31, p.119.
Takahashi, T., Broecker, W.S. and Bainbridge, A.E.: 1981, In "Carbon Cycle Modelling" (ed. B. Bolin) pp.271-286, SCOPE 16, J. Wiley & Sons, Chichester.
Tendal, O.S.: 1979, Galathea Report 12, p.8.
Turner, D.R., Dickson, A.G. and Whitfield, M.: 1980, Mar. Chem., 9, p.211.
Turner, D.R., Whitfield, M. and Dickson, A.G.: 1981, Geochim. Cosmochim. Acta, 45, p.855.
Whitfield, M.: 1979, Mar. Chem., 8, p.101.
Whitfield, M. and Turner, D.R.: 1982, In "Trace Metals in Seawater" (ed. C.S. Wong) Plenum Press, In press.

Wong, G.T.F., Brewer, P.G. and Spencer, D.W.: 1976, Earth Planet. Sci. Lett., 32, p.441.

THE GLOBAL CARBONATE-SILICATE SEDIMENTARY SYSTEM -- SOME FEEDBACK RELATIONS

R. M. Garrels
Dept. of Marine Science
Univ. of South Florida
St. Petersburg, Florida

R. A. Berner
Dept. of Geology & Geophysics
Yale University
New Haven, Connecticut

ABSTRACT

The steps required to make a box model of the important reservoirs and fluxes in the global geochemical cycle of carbon dioxide are detailed in a series of equations and diagrams. A summary schematic model is constructed that permits material transfer between various sedimentary reservoirs without changing the CO_2 content of the atmosphere or of the ocean. Then the actual present-day reservoirs and fluxes of various carbonates and silicates are presented, and discussed in terms of controls of atmospheric CO_2 fluctuation caused by inorganic processes.

INTRODUCTION

The sedimentary rock cycle is usually described in terms of the weathering and erosion of rocks on the continents, transport of the resultant materials to the oceans, deposition of new sediments with the same compositions as the old ones, and eventual uplift of the new sediments to make them available for weathering and erosion and thus complete the cycle. In this simplest view, the cycle could be classed as a cannibalistic steady-state system.

Every geologist knows that there are many important differences between the real world and this simplistic model. Yet it is useful to approach reality by first making schematic box models of steady state cannibalistic systems. These systems can then be modified, still schematically, to accord with observed deviations from the initial model. Next, estimates can be obtained for actual sedimentary reservoir sizes and for their fluxes in and out of the oceans. Finally, mathematical functions can be assigned to reservoir-flux relationships, and the effects of perturbation of the model assessed in terms of rates of change of the sizes of various reservoirs as a function of time (e.g., Garrels, Lerman and Mackenzie, 1976).

P. Westbroek and E. W. de Jong (eds.), Biomineralization and Biological Metal Accumulation, 73–87.

This sequence of model development will be illustrated by the global cycles of Ca-Mg carbonate and silicate rocks.

STEADY-STATE SYSTEM FOR Ca AND Mg CARBONATES AND SILICATES

A steady-state system for Ca and Mg silicates and carbonates is shown in Figure 1. The compositions of the rocks involved have been stylized.

Calcium silicate ($CaSiO_3$) stands for all calcium silicate minerals in continental rocks being eroded. The most important species is calcium feldspar in igneous and metamorphic rocks, so that the $CaSiO_3$ reservoir is represented chiefly by crystalline rocks exposed in the great shields of the continents. Magnesium silicate ($MgSiO_3$) represents both sedimentary and low grade metamorphic clay minerals, many of which are rich in magnesium, and igneous and metamorphic rock minerals such as olivines, pyroxenes, and amphiboles.

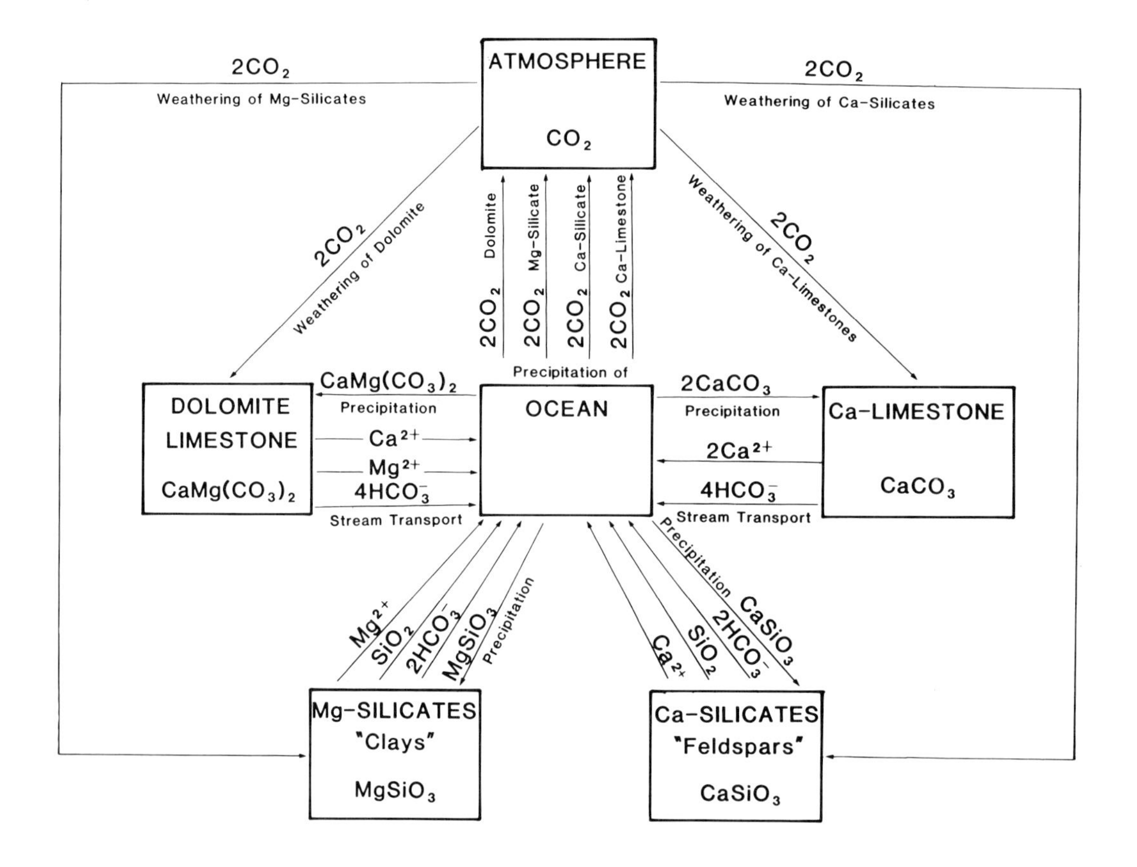

Figure 1. Idealized cannibalistic steady-state global cycling system for calcium and magnesium silicates and carbonates.

Almost all the calcium in sedimentary rocks is in carbonate minerals; only a small fraction resides in detrital plagioclase and in the interlayers of clays. One cannot give numbers to the masses of $MgSiO_3$ and $CaSiO_3$ in their reservoirs; these two silicates are important parts of the continental and oceanic crust. Their influence in the Ca-Mg cycles is measured by the areas over which they crop out, and the location of the outcrops with respect to climate.

The calcite ($CaCO_3$) reservoir represents all the carbonate minerals of sedimentary rocks made up of calcite or aragonite (both $CaCO_3$). Most of the $CaCO_3$ is in limestones; a little is found in sandstones and shales.

The dolomite reservoir represents the total mass of the mineral dolomite ($CaMg(CO_3)_2$), which is found chiefly in dolomitic limestones. A silica (SiO_2) reservoir (Fig. 2) refers to any mineral species of SiO_2, such as quartz or opal.

This model assumes that these five mineral reservoirs fit the cannibalistic steady-state model. The fluxes shown are simply the smallest integral coefficients that can be used to illustrate the chemical reactions involved. Each reservoir is weathered by reaction of CO_2. The products are carried to the ocean, where they are reprecipitated and returned to their reservoirs, returning the CO_2 used in weathering to the atmosphere. The steady state reactions are weathering and reprecipitation of:

a. dolomite $CaMg(CO_3)_2 + 2CO_2 + 2H_2O \rightleftharpoons Ca^{2+} + Mg^{2+} + 4HCO_3^-$

b. silica $SiO_{2(solid)} \rightleftharpoons SiO_2$ (dissolved)

c. calcite $CaCO_3 + CO_2 + H_2O \rightleftharpoons Ca^{2+} + 2HCO_3^-$

d. calcium silicate $CaSiO_3 + 2CO_2 + H_2O \rightleftharpoons Ca^{2+} + 2HCO_3^- + SiO_2$

e. magnesium silicate $MgSiO_3 + 2CO_2 + H_2O \rightleftharpoons Mg^{2+} + 2HCO_3^- + SiO_2$

In Figure 1 and succeeding diagrams, H_2O is not shown as a participant, although it is of course the medium in which cycling takes place.

INADEQUACIES OF THE SIMPLE STEADY STATE MODEL

I. $CaSiO_3$ Problems

Every geologist reading this article has already realized that there are several major flaws in the model that must be corrected. The most flagrant of these stems from the fact that weathered calcium silicates do not reprecipitate from the oceans, except in trivial amounts. Instead the calcium from the weathering of silicates is precipitated as $CaCO_3$. This behavior unbalances the system according to the reaction (which is obtained by summing the reactions for the weathering of Ca-silicate and the precipitation of $CaCO_3$):

$$2CO_2 + CaSiO_3 = CaCO_3 + SiO_2 + CO_2$$

Two CO_2 are used to weather $CaSiO_3$, only one is returned to the atmosphere when the Ca precipitates as $CaCO_3$. The SiO_2 carried to the ocean is available for the tests of organisms or for reactions with cations. Figure 2 shows schematically the effects of the weathering of calcium silicates.

II. The Disappearance of Dolomite

The second major alteration needed for the model is a change to take into account the disappearance of dolomite. Vinogradov, et al., (1957) showed that the Mg content of carbonate rocks is progressively less from Cambrian time to the Present, and that the Ca/Mg mol ratio

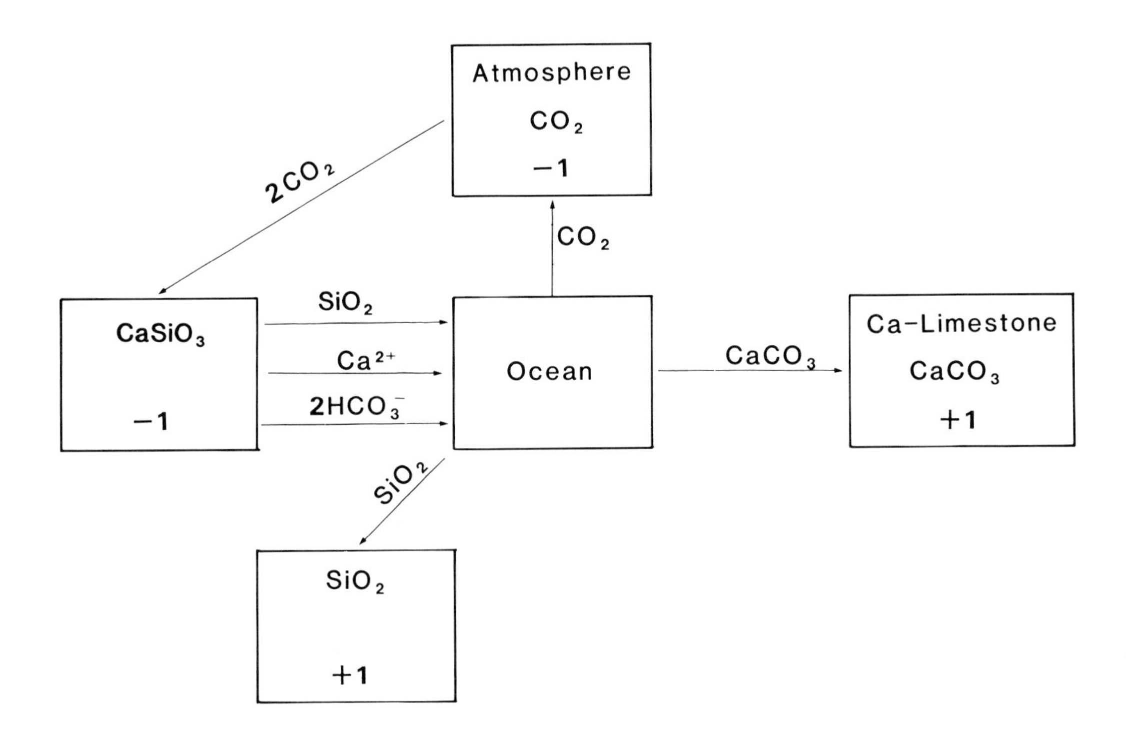

Figure 2. Diagram showing that the weathering of Ca-silicates, followed by precipitation of the Ca as $CaCO_3$, removes CO_2 from the ocean-atmosphere system.

has been about 40/1 since Cretaceous time. No significant masses of dolomite have been formed during the last 100 million years. Therefore the back fluxes of Mg and Ca from the ocean to dolomite must be removed. If the Mg is put into Mg silicates and the Ca removed as carbonate, the overall reaction is (which is obtained by summing the reactions for the weathering of dolomite and precipitation of calcite and Mg-silicate):

$$CaMg(CO_3)_2 + 2CO_2 + SiO_2 = CaCO_3 + MgSiO_3 + 3CO_2$$

This transfer has been incorporated into Figure 3. Note that there is a net addition of CO_2 to the atmosphere-ocean system when dolomite is converted to calcite and "clay". The removal of CO_2 from the ocean-atmosphere system as a result of the weathering of $CaSiO_3$ could be offset by the buffering action of the conversion of dolomite to clay and calcite. However, some other effects will be investigated before utilizing such feedbacks.

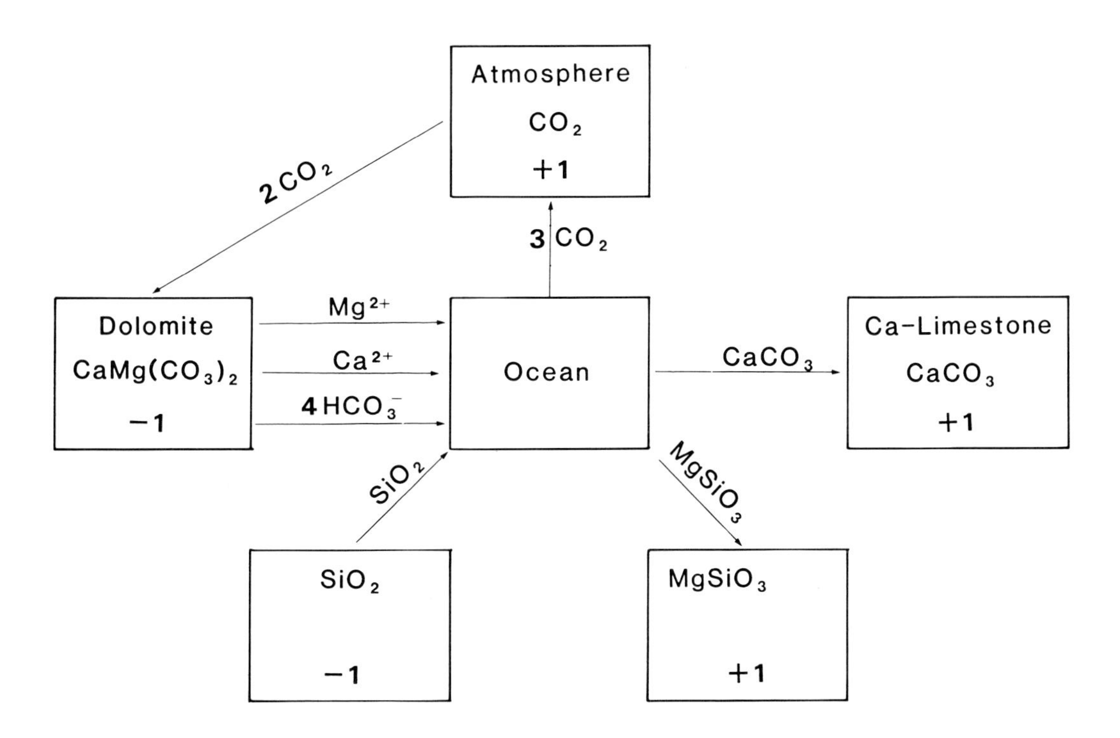

Figure 3. Diagram showing that the weathering of dolomite followed by the precipitation of the Ca as $CaCO_3$ and the Mg as "clay" ($MgSiO_3$), adds CO_2 to the atmosphere-ocean system.

III. Metamorphism of Carbonates

Another process that adds CO_2 to the atmosphere-ocean system is the metamorphism of carbonate minerals to silicate minerals. When subjected to high temperatures and pressures in the presence of a silica source such as quartz, carbonate minerals are converted to silicate minerals with the release of CO_2. This reaction takes place in many deep earth environments, such as during classical regional metamorphism of sediments, or during subduction of sediments down Benioff zones. The reactions involved are:

$$CaMg(CO_3)_2 + 2SiO_2 = CaSiO_3 + MgSiO_3 + 2CO_2$$

$$CaCO_3 + SiO_2 = CaSiO_3 + CO_2$$

Figure 4 shows the material transfers that result from metamorphism, and that the atmosphere-ocean system gains CO_2. The box representing

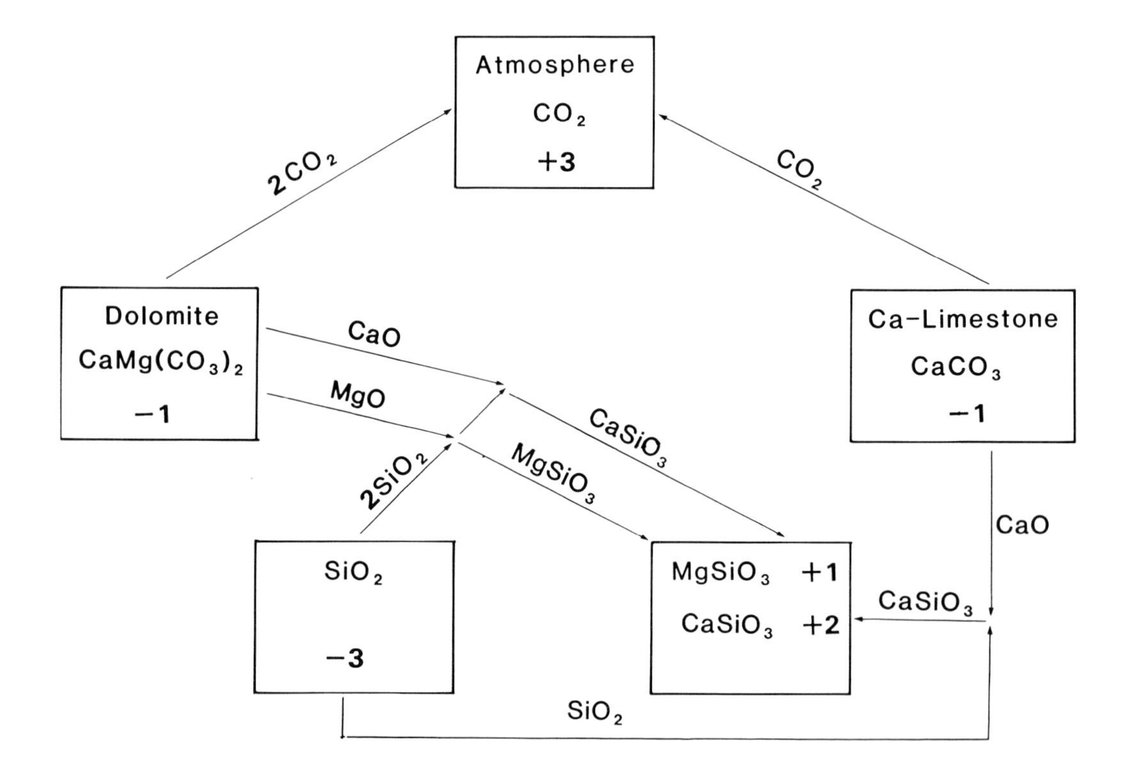

Figure 4. Diagram showing that metamorphism of carbonates to silicates adds CO_2 to the atmosphere-ocean system.

the ocean has been omitted from the diagram, because much of the metamorphic CO_2 return is directly from deep rocks to the atmosphere.

IV. Mg-Ca Exchange at Ridges

The processes discussed so far might have been considered adequate for construction of a schematic model of major effects a few years ago. But the recognition that the ocean continuously circulates through the hot basalts at the ocean ridges, with a total ocean circulation time of only about 10 million years (cf. Wolery and Sleep, 1976) made it evident that Mg-Ca relations must include Mg-Ca exchange at ridges. Berner and Garrels (in preparation), based on the work of Edmond, et al. (1979); Bischoff and Dickson, (1975); Mottl and others, (1978, 1979); conclude that the reactions of seawater with basalts at ridges are essentially removal of Mg as various silicates, and release of a corresponding amount of Ca. Figures 5 and 6 show the essence of the influence of this exchange on global Mg-Ca cycling.

$$CaSiO_3 + MgSiO_3 + 2CO_2 = CaCO_3 + SiO_2 + CO_2 + MgSiO_3$$

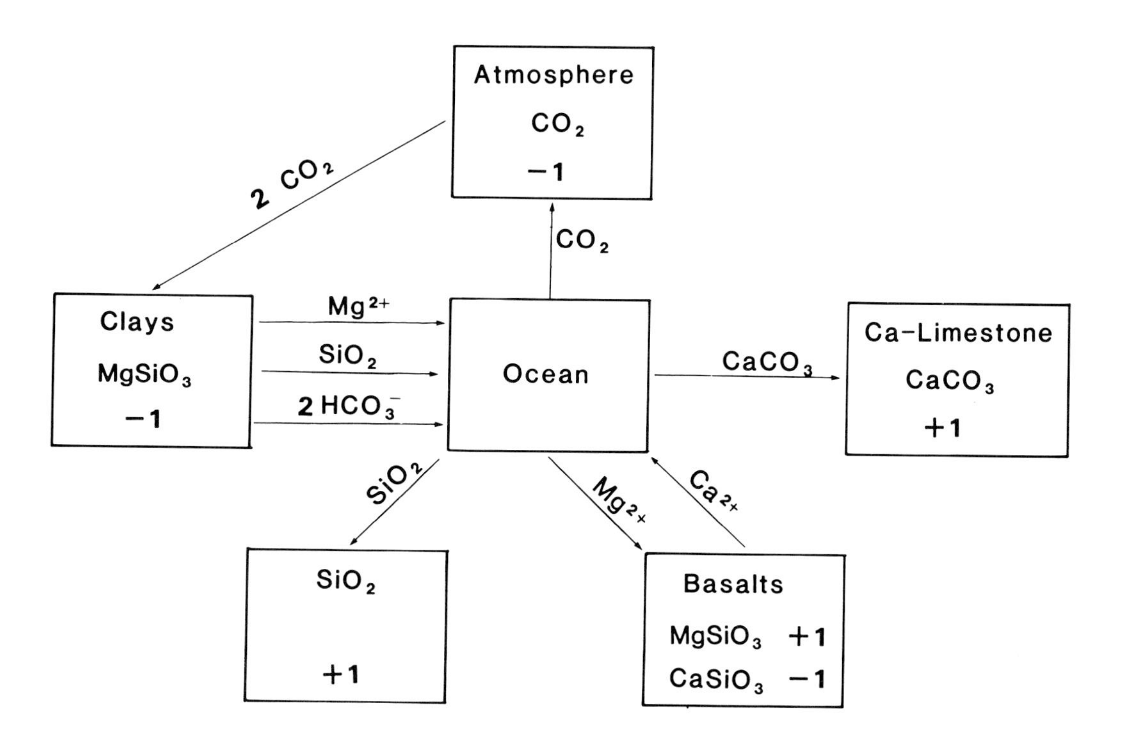

Figure 5. Diagram showing that transfer of Mg from weathering of clays to Ca by exchange at ridges results in depletion of atmosphere-ocean CO_2.

$$CaMg(CO_3)_2 + CaSiO_3 = MgSiO_3 + 2CaCO_3$$

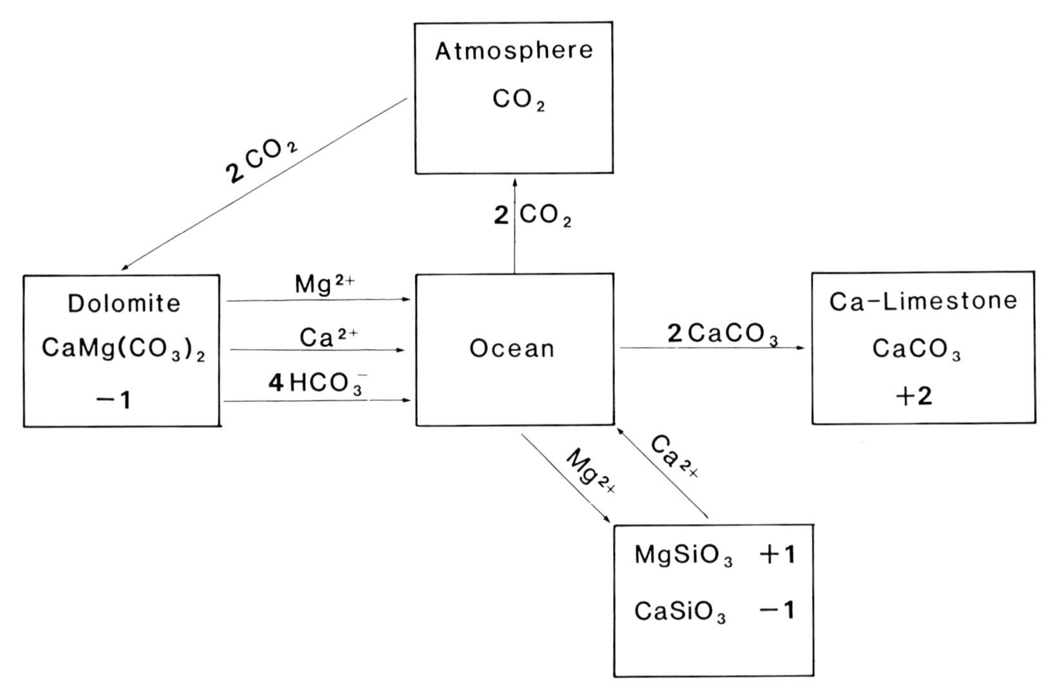

Figure 6. Diagram showing that if Mg exchanged for Ca at ridges is derived from dolomite, there is no change in atmosphere-ocean CO_2.

In Figure 6 the Mg exchanged at ridges comes from the weathering of dolomite. The Mg is exchanged for Ca at the ridge; the Ca then is removed from the ocean as $CaCO_3$. In this instance the exchange permits a balanced system, with no net CO_2 change in the atmosphere-ocean system.

V. Equilibrium between Atmosphere and $CaCO_3$-saturated Ocean

Although the ocean is supersaturated in its surface layers with respect to calcium carbonate, and undersaturated in the deeps, it can be assumed to be always close to saturation if considered *in toto*. This consideration puts a useful constraint on the carbonate-silicate modeling, and also buffers the additions and removals of CO_2 from the atmosphere. CO_2 is partitioned about 1/60 between atmosphere and ocean.

A SUMMARY DIAGRAM FOR Ca-Mg CARBONATE-SILICATE SYSTEM

The preceding discussion shows that there is a whole series of major erosional-depositional processes that must be balanced, or nearly balanced, if the atmosphere-ocean system is to maintain conditions suitable for the existence of life. The consumption of CO_2 by Ca-silicate weathering is such that, unopposed, it could lower CO_2 by a factor of 10 in half a million years (Holland, 1978).

A schematic steady-state system can be constructed by adding the various material transfer systems so as to have no net change of the CO_2 of the ocean-atmosphere system. Figure 7 shows a series of transfer reactions that fulfill this criterion of no net change. Figure 8 illustrates these transfer reactions graphically. The overall reaction of this chosen system transforms dolomite and Ca-silicate to Ca-limestone and Mg "clays". This system is by no means a unique solution for balance among the Ca and Mg carbonates. In fact, it balances CO_2 consumption by weathering of Ca-silicate against CO_2 production by transferring Mg in dolomite to Mg-precipitated silicates, whereas it now appears that in the real system Mg exchanges for Ca in silicates, and CO_2 loss by weathering of Ca-silicates may be balanced by CO_2 from metamorphic processes at depth. The value of such schematic "no ocean-atmosphere change" models is that they give the coefficients related to various assumed transfers, and permit testing of the effects of perturbations in the system.

A SYSTEM OF RESERVOIR CHANGES
IN WHICH DOLOMITE AND CA-SILICATE
ARE CHANGED TO CA-LIMESTONE AND CLAY,
WITHOUT CHANGING THE BALANCE
OF THE CO_2 IN THE ATMOSPHERE AND OCEAN

	Reaction	Process
1.	$3CaSiO_3 + 3CO_2 = 3CaCO_3 + 3SiO_2$	WEATHERING OF CA-SILICATE
2.	$CaMg(CO_3)_2 + SiO_2 + 2CO_2 = CaCO_3 + MgSiO_3 + 3CO_2$	WEATHERING OF DOLOMITE
3.	$CaMg(CO_3)_2 + 2SiO_2 = CaSiO_3 + MgSiO_3 + 2CO_2$	METAMORPHISM OF DOLOMITE
4.	$2CaCO_3 + 2SiO_2 = 2CaSiO_3 + 2CO_2$	METAMORPHISM OF CA-LIMESTONES
5.	$2MgSiO_3 + 2CaSiO_3 + 4CO_2 = 2CaCO_3 + 2MgSiO_3 + 2CO_2 + 2SiO_2$	WEATHERING OF CLAYS, MG-CA EXCHANGE AT RIDGES

OVERALL $2CaMg(CO_3)_2 + 2CaSiO_3 = 4CaCO_3 + 2MgSiO_3$

Figure 7. Balancing of material transfer equations.

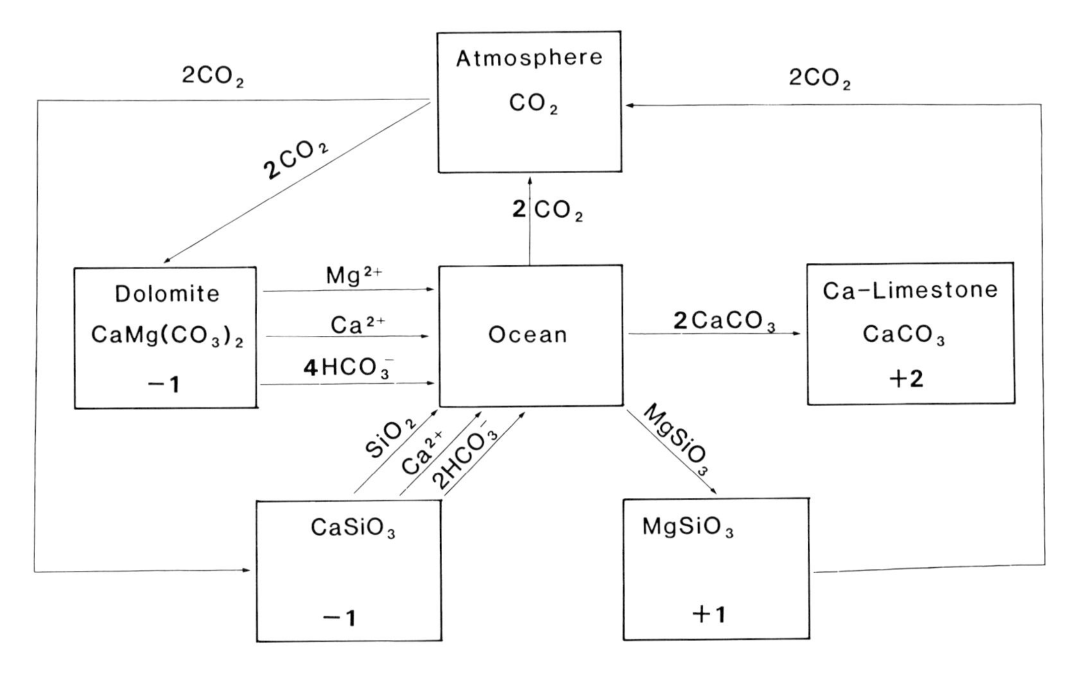

Figure 8. Constant atmosphere-ocean model showing conversion of dolomite and Ca-silicate to "clay" ($MgSiO_3$) and Ca-limestone.

REAL WORLD MODEL

Berner and Garrels (in preparation) have obtained real world numbers for reservoir sizes and fluxes to flesh out the schematic relations. Their model is shown in Figure 9. In this model 5.9×10^{18} mols of CO_2 are removed from the atmosphere-ocean system each million years as an end result of the weathering of Ca and Mg silicates. One reaction (in units of 10^{18} mols) is the consumption of 2.8 CO_2 by weathering of $CaSiO_3$ and conversion of the Ca to $CaCO_3$.

$$2.8CaSiO_3 + 5.6CO_2 \; = \; 2.8CaCO_3 + 2.8SiO_2 + 2.8CO_2 \tag{1}$$

The second is weathering of $MgSiO_3$ followed by exchange of Mg for Ca at ocean ridges, and precipitation of the Ca as $CaCO_3$

$$3.1H_2O + 3.1MgSiO_3 + 2CO_2 \longrightarrow 3.1Mg^{2+} + 6.2HCO_3^- + 3.1SiO_2 \tag{2}$$

$$3.1Mg^{2+} + 3.1CaSiO_{3(at\ ridge)} \longrightarrow 3.1Ca^{2+} + 3.1MgSiO_3 \tag{3}$$

$$3.1Ca^{2+} + 6.2HCO_3^- \longrightarrow 3.1CaCO_3 + 3.1CO_2 \quad (4)$$

The sum of the four reactions is simply

$$5.9CaSiO_3 + 5.9CO_2 = 5.9CaCO_3 + 5.9SiO_2$$

Thus 5.9 x 10^{18} mols of CO_2 per million years are removed by a combination of weathering of silicates, exchange at ridges, and precipitation of Ca-limestone. These 5.9 units are restored, in the model, by CO_2 from depth, attributed here entirely to metamorphism of carbonates to silicates.

The model serves to suggest some of the major reservoirs and fluxes involved in controls of atmospheric and oceanic CO_2, without necessarily specifying the nature of the processes that have resulted in the present system. It also demonstrates that the total CO_2 content of ocean and atmosphere (about 3 x 10^{18} mols of carbon) is of the same order of

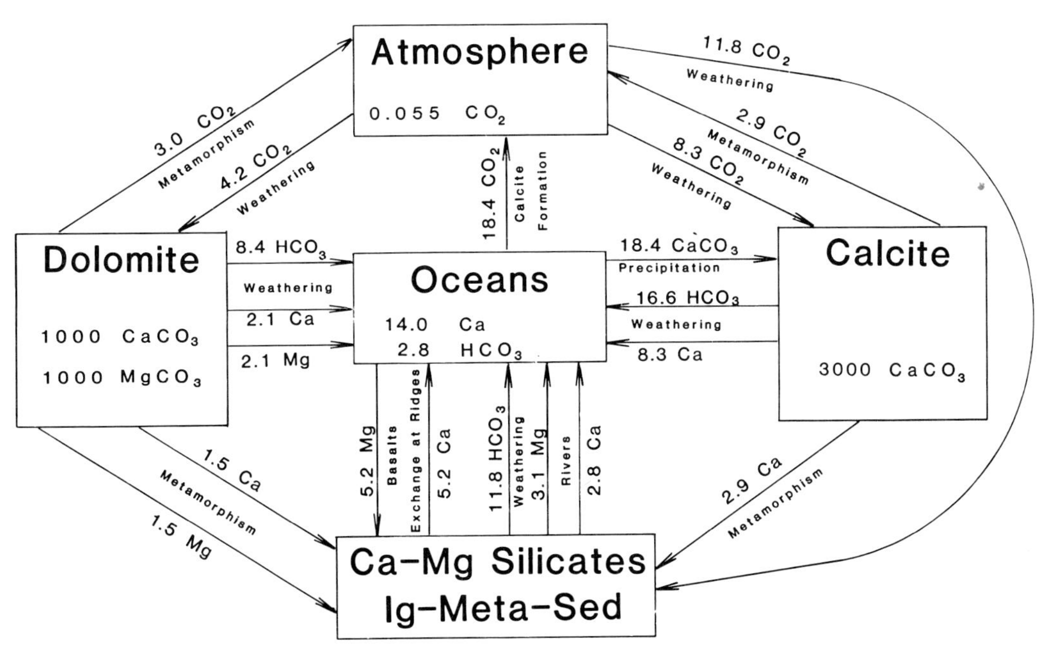

Figure 9. A balanced present-day reservoir-flux diagram for the Ca-Mg carbonate-silicate system (from Berner and Garrels, 1982).

magnitude as many of the million-year fluxes of carbon into and out of the ocean-atmosphere system.

The model also makes it clear that the CO_2 content of the atmosphere almost certainly must have fluctuated during earth history, and that there must be many negative feedbacks in the system that have prevented lowering of CO_2 to levels lethal to most photosynthetic organisms, or raising CO_2 to levels that would cause devastating greenhouse effects. There is also the implication that fluctuations of CO_2 levels over long time spans have been important to the evolution of life.

DISCUSSION

In view of the current controversy concerning the relative importance of organic versus inorganic processes in controlling the earth's surface environment (cf. Lovelock and Margulis, 1974) it seems appropriate to speculate on the interplay of such processes in that part of the Carbon Cycle modeled here. Life can be regarded as dominating the lighted realm, using solar energy to reverse the predictions of thermodynamics. The earth's internal energy tends to rule the realm of darkness of the subsurface. Materials that cycle into that region tend to react with appropriate increase of entropy.

It has been pointed out that the CO_2 content of the atmosphere is small compared to the million-year fluxes out of the atmosphere that go chiefly to the weathering of rocks, and the fluxes back into the atmosphere from precipitation and metamorphism of minerals.

The buffering effect of the carbonate system in the ocean perhaps can be regarded as a first line of defense against change of atmospheric CO_2. The current increase of atmospheric CO_2, generally estimated at about 25% over the last 100 years, would almost entirely disappear if the whole ocean were quickly mixed and equilibrated with the atmosphere.

An increase in the area of crystalline rocks containing abundant Ca-silicates should remove CO_2 from the ocean-atmosphere system (Fig. 2). As pointed out by Walker (1981), such a lowering would decrease the greenhouse effect, lower global temperatures, lessen precipitation, and thus decrease the rate of weathering; a strong negative feedback. Although the rate of weathering surely is heavily influenced by organisms, the rest of the system is essentially a series of inorganically dominated events. Increases in the area of crystalline rocks exposed to erosion are related to continental thickening and uplift, a function of the earth's internal processes. Whether there would also be a negative feedback to continental uplift in terms of increased metamorphism (Fig. 4) is unknown.

Walker (op cit) also pointed out that lowering of CO_2 by weathering processes could not result in a frozen earth because an ice cover would bring a halt to CO_2 consumption by weathering, while CO_2 would continue to come into the atmosphere as a result of deep earth metamorphic or volcanic processes. He pointed out that this feedback could also have compensated for a less luminous sun in the early days of the earth; the CO_2 level of the atmosphere would have risen to whatever level was required to melt the ice and start CO_2 consumption by weathering processes again.

Another likely long term CO_2 buffer is the dolomite-clay system. The chemical equation relating dolomite and clay can be represented by the reaction of dolomite with quartz to make the minerals calcite and the Mg-silicate talc:

$$H_2O + 3CaMg(CO_3)_2 + 4SiO_2 = Mg_3Si_4O_{10}(OH)_2 + 3CaCO_3 + 3CO_2$$

The equilibrium value for the CO_2 pressure at which talc, quartz, and dolomite are at equilibrium in the presence of water is not known accurately, but most calculations place it at a value somewhat higher than the CO_2 pressure of today's atmosphere -- at perhaps $10^{-2.5}$ to 10^{-3} atmospheres, or 3000 to 1000 ppmv CO_2 (Holland, 1978). As mentioned previously, dolomite seems to be disappearing slowly from the earth; the implication is that the effective CO_2 pressure in the ocean-atmosphere system has been less than the equilibrium value, and the Mg of dolomite has tended to go into Mg silicates. In most general terms, and with many caveats, this conversion tends to supply CO_2 to the ocean-atmosphere system. If, however, CO_2 pressure for any reason rises above the equilibrium pressure, there should be a tendency to convert Mg silicates back into dolomite, and, again in most general terms, subtract CO_2 from the atmosphere-ocean system. As shown in Figure 9, the mass of dolomite limestone is still substantial. Again the system considered seems to be dominated by inorganic equilibria. Whatever the reasons, organic or inorganic, for the specific control of CO_2 pressure, if that agent lowers CO_2 below the dolomite-clay equilibrium or raises it above that value, it must cope with the massive long-term mineral buffering of that system.

If we can suppose that CO_2 levels of the atmosphere are in the end controlled by a balance between the surface system that demands CO_2, and a deep system that yields it, we can surmise that as the earth gets older and older, radioactive heat diminishes, and the deep system becomes weaker and weaker. If dolomite were to disappear eventually, the system might fall back upon the $CaSiO_3$-$CaCO_3$ equilibrium, for which the basic specific reaction is:

water plus Ca-feldspar plus CO_2 = clay plus calcite

$$2H_2O + CaAl_2Si_2O_8 + CO_2 = Al_2Si_2O_5(OH)_4 + CaCO_3$$

The equilibrium CO_2 pressure for this reaction at room temperature is of the order of 10^{-6} x 10^{-7} atmospheres of CO_2, or about 1 ppmv. Such conditions would develop in a world in which dolomite limestones had been converted to clays, and Ca-silicates could precipitate from the oceans.

The relations that have been suggested for dolomite disappearance and for the possible future development of a Ca-silicate--$CaCO_3$ buffer, place the responsibility for such changes on changes in the strength of the heat generating system of the deep earth.

The preceding discussion is necessarily, at this stage of our knowledge, superficial. It is known that there are many reservoirs to consider, that there are unconsidered responses to these changes involving evolution of the biota.

SUMMARY

In summary, we have presented some of the interrelations of the massive sediment reservoirs of the Ca-Mg silicate-carbonate sedimentary system, as well as the sluggish exchanges between them. The suggestion has been made that these systems are important long-term buffers of the CO_2 content of the atmosphere - ocean system, and that the long-term controls of atmospheric CO_2 may be dominated by a slow diminution of CO_2 returned to the earth's surface as the earth's internal heat declines. Implicit in these suggestions is that the surface environment, dominated by photosynthesis, respiration, and decay of organisms, may in the end be dominated in its secular trends by happenings driven from the depths of the earth.

ACKNOWLEDGEMENT

This work was supported by National Science Foundation Grant # 12-2708-152.

REFERENCES

Berner,R.A.,and Garrels,R.M.:1982,In preparation.

Bischoff,J.L.,and Dickson,F.W.:1975,Earth Planet.Sci.Lett. 25,pp.385-397.

Edmond,J.M.,Measures,C.,McDuff,R.E.,Chan,L.H.,Collier,R.,Grant,B., Gordon,L.J.,and Corliss,J.B.:1979,Earth Planet.Sci.Lett. 46,pp.1-18.

Garrels,R.M.,Lerman,A.,and Mackenzie,F.T.:1976,Amer.Scien. 63,pp. 306-315.

Holland,H.D.:1978,Wiley Interscience, pp.351.

Lovelock,J.E.,and Margulis,L.:1974,Tellus 26,pp.1-10.

Mottl,J.J.,and Holland,H.D.:1978,Geochim.Cosmochim.Acta 42,pp. 1103-1115.

Mottl,J.J.,Holland,H.D.,and Carr,R.F.:1979,Geochim,Cosmochim.Acta 43,pp.869-884

Vinogradov,A.P.,Ronov,A.B.,and Ratinskii,V.M.:1957,Geochim.Cosmochim. Acta, 12,pp.273-276.

Walker,J.C.G.,Hays,P.B.,and Kasting,J.E.:1981,Jour.Geophys.Res. 86, pp.9776-9782.

Wolery,T.J.,and Sleep,N.H.:1976,Jour.Geol. 84,pp.249-275.

CALCIFICATION AND ATMOSPHERIC CO_2

Yasushi Kitano
Water Research Institute, Nagoya University, Chikusa-ku,
Nagoya 464, Japan

ABSTRACT

Two subjects will be discussed in this paper: (a) Chemical composition of the atmosphere of the earth. This subject concerns calcification in inorganic system, related to CO_2 in the primitive atmosphere before the origin of life. (b) Atmospheric CO_2 problem, that is now being faced, in connection with climatic change. This subject concerns calcification in biological system, related to CO_2 in the present atmosphere.

Almost all natural calcium carbonates are now formed in the marine biological system. Carbonate is a very valuable and important material for geochemical studies in the earth because of the very large amount of carbonate sediment that exists on the earth's surface and also because of the characteristic distribution of minor chemical elements and isotopes in natural carbonates. It is well known that the strontium content of calcareous fossils indicates the paleo-chemical composition of seawater and the $\delta^{18}O$ value indicates the paleo-temperature of seawater.

The physical and chemical properties of carbonates such as their crystal form and the chemical and isotopic composition of natural biogenic carbonates, are complex but interesting. The author presented a lecture on this subject as the presidential address to the Geochemical Society of Japan, on 13 February 1982 at Tokyo, which will be published in Geochemical Journal. In this paper, the author will concentrate on the geochemical significance of the large amount of carbonate sediment on the earth's surface and its relation to atmospheric CO_2.

Table 1 shows the inventory of carbon in the present earth's surface including the atmosphere, hydrosphere, biosphere and sedimentary rocks as presented by Rubey (1951) and Poldervaart (1955). From the table it is seen that sedimentary carbonate is the largest reservoir of CO_2 on the earth's surface and that there is a great difference in the amount of sedimentary carbonate between Rubey's and Poldervaart's estimates. Rubey's value was estimated from the geochemical balance of cal-

P. Westbroek and E. W. de Jong (eds.), Biomineralization and Biological Metal Accumulation, 89–98.

Table 1. Inventory of carbon (as CO_2) in atmosphere, hydrosphere, biosphere and sedimentary rock (10^{20} g)

	Rubey	Poldervaart
atmosphere	0.023	0.023
ocean and fresh water	1.3	1.3
living organisms and undecayed organic matter	0.14	0.14
(sedimentary rocks)		
carbonate	670	2240
organic carbon	250	250
coal, oil, etc.	0.27	0.27

cium entering and leaving seawater, and Poldervaart's value came from the observed amount of carbonate sediment existing on the earth's surface. The author has discussed these values and prefers the Poldervaart's estimate at this moment, for reasons that are discussed elsewhere (Kitano, 1980c).

If all of carbon estimated by Poldervaart was released to the earth's atmosphere, the partial pressure of CO_2 would be 45 atmospheres (or 15 atmospheres using Rubey's data).

Thus, the large amount of carbonate sediment existing on the earth's surface has a special significance in many important geochemical subjects. In this report, the following two geochemical subjects on carbonates are considered in connection with atmospheric carbon dioxide.

1. CHEMICAL COMPOSITION OF THE ATMOSPHERE OF THE EARTH

Table 2 shows the chemical composition of the present atmosphere on Mars, Earth and Venus. It is apparent from Table 2 that the chemical composition of the atmospheres on Mars and Venus is similar but that on Earth is completely different. But if all the carbon existing on the earth's surface (see Table 1) was released and added to the present atmosphere and if consideration is given to the fact that the evolution of oxygen was caused by the biological activity then the chemical composition of the hypothetical atmosphere on Earth becomes similar to that on Mars and Venus. The difference in the chemical composition of the atmosphere between Mars, Venus and Earth depends only on the exis-

Table 2. Chemical compositions of atmosphere in Mars, Venus and Earth

	Distance from sun (10^6 km)	Temperature of atmosphere (^oC)	Chemical composition of atmosphere (%)		
			CO_2	N_2	Ar
Mars	107	500	97	3	-
Earth	149	10	0.03	78	1
Venus	277	-60	95	2-3	1-2

tence of carbonate sediments and of organic carbon formed in the biological system of the earth. It should be noted that biological activity on the earth is related to the existence of liquid water on the surface of this planet (Kitano, 1980c).

These considerations make it possible to recognize that calcium carbonate formation in the sea removed CO_2 from the primitive atmosphere and led to the chemical composition of the present terrestrial atmosphere. These processes are connected with the birth and evolution of atmosphere, oceans and life on the earth.

It is well known that argon and carbon dioxide play very important roles in the history of the birth and evolution of the atmosphere and oceans. Generally geophysicists have considered argon and geochemists carbon dioxide in discussions of this subject (Hamano and Ozima, 1978a; 1978b; Kitano, 1975; 1980b; 1980c; Matsuo and Kitano, 1973; Ozima and Kudo, 1972; Ozima, 1975; Rubey, 1951).

Briefly it appears that when the earth's accretion occurred 4.6 x 10^9 years ago, the solid earth was covered with the dense primary atmosphere. The amount and chemical composition of this dense primary atmosphere are predictable from the cosmic abundance of chemical elements. A comparison of the volatile materials existing on the present earth's surface with the dense primary atmosphere indicates that the terrestrial atmosphere is not a remnant from the dense atmosphere. The estimated amount and chemical composition of the dense atmosphere are completely different from those of the volatile materials (Matsuo, 1980; Poldervaart, 1955; Rubey, 1951).

Thus it is now thought that the dense primary atmosphere was blown off the earth and that volatile gases then evolved as a secondary atmosphere from the earth's interior after the solid earth formed (Onuma, 1980). The volatile material thus became the source of the present atmosphere, hydrosphere and biosphere on the earth. Ozima and Kudo (1972) first suggested that information on the $^{40}Ar/^{36}Ar$ isotopic ratio when added to the ^{40}Ar inventory in the earth would impose a severe constraint on degassing models of the secondary atmosphere (Hamano and Ozima, 1978a; 1978b). This Ar inventory, including the isotopic ratio, made it appear that most (more than 85%) of the volatile materials were evolved by an early and sudden degassing from the solid earth, very likely before about 4.0 x 10^9 years (Ozima, 1975).

The chemical forms of the primitive volatiles which degassed from the interior of the earth were estimated as follows (Matsuo and Kitano, 1973; Matsuo, 1978; 1980; Kitano, 1980c). The amount of hydrogen gas, which was contained in the volatiles but escaped very rapidly from the primitive atmosphere to planetary space, was estimated from the Po_2 value in the chemical equilibrium system:

$$Mg_sSiO_4\ (\ell) \qquad 2\ MgO\ (s) + Si\ (\ell) + O_2 \tag{1}$$

where ℓ and s denote liquid and solid phases respectively. In that system Po_2 was measured to be $10^{-8.93}$ atmospheres at 2800°K, and in the system of basaltic magma Po_2 to be 10^{-8} at 1500°K (Katsura and Nagashima, 1974). Since the equilibrium constant in "$H_2O \rightleftharpoons H_2 + 1/2\ O_2$

system" was known, P_{H_2}/P_{H_2O} could be calculated by introducing the Po_2 value estimated above. Since P_{H_2O} is known, P_{H_2} is estimated. By using this P_{H_2} value, the chemical constituents of the secondary (primitive) atmosphere were estimated as:

$$H_2O > H_2 > CO_2\ (CO) > HCl > N_2 > SO_2\ (H_2S) \qquad (2)$$

It is supposed that these chemical forms were quenched at around 1500°K and therefore the chemical forms even at low temperatures were determined at the 1500°K temperature condition (Matsuo, 1978; 1980).

This scheme is supported by the following: (a) calcium carbonate with an age of 3.8 x 10^9 years has been found, and this was presumably formed in the inorganic system. This indicates that calcium carbonate was formed in the earth's surface before the origin of life in the sea (Allaardt, 1976; Moorbath *et al.*, 1973). The appearance of calcium carbonate would indicate the existence of carbon dioxide in the atmosphere (Kitano, 1980c; Shidlowski, 1978); (b) even when CH_4 existed in the primitive atmosphere it would be decomposed very quickly to CO_2 and CO under the primitive atmospheric conditions (Shimizu, 1974; 1978); and (c) hydrogen gas might escape from the primitive atmosphere, so that most would be lost in the first 10^7 years (Shimizu, 1974).

If the above suggestions are accepted, then the carbon dioxide shown in Table 1 (corresponding to about 40 atmospheres of Pco_2) should have been discharged into the terrestrial atmosphere about 4.0 x 10^9 years ago.

It has been proposed that, for the survival of even a very primitive life, the Pco_2 in the atmosphere must be below one atmosphere (Brancazio and Cameron, 1963; Rubey, 1951). Thus, most of carbon dioxide in the primitive atmosphere must have been removed throught the rapid formation of calcium carbonate in marine inorganic systems before the origin of life. The primitive volatile materials containing HCl gas, water vapor, H_2 gas, CO_2 gas, CO gas, N_2 gas etc. might be absorbed on the molten solid earth and might react with the melt to dissolve Ca, Mg, Na, K, Fe, Al etc. But under these high temperature conditions the chlorides of these metals might be hydrolized completely to HCl gas and their oxides. These reactions might occur in the melt successively and repeatedly. Finally HCl gas might be released from the melt to the atmosphere, and again the HCl gas might be dissolved in the melt. This dissolution and release might occur successively and repeatedly and other volatile materials such as CO_2 might also behave similarly.

Below 374°C, which is a critical temperature, liquid water appeared on the earth's surface. The dissolution and release of HCl gas to this water probably occurred repeatedly. Primitive seawater was originally 0.n N HCl solution (Rubey, 1951; Brancazio and Cameron, 1963). This primitive acid water would interact with basaltic rocks very actively and at these fairly high temperatures might dissolve Ca, Mg, Na, K, Fe, Al etc. from the rocks and become neutralized. The basaltic rocks would be weathered and transformed to clay minerals (Kitano, 1975).

The neutral solution (seawater) containing calcium and other ions could adsorb CO_2 from the atmosphere, and might precipitate calcium carbonates inorganically. The inorganic precipitation of carbonates would accelerate the removal of CO_2 from the atmosphere, while the seawater-clay interaction might keep the pH value of seawater slightly alkaline. It is thought that the formation of calcium carbonate, in other words the removal of CO_2 from the atmosphere might occur very rapidly under these fairly high temperature conditions. It seems possible, therefore, that the Pco_2 value of the atmosphere decreased to less than 1 atmosphere before the origin of life, 3.5 x 10^9 years ago (Kitano, 1980b; 1980c).

The formation of carbonates and the seawater-rock interaction make it possible to suppose that the major chemical composition of seawater approached that of the present seawater before the origin of life on earth.

2. ATMOSPHERIC CO_2 PROBLEM, THAT IS NOW BEING FACED, IN CONNECTION WITH CLIMATE CHANGE

It was reported that since 1850 human activities have increased the CO_2 content of the terrestrial atmosphere from 290 ppm to slightly more than 330 ppm (Bolin and Bishof, 1970; Broecker *et al.*, 1971; Keeling *et al.*, 1976; Kitano, 1980d; 1982). It is thought that the CO_2 content could reach twice the current CO_2 content by the year 2020 and at that time there will have been a rise in the average air temperature of about 2°C. This has produced a serious fear that a climatic change will occur and reduce some of the present granary regions to rather arid areas. Thus atmospheric CO_2 studies should be one of the most urgent research subjects internationally.

Until recently the CO_2 increase in the atmosphere was thought to be attributed to the burning of fossil fuels but it is now thought that the CO_2 increase is also due to the worldwide destruction of forests (Woodwell *et al.*, 1978). It has been shown that every year more trees are destroyed than are planted and forests do not act as a sink but rather as a source for CO_2 in the terrestrial atmosphere. The values of CO_2 discharged annually to the atmosphere from fossil fuels and from forests were estimated to be 18 x 10^9 ton and (3-6) x 10^9 ton-CO_2, respectively (Kitano, 1982).

The total ocean contains nearly 60 times (130 x 10^{12} ton-CO_2) as much CO_2 as the atmosphere (2.4 x 10^{12} ton-CO_2). Thus it was believed that only 1/60 value of CO_2 supplied to the atmosphere may remain in the atmosphere and the remaining 59/60 may be absorbed into seawater. But it was shown that about the half of the value of CO_2 released by the recent consumption of fossil fuels remains in the atmosphere, increasing the CO_2 content of the present atmosphere at the rate of 1 ppm per year. Thus, geochemists need to clarify the following subjects: (a) the estimate of the amount of CO_2 supplied to the atmosphere through the consumption of fossil fuels and also through the destruction of forests; (b) the destination of CO_2 supplied to the atmosphere through the above sources; and (c) the geochemical balance of CO_2 under both base line (natural) and the present day conditions.

The author has organized the research group on "The cycle of carbon dioxide through the phases of the atmosphere, hydrosphere and biosphere" in connection with the atmospheric CO_2 problem, sponsored by the Ministry of Education in Japan as a Special Project Research on Environmental Sciences (see Kitano, 1982).

It is apparent from Table 1 that the large reservoirs of CO_2 are in the carbonate sediments and organic material of soils. The atmosphere, oceans and living organisms are very important, however, as pathways from atmospheric CO_2 to carbonate sediment and organic carbon in soils, although the amounts of carbon in these systems are very small as compared with those in the carbonate sediment and the organic carbon in soils. Figure 1 shows the carbon dioxide balance on the surface of the earth. It indicates that CO_2 has been believed to be completely balanced between the following processes: (a) the removal of CO_2 from the atmosphere through photosynthesis and the supply of CO_2 to the atmosphere through respiration and decay of plants; (b) the removal of CO_2 from the atmosphere through adsorption to seawater and the supply of CO_2 to the atmosphere through release from seawater. It is important to consider whether or not the above belief is completely reasonable and acceptable, because the amounts of CO_2 exchanged in the above two processes are very large as compared with those supplied from the combustion of fossil fuels and the destruction of forests. Thus it should be questioned very carefully as to whether or not the amounts of carbonate sediment and organic carbon have ever changed and whether they have ever been constant.

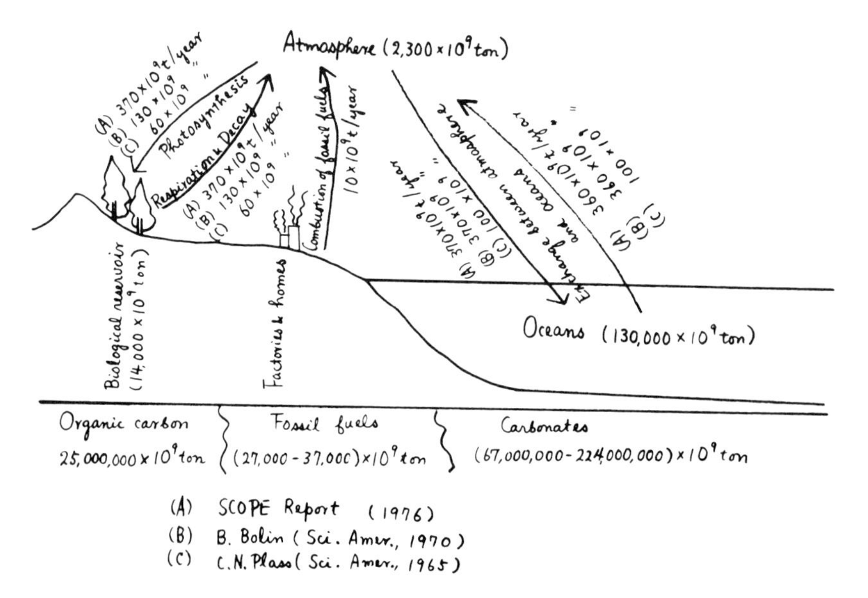

Figure 1. Carbon dioxide balance (as CO_2).

Thus, we return to the subject of the geochemical balance of CO_2 under the base line condition, in connection with the balance between the formation and dissolution of calcium carbonate sediments. Land water, such as river water, is the only important pathway for the discharge of major chemical constituents into seawater. Very large amounts of non-cyclic (non-marinogenic) chemical constituents with a completely different chemical composition from that of seawater have been discharged into the sea through land water such as rivers (Table 3). However, it is generally thought that the total volume and the major chemical composition of seawater have not changed for at least two billion years. In other words, the amounts of the individual chemical constituents contained in the entire seas have not changed for these two billion years. These facts should indicate that non-marinogenic chemical constituents discharged into seawater must be removed completely from seawater as authigenic chemical precipitates. The calcium ions discharged into seawater through land water must therefore be removed completely as calcium carbonates by the marine biological system (Kitano, 1975).

Table 3. Major chemical constituents discharged to seawater and contained in entire seawater

	Amounts of non-marinogenic chemical constituents discharged to seawater (10^{21}g/10^8 year)	Amounts of major chemical constituents contained in entire seawater (10^{21}g)
soluble SiO_2	43	0.08
HCO_3	190	0.91
SO_4^{2-}	33	3.7
Cl^-	0	26.1
Ca^{2+}	50	0.6
Mg^{2+}	12	1.9
Na^+	4	14.4
K^+	8	0.5

The author has already commented on the origin of these calcium ions in land water which are discharged into seawater. It was concluded that globally about 93% of the calcium ions in land water originated from the dissolution of limestone while the remaining 7% arose from the chemical weathering of silicate minerals in land areas. The calcium ions dissolved from limestone and transported to the sea through rivers are precipitated as calcium carbonate in the sea so that a steady state is maintained for both calcium and carbonate matter. But if calcium ions that are dissolved from silicate minerals are also precipitated as calcium carbonate in the sea, then there will be a loss of dissolved carbonate matter equivalent to the amount of extra calcium removed from seawater. In this case, calcium ions are in a steady state but dissolved carbonate matter is not.

The 7% of calcium ions discharged annually into seawater is estimated globally to be 0.0008×10^{15} moles. Therefore 0.0008×10^{15} moles of dissolved carbonate matter should be removed annually from seawater, if calcium ions dissolved from silicate are also precipitated as calcium carbonate. The total amount of dissolved carbonate matter in the entire sea is 3300×10^{15} moles, and the total amount of carbon dioxide in the atmosphere is 60×10^{15} moles (Kitano, 1980a).

Thus, if calcium ions, dissolved from both limestone and silicate minerals and transported into seawater, are precipitated as calcium carbonate in the sea, all of the carbonate matter dissolved in seawater and all of the carbon dioxide contained in the atmosphere will be removed completely in four million years.

This hypothesis is never accepted, because the degassing of juvenile CO_2 from the interior of the earth was already completed four billion years ago.

Thus the author would like to propose that the calcium ions dissolved from silicate minerals are removed from seawater through a silicate mineral-seawater interaction such as ion exchange reaction and Ca-montmorillonite formation,

$$Al_2Si_{2.4}O_{5.8}(OH)_4 + Ca^{2+} + SiO_2 + HCO_3^-$$

$$Ca_{0.17}Al_{2.33}Si_{3.67}O_{10}(OH)_2 + CO_2 + H_2O \qquad (3)$$

as proposed by MacKenzie and Garrels (1966) for magnesium and potassium ions. The amount of Ca-montmorillonite involved in the removal of this calcium annually is calculated to be 24×10^8 tons.

In conclusion therefore, it appears that 93% of calcium ions discharged to seawater through land water are precipitated as calcium carbonate in the sea and 7% as clay mineral. In this situation, both calcium and carbonate ions and calcium carbonate minerals are in a steady state. The amount of calcium carbonate sediment existing in the earth's surface is not changed under these base line (natural) conditions. In order to confirm this conclusion, however, the balance of organic carbon in the soil should be examined carefully and exactly.

As mentioned earlier, the CO_2 content of the atmosphere has increased due to the burning of fossil fuels and the worldwide destruction of forests. The influence of the CO_2 increase in the atmosphere on the rates of formation and dissolution of calcium carbonate and on the rates of photosynthesis and decomposition of organic materials should be studied urgently (Kitano, 1982). These studies are clearly related to the destination of the CO_2 discharged into the atmosphere from the above mentioned two sources. The amount of CO_2 absorbed from the atmosphere into seawater since the Industrial Revolution should be estimated from the measurements on the vertical distribution of dissolved carbonate matter in the open seawater. The decomposition of carbonates and organic materials in seawater and the adsorption of CO_2 from the atmosphere should also be examined carefully (Kitano, 1982). The transportation of particulate carbonates and organic materials from surface to deep waters should be examined through sediment trap observations. These studies are clearly of very great importance in understanding these systems.

REFERENCES

Allaardt, J.: 1976, "The Early History of the Earth", Windlye, B. (ed.), Wiley, London, pp. 117.
Bolin, B., and Bishof, W.: 1970, Tellus 22, pp. 431-442.
Brancazio, P.J., and Cameron, A.G.W. (eds.): 1963, "The Origin and Evolution of Atmospheres and Oceans", John Wiley & Sons, Inc., pp. 314.
Broecker, W.S., Li, Y.H., and Peng, T.H.: 1977, in "Impingement of Man on the Oceans", Hood, D.W. (ed.), Wiley-Interscience, New York, pp. 287-324.
Hamano, Y., and Ozima, M.: 1978a, in "Origin of Life", Noda, H. (ed.), Center Acad. Pub. Japan, pp. 29-33.
Hamano, Y., and Ozima, M.: 1978b, Adv. Earth Planet. Sci. 3, pp. 155-171.
Katsura, T., and Nagashima, S.: 1974, Geochim. Cosmochim. Acta 38, pp. 517-531.
Keeling, C.D., Bacastow, R.B., Bainbridge, A.E., Ekdahl, C.A. Jr., Guenther, P.R., Waslerman, L.S., and Chin, J.I.S.: 1976, Tellus 28, pp. 538-551.
Kitano, Y. (ed.): 1975, "Geochemistry of Water", Dowden Hutchinson & Ross, John Wiley & Sons, Inc., pp. 455.
Kitano, Y.: 1980a, Acta Oceanogr. Taiwanica 11, pp. 40-48.
Kitano, Y.: 1980b, in "Evolution of Materials", Kagaku Sosetsu, Chem. Soc. Japan (ed.), pp. 95-107.
Kitano, Y.: 1980c, "The History of Water and Earth", Nippon Hoso Shuppan Inc., Tokyo, pp. 222.
Kitano, Y.: 1980d, in "Chemistry of Earth and its Environment", Kitano, Y., and Matsuno, T. (eds.), Iwanami Shoten, Tokyo, pp. 105-127.
Kitano, Y. (ed.): 1982, "Cycle of CO_2 in the Phases of Atmosphere, Hydrosphere and Biosphere", Special Project Res. on Environmental Sci. (Min. Education, Japan), pp. 132.
MacKenzie, F.T., and Garrels, R.T.: 1966, Amer. J. Sci. 264, pp. 507-525.
Matsuo, S., and Kitano, Y.: 1973, Viva Origino 2, pp. 9-20.
Matsuo, S.: 1978, in "Origin of Life", Noda, H. (ed.), Center Acad. Pub. Japan, pp. 21-27.
Matsuo, S.: 1980, in "Evolution of Materials", Kagaku Sosetsu, Chem. Sco. Japan (ed.), pp. 83-94.
Moorbath, S., O'nions, R.K., and Pankhurst, R.J.: 1973, Nature 245, pp. 138.
Onuma, N.: 1980, in "Evolution of Materials", Kagaku Sosetsu, Chem. Soc. Japan (ed.), pp. 65-82.
Ozima, M., and Kudo, K.: 1972, Nature Phys. Sci. 239, pp. 22-24.
Ozima, M.: 1975, Geochim. Cosmochim. Acta 39, pp. 1127-1134.
Poldervaart, A.: 1955, Geol. Soc. Amer. Special Pap. 62, pp. 119-144.
Rubey, W.W.: 1951, Bull. Geol. Soc. Amer. 62, pp. 1111-1147.
Shidlowski, M.: 1978, in "Origin of Life", Noda, H. (ed.), Center Acad. Pub. Japan, pp. 3-20.
Shimizu, M.: 1974, Sym. Moon, Planetary Sci., Inst. Space and Aeronautical Sci., pp. 48-51.

Shimizu, M.: 1978, in "Origin of Life", Noda, H. (ed.), Center Acad. Pub. Japan, pp. 35-38.

Woodwell, G.M., Whittaker, R.H., Reiners, W.A., Linkens, G.E., Delwiche, C.C., and Botkin, D.B.: 1978, Science 199, pp. 141-146.

THE MODERN DISTRIBUTION AND GEOLOGICAL HISTORY OF CALCIUM CARBONATE BORING MICROORGANISMS

Susan E. Campbell
Department of Biology, Boston University, Boston, MA, U.S.A.

The ability of microbial endoliths to contribute to the cycling of the element carbon by transforming it from hard carbonate substrates into the biota or into solution had evolved by late Precambrian time, prior to the emergence of skeleton bearing metazoans. Throughout the Phanerozoic both inorganically and biogenically formed carbonate substrates were bored. Today, endoliths have representative species in all five kingdoms. They occur in fresh water and marine environments. Their vertical distribution ranges from alpine to the deep sea (to 4,000 m depth). Endoliths are affected by man's activities; a reduction in diversity accompanies certain types of pollution.

Microbial endoliths represent a heterogeneous group of organisms which have representatives in all five kingdoms (sensu Whittaker, see Margulis, 1974), and share a common ecological niche, i.e. they dissolve carbonate substrates as they penetrate them for nutrition and/or shelter. As a result, they leave within skeletons and rocks specific traces of their activity which generally conform to the surface morphologies of their bodies.

Today, microbial endoliths are widespread in marine, and to a lesser extent, fresh water environments. Their geological impact is most obvious in the destruction of carbonate rocky coasts (Schneider, 1976) and in sediment grain modification (Fig. 1) (Klement and Toomey, 1967). Although microbial endoliths are very small, the impact of their activity over time is enormous. They penetrate calcareous and phosphatic skeletal grains and inorganic carbonate substrates, such as limestones and dolostones, including ooids, beachrock and lithified stromatolites. Endoliths transfer carbon from its most abundant environmental pool ($CaCO_3$) into the biota and/or into solution (Golubic and Schneider, 1979). They have been involved in carbon budgeting of marine environments throughout the Phanerozoic (see Hessland, 1949; Gattrall and Golubic, 1970; Golubic *et al.*, 1975; James and Kobluk, 1978; Harris *et al.*, 1979; Golubic *et al.*, 1979). Recent results (Campbell, 1982, see Fig. 2) show that they evolved in the Precambrian prior to the skeleton bearing Phanerozoic organisms they bore.

P. Westbroek and E. W. de Jong (eds.), Biomineralization and Biological Metal Accumulation, 99–104.

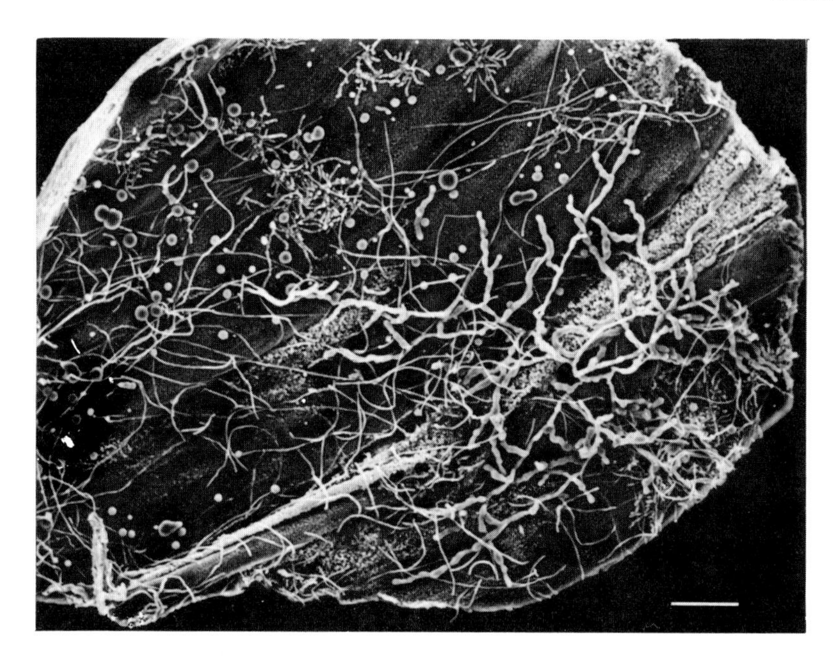

Figure 1. Resin-cast microbial borings within a 1 x 1.3 mm size shell fragment. Marine sediment, 30 m deep. Scale bar is 100 μm.

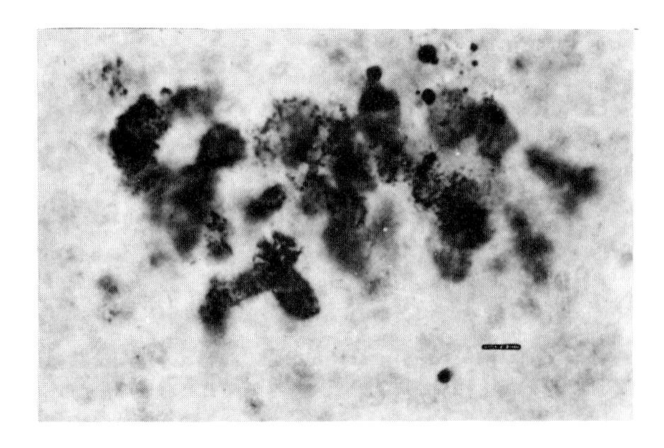

Figure 2. Fossil endolith borings in a late Precambrian (700-570 My old) silicified ooid. Scale bar is 10 μm.

Paleoenvironmental and paleobathymetric interpretations are made by comparing the resin casts of fossil borehole assemblages with modern ones to determine the ancient organismal types. The environmental distribution of modern endoliths is then used as a key for determination of ancient habitats (Golubic and Campbell, in press). This approach has its limits, due to evolutionary changes in organismal morphology and life style. However, fossil-to-fossil comparisons in conjunction with other accompanying environmental indicators can effectively overcome this problem. The most important basis for such studies is the distinction between the light dependent (photosynthetic) vs. light-independent (heterotrophic) endoliths accompanied by characterization of their specific borehole morphologies. While photoautotrophic endoliths (see Fig. 3) are light dependent and are active only in the photic zone, heterotrophs (see Fig. 4) are generally light independent and are active in surface sediments of deeper (aphotic) waters and other non-illuminated environments as well as illuminated ones. Today, the most frequent sources of organic carbon for the support of heterotrophic endolith activity is either the primary product of photosynthetic endoliths they attack (Fig. 4C and D) or the organic matrix laid down by skeleton forming animals (Fig. 4A and B). While photoautotrophic endoliths are restricted to the photic layer of the oceans which extends as deep as 200-370 m in clear seas (LeCampion *et al.*, in press), heterotrophs occur throughout the photic zone to the abyssal depths as great as 4,000 m (S.E. Campbell, unpublished; see Fig. 4A). In freshwater environments endolithic cyanophytes occur in lakes (Schneider, 1977) and endolithic lichens (Fig. 4D) penetrate carbonate rocks in alpine environments (Campbell, 1978).

The species representation and distribution of some modern endoliths may be affected by man's activities. Ongoing studies of modern coastal endoliths in relation to pollution gradients show that endolithic cyanophyte assemblages in the intertidal and supratidal zones exhibit reduced species diversity in heavily polluted harbors (Campbell, 1983a). While the influx of fresh-water into such wave and circulation protected marine environments may, in a large part, be responsible for these changes, comparison of two wave-exposed coasts subject to extremely high levels of organic pollution (Cortiou, France which is the sewage outflow of the entire city of Marseille and petrochemical effluent of Priolo-Melilli, Italy respectively) where salinity is near-marine, indicate that different types of pollutants result in different responses by the endolith community. A sharp reduction in species diversity (to two species) was found at Priolo-Melilli (Campbell, 1983b) whereas in Cortiou, species diversity was comparable to that of cleaner wave-exposed environments.

Figure 3. Photosynthetic prokaryotic (A-C) and eukaryotic (D-F) endoliths. A. *Hyella caespitosa*; B. *Solentia paulocellulare*; C. *Mastigocoleus testarum*; D. Conchocelis (*Porphyra*); E. *Eugomontia sacculata*; F. *Phaeophila dendroides*. Scale bars are 10 μm.

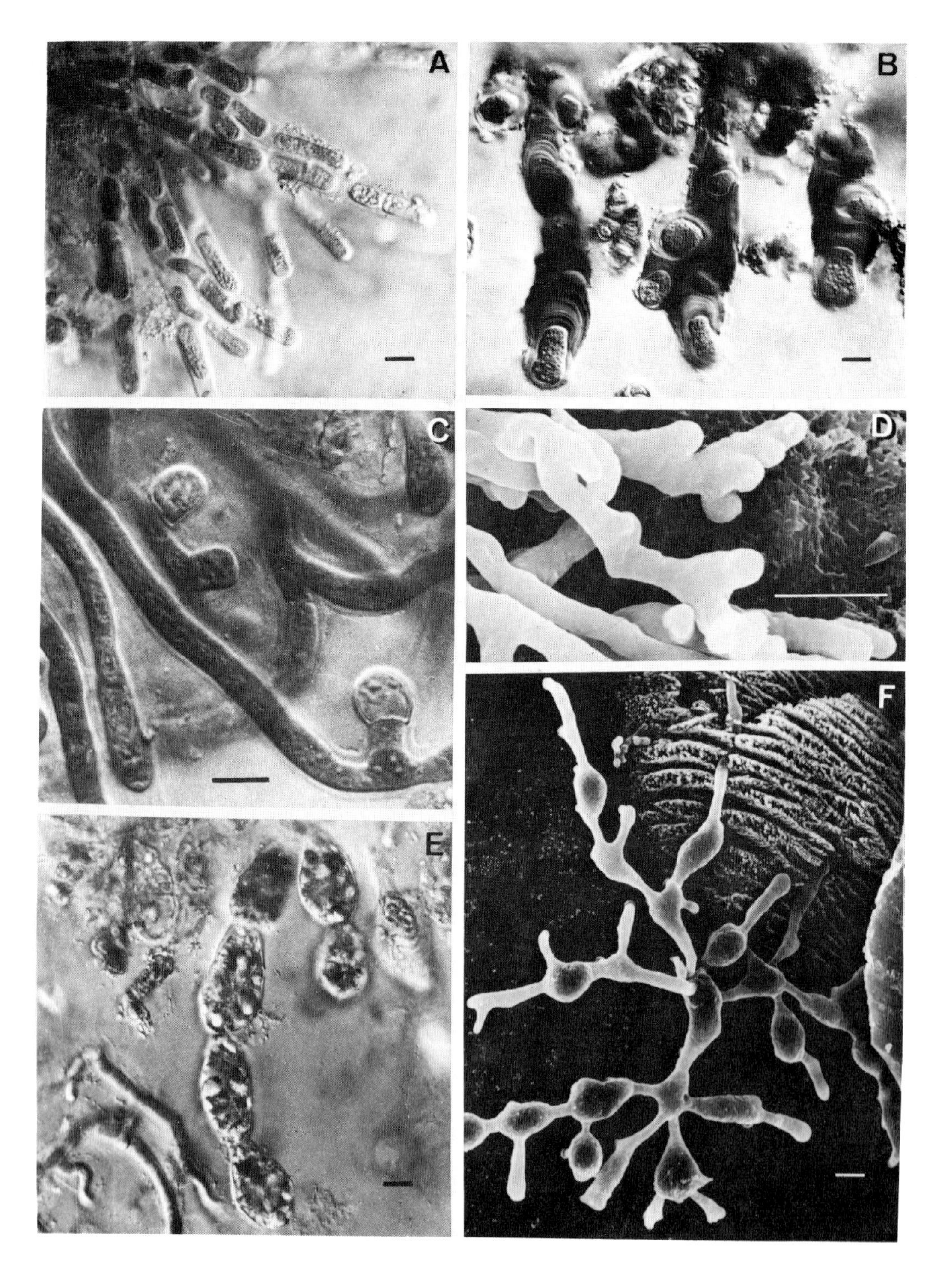
A
B
C
D
E
F

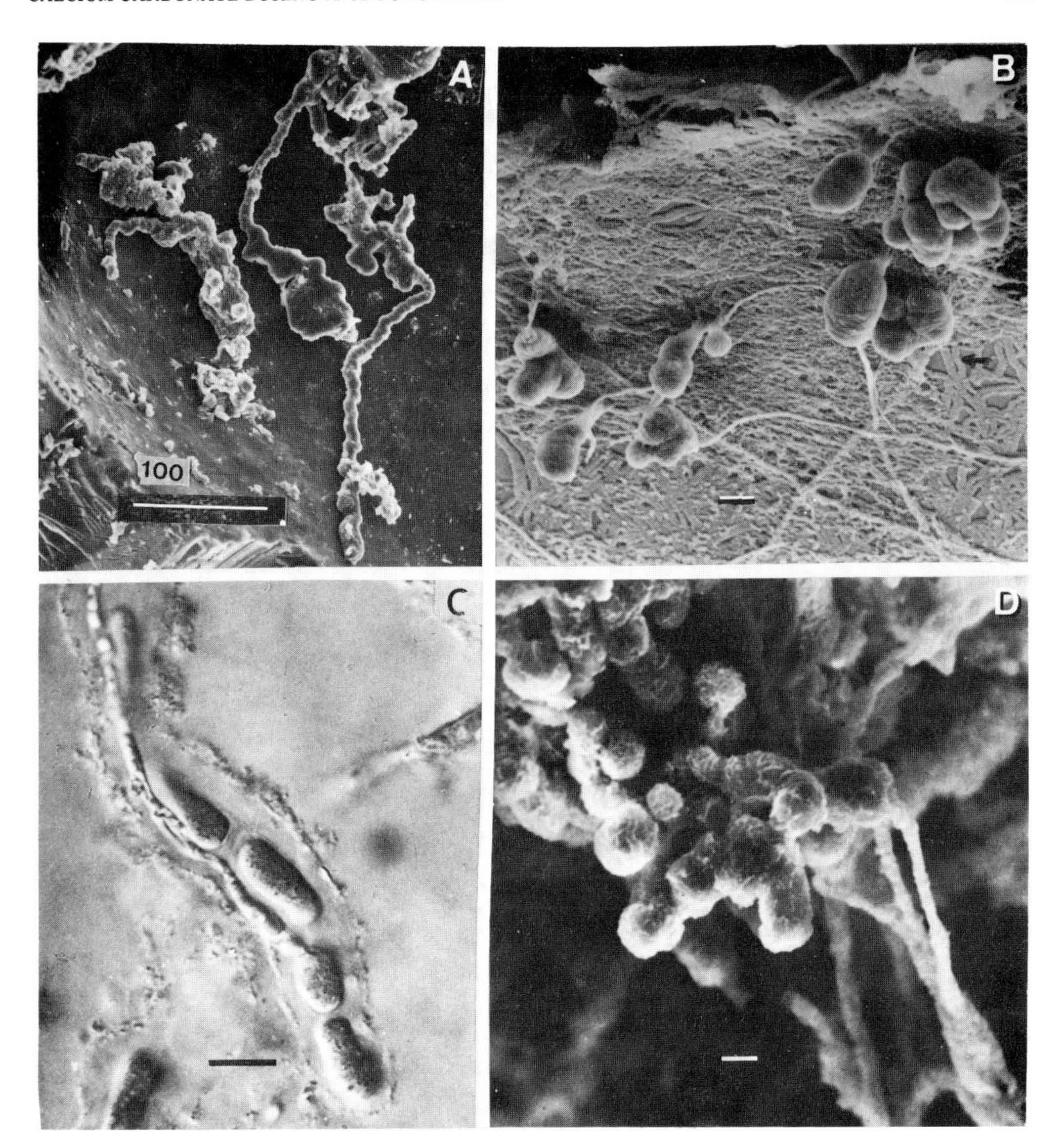

Figure 4. Heterotrophic (A-C) and symbiotic (D) endoliths. A. Deep sea endolith in a foraminiferal test, Atlantic ocean, 4,000 m depth; B. Sporangial clusters of endolithic lower fungi with a vermicular (probably prokaryotic) endolith (arrow); C. Endolithic fungus parasitizing on *Hyella caespitosa*; D. Endolithic lichen symbiosis. A-C. marine; D. freshwater subaerial environments. Scale bars 100 µm for A, and 10 µm for B-D.

Support for this research was provided by NASA grant NAGW-141, NSF grant EAR 8107686, Boston University grants 644-BI and 597-BI and the NATO Postdoctoral Fellowship Program.

REFERENCES

Campbell, S.E. (1978). 4th College Park Colloq. Chem. Evol.: Limits of Life, Abstr. w. Progr., p. 23.

Campbell, S.E. (1982). In preparation.

Campbell, S.E. (1983a). Thalassia Jugoslavica, in press.

Campbell, S.E. (1983b). XXIII Congress, Assemblée Pléniere de la CIESM, Cannes, December 2-11, 1982, in press.

Gattrall, M. and Golubic, S. (1970). In "Trace Fossils" (T.P. Crimes and J.C. Harper, eds.), Geol. Jour. Spec. Issue 3, pp. 167-178.

Golubic, S. and Campbell, S.E. (1983). In "Bioerosion: Case Histories in Earth Science" (R.G. Bromley, ed.), Springer, Heidelberg, Berlin, New York, accepted for publication.

Golubic, S., Hoffman, E.J. and Campbell, S.E. (1979). Amer. Assoc. Petrol. Geol., Ann. Meeting, Abstracts w. Progr., p. 94.

Golubic, S., Perkins, R.D. and Lukas, K.J. (1975). In "The Study of Trace Fossils" (R.W. Frey, ed.), Springer, New York-Heidelberg-Berlin, pp. 229-269.

Golubic, S. and Schneider, J. (1979). In "Biogeochemical Cycling of Mineral-Forming Elements" (P.A. Trudinger and D.J. Swaine, eds.), Elsevier, Amsterdam-Oxford-New York, pp. 107-129.

Harris, P.M., Halley, R.B. and Lukas, K.J. (1979). Geology 7, pp. 216-220.

Hessland, I. (1949). Bull. Inst. Univ. Uppsala 33, pp. 409-428.

James, N.P. and Kobluk, D.R. (1978). Sedimentology 25, pp. 1-35.

Klement, K.W. and Toomey, D.F. (1967). Jour. Sed. Petrol. 37, pp. 1045-1051.

LeCampion-Alsumard, T., Campbell, S.E. and Golubic, S. (1982). Jour. Sed. Petrol., in press.

Margulis, L. (1974). In "Handbook of Genetics" (R.C. King, ed.), pp. 1-41.

Schneider, J. (1976). Contr. to Sedimentol. 6, pp. 1-112.

Schneider, J. (1977). In "Fossil Algae" (E. Flügel, ed.), Springer, New York-Heidelberg-Berlin, pp. 248-260.

Part II

Aspects of Calcification

THE POSSIBLE ROLE OF MITOCHONDRIA AND PHOSPHOCITRATE IN BIOLOGICAL CALCIFICATION

A. L. Lehninger
The Johns Hopkins School of Medicine, Baltimore, Maryland, USA

ABSTRACT: Mitochondria from most vertebrate tissues and some invertebrates possess energy-dependent inner-membrane transport systems for both Ca^{2+} influx and efflux, whose rates are under regulation. Very large amounts of Ca^{2+}, together with phosphate, can be taken up by mitochondria both *in vitro* and *in situ* in intact cells, yielding electron-dense granules that are amorphous by x-ray analysis, prevented from crystallizing by inhibitor(s). Chemical studies suggest the inhibitor to be 3-phosphocitric acid, which was synthesized and found to be extremely active in stopping hydroxyapatite crystal growth and in preventing kidney calcification in animal models *in vivo*.

In my contribution to this Conference I shall review the main properties of the Ca^{2+} transport systems of mitochondria, the deposition of calcium phosphate granules in mitochondria, the possible participation of mitochondria in biological calcification mechanisms, as well as a trail of research which led us to phosphocitrate, a mitochondrial inhibitor of hydroxyapatite crystallization.

MITOCHONDRIAL CA^{2+} TRANSPORT

Various aspects of Ca^{2+} transport by mitochondria has been reviewed in some detail (1-7); the following represents a brief recapitulation. In the early 1960s it was found in our laboratory that isolated kidney and liver mitochondria can accumulate Ca^{2+} and phosphate from the medium in large amounts in a process that requires energy furnished either by mitochondrial electron transport or by ATP hydrolysis (8). These early experiments showed that Ca^{2+} accumulation proceeds at high rates and with a high affinity, but is prevented by respiratory inhibitors and uncoupling agents. The accumulation of Ca^{2+} was later found to be stoichiometric with electron transport in respiring mitochondria. Close to 2 Ca^{2+} ions are accumulated per pair of electrons passing through each of the three energy-conserving sites of the respiratory chain (9-11). The accumulation

P. Westbroek and E. W. de Jong (eds.), Biomineralization and Biological Metal Accumulation, 107–121.

of 2 Ca^{2+} ions is thus energetically equivalent to the phosphorylation of one molecule of ADP to ATP. Respiration-coupled Ca^{2+} uptake by mitochondria requires a counteranion which biologically is phosphate, but which may be replaced by several other anions, particularly those of weak aliphatic acids, such as acetate or butyrate (12). Ca^{2+} entry occurs electrophoretically in response to the negative-inside membrane potential that is generated by mitochondrial electron transport, with each Ca ion carrying 2 positive electric charges (13). Thus the inward transport of 2 Ca^{2+} ions responds to the generation of 4 negative charges inside the membrane as a pair of electrons passes through each of the three energy-conserving sites of the respiratory chain.

Mitochondria isolated from all the major tissues of the rat and other laboratory animals show almost identical Ca^{2+} transport properties. In fact, mitochondria from major tissues of representatives of all the vertebrate phyla show remarkably similar properties (14), as do mitochondria isolated from tissues of a number of insects, some crustacea (15), as well as the protozoan *Acanthameba castellani*. On the other hand, mitochondria from *Neurospora*, *S. cerevisiae*, and a dozen or more plant tissues, although capable of accumulating Ca^{2+}, have a rather low affinity for it (16).

THE CA^{2+} INFLUX CARRIER

Mitochondria contain 2 Ca^{2+} transport systems, one for influx, the other for efflux. Both require respiratory energy. The Ca^{2+} influx carrier, as indicated above, operates electrophoretically. The affinity of the Ca^{2+} carrier is extremely high, with a K_M of about 3 μM and a V_{max} that exceeds 1000 ng-atoms Ca^{2+} per min per mg mitochondrial protein. The Ca^{2+} influx carrier also can transport Mn^{2+} and Sr^{2+} at equal rates; it may also be involved in transporting Fe^{2+}, required in mitochondria to complete the synthesis of the cytochromes. The Ca^{2+} influx system is inhibited by the dye ruthenium red, as well as by lanthanide cations, such as La^{3+}, Sm^{3+}, Ne^{3+}, etc. Although a polypeptide of molecular weight 3000, imputed to be the Ca^{2+} carrier, has been isolated from heart mitochondria (17,18), it appears possible that specific phospholipids may also be required.

CA^{2+} EFFLUX

For a long time little was known about Ca^{2+} efflux from mitochondria; the nature of this process is only now being given intensive study (4,6, 7). Ca^{2+} cannot leave respiring mitochondria by the same transport process that brought it in, because of the positive-outside membrane potential, which imposes charge repulsion against Ca^{2+} efflux. Recent research indicates that Ca^{2+} efflux from liver mitochondria takes place in exchange for 2 H^+ coming in (19), on an independently regulated transport system, other than that involved in Ca^{2+} influx. In mitochondria from excitable tissues Ca^{2+} exchanges with Na^+ (5). The Ca^{2+} efflux system of liver mitochondria is insensitive to ruthenium red, but is inhibited by N-ethylmaleimide and by dibucaine. The Ca^{2+} efflux pathway has a rather high affinity, about 1 μM or less, but a relatively low V_{max} of <25 nmols/min/mg, in comparison to the maximum rate of Ca^{2+} influx, a point that will be examined further below.

MAINTENANCE OF CA^{2+} HOMEOSTASIS IN THE SUSPENDING MEDIUM (CYTOSOL) BY MITOCHONDRIA

Isolated respiring rat liver or heart mitochondria are able to maintain extramitochondrial free Ca^{2+} at remarkably constant concentrations, very close to those in the intact cell, through a virtually self-adjusting interplay of the Ca^{2+} influx and efflux systems. Ca^{2+} homeostasis maintained by mitochondria is an almost automatic result of the characteristic K_M and V_{max} values of the Ca^{2+} influx and efflux systems and the concentration range in which Ca^{2+} occurs in the cell (20-22). Suspensions of isolated liver mitochondria can maintain external Ca^{2+} concentrations of approximately 0.4 μM or less. When repeated small pulses of extra Ca^{2+} are added to such suspensions they will be taken up, with complete return of the Ca^{2+} concentration to its homeostatic level. Similarly, lowering of the Ca^{2+} level in the medium by addition of pulses of the Ca^{2+}-binding agent EGTA also results in readjustments of the Ca^{2+} influx and efflux systems, to restore the Ca^{2+} level of the medium (22).

MAINTENANCE OF INTRAMITOCHONDRIAL CA^{2+} HOMEOSTASIS

Recent research has indicated that mitochondria of some tissues may also maintain homeostasis of free Ca^{2+} concentration *within* the mitochondrial matrix compartment. Several enzymes in mitochondria require submicromolar concentrations of Ca^{2+}. Among these is the pyruvate dehydrogenase complex, which is brought into its active form by a specific Ca^{2+}-dependent phosphatase. Recent work indicates that the activity of this complex can be regulated by mitochondrial Ca^{2+} transport activity (23,24). In fact, it has even been suggested that it is the primary function of mitochondrial Ca^{2+} transport systems to regulate *intramitochondrial* Ca^{2+} concentration, rather than extramitochondrial Ca^{2+} (23). It is not inconceivable that mitochondria may help maintain Ca^{2+} homeostasis of *both* the extramitochondrial and intramitochondrial compartments.

REGULATION OF CA^{2+} EFFLUX

Ca^{2+} homeostasis in the extramitochondrial medium, as indicated above, can be maintained by the more or less automatic action of the Ca^{2+} influx and efflux systems. Normally, these two systems do not operate at high rates simultaneously; if they did they would waste respiratory energy through futile cycling of Ca^{2+} and effectively prevent oxidative phosphorylation. The Ca^{2+} turnover rate by the two systems is normally extremely low in respiring mitochondria *in situ* (4,19). Under such circumstances the maintenance of Ca^{2+} homeostasis would utilize only one or two percent of the total respiratory energy generated by mitochondria, the remainder going largely into oxidative phosphorylation. However, under some circumstances the Ca^{2+} concentration of the cytosolic compartment must be increased in order to activate certain Ca^{2+}-dependent functions. The question has arisen as to whether there are regulatory signals that can cause release of free Ca^{2+} from mitochondria, thus increasing the cytosolic Ca^{2+} concentration. We have found that Ca^{2+} influx does not appear to be under special regulation; rather, it is the Ca^{2+} efflux rate that can be modulated. We have been able to identify one such modulating signal, namely, the oxidized steady state of mitochondrial NADP (25,26). Whenever the $NADP^+$/NADPH ratio in mitochondria is relatively low, the Ca^{2+} efflux system is least active and Ca^{2+} tends to be retained by mitochondria. When the $NADP^+$/NADPH ratio is relatively high, Ca^{2+} efflux is stimulated some 10-15 fold. It has been postulated that regulation of Ca^{2+} efflux involves a closed feed-back loop between the cytosolic energy status and the rate of Ca^{2+} efflux, mediated by NADP redox state (26). However, other signals for releasing Ca^{2+} may occur but they are not yet well understood (6).

MASSIVE ACCUMULATION OF CALCIUM PHOSPHATE

Respiring kidney mitochondria can take up from the suspending medium almost 3000 nmols of Ca^{2+} per mg mitochondrial protein in only a few minutes; this amounts to over a third of their dry weight (8). Simple calculations showed that the concentrations of Ca^{2+} and phosphate so achieved within the mitochondrial matrix are in enormous excess of the solubility product for calcium phosphate and thus suggested that insoluble calcium phosphate salts must precipitate out in the mitochondrial matrix. This was confirmed by electron microscopy of massively-loaded mitochondria, which showed the presence of many large electron-dense granules within the matrix (27). These were subsequently identified in microincineration experiments as being largely calcium phosphate. Because the Ca/P ratio was between 1.5 and 1.7 it was suspected that the mitochondria might contain significant amounts of hydroxyapatite (Ca/P = 1.67). With the collaboration of Dr. Aaron Posner of the Hospital for Special Surgery, New York, we subjected desiccated Ca^{2+}-loaded mitochondria to x-ray analysis, to look for the characteristic diffraction patterns of hydroxyapatite. Surprisingly, there was no hint of the hydroxyapatite pattern; in fact, the mitochondrial deposits appeared to be amorphous (27). This was surprising, since at pH 7.4 the precipitation of Ca^{2+} and phosphate from a super-saturated solution will inevitably lead to the formation of calcium hydroxyapatite within 30 min at room temperature. We concluded that crystallization was in some way held in check within the mitochondrial matrix. Later investigations suggested that the calcium salt actually precipitated in the mitochondria may be tricalcium phosphate ($Ca_3(PO_4)_2$), possibly in a ultramicrocrystalline form that does not give an easily detected diffraction pattern (28); however, $CaHPO_4$ may also be involved.

Other studies have shown that ATP and pyrophosphate are accumulated together with Ca^{2+} and phosphate (29). The mitochondrial calcium phosphate granules contain significant amounts of ATP, as well as Mg^{2+}.

ELECTRON-DENSE CALCIUM PHOSPHATE GRANULES IN MITOCHONDRIA *IN SITU*

Electron-dense granules have been found not only in isolated mitochondria allowed to accumulate large amounts of Ca^{2+} *in vitro*, but also in the mitochondria of thin sections of intact tissues, particularly of bone-forming cells, such as chondrocytes (30,31). They are especially abundant in the neighborhood of healing bone fractures. Electron-dense mitochondrial granules, presumably of calcium phosphate, have been frequently observed in mitochondria of soft tissues after treatment of animals with certain drugs and toxic agents (reviewed in (2)). For example, heart mitochondria will accumulate such granules after acute administration of isoproterenol and kidney mitochondria will show such granules after treatment of the animal with mercuric chloride, which disturbs renal tubular function. Since, as indicated above, isolated mitochondria from all higher animal tissues seem to have the capacity for massive accumulation of calcium and phosphate, it can be expected that such mitochondrial granules will be detectable in cells *in situ* after treatments that are known to cause pathological calcification. These relationships and observations raise questions concerning the regulation of the accumulation of calcium phosphate by mitochondria in intact cells. In particular, how is it that calcium phosphate deposition in the mitochondria of soft tissues is normally held in check, and becomes overt only after certain drugs or during specific pathological changes?

CALCIUM AND PHOSPHATE IN BODY FLUIDS; KIDNEY STONES AND NEPHROCALCINOSIS

The concentration of free Ca^{2+} in blood plasma and interstitial fluid is approximately 1 mM; similarly, the concentration of free phosphate is about the same. At pH 7.4, if the laws of chemical equilibrium hold, Ca^{2+} and phosphate should precipitate out in the body fluids and form hydroxyapatite, since $\sim$1 mM Ca^{2+} and phosphate will greatly exceed the solubility product of hydroxyapatite. Indeed, if it were not for some inhibitory effects on this process, we should all, in the words of the Bible, turn into stone. Thus body fluids may contain inhibitors of calcium phosphate precipitation and crystallization. This came out clearly in clinical observations by Howard (32), who long ago found that individuals repeatedly forming calcium stones in their kidneys appear to be deficient in some chemical factor in their urine which prevents formation of kidney stones in normal individuals. The possibility that normal urine and mitochondria may both contain factors inhibitory to precipitation of insoluble calcium salts thus brought two lines of research together.

Efforts to isolate such inhibitory substances began on normal urine as starting material. Early experiments in the 1960s strongly suggested that the urinary inhibitor was a short peptide. However, it was ultimately found that it was not the peptide that was active but some substance that migrated with it during column chromatography. The task of concentrating the inhibitor proved to be puzzling and difficult; however, other developments occurred in the meantime which helped to clarify some of the logic in pursuing this work (2,32).

STEPS IN THE FORMATION OF HYDROXYAPATITE

Physical-chemical studies of the precipitation and crystallization of calcium phosphate and hydroxyapatite by a number of investigators provided one clue. In brief, such studies indicated that the formation of crystalline hydroxyapatite occurs in at least two major stages. In the first, Ca^{2+} and phosphate precipitate out to form a sparingly soluble form of calcium phosphate, which was for a long time thought to be amorphous tricalcium phosphate ($Ca_3(PO_4)_2$), but which more recent evidence suggests may be brushite ($CaHPO_4 \cdot 2H_2O$). This intermediate insoluble form of calcium phosphate constitutes an obligatory step in hydroxyapatite formation. In the second major stage a complex process of re-solution, re-precipitation, and partial hydrolysis occurs, with the formation of hydroxyapatite, which is far less soluble than the intermediate tricalcium phosphate or brushite. Although the overall process of hydroxyapatite crystallization is for all practical purposes irreversible, it is held in check in body fluids by two factors. The first appears to be that the obligatory intermediate tricalcium phosphate and/or brushite, does not precipitate readily at the concentrations of Ca^{2+} and phosphate present in body fluids; therefore hydroxyapatite cannot form. The second factor is the presence of inhibitors in body fluids. If Ca^{2+} and phosphate are mixed in the test tube at pH 7.4 and 37° in concentrations high enough to precipitate insoluble calcium phosphate, the latter will spontaneously crystallize as hydroxyapatite in an hour or less. However, if a small amount of normal human urine is added to such a mixture it will inhibit both the precipitation of insoluble calcium phosphate, as well as its conversion into hydroxyapatite. This is not a new discovery. The inhibitory action of urine on calcification has long been known to masons and cement workers: urine will prevent concrete from setting and hardening.

MITOCHONDRIAL CALCIUM PHOSPHATE PRECIPITATION AS AN EARLY STEP IN BONE CALCIFICATION

These properties suggested the possibility that mitochondria of bone-forming cells may participate in the formation of bone mineral (2). It had been suspected for a long time that bone-forming cells must "do something" to Ca^{2+} and phosphate to prepare it for extracellular deposition in the collagen nidus. What bone-forming cells might do to Ca^{2+} and phosphate is to take them up from the surrounding extracellular fluid into the cytosol and then concentrate them in the mitochondria at the expense of energy generated by mitochondrial electron transport. In the mitochondrial matrix Ca^{2+} and phosphate would be brought to a high enough concentration to precipitate them out as amorphous calcium phosphate, the obligatory precursor of hydroxyapatite. The calcium phosphate granules in the matrix would be prevented from crystallizing into hydroxyapatite by an inhibitor. Such granules or "micropackets" of amorphous calcium phosphate might depart from the mitochondria, pass through the cytosol to the plasma membrane, and thence be ejected into the neighboring extracellular fluid, in the neighborhood of the calcifying collagen nidus. We suggested further that the tightly bound inhibitor that stabilized and protected the micropacket, as it passed from the mitochondria to the extracellular site of mineralization, would then be destroyed by enzymatic action, thus allowing the insoluble calcium phosphate to mature into hydroxyapatite crystals (2). Some circumstantial evidence consistent with this idea has arisen from work in other laboratories (30,31). Thus electron-dense particles of calcium phosphate are prominent in mitochondria of bone-forming cells. Also, it has been found that when experimental animals are given mercuric bichloride, which causes deposition of calcium phosphate in kidney mitochondria, such granules have been found to move out through the mitochondrial membrane into the cytosol (33). Moreover, granules of calcium phosphate have been detected in the extracellular space near sites of bone mineralization (34). Such granules are not surrounded by membranes. The calcium phosphate crystals once observed within membrane-surrounded "matrix vesicles", which were postulated to be the precursors of bone mineral now appear to be artifacts of tissue preparation (34). These pieces of evidence in support of a possible role of mitochondria in normal bone calcification, although circumstantial, are interesting and deserve further study.

THE SEARCH FOR MITOCHONDRIAL CALCIFICATION INHIBITORS

Preliminary experiments showed that extracts of rat liver mitochondria contained materials inhibitory to calcification of rat rib cartilage in a bioassay system used in study of urinary inhibitors (32). They led to a search for the best sources of such inhibitors. It came to our attention, through a personal communication from Dr. Dorothy Skinner, that terrestrial or fresh water crustacea might be a rich source, on the grounds that these organisms salvage and store Ca^{2+} from their exoskeletons in the hepatopancreas during ecdysis, to be mobilized again for calcification of the new shell. After some experiments with land crabs, we changed to the common blue crab *Callinectes sapidus*, obtained from the brackish water of the upper Chesapeake Bay. We found that the hepatopancreas of this organism, even during intermoult, is extremely rich in Ca^{2+}. Moreover, isolated crab hepatopancreas mitochondria are extremely active in accumulating Ca^{2+}, up to 6000 ng-ions per mg protein (35). The cytoplasm in these cells is profuse in electron-dense granules, which were isolated and found to be primarily calcium phosphate, with about 8 percent ATP + ADP (36). Most important, the calcium phosphate in the mitochondria and in the cytoplasmic granules was again amorphous by x-ray analysis (37), suggesting that the unusually large deposits of calcium phosphate in the crab hepatopancreas contain inhibitor(s) in very large amounts. This was verified by assay of crab liver extracts.

Efforts to concentrate and purify the inhibitory factors were then focused on the crab hepatopancreas as source. After many efforts chromatographic and electrophoretic procedures were developed which yielded considerable purification of the inhibitory substance, which appeared to have about the same chromatographic properties when derived from crab liver, rat liver mitochondria, or normal urine. However, further purification became most difficult because of the highly acidic character of the inhibitor. A series of studies was then carried out on the proton NMR spectra of the concentrated material, which gave important clues to some aspects of its structure. Mass spectrometry of the compound in derivatized form yielded a characteristic set of fragments, which, taken together with the proton NMR data, pointed to several possible structures, assuming a molecular weight below 400 (38). One of these putative structures for the inhibitor, 3-phosphocitric acid, was of special interest since it could be derived from biologically occurring metabolites. Citrate, which originates in mitochondria during Krebs cycle activity, is present in large amounts in bone. Moreover, mitochondria are the site of ATP production. Thus phosphocitrate appeared to be a reasonable structure for the inhibitor. Because further purification of the small amount of the natural inhibitor was yielding diminishing returns, we decided to take the Monte Carlo approach and try to synthesize 3-phosphocitric acid by organic-chemical methods, on the gamble that it represents the correct structure. We thought this would be a rather simple one-month project; however, it required several months. The compound was ultimately synthesized and characterized (38). The most difficult reaction in the synthesis, which still causes trouble because of variable yields, is the last step, the selective removal of the blocking ethyl groups from the 3 carboxyls of triethyl 3-phosphocitrate, without disturbing the phosphoester group. Two of the three ethyl groups hydrolyze quite readily, but the third is rather resistant. Reactions capable of removing the latter also cause significant loss of the phosphate group, with the result that the yields are quite variable, depending on the precise conditions used in the hydrolysis. In any case, the compound was finally obtained in pure form as its sparingly soluble Ca^{2+} salt. The mass spectrum of 3-phosphocitric acid closely resembled that of the purified natural inhibitor (38).

INHIBITORY ACTIVITY OF SYNTHETIC 3-PHOSPHOCITRATE

Synthetic 3-phosphocitrate was then put to the real test. When examined in the cartilage calcification assay it was found to be highly active. However, because of a possible permeability barrier, it was not possible to assess its absolute activity by this method. A more quantitative assay involved its capacity to inhibit the growth of hydroxyapatite seed

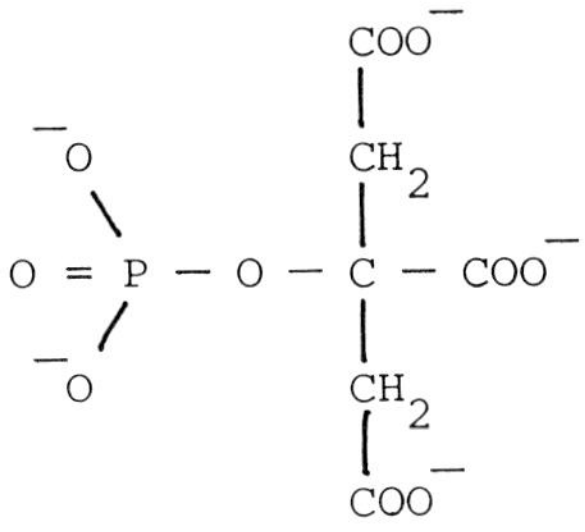

```
                          COO⁻
                           |
     ⁻O                   CH₂
        \                  |
     O = P  —  O  —  C  —  COO⁻
        /                  |
     ⁻O                   CH₂
                           |
                          COO⁻
```

Figure 1. 3-phosphocitrate

crystals in a supersaturated solution of calcium phosphate. Phosphocitrate was found to be extremely active in this assay, far more active than any other known inhibitor for this process (38). Although our identification of 3-phosphocitrate as the urinary and mitochondrial inhibitor is still only provisional, the fact that the synthetic compound is extremely active in inhibiting calcification is the most convincing evidence that it or a closely related compound is the true biological inhibitor.

Phosphocitrate is not only far more active than other compounds tested but its activity appears to depend rather strikingly on specific features of its structure (38). Free citrate is only about 1/300 as active as 3-phosphocitrate, indicating the importance of the 3-phosphate group. If all three carboxyl groups are esterified with ethanol the compound is totally inactive in inhibiting hydroxyapatite crystal growth. Similarly, monoethyl 3-phosphocitrate is also inactive. Therefore it appears that all three carboxyl groups must be free and the phosphate group must be present. At pH 7.4 3-phosphocitrate has a maximum of five closely-spaced negative charges (Figure 1). From models and also thermodynamic considerations it is likely that the three-dimensional conformation of phosphocitrate is highly constrained. The five negative charges would repel each other maximally and force the molecule into a conformation that may possibly fit specific growth points on the hydroxyapatite crystal lattice.

Phosphocitrate does not appear to function simply to bind Ca^{2+}. In fact, titrations of phosphocitrate with Ca^{+} and H^{+} show it to be a relatively poor chelating agent. Phosphocitrate appears to act more like an inhibitor of crystal growth. For example, 0.3 μM phosphocitrate can completely prevent hydroxyapatite crystal growth in a 1.7 mM calcium phosphate solution. It would appear that one molecule of phosphocitrate can prevent literally thousands of molecules of calcium phosphate from crystallizing. Other biologically occurring compounds with significant activity are pyrophosphate and 2,3-diphosphoglycerate; ATP is much less active.

Many crystallization inhibitors are known in industry. For example, ice crystals are kept from forming in ice cream by the presence of various inhibitors; one such is gelatin. Similarly, exposed photographic emulsions are kept from forming large blotches of reduced silver by suitable inhibitors, thus making possible fine-grain emulsions. Biologically occurring crystal growth inhibitors are also known, in particular the "anti-freeze" protein of the blood plasma of Antarctic fish and the Atlantic flounder. This compound, a simple polypeptide containing sugar substituents at every third amino acid residue along the chain, functions by keeping the water phase of blood plasma in a metastable state below its normal freezing point.

Some crystal growth inhibitors appear to function by binding to the crystal lattice at specific growth points, thus preventing access of additional lattice units. Such an action of phosphocitrate appears likely from some preliminary experiments (38). Phosphocitrate was prepared with the citrate portion labeled with ^{14}C. On adding the labeled phosphocitrate to suspensions of seed crystals of hydroxyapatite we found that it is almost completely and instantaneously bound to the crystals. Subsequently no further crystal growth occurs for many minutes or hours, depending upon the conditions. Ultimately, however, crystal growth resumes and will eventually reach its normal rate unless a second addition of labeled phosphocitrate is made. This again binds very tightly and prevents further crystallization for an extended period.

IN VIVO TESTS OF THE ACTION OF PHOSPHOCITRATE ON MITOCHONDRIAL CALCIFICATION

Although synthetic phosphocitrate is extremely effective in inhibiting the growth of hydroxyapatite seed crystals in the non-biological, i.e. physical-chemical assay described above, we also wished to test phosphocitrate in a biological calcification assay system. Three different types of biological assays have been employed. 1. Inhibition of induced kidney calcification in small animals. 2. Inhibition of the calcification of rat rib cartilage, detected by histochemical staining. 3. Inhibition of calcification by bone-forming cells maintained in culture.

The action of phosphocitrate has been tested in three different experimental models of nephrocalcinosis in mice and rats (39,40). In the first, mice on a high phosphate diet were given doses of excess parathyroid hormone daily over a period of 4-5 days. This regimen causes calcification of the kidneys, which usually results in complete renal failure within about a week. The earliest observable evidence of calcification is the deposition of calcium phosphate granules in the matrix of the mitochondria of tubule cells. Subsequently, calcium phosphate deposits also form in the cytosol; eventually, insoluble calcium salts also appear in the lumen of the tubules. This course of events can be followed by both optical and electron microscopy. We have also measured the total calcium content of the kidney, which may increase as much as 100-fold over this relatively short period of treatment.

When animals on this regimen also receive daily small doses of phosphocitrate intraperitoneally one hour prior to the administration of the parathyroid hormone, the dramatic increase in calcium content of the kidneys is almost completely prevented. In addition, the appearance of calcium phosphate deposits in the mitochondrial matrix is also almost completely suppressed; no calcification occurred in the cytosol or in the tubules. As a consequence, most of the phosphocitrate-treated animals will survive the rather drastic 4-day regimen, but the control animals usually do not. When equivalent amounts of either free citrate or phosphate or both were given, instead of phosphocitrate, absolutely no inhibitory effects on renal calcification induced by the hormone were observed. It was therefore concluded that phosphocitrate *per se* is very active *in vivo* in prevention of mitochondrial calcification in renal tubule cells and thus appears to prevent subsequent calcification in the cytosol and tubules (39).

In the second animal model (39), renal calcification was induced by daily infusion of calcium gluconate, which raises the free serum calcium level several-fold. These experiments were carried out in large rats, which developed calcification lesions, first in the mitochondria and later in the cytosol, as in the case of parathyroid hormone administration to mice; this regimen also resulted in a very large increase in the kidney total calcium content, measured by atomic absorption analysis. Phosphocitrate given to such animals in small quantities daily, one hour prior to the infusion of calcium gluconate, again almost completely suppressed calcification of the mitochondria (39). Measurements of the serum calcium concentration showed a rise from about 1 mM to nearly 5 mM during the period, but this increase was not diminished by simultaneous administration of phosphocitrate, indicating that the action of phosphocitrate in suppressing kidney calcification is not exerted simply through the lowering of serum calcium concentration, but probably through some direct action of the phosphocitrate on the cells of the renal tubules. Further experiments on the dosage of phosphocitrate showed that maximal effects in suppression of calcification were given by 0.1 μmol phosphocitrate per rat, corresponding to about 0.3 μmol per kg body weight or 0.08 milligrams per kg. This amount of phosphocitrate, if equally distributed in the total body water of the rats, would yield a concentration of approximately 4×10^{-7} M. In contrast, the calcium concentration in the extracellular fluids of the calcium gluconate-treated animals was calculated to be between 3 to 5 mM.

Further studies were carried out on a third animal model of nephrocalcinosis, one closely related to a human medical problem (40). Adult rats were subjected to unilateral nephrectomy. After recovery from the surgery the animals were placed on a high phosphate diet. On this regimen the animals gradually become uremic and the remaining kidney undergoes calcification as a consequence of the increased excretory work load and the elevated phosphate concentration in the blood plasma. When such animals were given phosphocitrate daily by the intraperitoneal route, uremia and calcification of the remaining kidney were largely suppressed. Many of the same observations made on the calcium gluconate and parathyroid hormone models also prevailed. Because renal calcification develops, but much more slowly, in humans who have lost a kidney, the possibility that phosphocitrate may be useful in prevention or therapy of this condition arises (40). Further animal studies have shown that phosphocitrate may be given intraperitoneally in the form of its monoethyl ester, with full retention of activity. Although monoethyl phosphocitrate is inactive in the non-biological hydroxyapatite crystal growth assay, presumably it is active in animals because the ethyl group is removed by enzymatic action.

OTHER BIOLOGICAL TESTS OF PHOSPHOCITRATE ACTION

All the above biological assays of phosphocitrate activity were carried out in systems in which abnormal or pathological calcification processes were induced in the kidney or in rib cartilage. The question arose as to whether phosphocitrate inhibits the normal process of bone calcification. In one test (41) phosphocitrate was added to bone cells cultured *in vitro*. Calcification occurs in such cultures and it is believed to be equivalent to that normally taking place in epiphyseal growth plates. It was a most interesting and potentially important finding that phosphocitrate, even when added in quite high concentrations, does not inhibit calcification in such *in vitro* bone cell cultures. This finding suggests that phosphocitrate does not inhibit normal calcification processes such as occur in hard tissues but may act primarily to suppress pathological calcification of normally soft tissues, as in nephrocalcinosis. To follow up this important idea, specimens of the epiphyseal disks of the long bones of normal mice and rats receiving daily doses of phosphocitrate have been collected and subjected to pathological examination. Even after rather long periods of administration there has been no evidence that phosphocitrate has altered the histology or staining characteristics of the epiphyseal growth plates in these animals (Walker, W.G. *et al.*, unpublished). Obviously, more comprehensive tests are required before any firm conclusions can be drawn. Nevertheless, the concept that phosphocitrate interferes only with pathological or metastatic calcification seems appropriate to consider. It is now well known that most soft tissues of the body may potentially undergo calcification after certain insults or challenges (42). Abnormal calcification occurs in arteriosclerosis, for example.

OTHER CALCIFICATION INHIBITORS

Phosphocitrate is not the only biological agent capable of suppressing abnormal calcification. One such may be statherin (43), a polypeptide found in saliva, which has been suggested to suppress the formation of dental calculi, which have a mineral base of hydroxyapatite. Another calcification inhibitor is a glycopeptide recently isolated from kidney cell culture medium (44), which inhibits the precipitation and crystallization of calcium oxalate, but apparently has no action on precipitation of calcium phosphate. Thus there may be different kinds of calcification inhibitors, possibly differing in their chemical nature, site of synthesis, and distribution, and in the type of calcium deposition process that they influence. The biochemistry and physical chemistry of such reagents and how they function in controlling biological calcification is a matter of great interest. Presumably the concentration and life-time of such agents in the tissues and body fluids must be under stringent control in order to suppress pathological calcification of soft tissues without harming the normal patterns of mineralization during growth and development of the skeletal system and teeth. We have still much work to do on phosphocitrate, particularly on the mechanism of its enzymatic synthesis and degradation, as well as the regulation of these processes. We must also learn in what portions of which cells this and other calcification inhibitors are synthesized and, ultimately, how they can control calcium phosphate deposition at the molecular level.

CALCIUM CARBONATE ACCUMULATION BY MITOCHONDRIA

Finally, I want to point out that isolated mitochondria not only can accumulate insoluble calcium phosphate salts and keep them in an amorphous state, but they are also capable of accumulating calcium carbonate, as we showed some years ago (45). Nearly all the CO_2 formed in aerobic animal cells is generated in the mitochondria through oxidations of the Krebs citric acid cycle. The concentration of bicarbonate in the mitochondrial matrix *in vivo* has been estimated to be at least 60 mM. Such a high concentration of bicarbonate at a matrix pH of $\sim$7.8 can be expected to result in the deposition of calcium carbonate if the Ca^{2+} level in the matrix is sufficiently high. We have found, with $^{45}Ca^{2+}$ and ^{14}C-labeled bicarbonate buffer, that both can be accumulated in rather large and equimolar amounts by respiring rat liver mitochondria, indicating that the accumulated form is calcium carbonate. When equal concentrations of bicarbonate and phosphate (2 mM) are offered to respiring rat liver mitochondria in the presence of a given concentration of Ca^{2+}, calcium phosphate will be accumulated in preference to calcium carbonate. On the other hand, in the presence of intracellular, i.e. cytosolic concentrations of phosphate (1.0 mM) and bicarbonate (25 mM), in equilibrium with 5% CO_2 in the gas phase, calcium carbonate will accumulate in the matrix. However, dense granules of $CaCO_3$ are not visible after standard methods of fixation, which appear to cause immediate loss of the $CaCO_3$. Such experiments suggest that insoluble calcium carbonate may be deposited in mitochondria of some tissues *in vivo*.

Thus calcium phosphate and calcium carbonate, the two most abundant minerals of hard tissues in the animal and plant kingdoms, may be concentrated and precipitated in mitochondria at the expense of energy generated by respiration. Moreover, as will be discussed by Dr. Volcani in another paper in this volume, silicon, another ubiquitous component of hard tissues, also can become bound or concentrated by mitochondria. While the evidence that mitochondria participate in biological mineralization with Ca^{2+} is still circumstantial, they appear to have the potential to do so with respect to rate and capacity, and could participate in a wide variety of different biomineralization processes (46).

REFERENCES

1. Lehninger, A.L., Carafoli, E., and Rossi, C.S.: 1967, Adv. Enzymol. 29, pp. 259-320.
2. Lehninger, A.L.: 1970, Biochem. J. 119, pp. 129-138.
3. Bygrave, F.L.: 1977, Curr. Top. Bioenerg. 6, pp. 259-318.
4. Fiskum, G. and Lehninger, A.L.: 1979, Fed. Proc. 38, p. 304.
5. Carafoli, E. and Crompton, M.: 1978, Curr. Top. Membrane Transp. 10, pp. 151-216.
6. Saris, N. and Åkerman, K.E.O.: 1980, Curr. Top. Bioenerg. 10, pp. 104-179.
7. Nicholls, D.G.: 1981, TIBS 6, pp. 36-38.
8. Vasington, F.D. and Murphy, J.V.: 1962, J. Biol. Chem. 237, pp. 2670-2677.
9. Rossi, C.S. and Lehninger, A.L.: 1963, Biochem. Z. 338, pp. 698-713.
10. Rossi, C.S. and Lehninger, A.L.: 1964, J. Biol. Chem. 239, pp. 3971-3980.
11. Chance, B.: 1965, J. Biol. Chem. 240, pp. 2729-2748.
12. Lehninger, A.L.: 1974, Proc. Natl. Acad. Sci. USA 71, pp. 1520-1524.
13. Fiskum, G., Reynafarje, B., and Lehninger, A.L.: 1979, J. Biol. Chem. 254, pp. 6288-6295.
14. Carafoli, E. and Lehninger, A.L.: 1971, Biochem. J. 122, pp. 681-690.
15. Chen, C.-H. and Lehninger, A.L.: 1973, Arch. Biochem. Biophys. 154, pp. 449-459.
16. Chen, C.-H. and Lehninger, A.L.: 1973, Arch. Biochem. Biophys. 157, pp. 183-196.
17. Jeng, A.Y. and Shamoo, A.E.: 1980, J. Biol. Chem. 255, pp. 6897-6903.
18. Jeng, A.Y. and Shamoo, A.E.: 1980, J. Biol. Chem. 255, pp. 6904-6912.
19. Fiskum, G. and Lehninger, A.L.: 1979, J. Biol. Chem. 254, pp. 6236-6239.
20. Nicholls, D.G.: 1978, Biochem. J. 176, pp. 463-474.
21. Becker, G.L.: 1980, Biochim. Biophys. Acta 591, pp. 234-239.
22. Becker, G.L., Fiskum, G., and Lehninger, A.L.: 1980, J. Biol. Chem. 255, pp. 9009-9012.
23. Denton, R.M., McCormack, J.G., and Edgell, N.J.: 1980, Biochem. J. 190, pp. 107-117.
24. Hansford, R.G.: 1981, Biochem. J. 194, pp. 721-732.
25. Landry, Y. and Lehninger, A.L.: 1976, Biochem. J. 158, pp. 427-438.
26. Lehninger, A.L., Vercesi, A., and Bababunmi, E.A.: 1978, Proc. Natl. Acad. Sci. USA 75, pp. 1690-1694.
27. Greenawalt, J.W., Rossi, C.S., and Lehninger, A.L.: 1964, J. Cell Biol. 23, pp. 21-38.
28. Betts, F., Blumenthal, N.C., Posner, A.S., Becker, G.L., and Lehninger, A.L.: 1975, Proc. Natl. Acad. Sci. USA 72, pp. 2088-2090.

29. Carafoli, E., Rossi, C.S. and Lehninger, A.L.: 1965, J. Biol. Chem. 240, pp. 2254-2261.
30. Martin, J.H. and Matthews, J.L.: 1969, Calcif. Tiss. Res. 3, p. 184.
31. Brighton, C.T. and Hunt, R.M.: 1976, Fed. Proc. 35, pp. 143-147.
32. Howard, J.E.: 1976, J.H. Med. Journal 139, pp. 239-252.
33. Trump, B.F., Dees, J.H., Kim, K.H., and Sahaphong, S.: 1972, In: Urolithiasis: Physical Aspects (Finlayson, B., Hench, L.L., and Smith, L.H., Eds.), Natl. Acad. Sci., Wash., D.C., pp. 1-42.
34. Landis, W.J. and Glimcher, M.J.: 1982, J. Ultrastructural Res. 78, pp. 227-268.
35. Chen, C.-H., Greenawalt, J.W., and Lehninger, A.L.: 1973, J. Cell Biol. 61, pp. 301-315.
36. Becker, G.L., Chen, C.-H., Greenawalt, J.W., and Lehninger, A.L.: 1974, J. Cell Biol. 61, pp. 316-326.
37. Becker, G.L., Termine, J.D., and Eanes, E.D.: 1976, Calcif. Tiss. Res. 21, pp. 105-113.
38. Tew, W.P., Mahle, C., Benavides, J., Howard, J.E., and Lehninger, A.L.: 1980, Biochemistry 19, pp. 1983-1988.
39. Tew, W.P., Malis, C.D., Howard, J.E., and Lehninger, A.L.: 1981, Proc. Natl. Acad. Sci. USA 78, pp. 5528-5532.
40. Gimenez, L., Walker, W.G., Tew, W.P., and Herman, J.R.: 1982, Kidney International, in press.
41. Reddi, A.H., Meyer, J.L., Tew, W.P., Howard, J.E., and Lehninger, A.L.: 1980, Biochem. Biophys. Res. Communs. 97, pp. 154-159.
42. Selye, H.: 1962, Calciphylaxis, Univ. of Chicago Press.
43. Schlesinger, D.H. and Hay, D.I.: 1977, J. Biol. Chem. 252, pp. 1689-1695.
44. Nakagawa, Y., Margolis, H.C., Yokoyama, S., Kezdy, F.J., Kaiser, E.M., and Coe, F.L.: 1981, J. Biol. Chem. 256, pp. 3936-3944.
45. Elder, J.A. and Lehninger, A.L.: 1973, Biochemistry 12, pp. 976-982.
46. Lehninger, A.L.: 1977, Horizons in Biochem. and Biophys. 4, pp. 1-30.

A COMPARISON OF THE ROLES OF CALCIUM IN BIOMINERALIZATION AND IN CYTOSOLIC SIGNALLING

Robert H. Kretsinger

Department of Biology, University of Virginia
Charlottesville, VA, 22901, U.S.A.

Calcium serves at least two very significant and distinct functions in Biology--one as a structural component, the other in information transfer. We are seeking unifying principles or a general conceptual framework within which we can study these phenomena. The following observations should be relevant.

CHEMISTRY OF CALCIUM

Coordination Geometry

The outer two electrons of the calcium atom are easily dissociated. The resulting Ca^{2+} ion is strongly electropositive. Its bonds are entirely electrostatic with essentially no covalent character. Of atoms found in biopolymers, oxygen is usually the most electronegative and is much more frequently observed as a calcium ligand than are nitrogen or sulfur. In inorganic salts calcium often complexes with halides. Of particular interest, in fluoroapatite, $[Ca_5(PO_4)_3F]$, F^- replaces the OH^- ion of hydroxyapatite.

The positive charge of the Mg^{2+} ion (ionic radius 0.65 A) is even more concentrated than is that of calcium. It has a slightly greater affinity for electronegative ligands. The Mg^{2+} ion is almost always six coordinate with oxygen. Five and seven coordinated Mg^{2+} are rare. With a Mg-O distance about 2.00 A the liganding oxygen atoms are in stabilizing van der Waals contact with one another at 2.82 A. For instance Mg $(H_2O)_6{}^{2+}$ is quite stable; aquomagnesium dissociates slowly.

In contrast, the Ca^{2+} ion (ionic radius 0.99 A) can accommodate eight oxygen ligands in square antiprism configuration. With a Ca-O distance of 2.40 A, the O-O distance is 2.77 A. One might anticipate that calcium would always be eight coordinate with oxygen so long as the geometry of the organic molecule did not place steric constraints on the disposition of the oxygen ligands. However, rather surprisingly there

P. Westbroek and E. W. de Jong (eds.), Biomineralization and Biological Metal Accumulation, 123–131.

is quite a variation in Ca-O oxygen bond lengths (2.3 to 2.6 A). The energy minimum appears to be broader than for the Mg-O bond. There are as many examples of seven coordinate calcium compounds as of eight (Kretsinger and Nelson, 1976). Aquocalcium exists as a mixture of six, seven, and eight (or higher) coordinate. Mean coordination number decreases with increasing calcium concentration (Hewish et al, 1982).

Crosslinking Ability

The Ca^{2+} ion is an excellent crosslinker of biopolymers that have oxygen ligands (Williams, 1974). Its phosphate and carbonate salts are insoluble. It is still not possible to predict solubility products and rate constants from first principles. Yet, in a qualitative sense one attributes this crosslinking ability, relative for instance to that of magnesium, to variability in bond lengths and angles and in coordination number. Further the affinity of any ligand in aqueous solution must be considered relative to that of water. Many proteins are stabilized by calcium and/or aggregated and precipitated.

These solubilities determine the nature of our planet. Calcium is the sixth most abundant element in the earth's crust. Carbon is a relatively rare element. About 99% of the earth's carbon is in $CaCO_3$. Nearly all of these calcite sediments are of biogenic origin.

Of particular relevance to mammals, hydroxyapatite is the least soluble calcium phosphate above pH 7.0. With $K_s(Ca_5(PO_4)_3OH) = 10^{-58.5}$ M^9 a solution is saturated at the following pairs of concentrations for total phosphate (H_3PO_4, $H_2PO_4^-$, HPO_4^{2-}, and PO_4^{3-}) and calcium, calculated at pH 7.2.

$[Ca^{2+}]$ M	$10^{-7.0}$	$10^{-6.0}$	$10^{-5.0}$	$10^{-4.0}$	$10^{-3.0}$	$10^{-2.0}$
[Total Phosphate] M	$10^{-0.1}$	$10^{-1.7}$	$10^{-3.4}$	$10^{-5.1}$	$10^{-6.7}$	$10^{-8.4}$

In any body fluid this solubility product cannot be exceeded for an extended period without calcium phosphate precipitation.

Dehydration Rate

Eigen and Wilkins (1965) determined the dehydration rate of calcium to be 10^8 sec^{-1}, while that of magnesium is $10^{4.5}$ sec^{-1}. Most Ca^{2+} or Mg^{2+} ion binding reactions have k_{on} rates of 10^8 sec^{-1} M^{-1} or $10^{4.5}$ sec^{-1} M^{-1} respectively. The dehydration is rate limiting. Eigen suggested that this rapid reaction rate might account for calcium's functioning as a second messenger instead of magnesium. I doubt that this speed of reaction was crucial $3 \cdot 10^9$ years ago when this evolutionary decision was made. All eukaryotes, and possibly some prokaryotes, use calcium as a cytosolic messenger.

The relatively slow speed of the magnesium dehydration may be due to the snug fit of the oxygen atoms in $Mg(H_2O)_6^{2+}$. The oxygen atoms of aquocalcium may not fit so well and hence are more readily exchanged.

Most calcium binding proteins within the cytosol, so called calcium modulated proteins, bind calcium with about 10^4 higher affinity than they do magnesium. Hence under the conditions that obtain in the resting cell, (pCa = 7.5, pMg = 2.8), the low divalent cation affinity proteins, such as troponin (sites 3 and 4) are predominately in the apo form; while the high affinity proteins, such as parvalbumin, are in the magnesium bound form. When second messenger calcium enters a muscle cell it first binds to troponin; then, after magnesi-parvalbumin has dissociated the calcium leaves troponin and binds to apo-parvalbumin (Robertson et al, 1981 and Gillis et al, 1982). It remains to be seen how important these reaction rates will be in Biology.

Supersaturation and Crystal Formation

There is no question about the importance of reaction rates in the formation of calcium salts. In over simplification consider a bone, or a tooth, in contact with serum, or saliva; or consider a shell in contact with the ocean. If the saliva is undersaturated why doesn't the tooth dissolve; if supersaturated why doesn't the calcium phosphate precipitate from the saliva? Part of the answer is that many biominerals have monomolecular surface films of carbohydrate or of protein.

Of more general interest consider the addition of ion pairs to water within a closed system. At saturation, more energy would be released if the ions were to bind to a crystal than would be released upon dissolution. However, no crystal seed yet exists so the added ions continue to dissolve. The solution is supersaturated. It is then a question of probability as to when the appropriate ionic collisions might occur to form a crystal seed. Some supersaturated solutions are stable for decades, others for only seconds.

In the calcium phosphates, a more soluble crystal form, octacalcium phosphate, is actually most likely to precipitate from supersaturated solutions. Additional octacalcium phosphate grows on this seed. However, the lowest free energy of the system is finally reached when the $Ca_8H_2(PO_4)_6 \cdot 5H_2O$ "ripens" to hydroxyapatite, $Ca_5(PO_4)_3OH$. This, in rough outline, is what happens in bone deposition when amorphous precipitate matures to apatite (Brown, 1962).

Two additional factors are probably introduced in most biomineralization processes. Near the site of potential mineral deposition the solubility product has been reached or slightly exceeded. A polyglycan or a protein such as collagen might possess a periodicity and (partial) charge distribution such that it functions as a seed. This theory of epitaxis has yet to be proven but there is strong inferential evidence in its support.

Conversely, organic molecules can retard crystal formation. All one need do to prevent precipitation of a supersaturated solution is to remove seeds as they appear. In particular it has been suggested that phosphocitrate in the urine and the proteins, statherin (Schlesinger and Hay, 1977) and proline rich phosphoprotein (Bennick et al, 1981) in saliva serve this retarding function.

Calcium in the Environment

In 1935 McLean and Hastings observed that "The concentration of free calcium ion in the serum is one of Nature's physiological constants." They used the frog heart assay to distinguish the concentration of the free Ca^{2+} ion from total calcium. The concentration of total calcium in the serum is 2.5 mM; half is bound and half is free Ca^{2+} ion. This is close to the ocean concentration of calcium, about 8.0 mM.

The calcium concentrations in most extracellular fluids in eukaryotes are in these ranges. The usual interpretation is that life evolved in a millimolar calcium environment, grew accustomed to it, and then in the metazoa evolved elaborate mechanisms to maintain this milieu. Several questions are left unanswered. Was the prebiotic ocean, four billion years ago, millimolar in calcium; and in any case was it the site of formation of the first cells? Our present ocean is nearly saturated in calcium, especially near the surface where HCO_3^- concentration is highest.

In my concluding chapter I will argue that all cells have evolved calcium extrusion mechanisms because calcium is deleterious to cytosolic function. Given this pumping ability, why are animals so sensitive to slight increases or decreases in serum calcium?

BIOLOGY OF CALCIUM

Cytosolic Concentration of Calcium

The concentration of free Ca^{2+} ion within the cytosol of all "resting" eukaryotic, and probably prokaryotic, cells is about $10^{-7.5}$ M. Three techniques have proven to be particularly valuable: the injection of the calcium specific photoprotein, aequorin; the injection of calcium specific dyes, such as arsenazo III; and the insertion of calcium specific microelectrodes. All of these techniques are difficult to calibrate and subject to errors up to a factor of ten. Nonetheless, many different cells have been measured by different techniques and by different groups. Further, purified calcium modulated proteins, found in the cytosol, have calcium affinities consistent with low cytosolic calcium.

One cannot inject bacteria but two sorts of experiments indicate that at least some bacteria also maintain $[Ca^{2+}]_{in}$ between 10^{-7} and 10^{-8} M. Under special conditions protoplasts, derived from bacteria whose cell walls have been removed, can be lysed and made to reseal inside out. Their calcium pumps are still active and they reduce the Ca^{2+} ion concentration below 10^{-7} M in the medium (Lee et al, 1979).

B. subtilis change their direction of swimming by reversing the sense of rotation of their flagella. Counterclockwise rotation causes forward swimming; clockwise produces tumbling. In the presence of the calcium ionophore, A23187, the sense changes from counterclockwise, to inactive, to clockwise as [Ca^{2+}] in the medium is raised from 10^{-8} to $10^{-6.5}$ M. This may be the controlling factor within the intact bacterium (Ordal, 1977).

One can consider the maintenance of cytosolic Ca^{2+} ion concentrations less than 10^{-7} M to be a universal and necessary condition of life.

Calcium Extrusion

All cells have specific calcium extrusion mechanisms in order to maintain these low concentrations. They seem to be of three general sorts: Ca-ATPase, Na^{+}-Ca^{2+} exchange, and H^{+}-Ca^{2+} exchange.

Calcium pumping Mg-ATPase enzymes have been identified in a wide variety of animal tissues. The best characterized ones come from erythrocytes, from the plasma membrane of cardiac cells, and from the sarcoplasmic reticulum of skeletal muscle. Although this is an area of intense investigation, we still don't know whether they are homologous to one another nor do we know their basic stoichiometries. Of course there are many ATPases that are activated by calcium, such as those involved in muscle contraction and ciliary motion, that are not calcium pumps.

Most bacteria are considered to pump calcium by exchanging it for protons. A similar mechanisms exists in mitochondria, which appear to have evolved from symbiotic bacteria. The energy for this pumping derives from oxidative phosphorylation. Whether the proton gradient is an obligatory intermediate is debated. There is one report of an ATPase type pump in S. faecalis. Kobayashi et al (1978) suggested "that calcium entry is detrimental but unavoidable, and that the function of the ATP-linked pump is to prevent the accumulation of calcium in the cytoplasm."

Most eukaryotic cells also contain Na^{+} ion extrusion mechanisms. These Na-ATPases are in many ways similar to the Ca-ATPase pumps. Particularly in nervous tissue the energy used to extrude calcium comes from the sodium gradient. The outflow of one Ca^{2+} ion is coupled to the inward flow of three Na^{+} ions down their concentration gradient.

The detailed mechanisms for none of these pumps is known; their ubiquity is well established.

Calcium as a Cytosolic Messenger

All cells respond somehow to stimuli appropriate to the cell type --muscles to acetylcholine, the retina to photons, eggs to sperm. Many of these stimuli never enter the target cell; and even if they do, they do not interact directly with the enzymes or structural proteins that are responsible for the cell's response. The stimulus is coupled to the responding protein(s) via a "second messenger", two of which, the Ca^{2+} ion and cyclic adenosine monophosphate, have been well studied. Others may function in special situations.

The same techniques used to determine $[Ca^{2+}]$ in larger cells such as muscle, neurones, eggs, and salivary gland have been applied to follow the change in $[Ca^{2+}]$ following stimulation. They do not yet have adequate spatial or temporal resolution to describe the diffusion path in detail. Following stimulation the Ca^{2+} ion concentration rises rapidly from $10^{-7.5}$ M to about $10^{-5.5}$ M and then returns to the resting level within, at most, a few seconds. This pulse of Ca^{2+} ions has been measured with both aequorin (Blinks, 1978) and arsenazo III (Miledi et al, 1980). Although the concentration of Ca^{2+} just next to the site of entry may exceed $10^{-5.5}$ M, it seems that the molecule that detects the rise in $[Ca^{2+}]$ must have a $pK_d(Ca^{2+})$, under cytosolic conditions, in the range of 6.0 to 7.0 in order to switch on, then off, during the rise, then fall, of the calcium pulse.

I have argued that Ca^{2+} ion levels cannot be too high, too long within the cytosol without precipitating phosphate, normally about 10^{-3} M, or other phosphate containing compounds. Further it appears that only proteins, not carbohydrates or lipids, have the affinity and the specificity to coordinate calcium at $10^{-5.5}$ M in the presence of cytosolic concentrations of Mg^{2+} ion, about $10^{-2.8}$ M.

Homology of Calcium Modulated Proteins

Calcium modulated proteins are the targets of calcium functioning as a second messenger. They are defined by two characteristics. They are present in the cytosol or on membranes facing the cytosol. Their affinities for calcium, under cytosolic conditions, are such that in the resting cell they are in the apo or in the magnesium form; in the stimu- lated cell they have bound calcium. They are the switches activated by calcium.

Amino acid sequences are known for several of these calcium modulated proteins and crystal structures for two, parvalbumin and intestinal calcium binding protein, have been determined. They have all evolved from a common precursor, the so called "EF hand domain". The most ubiquitous of them, calmodulin, is found in all, or most, eukaryotic cells.

The EF hand domain consists of about thirty amino acids--ten in a length of α-helix known as the E helix, ten in a loop about the calcium ion, and another ten in α-helix F. This domain is easily recognized in an amino acid sequence. Residues on the sides of the two α-helices facing the inside of the molecule have hydrophobic side chains at positions 2, 5, 6, 9, 22, 25, 26 and 29. Residues at positions 10, 12, 14, 18, and 21 in the calcium binding loop have oxygen containing side chains; they coordinate the calcium ion (Kretsinger, 1980).

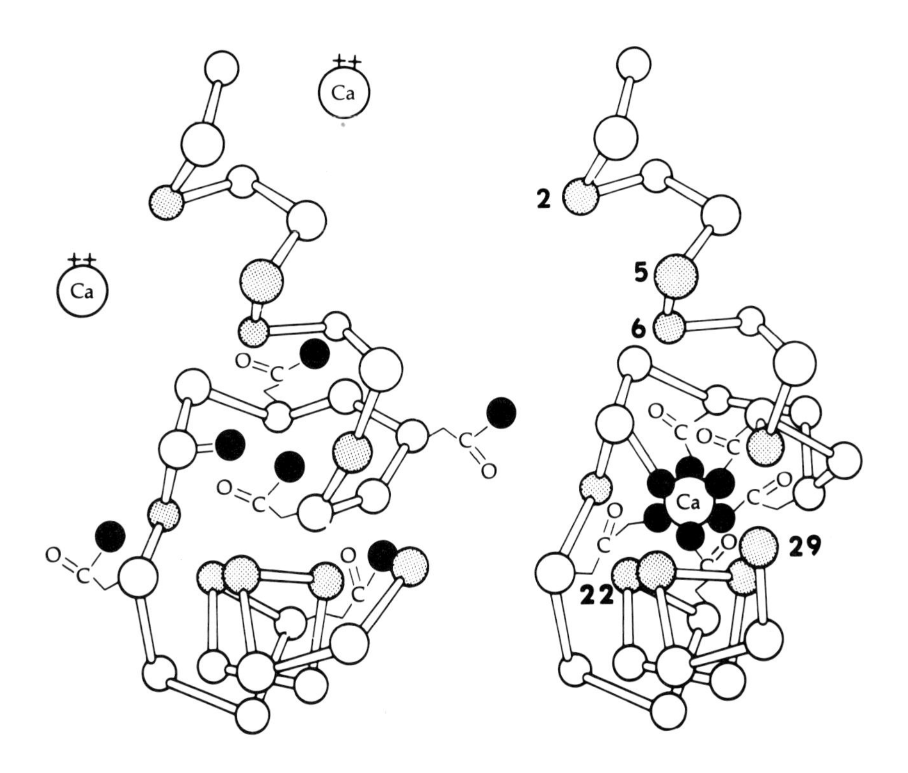

The EF hand domain undergoes a conformational change upon binding calcium, figure 1. In retrospect, this helix-loop-helix conformation seems ideally suited to transmitting the information, and energy, content of a change in calcium concentration. It is not clear, however, why it is so generally and uniquely employed.

Six proteins--calmodulin, troponin C, myosin light chain, parvalbumin, S-100, and intestinal calcium binding protein--are known to contain four, three, or two EF hands within their single polypeptide chains. None of these proteins are themselves enzymes or structural proteins. They are inferred to transmit, in their calcium bound forms, information to their ultimate targets. The highly conserved calmodulin interacts with at least eighteen proteins--phosphodiesterase, adenylate cyclase, guanylate cyclase, myosin light chain kinase, NAD^+ kinase, phosphorylase kinase, glycogen synthase kinase, protein kinase(s), actomyosin, membrane Ca-ATPase, dynein ATPase, phospholipase A_2, phosphoprotein phosphatase 2B, N-methyl transferase, O-methyl transferase, succinate dehydrogenase, spectrin, calcineurin, the gap junction protein, the K^+ channel protein and microtubules.

Other cytosolic proteins--such as aequorin, calcium protease, villin, synexin, gelsolin, spasmin, invertebrate sarcoplasm calcium binding protein--appear to be calcium modulated. Whether they too contain EF hand domains is not yet known.

There are numerous extracellular calcium binding proteins. Among them those involved in blood clotting and, like osteocalcin, in bone formation, contain the unique amino acid γ-carboxy glutamic acid (Hauschka et al, 1982). None of these proteins appear to be calcium modulated; none of them contain EF-hand domains.

We still must understand the structural changes induced by calcium binding and in particular the evolutionary success of the EF hand domain proteins.

Organic Matrices and Biomineralization

Most calcium biominerals are formed in contact with organic matrices. Several possibilities must be explored. There have been reports of calcium phosphate precipitates being formed within the mitochondria of cells associated with mineralization, such as the egg laying gland of the chicken (Lehninger, 1977). However, the suggestion that these seeds might then be transported outside of the cell has not been confirmed. It seems likely that, at least in most cases, the initial precipitation of calcium phosphate or calcium carbonate occurs extracellularly.

Proteins, polyglycans, or small molecules might function in one of, at least, two ways. In the immediate vicinity of the area to be mineralized the solution is supersaturated but stabilized by a "seed-chelator". Mineralization would be initiated or controlled by an extracellular enzyme that removes this putative seed-chelator. Although there exist such chelators, I know of no evidence that they actually function in this regulatory capacity.

In an epitaxis model the medium is slightly undersaturated. The biopolymer serves as a crystal template. The conformation, periodicity, and reactivity of its subunits bear some resemblance to the surface of the nascent crystal. In effect this organic seed prevents the solution from becoming supersaturated. Further if its affinities for Ca^{2+} and/or for PO_4^{3-} ions were high enough, it might reduce the effective solubility product within its immediate environment. Several unit cell thicknesses of distorted crystal might be deposited before the macroscopic solubility product would obtain.

The results of Berthet-Colominas et al (1979) describing the deposition of amorphous calcium phosphate, then hydroxyapative, on collagen in developing turkey tendon support this epitaxis model. However, the geometric relationship, if any, between the collagen structure and that of either octacalcium phosphate or hydroxyapatite has yet to be established.

CONCLUSION

One would like to bring answers to such an important and exciting field as biomineralization. Failing that, I hope to have posed meaningful questions. These basic chemical and biological properties of calcium surely must be included in any comprehensive model.

REFERENCES

Bennick, A., McLaughlin, A.C., Grey, A.A. & Madapallimattam, G. "The Location and Nature of Calcium-binding Sites in Salivary Acidic Proline-rich Phosphoproteins" J. Biol. Chem. (1981) 256 4741-4746.

Berthet-Colominas, C., Miller, A. & White, S.W. "Structural Study of the Calcifying Collagen in Turkey Leg Tendons" J. Mol. Biol. (1979) 134 431-445.

Blinks, J.R. "Applications of Photoproteins in Experimental Biology" Photochem. Photobiol. (1978) 27 423-432.

Brown, W.E. "Octacalcium Phosphate and Hydroxyapatite" Nature (1962) 196 1048-1055.

Eigen, M. & Wilkins, R.G. "The Kinetics and Mechanism of Formation of Metal Complexes" Adv. Chem. (1965) 49 55-80.

Gilles, J.-M, Thomason, D.B., LeFevre, J. & Kretsinger, R.H. "Parvalbumin and Muscle Relaxation: A Computer Simulation Study" J. Muscle Res. Cell Motil. (1982) in press.

Hauschka, P.V., Carr, S.A. & Biemann, K. "Primary Structure of Monkey Osteocalcin" Biochemistry (1982) 21 638-642.

Hewish, N.A., Neilson, G.W. & Enderby, J.E. "Environment of Ca^{2+} Ions in Aqueous Solvent" Nature (1982) 297 138-139.

Kobayashi, H., Van Brunt, J. & Harold, F.M. "ATP-linked Calcium Transport in Cells and Membrane Vesicles of Streptococcus faecalis" J. Biol. Chem. (1978) 253 2085-2092.

Kretsinger, R.H. "Structure and Evolution of Calcium-Modulated Proteins: C. R. C. Crit. Rev. Biochem. (1980) 8 119-174.

Kretsinger, R.H. & Nelson, D.J. "Calcium in Biological Systems" Coord. Chem. Rev. (1976) 18 29-124.

Lee, S.-H., Kalra, V.K. & Brodie, A.F. "Resolution and Reconstitution of Active Transport of Calcium by a Protein(s) from Mycobacterium phlei" J. Biol. Chem. (1979) 254 6861-6864.

Lehninger, A.L. "Mitochondria and Biological Mineralization Processes: an Exploration" Horizons Biochem. Biophys. (1977) 4 1-30.

McLean, F.C. & Hasting, A.B. "Clinical Estimation and Significance of Calcium-Ion Concentrations in the Blood" Am. J. Med. Sci. (1935) 189 601-613.

Miledi, R., Parker, I. & Schalow, G. "Transmitter Induced Calcium Entry across the Post-Synaptic Membrane at Frog End-Plate Measured Using Arsenazo III" J. Physiol. (1980) 300 197-212.

Ordal, G.W. "Calcium Ion Regulates Chemotactic Behaviour in Bacteria" Nature (1977) 270 66-67.

Robertson, S.P., Johnson, J.D. & Potter, J.D. "The Time-Course of Ca^{2+} Exchange with Calmodulin, Troponin, Parvalbumin, and Myosin in Response to Transient Increases in Ca^{2+}" Biophys. J. (1981) 34 559-569.

Schlesinger, D.H. & Hay, D.I. "Complete Covalent Structure of Statherin, a Tyrosine-rich Acidic Peptide Which Inhibits Calcium Phosphate Precipitation from Human Parotid Saliva" J. Biol. Chem. (1977) 252 1689-1695.

Williams, R.J.P. "Calcium Ions: Their Ligands and Their Functions" Biochem. Soc. Symp. (1974) 39 133-138.

HORMONAL REGULATION OF CALCIFICATION, WITH PARTICULAR REFERENCE TO THE HORMONAL CONTROL OF EGGSHELL FORMATION IN BIRDS AND SHELL GROWTH IN MOLLUSCS

J. Joosse
Department of Biology, Free University, P.B. 7161,
1007 MC Amsterdam, The Netherlands

A short introduction to the functions of the hormones involved in blood calcium homeostasis in mammals, is followed by a comparative endocrinological view on the formation of the eggshell in birds and the growth of the shell in a gastropod. There are some interesting parallels. Both structures consist mainly of calcium carbonate. In their formation calcium binding proteins and carbonic anhydrase play an important role. Control of the formation of these structures is not exerted by hormones involved in calcium homeostasis, but by the reproductive hormones in female birds and by the growth hormone in the gastropod. In both cases actual secretion is triggered by local sensory input and is probably mediated by local nervous elements.

1. INTRODUCTION

In all organisms calcium plays an important role in the control of all kinds of cellular activities. Therefore the ionic concentration of this element in the cytosol is carefully controlled (cf. Lehninger, this issue). Advanced multicellular organisms, such as the vertebrates, have extended this control to the extracellular fluids. Their cells, particularly the nerve and muscle cells, are adapted to this situation and have a low tolerance to changes in the calcium concentration of the blood. Therefore, control of calcium levels has to be exerted throughout the lifetime of these animals. However, also in less advanced groups, e.g. in molluscs and crustaceans, blood calcium levels show a high degree of stability.

For animals living in sea water the control of blood calcium levels is no problem since calcium is always available in their environment. Generally the same is true for animals living in freshwater habitats. Terrestrial animals, however, do not have continuous access to calcium sources in their environment. They depend on calcium present in food and drinking water. Particularly for these animals it is benificial to build up internal calcium stores in the form of calcium phosphate and carbonate. In many animals these stores also serve other purposes, e.g. in vertebrates they form an important component of the internal skeleton, and in

P. Westbroek and E. W. de Jong (eds.), Biomineralization and Biological Metal Accumulation, 133–144.

many molluscs the calcareous shell serves as an external skeleton.

For long term homeostatic control of calcium in blood, hormones are particularly suitable. In principle the activity of the hormone producing cells and the calcium concentration in the extracellular space could form a simple feedback loop. However, interaction of these hormones with cells involved in formation or resorption of calcium in stores is apparently insufficient. The actual control scheme is much more complicated and includes uptake and removal of calcium from the body and protection of the other functions of the stores, e.g. bones in vertebrates.

It is the aim of the present paper to present a comparative endocrinological view on the hormones involved in control of calcium levels in the blood and in bone formation of mammals, in eggshell formation in birds, and in shell formation in molluscs. Most of the attention will be given to the latter two aspects.

For general information on hormones involved in calcium metabolism in mammals, see Talmage et al. (1975) and Norris (1980); in birds, see Sturkie (1976) and Epple and Stetson (1980); and in molluscs, see Joosse and Geraerts, in press.

2. ENDOCRINE CONTROL OF BLOOD CALCIUM LEVELS AND BONE FORMATION IN MAMMALS

Bone is the major component of the mammalian skeleton. The skeleton supports the soft tissues, protects vital organs such as the brain and is used for the attachment of skeletal muscles. Moreover, it contains the bone marrow in which blood cells are produced, and it is used as a storage site for minerals, particularly calcium.

Bone consists of an organic matrix of proteinaceous collagen fibres. The inorganic material in bone is mainly calcium phosphate in the form of hydroxyapatite. Most of this is represented by very small crystals of 30 x 3 nm size. The combination of fibres and crystals is the physicochemical basis of the hardness and rigidity of bone.

Bone is a living tissue. It has a microscopic honeycomb-like appearance, due to the presence of numerous small chambers (lacunae), each of which contains an osteocyte bone cell. Numerous fine channels (canaliculi) radiate from the lacunae through the surrounding mineral substance and these provide a pathway through which dendrite-like extensions of the osteocytes maintain contact with other cells and with the blood vascular system for transport of nutrients and metabolites. Large bones are traversed by wider channels, the Haversian canals, containing blood vessels and nerves, which branch through the bone. Externally bone is surrounded by a tough periosteum. All internal channels and spaces are lined by the endosteum.

It should be stressed that bone is a dynamic tissue. Particularly in growing animals, but to a lesser extend also in adults, there is a continuous activity in modelling and remodelling bone in relation to changes in the function of the skeleton and the various activities of the individual.

Two cell types are involved in bone dynamics. Osteoblasts are responsible for bone formation. They secrete the collagen fibres. In the

EM picture these cells show the structural characteristics of protein synthesis and secretion. Sometimes osteoblasts get surrounded by bone matrix and then change into osteocytes. Osteoclasts are involved in bone resorption. They are motile, multinuclear, irregularly branched cells, often found in small holes in the surface of bone. The active part of the cell membrane shows numerous foldings (ruffled border). Osteoclasts secrete enzymes for the breakdown of collagen. Furthermore, they contain many lysosomes which may be involved in final intracellular digestion of the organic material.

Osteoblasts and osteoclasts are found in adjacent positions. This makes sense in view of their roles in the remodelling of bone, but from the physiological point of view it raises the problem that it also implies that crystal formation and mobilization occur in the same micro-environment. Moreover, it is difficult to understand that the activities of these cells differ depending on their position in the skeleton, especially at locations where breakdown and formation of new bone occur simultaneously. The control of these processes is still poorly understood.

Hormones play an important role in control of bone formation. Of great importance are two hormones which seem to be primarily involved in the homeostatic control of blood calcium levels. Parathyroid hormone (PTH) is produced by the parathyroid glands. It is more actively released when calcium levels in the blood decrease. PTH stimulates the mobilization of calcium from bone, activates the reabsorption of calcium from prourine in the kidney, and stimulates the uptake of calcium by the small intestine. All these processes result in a rise of calcium in the blood. Bij negative feedback action of the blood calcium titre the release of PTH decreases.

Similarly the release of thyrocalcitonin (TCT), produced by the C-cells in the thyroid of mammals, is stimulated when calcium levels in the blood tend to be too high. TCT decreases the mobilization of calcium from bone. It inhibits the effects of PTH. An important role of TCT seems to be the protection of the skeleton from excessive destruction in meeting the calcium requirement during pregnancy and lactation (Norris, 1980).

The ways in which PTH and TCT act on the bone cells is still a matter of debate. PTH stimulates the activity of osteoclasts and may increase their number. The prostaglandin PGE_1 is in some way involved in these processes. PGE_1 and PTH have identical effects, but seem to use different receptors (Herrmann-Erlee et al., 1980). Several investigators have suggested that at low, physiological, levels, PTH acts primarily on osteoblasts by stimulating the transport of calcium ions from the bone surface to the blood. The major site of action of TCT might be on the osteoclasts by an intervention of the action of PTH and/or prostaglandins.

A third hormone involved in calcium metabolism is 1,25 dihydroxycholecalciferol (1,25-$(OH)_2$-D_3), a metabolite of vitamin D_3. This substance is produced by the kidney from a precursor formed in the liver. It acts on the intestine, the bone cells and the kidney. Its production is stimulated by PTH. In the cells of the intestinal mucosa 1,25-$(OH)_2$-D_3 induces the synthesis of a calcium binding protein (CaBP) and, in this way, enhances the uptake of calcium from the intestinal lumen (Kretsinger, this issue). Furthermore, it increases the sensitivity of the bone cells and the kidney to PTH.

From the above it is clear that PTH, TCT and 1,25-$(OH)_2$-D_3 primarily function in the homeostatic control of blood calcium levels. This raises the question how they are involved in bone modelling and remodelling. This is not clearly understood. It should be stressed, however, that other hormones are also involved in the growth and maintenance of the skeleton. Growth hormone and thyroid hormone are important for skeletal growth. The raise in gonadal hormones at puberty determines the end of skeletal growth by inducing the closure of the epiphysial plates in the limb bones. This picture is still incomplete, as no attention is given here to the anions, particularly phosphate, in relation to these processes.

In conclusion it can be stated that in mammals in the control of calcium metabolism a number of hormones is interacting. PTH, TCT and 1,25-$(OH)_2$-D_3 titers are controlled by negative feedback mechanisms from blood calcium levels and by mutual interactions. Prostaglandins act locally. This complex of hormones seems to function independent of a neuroendocrine control by the nervous system. The latter influence is exerted via growth hormone, thyroid hormone and the gonadal hormones.

3. CONTROL OF CALCIUM METABOLISM AND EGGSHELL FORMATION IN BIRDS

3.1. Calcium metabolism

In birds control of calcium also involves PTH, CT and 1,25-$(OH)_2$-D_3. Calcitonin (CT) is produced in separate glands, the ultimobranchial bodies. The role of calcitonin may differ from that in mammals. Possibly in birds CT controls primarily phosphate metabolism (Ringer and Meijer, 1976).

3.2. Eggshell function, structure and formation

The formation of the egg occurs in the oviduct (for literature, see Sturkie, 1976). After ovulation the large oocyte is fertilized in the funnel-like infundibulum. Albumen is secreted by the magnum. The inner and outer shell membranes are secreted by the isthmus. These membranes show stretching. The resulting "space" within the inner membrane is filled up by entry of water and salts ("plumping"). The swollen egg is meanwhile and afterwards surrounded by the shell, secreted by the shell gland which is the uterine part of the oviduct. A cuticle is added just before the egg is laid.

Besides its function as a protecting capsule, the eggshell has a role in gas exchange and forms a calcium source for the embryo. Together with the shell membrane it prevents infection of the egg.

In contrast to bone, the eggshell is a dead structure. Besides numerous organic matrix components (e.g. ovocalcin; Krampitz, 1980), some of which can bind calcium, the shell consists mainly of columns of calcite, a crystal form of calcium carbonate (Simons, 1971). The columns are attached to the outer shell membrane by hemispherical structures known as mammillary knobs. The cores of these knobs consist of a protein-mucopolysaccharide complex. From the outer membrane anchoring fibers run into the knobs. The cores are laid down before the egg enters the shell gland. The columnar part of the shell is called the palisade layer.

The shell has numerous pores, about 1 per mm^2. It is through these pores that exchange of oxygen, carbon dioxide and water vapour (evaporation) can occur. The pores connect the air filled spaces in the fibrillous egg membranes with the ambient air. The respiratory function of the shell is of increasing importance during embryonic development. For the consequences of changes in barometric pressure for the effective conductance of eggshells, see Visschedijk and Rahn (1981).

Shell formation starts in the isthmus with the appearance of great numbers of small granules in the outer shell membrane. Later on these granules decrease in number but increase in size. Upon arrival of the egg in the shell gland region of the oviduct the deposition of the mammillary cores and the nucleation of the crystals have already started. Thus crystal growth and matrix deposition occur in the shell gland lumen. It is not clear how the pores are formed during shell formation.

In the chicken the egg remains in the shell gland for approximately 20 hr. During the greater part of this period calcium moves across the shell gland at a rate of 100-150 mg/hr, which is per hour 5x the total amount of calcium in the blood. This calcium, derived from the blood, is transported to the lumen of the gland, possibly mediated by a CaBP in the shell gland cells (cf. Sturkie et al., 1976). The shell gland fluid is very rich in calcium, but the greater part may be protein bound, so it is not clear whether calcium transport through the mucosa of the shell gland is active.

3.3. Endocrine control of eggshell formation

The female reproductive tract is activated by estrogen and progesterone, synchronously with the development of ripe oocytes in the ovary. For the shell gland this activation includes the full development of the mucosa cells and the tubular glands except for the release of the products.

The various physiological activities involved in shell formation can be grouped as follows.

a. The secretory activity of the shell gland starts as soon as an egg (or any other object) enters the uterus. Sensory and nervous mechanisms are presumably involved in this activation.

b. The uptake of calcium by the shell gland decreases the levels of both the ultrafilterable and non-ultrafilterable calcium fractions in the blood. This increases PTH release, which causes calcium-mobilization from stores in medullary bone (see below). Recently, Hertelendy (1980) suggested that not PTH, but prostaglandins (particularly PGE_2) are involved in calcium mobilization during eggshells formation.

c. Bone consists of calcium phosphate, eggshell of calcium carbonate. Where does carbonate come from? It appears that the epithelium and the tubular glands of the shell gland contain high concentrations of the enzyme carbonic anhydrase (Gay and Mueller, 1973). In the gland bicarbonate is produced from metabolic CO_2 and H_2O. The subsequent decrease in the pH and the bicarbonate concentration in the blood during eggshell formating (metabolic acidosis) causes a compensatory increase in respiratory rate. The phosphate from the mobilized bone is, at least partly, excreted by the kidney. Activation and inactivation of carbonic anhydrase probably does not involve hormones.

d. The skeleton of birds is kept at a minimum weight to reduce metabolic costs of flight. Therefore it is not very suitable to function as a calcium reservoir for eggshell formation. Therefore female birds have to take anticipating measures for egg laying. Such measures cannot be taken by homeostatic hormones such as PTH, but need the coordinating action of the reproductive hormones.

In the pullet, about 10 days prior to the start of laying, the retention of dietary calcium and phosphate raises and the skeletal weight increases about 20%, indicating that most of the additional mineral is incorporated into bone (Taylor, 1970). These changes can be induced by the combined action of estrogen and androgen.

The increase in skeletal weight is due to the formation of a new type of bone which occurs in female birds only. This bone is particularly found in the marrow cavity of femur, tibia and humerus and is therefore called medullary bone. It grows from the endosteal surface of the cortex in the form of interlacing spicules between the existing vascular system which is left intact. As a result medullary bone has a relatively large surface area and a rich blood supply. Calcium mobilization and deposition in medullary bone is cyclic in relation to the egg laying cycle. It has been shown that the osteoclasts increase in number during mobilization. However, recent observations in the chick showed that the activity of both osteoclasts and osteoblasts is raised during egg shell formation (Touw et al., 1981). Like the effects on the intestinal calcium retention, medullary bone formation is caused by the synergistic action of estrogens and androgens.

e. The production of the very large yolky egg cells is sustained by the synthesis of specific proteins by the liver. The synthesis and release of these vitellogenins and their uptake by the oocytes in the ovary are induced by estrogens. The vitellogenins (phosphovitin) have a high calcium binding capacity and contribute to calcium transport by the blood during shell formation.

f. Like in mammals 1,25-dihydroxycholecalciferol (a vitamin D_3 metabolite) plays a role in calcium metabolism of birds. During egg laying the production of this hormone is increased by a direct action of estrogens on the liver to produce 25-OH-D_3 (Nicholson et al., 1979). With this compound as a precursor the raised PTH levels stimulate the kidney to produce 1,25-$(OH)_2$-D_3. This hormone increases the CaBP content of the intestine (Bar and Hurwitz, 1979) and of the shell gland (Bar and Norman, 1981) via an increased capacity of this organ to bind it to specific receptors (Coty, 1980).

In conclusion eggshell formation in birds is started by the arrival of the egg in the isthmus and shell gland region of the oviduct. This activation is presumably mediated by sense and nerve cells in the wall of the oviduct. Mobilization of calcium during eggshell formation is probably mediated by raised PTH levels. In anticipation to as well as during egg laying, intestinal calcium retention and the formation of medullary bone is induced and maintained by estrogens and androgens. PTH mobilizes only medullary bone. In addition 1,25-$(OH)_2$-D_3 enhances intestinal calcium retention and egg shell calcium secretion by the induction or raise of the CaBP content in these organs. Carbonic anhydrase is crucial for the production of the carbonate component of the shell from metabolic CO_2.

4. CONTROL OF EGGSHELL AND SHELL FORMATION IN MOLLUSCS

4.1. Eggshell formation in terrestrial gastropods

In gastropods shell formation starts while the embryo is still in the egg. Embryos in eggs, developing in an aquatic environment, have direct access to calcium of the surrounding water. However, development in a terrestrial environment makes it necessary that calcium is made available by the parent. A number of provisions are known, varying from the deposition of calcium salt granules on the egg surface to a solid eggshell, comparable to the avian one, secreted by the uterus of the parent and consisting of calcite crystals with small holes for gas exchange (Tompa, 1976, 1980). This variability among the gastropods is of phylogenetic interest.

As yet it is not known whether hormones are involved in calcium carbonate secretion by the female tract of terrestrial gastropods. It is to be expected that in stylommatophoran snails and slugs the female gonadotrophic Dorsal Body Hormone as well as gonadal hormones, will be involved (cf. Joosse and Geraerts, in press). It is interesting to note that during egg laying of *Helix aspersa* bound and unbound blood calcium levels are enhanced by about 70% (Tompa and Wilbur, 1977).

4.2. Shell formation in molluscs

The most obvious specialization of molluscs is their ability to form a calcium carbonate shell. This shell encloses and protects the body, is used for the attachment of muscles and functions as a calcium store In terrestrial molluscs the shell is important for protection against desiccation.

The molluscan shell is a dead structure, secreted by the mantle epithelium (cf. Wilbur, 1976). The outer part, formed first, consists of the organic periostracum. It is secreted by cells in a specialized area of the mantle edge, the periostracal groove, and becomes tanned. Against the periostracum the outer crystalline layer of calcium carbonate is deposited. Later on the inner crystalline layer, of variable thickness, is deposited against the outer one. For calcification, calcium bicarbonate and CO_2 pass through the mantle epithelium to the extrapallial fluid between the mantle and the inner shell surface. The mantle secretes organic material, largely protein, into the extrapallial fluid or directly onto the inner shell surface. This provides the organic matrix of shell. Crystal nucleation occurs on the organic matrix or on the surface of crystals previously deposited.

4.3. Shell growth - body growth

In molluscs there is an intimate relation between body and shell. The inner surface of the shell is in contact with the mantle epithelium, leaving a narrow space between the two, filled with the extrapallial fluid. The mantle edge is always in contact with the shell edge, except when the animal shows mantle retraction during adverse conditions. In this situation shell growth (=increment) does not occur. Shell growth is also arrested during starvation or after damage of the shell. In order to keep the relation between shell and body intact, growth of the two has to be closely interrelated. In studies on the freshwater snail Lymnaea stagnalis, we were able to show that this interrelation is controlled at the endocrine level (Geraerts, 1976; cf. Joosse and Geraerts, in press).

4.4. Endocrine control of body and shell growth

In invertebrates the majority of the hormones is produced by neuroendocrine neurons. In L. stagnalis the Light Green Cells (LGC) in the cerebral ganglia produce a hormone that stimulates the proportional growth of all body parts and therefore is called growth hormone (Geraerts, 1976). In the absence of the hormone body and shell growth cease. Injections of growth hormone extracts in intact animals stimulates the activity of the enzyme ornithine-decarboxylase (Dogterom and Robles, 1980), which is comparable to the effect of growth hormone in vertebrates.

The role of growth hormone in shell formation has been studied in detail by Dogterom and co-workers. He showed that the hormone stimulates the synthesis and release of periostracum material (Dogterom and Jentjens, 1980). Moreover, the hormone maintains a high concentration of calcium and of a specific CaBP in the mantle edge (Dogterom and Doderer, 1981). At the moment purification of this protein is in progress (see Doderer, this issue). In the absence of the hormone the incorporation of injected labelled calcium into the shell edge was significantly lower, but in the remaining part of the shell there was no difference in incorporation of calcium (Dogterom et al., 1979). This suggests that the hormone stimulates the formation of the outer crystalline layer. The formation of the inner crystalline layers presumably depends on the availability of calcium and on metabolic activity.

The effect of the growth hormone on bicarbonate movements was also studied (Dogterom and Van der Schors, 1980). In the absence of the growth hormone, bicarbonate levels in the blood were 20-30% lower compared to controls. Injection of labelled bicarbonate in these operated animals resulted in a 90% lower incorporation of the label in the shell edge. However, the deposition of carbonate in the shell was also lower. In the mantle and in the mantle edge carbonic anhydrase is present in the basal parts of the cells (Boer and Witteveen, 1980). This enzyme was still there in animals without growth hormone, but its activity may have decreased.

From these results it is concluded that the primary actions of the growth hormone of Lymnaea in the stimulation of shell enlargement are on the synthesis of periostracum material and on the maintenance of a high concentration of CaBP in the mantle edge. The actual secretion of periostracum material, the transport of calcium and the production and transport of bicarbonate will occur only when the mantle edge is in proper contact with the shell edge. Only in this situation the periostracum can be fixed in the right plane and secreted with the appropriate rate. These secretory processes are considered to be under local sensory and nervous control. The mantle edge of Lymnaea is richly innervated. The details of this control mechanism are still unknown.

Whether the induction of the CaBP by the growth hormone depends on a Vitamin D_3 metabolite is not known. Evidence has been presented by Weiner, 1979) that Vitamin D_3 is metabolized in snails.

Eggshell formation in birds and shell formation in molluscs shows interesting parallels. First, in the formation of both structures a CaBP and carbonic anhydrase are involved. Second, the formation of both structures is not controlled by hormones involved in calcium homeostasis, but by the reproductive hormones in birds and by the growth hormone in the mollusc. Third, the activation of the secretory process is locally determined and probably involves the action of the nervous system.

However, there are also differences. In the molluscs the periostracum is formed as a basis for the start of calcification. Moreover, the secondary inforcement of the molluscan shell seems to be independent of the growth hormone. Finally, in the mollusc there is no need for a calcium store, such as the medullary bone in female birds. In the mollusc shell enlargement is a relatively slow process and is stopped as soon as conditions are less favourable.

4.5. Control of blood calcium levels in Lymnaea

The growth hormone of Lymnaea does not influence the haemolymph calcium concentration (Dogterom, 1980). This raises the question whether other factors are involved in the control of calcium metabolism in molluscs.

The blood calcium concentration of L. stagnalis is strongly regulated and even maintained in specimens kept in calcium-free water (Greenaway, 1971a). This indicates a well established control mechanism (De With, 1977). In contrast to this reaction of the snail to environmental changes in calcium, the blood calcium levels appeared to change to a significantly lower and again constant level after 24 hr of food deprivation. De With and Sminia (1980) and De With et al. (this issue) have put forward a hypothesis to explain this phenomenon. This hypothesis assumes that the haemolymph is saturated with respect to $CaCO_3$ and, hence, that the product of the concentrations of Ca^{2+} and CO_3^{2-} is constant. The equilibrium is CO_3^{2-} directed, i.e. the haemolymph concentration of Ca^{2+} is constant because of the fact that the concentration of CO_3^{2-} is strongly regulated (CO_2-metabolism, acid-base balance). To remain saturated the haemolymph must be in equilibrium with a solid-$CaCO_3$ phase. In L. stagnalis this phase could consist of the $CaCO_3$ spherules inside the numerous calcium cells. These spherules are located inside vacuoles; the membrane of these vacuoles can fuse with the cell membrane and, in this way, forms "pores", via which exchange of materials seems possible (Sminia et al., 1977).

This theory has to be adjusted as in L. stagnalis, depending on the feeding condition, two different time-independent values of the ionic product $Ca^{2+} \times CO_3^{2-}$ can be found (see De With et al., this issue).

Acknowledgements. The author is highly indebted to Dr. M.P.M. Hermann-Erlee, Dr. P.C.M. Simons and Dr. A.K.J. Visschedijk for helpful discussions and information about the vertebrate aspects, to Dr. N.D. de With and Mr. A. Doderer for their encouragement and help in the preparation of the manuscript, and to Ms. Thea Laan for typing the manuscript.

5. REFERENCES

Bar, A., and Hurwitz, S.: 1979, The interaction between dietary calcium and gonadal hormones in their effect on plasma calcium, bone, 25-hydroxycholecalciferol-1-hydroxylase, and duodenal calcium-binding protein, measured by a radio-immunoassay in chicks. Endocrinology 104, pp. 1455-1460.

Boer, H.H., and Witteveen, J.: 1980, Ultrastructural localization of carbonic anhydrase in tissues involved in shell formation and ionic regulation in the pond snail *Lymnaea stagnalis*. Cell Tiss. Res. 209, pp. 383-390.

Coty, W.A.: 1980, A specific, high affinity binding protein for 1α, 25-dihydroxy vitamin D in the chick oviduct shell gland. Biochem. Biophys. Res. Comm. 93, pp. 285-292.

Dogterom, A.A.: 1980, The effect of the growth hormone of the freshwater snail *Lymnaea stagnalis* on biochemical composition and nitrogenous wastes. Comp. Biochem. Physiol. 65B, pp. 163-167.

Dogterom, A.A., and Doderer, A.: 1981, A hormone dependent calcium-binding protein in the mantle edge of the freshwater snail *Lymnaea stagnalis*. Calcif. Tissue Int. 33, pp. 505-508.

Dogterom, A.A., and Jentjens, Th.: 1980, The effect of the growth hormone of the pond snail *Lymnaea stagnalis* on periostracum formation. Comp. Biochem. Physiol. 66A, pp. 687-690.

Dogterom, A.A., and Robles, B.R.: 1980, Stimulation of ornithine decarboxylase activity in *Lymnaea stagnalis* after a single injection with molluscan growth hormone. Gen. Comp. Endocrinol. 40, pp. 238-240.

Dogterom, A.A., and Schors, R.C. van der: 1980, The effect of the growth hormone of *Lymnaea stagnalis* on (bi)carbonate movements, especially with regard to shell formation. Gen. Comp. Endocrinol. 41, pp. 334-339.

Dogterom, A.A., Loenhout, H. van, and Schors, R.C. van der: 1979, The effect of the growth hormone of *Lymnaea stagnalis* on shell formation. Gen. Comp. Endocrinol. 39, pp.63-68.

Epple, A., and Stetson, M.H. (eds.): 1980, Avian Endocrinology. Academic Press, New York.

Gay, C.V., and Mueller, W.J.: 1973, Cellular localization of carbonic anhydrase in avian tissues by labelled inhibitor autoradiography. J. Histochem. Cytochem. 21, p. 693.

Geraerts, W.P.M.: 1976, Control of growth by the neurosecretory hormone of the light green cells in the freshwater snail Lymnaea stagnalis. Gen. Comp. Endocrinol. 29, pp. 61-71.

Greenaway, P.: 1971, Calcium regulation in the freshwater mollusc, Lymnaea stagnalis (L.) (Gastropoda: Pulmonata). I. The effect of internal and external calcium concentration. J. exp. Biol. 54, pp. 119-214.

Hermann-Erlee, M.P.M., Meer, J.M. van der, and Hekkelman, J.W.: 1980, In vitro studies of the adenosine 3',5'-monophosphate (cAMP) response of embryonic rat calvaria to bovine parathyroid hormone -(1-84) [bPTH-(1-84)], bPTH-(1-34) and bPTH-(3-34) and the loss of cAMP responsiveness after prolonged incubation. Endocrinology 106, p. 2013.

Hertelendy, P.: 1980, Prostaglandins in avian endocrinology. In: A. Epple and M.H. Stetson, Avian Endocrinology, pp. 455-479. Academic Press, New York.

Joosse, J., and Geraerts, W.P.M.: in press, Endocrinology. In: Wilbur K.M. and Saleuddin, A.S.M. (eds.), Biology of Mollusca, Vol. Physiology. Academic Press, New York.

Krampitz, G.: 1980, Molekularmechanismen der Biomineralisation: Biochemische Untersuchungen an Eischalen. Praktische Tierarzt 61, pp. 250-253.

Nichelson, R.A., Akhtar, M., and Taylor, T.G.: 1979, The metabolism of cholecalciferol in the liver of Japanese quail (Coturnix coturnix japonica) with particular reference to the effects of oestrogen. Biochem. J. 182, pp. 745-750.

Norris, D.O.: 1980, Vertebrate Endocrinology. Lea & Febiger, Philadelphia.

Simons, P.C.M.: 1971, Ultrastructure of the hen eggshell and its physiological interpretation. Pudoc, Wageningen, Netherlands.

Sturkie, P.D. (ed.): 1976, Avian Physiology. Springer Verlag, New York, Heidelberg, Berlin.

Talmage, R.V., Owen, H., and Parsons, J.A.: 1975, Calcium Regulating Hormones. American Elsevier Publ. Co., New York.

Taylor, T.G.: 1970, The role of the skeleton in egg-shell formation. Ann. Biol. Anim. Biochim. Biophys. 10, pp. 83-91.

Touw, J.J.A, Velde, J.P. van de, and Vermeiden, J.P.W.: 1981, Remodeling of medullary bone during the egg laying cycle. Calcified Tissue Intern. 33, suppl, abstract nr. 109.

Tompa, A.S.: 1976, A comparative study of the ultrastructure and mineralogy of calcified land snail eggs (Pulmonata: Stylommatophora). J. Morph. 150, pp. 861-888.

Tompa, A.S.: 1980, Studies on the reproductive biology of gastropods: Part III. Calcium provision and the evolution of terrestrial eggs among gastropods. J. Conch. 30, pp. 145-154.

Tompa, A.S., and Wilbur, K.M.: 1977, Calcium mobilization during reproduction in snail *Helix aspersa*. Nature 270, pp. 53-54.

Visschedijk, A.H.J., and Rahn, H.: 1981, Incubation of chicken eggs at altitude: theoretical consideration of optimal gas composition. British Poultry Science 22, pp. 451-460.

Weiner, S., Noff, D., Meyer, M.S., Weisman, Y., and Edelstein, S.: 1979, Metabolism of cholecalciferol in land snails. Biochem. J. 184, pp. 157-161.

Wilbur, K.M.: 1976, Recent studies of invertebrate mineralization. In: Watabe, N. and Wilbur, K.M.: The mechanisms of mineralization in the invertebrates and plants. Univ. of South Carolina Press, Columbia (U.S.A.), pp. 79-108.

With, N.D. de: 1977, Evidence for the independent regulation of specific ions in the haemolymph of *Lymnaea stagnalis* (L.). Proc. Kon. Ned. Akad. Wet. C80, pp. 144-157.

With, N.D., and Sminia, T.: 1980, The effects of the nutritional state and the external calcium concentration on the ionic composition of the haemolymph and the activity of the calcium cells in the pulmonate freshwater snail *Lymnaea stagnalis*. Proc. Kon. Ned. Akad. Wet. C83, pp. 217-227.

PARTIAL PURIFICATION OF A HORMONE DEPENDENT CALCIUM-BINDING PROTEIN FROM THE MANTLE EDGE OF THE SNAIL *LYMNAEA STAGNALIS*

Albert Doderer, Department of Biology, Free University, Amsterdam, The Netherlands

Shell and body growth of the pulmonate freshwater snail *Lymnaea stagnalis* are under control of a neurohormone (molluscan growth hormone) produced by the neurosecretory Light Green Cells (LGC) located in the cerebral ganglia (Geraerts, 1976).

With respect to the process of shell growth (=enlargement) this neurohormone maintains a high concentration of calcium and of a calcium-binding protein (CaBP) in the mantle edge. Furthermore the hormone stimulates periostracum formation and calcium deposition at the shell edge. Matrix formation and calcification at the inner surface of the shell are not influenced by the hormone (Dogterom et al., 1979; Dogterom and Jentjens, 1980; Dogterom and Doderer, 1981).

It has been suggested that the CaBP plays an essential rôle in the process of shell formation. As part of a research program on the chemical structure and properties of this CaBP as well as its cellular origin, we started to purify this protein to homogeneity.

After gelfiltration of a 10% (W/V) homogenate of mantle edge tissue in 40% ethanol on a Bio-gel P2 column, the excluded volume contained all the calcium binding activity (as measured by the method of Dogterom and Doderer, 1981).
This crude fraction was further purified by means of partition chromatography on a Waters Asc. C18 Sep-Pak cartridge; elution with 40% methanol. The elution fluid contained 100% of the initial calcium binding activity; the protein concentration however, diminished by 50 µg to 136 µg (table 1).

After lyophilization the active fraction was dissolved in 200 µl buffer and high performance gel permeation chromatography (HPGPC) was carried out using a TSK column (Blue Column, LKB, Upsala, Sweden). Elution was performed with a 100 mM triethylamine-acetic acid buffer (pH 4.0), at a flow rate of 1 ml/min using a Waters 6000 A solvent delivery system. Absorption was measured at 280 nm and fractions of 500 µl were collected. The calcium binding activity was found in the fractions 24 and

P. Westbroek and E. W. de Jong (eds.), Biomineralization and Biological Metal Accumulation, 145–147.

	DPM ^{45}Ca BOUND*	UG PROTEIN*	DPM/UG PROTEIN	RECOVERY
HOMOGENATE	43650	223.2	195.56	100%
P2 COLUMN	43856	186.8	234.78	100%
C18 SEP PAK	42120	136.4	308.80	97%
TSK (HPGPG)	41468	0.69	60098.55	95%

Table 1. Calcium binding activity, protein concentration and recovery of the calcium-binding protein after the successive purification procedures (*expressed per mantle edge equivalent).

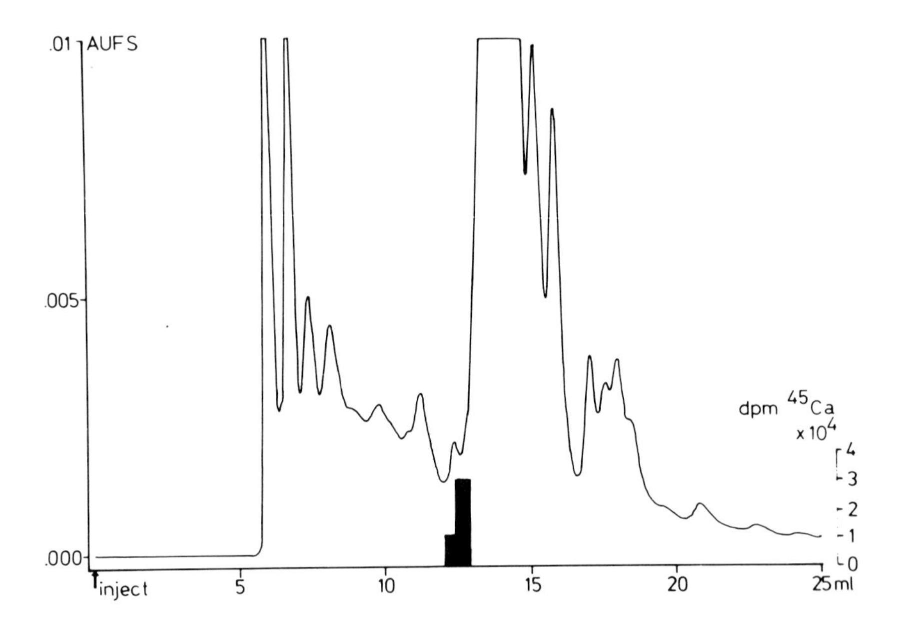

Fig. 1. Elution pattern (E 280) and calcium binding activity of a P2 and C18 purified mantle edge homogenate processed through a TSK column (HPGPG).

25, corresponding with a small absorption peak in the elution pattern (cf. Fig. 1).
Base line separation was obtained after reinjection of these two pooled fractions onto the same column. The protein concentration of the active fraction was 0.69 µg, corresponding with a 200-300 fold concentration of the calcium-binding protein. The recovery was 95% (cf. table 1).

Preliminary studies, using partition chromatography on a CN-column as next step, were very promising and suggest that the protein can be purified to homogeneity within reasonable time. Thereafter, studies on the amino-acid composition of the CaBP will be started.

References:

Dogterom,A.A., Van Loenhout, H., Van der Schors, R.C.: The effect of the growth hormone of *Lymnaea stagnalis* on shell calcification, Gen. Comp. Endocrinol. 39: 63-68, 1979.
Dogterom, A.A., Jentjens, T.: The effect of the growth hormone of the pond snail *Lymnaea stagnalis* on periostracum formation, Comp. Biochem. Physiol. 66A: 687-690, 1980.
Dogterom, A.A., Doderer, A.: A hormone dependent calcium-binding protein in the mantle edge of the freshwater snail *Lymnaea stagnalis*, Calcif. Tissue Int. 33: 505-508, 1981.
Geraerts, W.P.M.: Control of growth by the neurosecretory hormone of the Light Green Cells in the freshwater snail *Lymnaea stagnalis*. Gen. Comp. Endocrinol. 29: 61-71, 1976.

STUDIES ON THE CONSTANCY OF THE VALUE OF THE IONIC PRODUCT $Ca^{2+} \times CO_3^{2-}$ IN THE HAEMOLYMPH OF THE FRESHWATER SNAIL *LYMNAEA STAGNALIS*

N.D. de With, G.J. van der Wilt and R.C. van der Schors
Department of Biology, Free University, Amsterdam,
The Netherlands

Many molluscan species are able to maintain a more or less constant haemolymph Ca concentration despite large fluctuations in the availability of this element in the milieu (cf. De With, 1977). This ability is mainly attributed to the activity of a buffer system capable to remove excess Ca from the haemolymph or to release it in case of shortage. The involved deposits of Ca, mostly in the form of $CaCO_3$, are found in the shell and the soft body parts. Because of the involvement of $CaCO_3$ important interactions occur between Ca^{2+} and CO_3^{2-} metabolism. The latter metabolism is also connected with acid-base balance.

A model to explain the phenomena with respect to the Ca concentration in the haemolymph of *L. stagnalis* has been presented by De With and Sminia (1981). Crucial in this model is the assumption that the haemolymph is saturated with respect to some $CaCO_3$ deposit. Accordingly, the value of the product of the activities of Ca^{2+} and CO_3^{2-} in the haemolymph (IP = ionic product) should be constant and agree with that of a solubility product (SP) of one of the crystal forms of $CaCO_3$. The latter requirement is unfortunately largely theoretical as these values, under chemically non-ideal conditions, are unpredictable (cf. Simkiss, 1976).

This paper presents the results of some studies on the constancy of the value of the IP in the haemolymph of *L. stagnalis* and concerns effects of the feeding regimen and of hypercapnia.

The IP value in the haemolymph was determined as follows: in each sample the pH, (CO_2 tot.) and ($Ca_{tot.}$) were determined (cf. De With and Sminia, 1980). aCO_3^{2-} was calculated from pH and (CO_2 tot.):

$$aCO_3^{2-} = \frac{K_1K_2}{(10^{-pH})^2 + 10^{-pH}K_1(1/f_{HCO_3^-}) + K_1K_2(1/f_{CO_3^{2-}})} \; CO_{2\,tot.}$$

$K_1=10^{-6.387}$, $K_2=10^{-10.381}$; cf. Plummer et al., 1976.

P. Westbroek and E. W. de Jong (eds.), Biomineralization and Biological Metal Accumulation, 149–153.

Neglecting protein-bound Ca (0-6%; Van der Borght and Puymbroeck, 1964), $Ca_{tot.}$ in the haemolymph was assumed to be the sum of the concentrations of free Ca^{2+} and the ion-pairs $CaHCO_3^+$ and $CaCO_3^0$, hence:

$$aCa^{2+} = \frac{Ca_{tot.}}{1/f_{Ca^{2+}} + 1/f_{CaHCO_3^+} + aHCO_3^- (1/K_3) + aCO_3^{2-} (1/K_4)}$$

(for mM use K_3=113 and K_4=0.77; cf. Plummer et al., 1976).
The individual ion activity coefficients were calculated using the extended Debye-Hückel equation.

The values of the parameters in fed and starved (7 days) snails, adapted to the local tap water (Ca^{2+}= 2.3 mM, HCO_3^-=3.2 mM) are shown in Table I. In the haemolymph of starved snails, relative to the situation in fed snails, the pH was higher whilst the concentrations of $CO_{2\ tot.}$, $Ca_{tot.}$, aCO_3^{2-}, aCa^{2+}, and accordingly the IP value, were lower. The time course of the changes in the IP value, following a change in the feeding regimen, is shown in Fig. 1. The IP values interchange; adjustment being completed in about 24 hours. From these results it is concluded that the IP in the haemolymph shows 2 different time-independent values.

Table I. The composition of the haemolymph of fed and starved specimens of *L. stagnalis*. Means and standard deviations (n=15).

	starved	fed
pH	7.94±0.04	7.87±0.05
$CO_{2\ tot.}$ (mM)	13.6±0.8	20.6±1.4
$Ca_{tot.}$ (mM)	3.51±0.22	4.16+0.24
Carbonate activity (μM)	38±3	49±7
Calcium activity (mM)	1.47±0.09	1.69±0.09
IP (10^{-9} M^2)	56±7	84±13

In the experiments on the effects of hypercapnia haemolymph samples were collected from snails exposed for at least 4h to tap water gasified with different CO_2 tensions in air. This time is sufficient for the snails to reach a constant haemolymph pCO_2. The normal pCO_2 in the haemolymph is about 6 Torr in starved and about 11 Torr in fed snails. Hypercapnia caused large, CO_2 tension dependent changes in the haemolymph with respect to pH and the concentrations of $CO_{2\ tot.}$ and $Ca_{tot.}$ (Table II).

In fed as well as in starved snails the values of the IP increased with the applied CO_2 tension (Fig. 2). The characteristic difference between fed and starved snails, however, persisted.

Fig. 1. Time course of the change of the value of the IP ($\times 10^{-9}$ M^2) in the haemolymph of *L. stagnalis* following a change in the feeding regimen. o, starving, formerly fed, •, feeding, formerly starved specimens. Each point represents the mean of 5 observations.

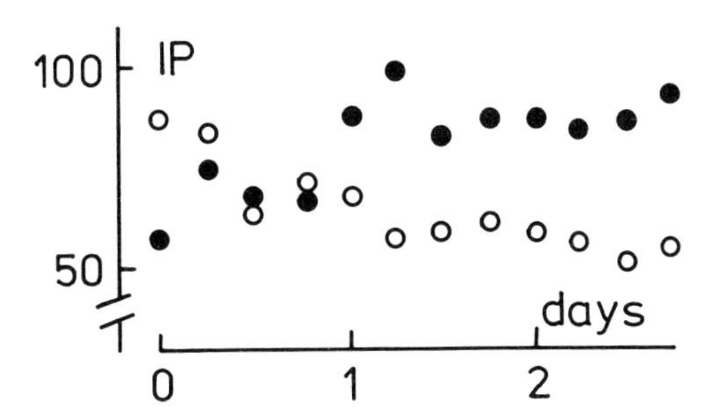

Table II. The effects of the external pCO_2 (Torr) on the pH and on the concentrations of $CO_{2\ tot.}$ and $Ca_{tot.}$ in the haemolymph of starving (A) and fed (B) specimens of *L. stagnalis*. Means, standard deviations and number of observations.

pCO_2	pH	$CO_{2\ tot.}$	$Ca_{tot.}$
A			
0	7.82±0.03(7)	14.4±0.4(7)	3.47±0.13(7)
7.6	7.75±0.02(4)	14.5±1.0(4)	4.12±0.18(5)
15.2	7.60±0.04(5)	19.8±1.2(5)	5.36±0.81(5)
38	7.38±0.02(5)	27.6±1.0(5)	7.44±0.31(5)
76	7.24±0.02(7)	36.8±1.8(7)	11.23±0.55(7)
114	7.12±0.04(5)	43.0±5.2(5)	13.23±1.61(5)
B			
0	7.83±0.07(6)	23.3±0.9(6)	4.04±0.24(6)
7.6	7.77±0.03(5)	23.8±1.3(5)	4.16±0.21(5)
15.2	7.66±0.02(5)	28.6±1.8(5)	5.09±0.55(5)
38	7.48±0.01(5)	37.4±1.4(5)	6.99±0.57(5)
76	7.30±0.03(7)	44.1±2.5(7)	10.30±0.64(7)
114	7.25±0.03(5)	52.0±3.7(5)	13.02±0.66(5)

Fig. 2. Relation between the value of the IP (10^{-9} M^2) in fed (o) and starved (•) specimens of *L. stagnalis* and the external pCO_2 (Torr). Each point represents the mean of at least 5 observations.

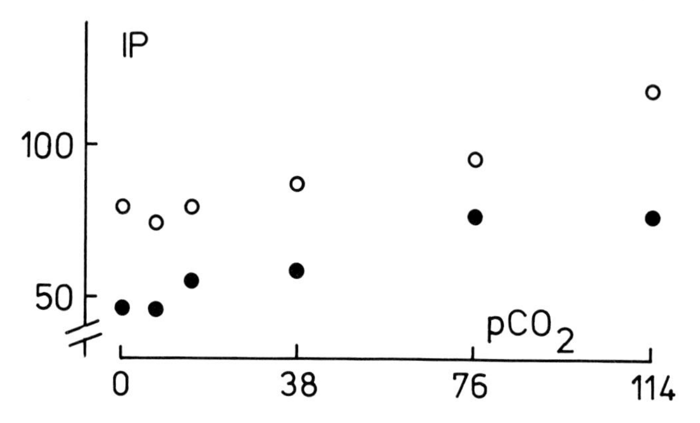

The values of the IP's in the haemolymph of fed and starved snails at CO_2 tenstions $\leqslant$ 38 Torr were relatively constant, despite considerable changes in ($CO_{2\ tot.}$) and ($Ca_{tot.}$) (Table II). In this range the IP's in the haemolymph of fed as well of starved snails behaved as SP's. The interpretation of the increase of the IP values at higher CO_2 tensions is not straightforward at the moment. It might represent a real phenomenon (with serious theoretical implications) or might result from the use of inappropriate values for the constants needed in the calculations (especially the values of K_3 and of K_4, presented in the literature differ widely).

When the IP's in the haemolymph of fed and starved snails represent SP's and their constancy does not merely result from other physiological mechanisms, which and where are the involved $CaCO_3$ solid phases? What is the explanation of the finding that in the haemolymph of *L. stagnalis* 2 completely different values of the IP can be found? These questions are largely unanswered.

In *L. stagnalis* solid $CaCO_3$ is mainly present in 2 forms, in the shell and inside the calcium cells. In the latter cells $CaCO_3$ is present in the form of spherules in special vacuoles. These vacuoles are peculiar in this respect that their membranes can fuse with the cell membrane and, in this way, can form "pores". Via these "pores" exchange of materials between vacuole and haemolymph seems possible (cf. Sminia et al., 1977). Preliminary studies have demonstrated that the percentage of calcium cells possessing "pores" can differ considerably under different physiological conditions. Whether shell and calcium $CaCO_3$ are involved in the regulation of the haemolymph Ca concentration by maintaining a constant value of the IP, and whether changes in the percentage of calcium cells having "pores" might explain the difference in the IP values between fed and starved snails are subjects for future investigations.

Literature:

Borght, O. van der and Puymbroek, S. van - Active transport of alkaline earth ions as physiological base of the accumulation of some radionuclides in freshwater molluscs. Nature, London, 204, 533-5 (1964).

Plummer, L.N., Jones, B.F. and Traesdell, A.H. - WATEQF - A fortran IV version of WATEQ, a computer program for calculating chemical equilibrium of natural waters. K.s. Geol. surv. Water-Resour. Investig. 76-13, 63 p. (1976).

Simkiss, K. - Intracellular and extracellular routes in biomineralization. In: Calcium in biological systems (Symposia of the Society for Experimental Biology, 30), Cambridge University Press, Cambridge, U.K., 423-44 (1976).

Sminia, T., With, N.D. de, Bos, J.L., Nieuwmegen, M.E. van, Witter, M.P. and Wondergem, J. - Structure and function of the calcium cells of the freshwater pulmonate snail *Lymnaea stagnalis*. Neth. J. Zool. 27, 195-208 (1977).

With, N.D. de - Evidence for the independent regulation of specific ions in the haemolymph of *Lymnaea stagnalis*. Proc. K. Ned. Akad. Wet. (Ser.) C 80, 241-8 (1977).

With, N.D. de and Sminia, T. - The effects of the nutritional state and the external calcium concentration on the ionic composition of the haemolymph and on the calcium cells in the pulmonate freshwater snail *Lymnaea stagnalis*. Proc. K. Ned. Akad. Wet. (Ser.) C 83, 217-27 (1980).

MINERALIZATION REACTIONS INVOLVING CALCIUM CARBONATES AND PHOSPHATES

G.H.Nancollas, K.Sawada, and E.Schuttringer
Chemistry Department, State University of New York at Buffalo, Buffalo, N.Y. 14214.

ABSTRACT

The calcium carbonates and phosphates may precipitate as different polymorphs and phases under conditions typical of those in vivo. A constant composition method has been used to study the kinetics of crystallization in which the activities of ionic species in the supersaturated solutions are maintained constant by the potentiometrically controlled addition of reagents. For calcite, the rate of reaction is proportional to the square of the supersaturation, expressed in terms of the activities of free calcium and carbonate ions. The constant composition study of the dissolution of this salt indicates the participation of both transport and surface processes over a range of pH. At lower pH values, the reaction is controlled by the rate of formation of carbon dioxide. In the growth experiments, metastable phases may persist for considerable periods and kinetic considerations are of paramount importance in understanding the mechanisms of the reactions. Thus the calcium phosphates which form in solution are markedly dependent on pH, supersaturation, ionic strength, temperature, and the nature of the solid phases already present.

Crystallization reactions involving the calcium phosphates and calcium carbonate polymorphs play an important part in the formation of mineral phases in biological systems. Thus hydroxyapatite ($Ca_5(PO_4)_3OH$, hereafter, HAP) formation in aqueous solutions is of particular interest since this phase is considered to be the model compound for tooth and bone minerals (Nancollas, 1979). Calcium carbonate has probably received the greatest attention in connection with its precipitation and dissolution in natural waters. The reaction is very sensitive to the presence of magnesium ions which are normally present at relatively high concentrations in biological systems (Taft, 1967; Kitano, 1965; Bischoff and Fyfe, 1968) and which have been shown to reduce the rate of calcite crystal growth in sea water systems (Pytkowicz, 1965). In sea water, the effect of various ionic constituents upon the rate of calcium carbonate crystallization may explain the existence of metastable polymorphs of the salt. Clearly, kinetic considerations are of paramount importance in determining the nature of the mineral phases which precipitate in vivo.

P. Westbroek and E. W. de Jong (eds.), Biomineralization and Biological Metal Accumulation, 155–169.

Almost all natural calcium apatites contain carbonate ions in their lattices. It is therefore also necessary to take into account the influence of carbonate ion on calcium phosphate precipitation and that of phosphate ion upon the rate of calcium carbonate crystallization. In the latter case, traces of orthophosphate ions have been shown to markedly reduce the rate of calcite growth (Nancollas et.al., 1981). In many systems, mixed phases may form by the crystallization of one phase upon another in metastable supersaturated solutions. In the light of the appreciable carbon dioxide and calcium ion concentrations in the environemnt, calcium carbonate may almost be regarded as an ubiquitous solid phase upon which other minerals may grow. Thus the exclusive crystallization of HAP upon calcite crystallites has been demonstrated in a constant composition study in solutions of calcium phosphate at very low supersaturation (Koutsoukos, 1981). Although it is not possible to confirm unequivocally the true epitaxial growth of these mixed phases, the presence of the calcite substrate allows nucleation and growth of the new phase at supersaturations considerably lower than those required for homogeneous precipitation.

It is now quite well established that the mineralization of calcium carbonate and phosphate involves the formation of a number of precursor phases which may subsequently dissolve during the reprecipitation of the thermodynamically more stable phases. Thus, during the precipitation of calcium carbonate, the homogeneously nucleated particles are thought to be amorphous calcium carbonate hydrates (Packter, 1975) which are visible for a few minutes and gradually dissolve and grow on heteronucleated crystals. In the crystallization of calcium phosphates, many studies have been made in solutions supersaturated with respect to at least four calcium phosphate phases, in order to increasing solubility: HAP, tricalcium phosphate ($Ca_3(PO_4)_2$, TCP) octacalcium phosphate ($Ca_8H_2(PO_4)_6.5H_2O$,OCP) and dicalcium phosphate dihydrate ($CaHPO_4.2H_2O$, DCPD). The stoichiometric molar ratio of calcium to phosphate calculated from changes in concentration of the lattice ions during the precipitation is frequently in the range 1.45 ± 0.05, considerably lower than the value, 1.67, required for the thermodynamically most stable HAP. It is usually assumed that the first formed phase is an unstable amorphous precurso, ACP, consisting of a highly hydrated cryptocrystalline phosphate (Eanes et.al., 1966). The ACP usually takes the form of chain-like aggregates of spherolytic rather than monodispersed particles and it is still not established whether this phase is a single stoichiometric unit. X-ray radial distribution measurements made by Betts et.al.,(1981) suggest that the ACP contains a basic structural unit consisting of a spherical cluster of about 9.5 Å in diameter with a stoichiometry of $Ca_9(PO_4)_6$. Adsorbed phosphate ions or occluded calcium ions were suggested as accounting for the lower and higher Ca/P ratios, respectively. An alternative theory is that ACP contains OCP-like structures. This is based on the idea that the ribbons first formed in the enamel matrix, which are best described as single unit cell thicknesses of OCP, are prototypes for the simplest particles in ACP. The conversion of OCP to HAP may readily be understood from the similarity of their crystal lattices (Brown, 1973), and the addition of further layers of OCP or apatite to these units would provide crystallites of various thicknesses and Ca/P ratios.

In studies aimed at modelling biological systems, experiments are frequently made to determine spontaneous precipitation thresholds by mixing solutions of salts containing lattice ions and observing the appearance of the first solid phase.

In such systems it is impossible to eliminate foreign substances or dust particles that can act as sites for heterogeneous nucleation. As indicated above, thermodynamic interpretations of the results may ignore important kinetic considerations which determine the nature of the phases which form initially. In view of this fact, highly reproducible seeded growth techniques were applied to characterize calcite (Nancollas and Reddy, 1971) and calcium phosphate (Nancollas and Mohan, 1970) crystallization. Unlike the spontaneous precipitation studies, these experiments allow reliable measurements of the crystal growth rate in the presence and absence of additives and crystal growth occurs on well defined surfaces of known morphology. In biological systems, precipitation invariably takes place on a surface already present either of the mineral itself or of an organic substrate offering available sites for adsorption of lattice ions. Seeded crystal growth methods may therefore simulate in vivo conditions much more closely than spontaneous precipitation studies.

In the seeded growth techniques which have been previously established, the pH may either be allowed to decrease in solution during the reaction or it may be maintained by the pH-stat controlled addition of base. In both cases the concentrations of calcium, carbonate, and phosphate ions change rapidly with time so that precursor phases may form and subsequently dissolve during the reactions. In this paper, a method is discussed by which all solution concentrations are maintained constant during the crystal growth reactions so that the kinetics of crystallization and dissolution can be studied over wide ranges of super-and under-saturation, respectively. The method is particularly useful for investigating the mechanism of crystal growth and dissolution since the extents of reaction can be varied appreciably. In the case of crystal growth, factors such as secondary nucleation and the influence of foreign ions can be studied under highly reproducible conditions. The incorporation of trace impurities into the growing crystals and their influence on the subsequent rate of crystallization can also be investigated. The present paper will discuss some recent results obtained for the crystallization and dissolution of calcium carbonate and the crystallization of calcium phosphate.

EXPERIMENTAL METHODS

Reagent grade chemicals and triply distilled, CO_2-free water were used and the calium phosphate crystallization experiments were conducted in a nitrogen atmosphere. Constant composition growth and dissolution experiments were made in a water jacketed Pyrex cell with water thermostated at the desired temperature ($\pm$ 0.05 K) circulating in the jacket. Stable super-and under-saturated solutions of calcium carbonate and calcium phosphate were prepared by mixing solutions containing the lattice ions and bringing to the desired pH by the controlled addition of potassium hydroxide solution. The resulting supersaturated solutions were stable for periods of several hours prior to the initiation of the crystallization by the introduction of seed crystals.

Seed crystals of HAP and calcite were prepared as described previously (Nancollas and Mohan, 1970; Nancollas and Reddy, 1971). Calcium and phosphate determinations were made by a combined method using an atomic absorption spectrometer (Perkin-Elmer Model 503) with a precision for calcium

($\pm$ 0.2%) and for phosphate as vanadomolybdate ($\pm$ 2%) (Tomson et.al., 1977). Specific surface areas (SSA) of seed as well as of the grown material were measured ($\pm$1%) by a single point BET nitrogen adsorption using 30% nitrogen/helium mixture (Quantasorb II, Quantachrome, Greenvale, N.Y.) The seed slurries were aged for at least four months before use and the solid phases were characterized by x-ray powder diffraction (Phillips XRG 3000 x-ray diffractometer) and by scanning electron microscopy (ISI Model Super II).

Following the addition of seed crystals of calcite or HAP to the calcium carbonate or phosphate supersaturated solutions, the pH was maintained constant by the simultaneous addition of titrant solutions containing the lattice ions. The addition of titrant was monitored and, in addition, samples were withdrawn from the supersaturated solutions, filtered, (0.22 um filter, Millipore, Bedford, Mass.) and analyzed for calcium and phosphate. The solid phases, withdrawn from time to time, were also air or freeze-dried and examined by scanning electron microscopy, SSA, infrared, and x-ray analysis. In addition, background electrolyte (usually potassium chloride or potassium nitrate) was also added to the cell and titrant solutions in order to maintain the ionic strength at a constant value. This was necessary since changes in activity coefficients could also trigger the addition of reagent titrant solutions. Using this technique, it was possible to study the rates of crystallization, as given by the rates of titrant solutions added with a precision of approximately 3%. Under conditions of constant composition, the stoichiometry of the precipitating phases was uniquely given by the stoichiometric molar ratios of the titrant solutions necessary for maintaining the ion activities.

In order better to define the fluid dynamics in the calcite dissolution experiments, a single crystal of Iceland Spar was centrally set in an inert self-setting acrylic resin machined into a disc shape shown in Fig.1 which has been found to meet the requirements of the hydrodynamic boundary layer calculations of Levich, (1962) (Gregory and Riddiford, 1956). The demonstrated laminar flow at such rotating disc surfaces enables calculations to be made of the thickness of the Nernst diffusion layer and the dependence of rate of reaction upon the disc rotation speed. The exposed calcite surface (about $1cm^2$) was carefully polished with 6μm and 1μm diamond dust. The final polish before each growth experiment was made with 0.05μm gamma alumina. Before use in a rate experiment, the disc was washed in an ultrasonic cleaner with distilled water for approximately 5 min. This procedure was repeated three times and surface areas were determined geometrically by photographic reproduction.

RESULTS AND DISCUSSION

To analyze the kinetic data in terms of the free ionic species, it is necessary to take into consideration the formation of ion-pairs and complexes. The concentrations of ionic species in the supersaturated solutions at any time were calculated from mass balance, electroneutrality, proton dissociation constants for phosphoric and carbonic acids and the formation of the ion-pairs, $CaH_2PO_4^+$, $CaHPO_4$, $CaPO_4^-$, $CaHCO_3^+$ and $CaCO_3$. The computations were made by successive approximations for the ionic strength and activity coefficients, y_z, of z-valent ions were calculated using the modified Debye Hückel equation proposed

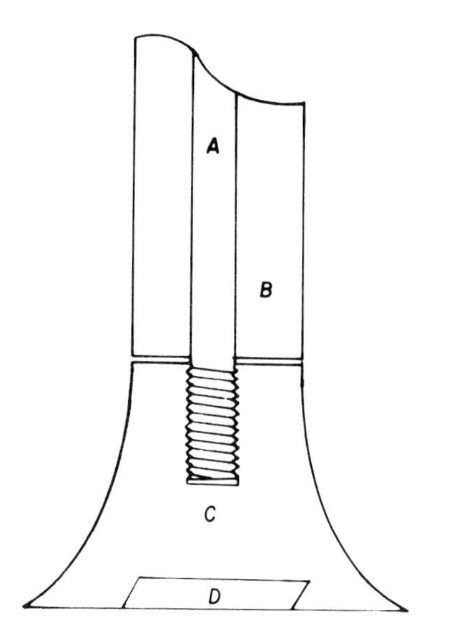

Fig. 1. Rotating Disc.
A, Metal spindle;
B, Teflon sleeve;
C, inert acrylic resin;
D, single crystal of Iceland Spar from Mexico.

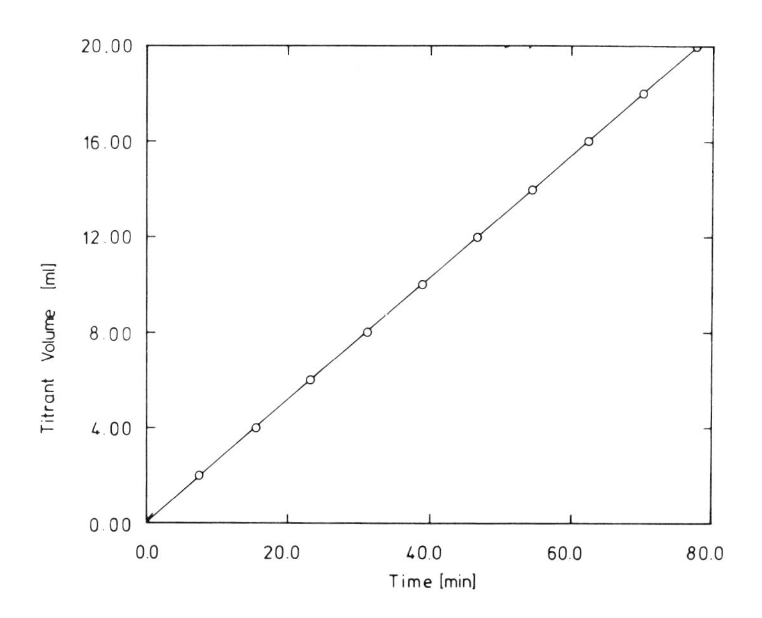

Fig. 2. Constant composition experiments. Growth of calcite; plots of mixed titrants addition as a function of time. $T_{Ca} = 2.500 \times 10^{-3}$M, $T_C = 0.951 \times 10^{-3}$M, 99.6 mg seed L^{-1}. pH = 8.50

by Davies (1962). The calcium carbonate system is complicated by the relatively slow gas/solution equilibrium $CO_2 + H_2O = H^+ + HCO_3^-$. In the crystallization work, this problem was avoided by keeping to a minimum the volume of the gas space above the solution in the crystallization cell and the effective isolation of the system was verified by the constancy of the pH of the supersaturated solutions for periods of hours in the absence of inoculating seed crystals. In contrast, it will be shown that in the case of dissolution experiments at relatively low pH, the formation of CO_2 may control the rate of the reaction.

Calcium Carbonate Crystallization:

The rate of crystallization of a number of sparingly soluble 2-2 electrolytes has been shown to be proportional to the square of the relative supersaturation (Nancollas, 1979). For calcium carbonate, the kinetic equation can be written in terms of the activities of the ionic species:

$$\text{Rate} = -dn/dt = k_g K_{so} \sigma^2 \quad (1)$$

$$\sigma = \left[(\{Ca^{2+}\}\{CO_3^{2-}\})^{1/2} - K_{so}^{1/2}\right] / K_{so}^{1/2}$$

In these equations, the braces enclose the activities of the calcium and carbonate ions, K_{so} is the thermodynamic solubility product of the calcium carbonate phase, n the moles of calcium carbonate precipitated, k_g the rate constant, and s a function of the surface area of the seed crystals.

Following the inoculation of metastable supersaturated solutions of calcium carbonate with calcite seed crystals, a typical plot of the volume of mixed titrants added as a function of time in order to maintain the solution supersaturation, is shown in Fig.2. The rate plots are strikingly linear when the data are corrected for changes in surface area of the crystals during growth, and it is interesting to note that in 60 min of reaction shown in Fig.2, the calcite surface area available for growth increased more than fivefold. A typical kinetic plot of the data, following Equation 1 is shown in Fig.3 from which it can be seen that the rate of crystallization is proportional to the square of the supersaturation, σ. The constant composition crystallization experiments with calcite seed have been made over a range of molar ionic calcium/carbonate ratios ranging from 14-134, σ from 0.93 to 5.5 and the ionic strengths and seed concentrations have been varied by factors of 40X and 5X, respectively (Kazmierczak et.al., 1982). Despite these wide variations, the rate constant for each seed preparation was constant to within $\pm$ 12%. These results provide striking justification for the use of activity coefficients in rate equation (1).

Crystallization of calcite in the presence of low levels of magnesium ion was markedly inhibited with the formation of magnesian calcite. The distribution coefficient of magnesium ion between solution and the growing calcite phase $(Mg)/(Ca)^{solid}/(Mg)/(Ca)^{solution}$ was shown to be 0.028 (Sawada and Nancollas, unpublished results). The retardation of calcite growth by the magnesium ion could be interpreted in terms either of the specific interaction of magnesium ion with the surface of the crystal or the increased solubility of magnesian calcite induced by the incorporation of Mg^{2+} having a smaller radius than Ca^{2+} thus producing lattice strain. In contrast, the influence of magnesium ion on the

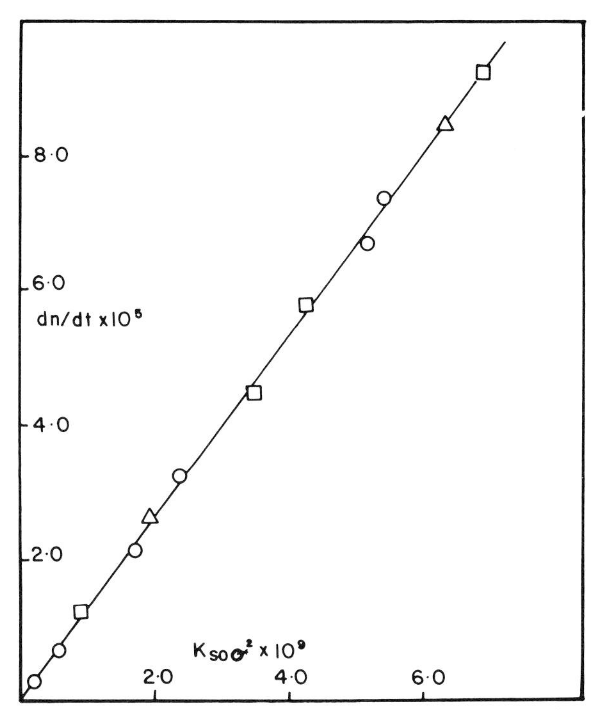

Fig. 3. Plot of dn/dt as a function of $K_{so}\sigma^2$. T_{Ca} = 2.49 x 10^{-3}M, T_C= 0.975 x 10^{-3}M, pH = 9.38, 56 mg seed L^{-1}

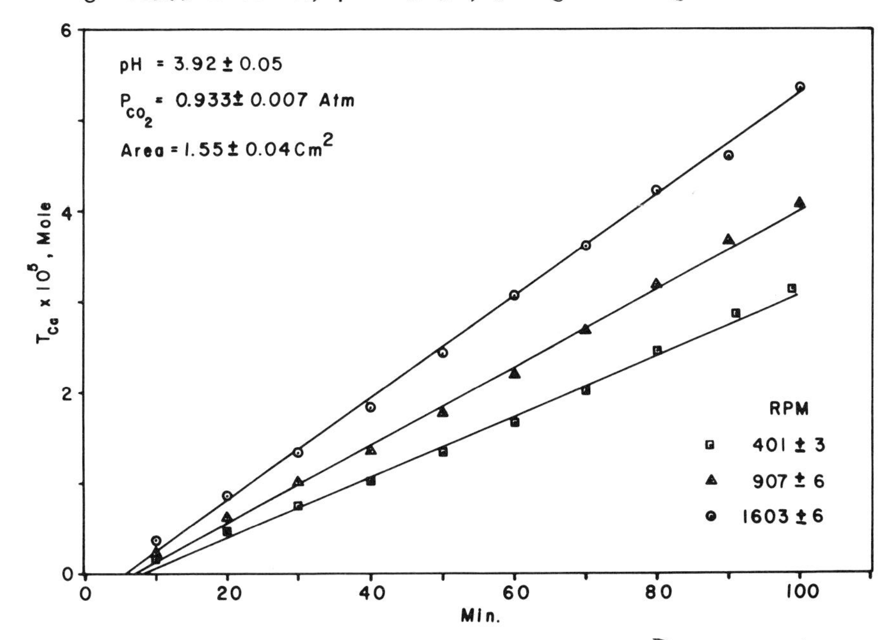

Fig. 4. Dissolution of calcite. Plots of T_{Ca} as a function of time.

crystallization of aragonite and vaterite polymorphs was very small with little or no incorporation of Mg^{2+} in the crystal lattices. Indeed, at higher concentrations of magnesium ion ($6-12 \times 10^{-4}$M) the addition of calcite seed crystals to supersaturated solutions of calcium carbonate resulted in the spontaneous precipitation of aragonite in experiments at 70°C (Nancollas and Sawada, 1982).

Crystallization experiments of calcite, aragonite, and vaterite, in the presence of hydroxyethylidene 1-1 diphosphonic acid, HEDP, indicated that at a concentration as low as 10^{-7} M, this additive markedly inhibited the seeded crystal growth of calcite and aragonite. The influence of HEDP on the crystallization of vaterite seed was particularly interesting since a concentration of 2.7×10^{-6}M had little effect on the rate. The addition of vaterite seed crystals to supersaturated solutions of calcium carbonate at 70°C containing HEDP was shown to proceed with the exclusive formation of vaterite despite the fact that this was the most thermodynamically unstable polymorph. The possible stabilization of this phase by the presence of inhibitors may be of particular significance in biological mineralization reactions. By inhibiting the formation of aragonite and calcite, the crystallization of vaterite was favoured and, at 70°C, the reaction took place with a rate which was appreciably greater than that of either calcite or aragonite. It is clear that the nature of the phase which forms may be dependent upon the seed substrate which is added to the solution and also the possible presence of traces of growth inhibitors which may selectively prevent the precipitation of some calcium carbonate polymorphs and influence the factors governing their interconversion.

Calcium Carbonate Dissolution:

Since the formation of biological minerals may involve concomitant dissolution of precursor phases, it is also important to understand the mechanism of this process. In addition, the fluctuating concentrations of biological fluids may lead to the partial dissolution of minerals which may have precipitated in other parts of the organisms. The dissolution process has usually been regarded as being determined by the transport of solute species from the surface to the bulk of the solution. The rate of reaction would then be expected to be linearly related to the undersaturation. However, for a number of sparingly soluble salts such as the divalent metal sulfates, the rate of dissolution is second order in undersaturation and independent of stirring rate (Liu and Nancollas, 1975; Campbell and Nancollas, 1969). The evidence points to a contribution from a surface reaction in the dissolution process. A similar conclusion was reached by Christoffersen et.al., (1978) for the dissolution of HAP in water at constant pH.

Experiments involving the dissolution of crystals in stirred suspensions suffer from the disadvantage of an ill-defined fluid dynamics. In contrast, when transport occurs to or from a rotating disc, an exact solution of the equations governing mass transport can be achieved (Levich, 1962). The use of a rotating disc system has permitted the evaluation of steady state concentrations of species within the diffusion layer for a given experimental rate of dissolution (Jones et al, 1972). This treatment was based on a diffusion controlled process in the solution phase, assuming that all chemical equilibria were rapid compared to the rate of diffusion (Brunner, 1904). In the dissolution of calcium carbonate, however, the formation of carbon dioxide may be rate determining under certain conditions of concentration.

In the rotating disc studies, laminar flow ensures that the whole surface of the disc is uniformly accessible and the thickness of the diffusion boundary layer may be calculated (Levich, 1962). As the disc rotates, the ambient fluid is drawn towards the face and forced away radially over the surface by the centrifugal force developed by the surface drag. It is one of the few systems for which an explicit solution of the stationary form of the Navier-Stokes equation is possible (Goldstein, 1938). Solution of the equation for transport normal to the rotating disc led to expressions for the thickness of the boundary layer, δ as a function of the characteristic velocity of the rotating disc, ω (Levich, 1962; Gregory and Riddiford, 1956) given by equation 2.

$$\delta = 1.805\ D^{1/3}\nu^{1/6}\left[0.8934 + 0.316\ (D/\nu)^{0.36}\right] \qquad (2)$$

In equation 2, D is the diffusion coefficient and ν is the kinematic viscosity. Provided that the volume of the solution is large enough to reduce the effects of the vessel walls to a minimum, and the diameter of the disc is much larger than the thickness of the boundary layer, the solution may be applied to a system of finite size. The important feature of the rotating disc is that for laminar flow, there is a region, δ, adjacent to the solid surface wherein the diffuse transport of matter is important and independent of the position of the rotational axis.

Studies of the dissolution of calcite (Icelandic Spar) suggested that both surface and mass transport were important in controlling the rate. The experiments were made at a constant undersaturation maintained through the addition of acidified background electrolyte, controlled by a pH glass electrode as described previously. Typical plots of the rate of dissolution at a pH of 3.92 are shown in Fig.4. It can be seen that the rate of dissolution was constant for at least 100 min during which time an infinitesimal amount of the surface layer was removed through dissolution (less than 0.15um min^{-1}). The rate of dissolution calculated from the slopes of the curves in Fig.4 are plotted as a function of pH in Fig.5. At lower pH values, the rates of reaction were much more sensitive to changes in rotational speeds than at higher pH. This indicates that the dissolution mechanism is dependent on the mass transport from the calcite interface at lower pH whereas at higher values, surface control processes may be important.

The net dissolution flux per unit area may be written J_d/s. Since only the lattice ions calcium (total molar concentration T_{Ca}) and carbonate (total molar concentration T_c) are transported in the diffusion layer, $J_d = J_{TCa} = J_{Tc}$. The total fluxes of calcium and carbonate species may be written in terms of the individual fluxes by equation 3:

$$J_{T_{Ca}} = J_{Ca} + J_{CaHCO_3} + J_{CaOH} + J_{CaCO_3} \qquad (3)$$

and
$$J_{T_c} = J_{CO_3} + J_{HCO_3} + J_{CO_2} + J_{CaHCO_3} + J_{CaCO_3} \qquad (4)$$

In equations 3 and 4, the charges have been omitted for clarity. The concentration of soluble species were calculated by successive approximations as

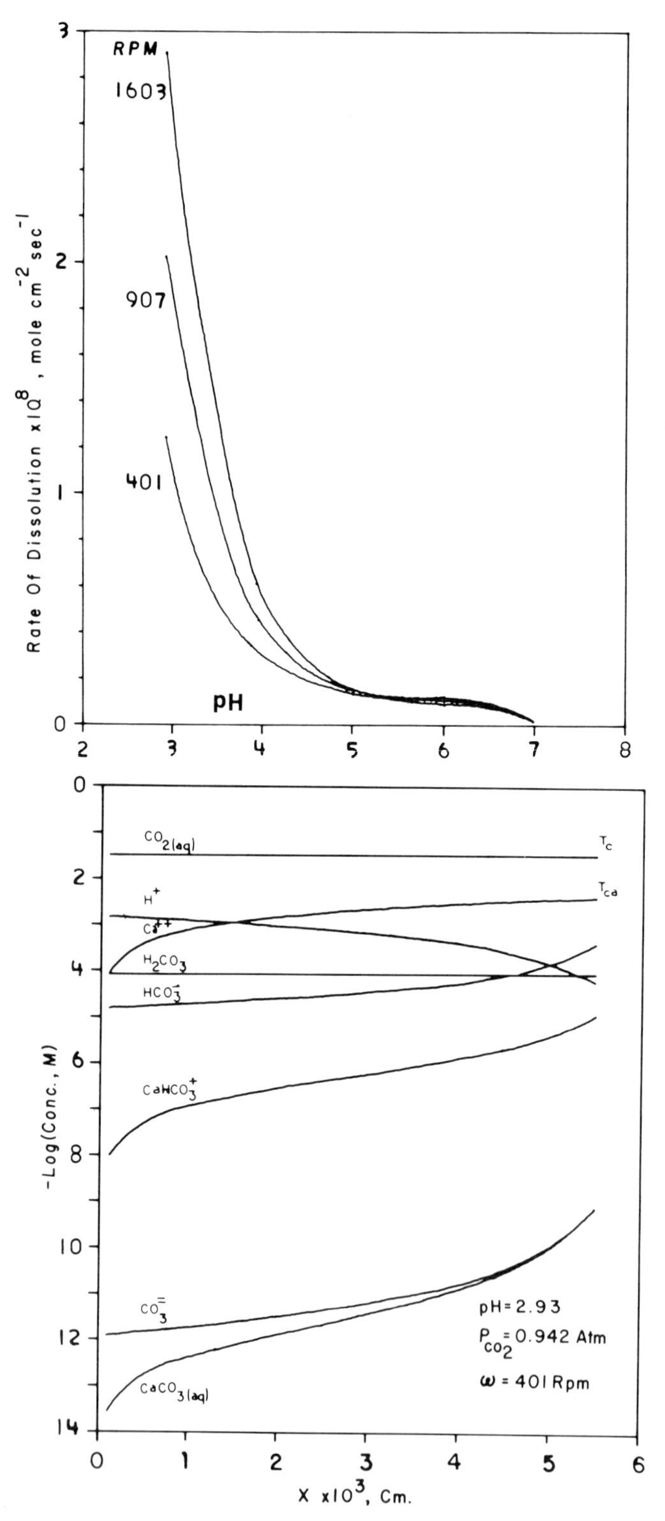

Fig.5. Rate of calcite dissolution as a function of solution pH

Fig.6. Species concentrations within the diffusion layer during calcite dissolution

described previously. Since no net charge is transported in the diffusion layer during dissolution,

$$2J_{Ca} + J_{CaHCO_3} + J_{CaOH} + J_H - 2J_{CO_3} - J_{HCO_3} - J_{OH} = 0$$

In Fig.6, the concentrations of ionic species are plotted as a function of the distance of penetration into the diffusion layer, X. The relative thickness of the latter was calculated from the Levich equation assuming, as a first approximation, a diffusion coefficient for the hydrogen ion. In this way, the calculation refers to the maximum possible penetration since the proton has the largest value for the diffusion coefficient. For the system as a whole, the mutual diffusion coefficient is, in all likelihood, less than that of a proton.

The concentrations of the species within the diffusion layer were obtained by applying the Nernst-Planck equation together with the mass and charge balance relationship by means of an iterative solution of their steady state concentrations at finite intervals from the bulk solution. The concentration profiles describe the magnitude and direction of molecular diffusion throughout the diffusion layer, and their curvature generated rates of homogeneous reaction, through Fick's Second Law, at finite distances in the diffusion layer. In Fig.7, the rates of carbon dioxide formation are plotted as a function of distance into the diffusion layer at three rotation speeds. The points marked X refer to kinetic limits of CO_2 formation by the reaction (Kern, 1960),

$$H^+ + HCO_3^- = CO_2(aq) + H_2O \ .$$

It can be seen that at the relatively low pH (2.92 ± 0.05), the rate of dissolution is governed by the rate of formation of carbon dioxide. The latter is strongly coupled to the transport phenomena and the mass action relationships involving CO_2 are no longer valid. Above a limiting pH of 4-5, such homogeneous kinetic considerations were found to have no influence on the validity of the computational model.

It can be seen in Fig.7 that the distance from the dissolving surface at which the formation of carbon dioxide becomes rate determining depends markedly upon the rotational speed. The distance is also dependent upon the bulk solution pH. The inward diffusion of H^+ and outward diffusion of HCO_3^- (Fig.6) control the formation of carbon dioxide at each point in the diffusion layer. It reaches its kinetic limit (X in Fig.7) at a distance within the diffusion layer, dependent upon the instantaneous concentrations of these reacting ions. The hypothetical maxima shown in Fig.7, beyond the kinetic limits, refer to the positions of optimum concentrations of H^+ and HCO_3^-. This model implies that the solution is supersaturated with respect to carbon dioxide under these conditions. Dissolution experiments at lower pH values for which the carbon dioxide supersaturation is increased, bring these points closer to the crystal surface where carbon dioxide gas bubble nucleation will take place.

In the interpretation of kinetic data for the dissolution of calcium carbonate polymorphs it is clearly important to take into account the possibility that the rate of reaction is controlled not by transport or surface processes but rather by the rate of formation of carbon dioxide. Thus at pH <8 the rate constant, k, for the

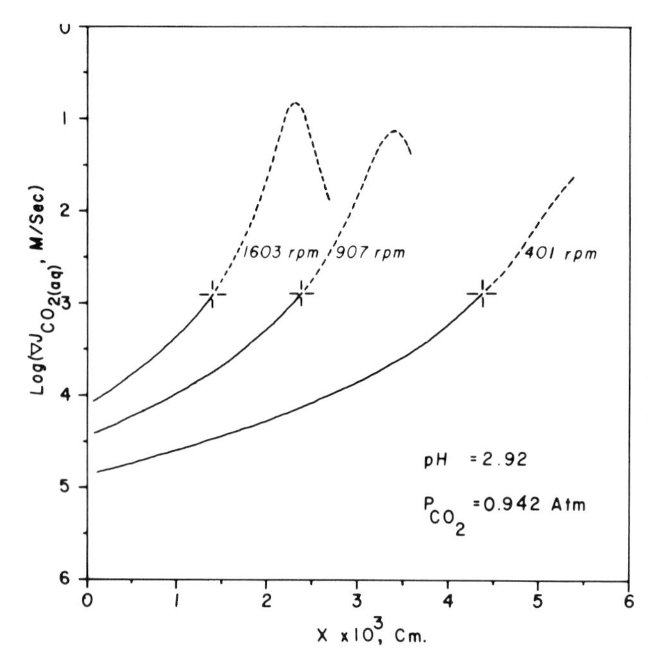

Fig. 7. Rate of CO_2 formation as a function of distance within the diffusion layer

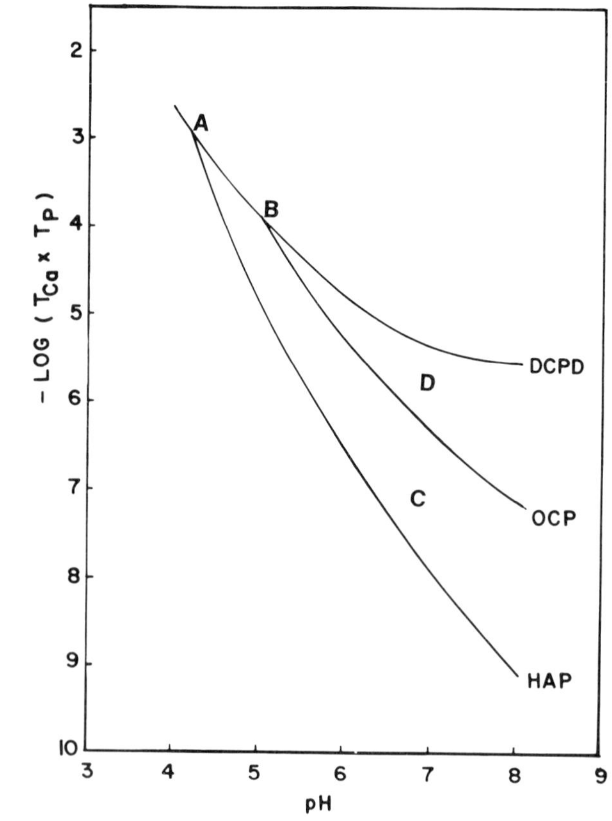

Fig. 8. Solubility isotherms for calcium phosphates at 37^0C. Calculated at $I = 0.5M$

direct hydration of carbon dioxide is $3.665 \times 10^{-1} s^{-1}$ for the forward and $13.7 s^{-1}$ for the reverse reactions (Edsall,1968). Conditions may therefore be encountered for which the rate of reaction $dT_{Ca}/dt > k_1[CO_2]$. In the corresponding studies of the kinetics of crystal growth of calcite, the rate of reaction is much less than the rate of hydration of carbon dioxide both in calcium bicarbonate (House, 1981) and calcium carbonate supersaturated solutions (Nancollas et.al., (1981); Kazmierczak et.al., (1982).)

Calcium Phosphate Crystallization:

Typical solubility isotherms for the important calcium phosphate phases which may precipitate at 37°C in the system $Ca(OH)_2$-H_3PO_4-KNO_3-H_2O are shown in Fig.8 (Shyu et.al., 1982). It can be seen that although HAP is the most stable phase under many conditions, it becomes less stable than DCPD if the solution is sufficiently acid. Calcium phosphate phases which are exposed to a solution at a pH lower than that corresponding to the singular point (A, Fig.8) may be expected to become covered by a surface coating of a more acid calcium phosphate. The apparent solubility behaviour may therefore be quite different from that of the phase itself and this has probably been the source of the widely discrepant data in the literature for the solubility of calcium phosphate phases. This was emphasized by Brown (1973) who suggested that this uncertainty may have led many workers to conclude that the properties of calcium phosphate phases in general and HAP in particular, are rather variable. In considering solubility diagrams such as that in Fig.8 and also those for the calcium carbonates, it is essential to establish that the systems are at equilibrium in order to be able to use the thermodynamic solubility products which refer to the free energies of single solid phases. The position of the singular points A and B in Fig.8 at which two solid phases are in equilibrium with the aqueous solution is clearly markedly dependent upon pH, the calcium/phosphate molar ratios, and the ionic strengths of the solutions. Where multiple charged ions are involved, the importance of introducing activity coefficient corrections becomes increasingly apparent. Thus at ionic strengths of about 0.1M, the calculated ionic products for HAP may be in error by as much as a factor of 10^6 if activity coefficients are neglected.

It can be seen in Fig.8 that at physiological pH, at least three calcium phosphate phases may form during conventional crystallization experiments. The formation of apatite and the growth of HAP seed crystals in such metastable supersaturted solutions is therefore much more complex than at low pH where only DCPD may be kinetically stabilized. In addition, different calcium phosphate phases may be stabilized or destabilized by the presence of various cations and anions which may not be significantly incorporated into the calcium phosphate crystal lattice. As discussed previously, the overall precipitation reaction may involve not only the formation of different calcium phosphate phases but also the concomitant dissolution of thermodynamically unstable phases such as OCP which may form rapidly in the initial stages of the reaction.

The problems involved in interpreting the kinetics of the precipitation are a direct result of the fact that in such conventional crystallization experiments, the concentrations fall rapidly. If attempts are made to conduct conventional crystallization experiments at points corresponding to C in Fig.8 in which the solutions are supersaturated only with respect to HAP, the resulting rate of crystallization is so slow that it is impossible to quantify changes in calcium and phosphate concentration in order to calculate the rates of reaction and to determine the stoichiometry of the precipitating phases. The constant composition method has enabled studies to be made of the exclusive formation of HAP on HAP and other seed crystals in solutions corresponding to C(Fig.8). In these cases, the titrant solutions are used to replace calcium, phosphate, and hydroxide ions following the addition of well characterized HAP seed crystals. The rate of crystallization , given by the rate of titrant addition, was shown to be linear for long periods of time provided that corrections were made for decreases in specific surface area of the HAP seed crystals in the early stages of the crystallization reactions (Koutsoukos et.al., 1980).

The exclusive growth of HAP without participation of precursor phases, is of particular interest since the conditions model those frequently encountered in vivo in which the lattice cation may be complexed by other components present in the biological fluids, resulting in relatively low supersaturations with respect to the phosphate salts of interest. In the case of calcium phosphates, the constant composition studies have provided direct evidence for OCP as a precursor for HAP formation at supersaturations corresponding to D(Fig.8) while studies at very low supersaturation C(Fig.8) have shown that macroscopic quantities of stoichiometric, highly crystalline HAP can be formed. The results suggest a model for calcium phosphate precipitation in which OCP, formed as a precursor phase, hydrolyzes partially or completely to HAP, depending upon the rate of precipitation. This mechanism would apply in cases in which the supersaturation was represented by points such as D in Fig.8. It is possible that only a fraction of the OCP molecules transform to HAP leading to the variable stochiometric ratios often reported in the literature.

The presence of magnesium ion has been shown to reduce the rate of crystallization of calcium phosphate while stabilizing certain precursor phases (Betts et.al., 1981). In the presence of 1mM of this ion, the hydrolysis of OCP was dramatically inhibited in boiling water. The relatively large concentration of magnesium ion in blood serum is thought to be the principal component which prevents the precipitation of HAP in vivo. In contrast to the influence of magnesium ion, another alkaline earth cation, strontium, is incorporated into the crystallizing HAP lattice at very low supersaturations with respect to calcium phosphate (eg. point C Fig.8). (Koutsoukos and Nancollas, 1982). Solid phases were formed with considerably lower Sr/Ca molar ratios than those in the supersaturated solutions. Changes in the unit cell lattice parameters and infrared spectra were linearly related to the strontium content of the precipitated solid phases. The presence of strontium ion in the crystallizing solutions also markedly reduced the size of the developing calcium/strontium apatite crystallites. The constant composition method may be useful for preparing solid solutions of substituted apatites and for modelling biological mineralization processes which take place in such mixed electrolyte solutions.

ACKNOWLEDGEMENTS

We thank the National Institute of Dental Research (Grant DE03223), the National Institutes of Arthritic, Metabolic, and Digestive and Kidney Diseases

(Grant R01AM19048) and the Donors of the Petroleum Research Fund administered by the American Chemical Society for grants in partial support of this research.

REFERENCES

Betts,F., N.C.Blumenthal and A.S.Posner. 1981. J.Crystal Growth, 53, 63.
Bischoff,J.L. and W.S.Fyfe. 1968. Am.J.Sci. 266, 65.
Brown,W.E. 1973. In Environmental Phosphorus Handbook, E.J.Griffith, A.Beeton, J.M.Spencer and D.T.Mitchell, eds. New York, Wiley, pp.203.
Brünner,E. 1904. Z.Physik.Chem. 47, 56.
Campbell,J.R., and G.H.Nancollas. 1969. J.Phys.Chem., 73, 1735.
Christoffersen J., M.R.Christoffersen and N.Kjaegaard 1978. J.Crystal Growth, 43, 501.
Davies, C.W. 1962. Ion Association, Butterworths, London.
Eanes,E.D., I.H.Gilessen and A.S.Posner. 1966. In Crystal Growth, H.S.Peiser, ed, Oxford: Pergamon Press, pp. 373-376.
Edsall,J.T. 1968. Carbon Dioxide, Carbonic Acid and Bicarbonate Ion: Physical Properties and Kinetics of Interconversion, NASA SP188 report.
Gregory,D.P. and A.C.Riddiford. 1956. J.Chem.Soc., 3756.
House,W.A. 1981 J.Chem.Soc.Faraday, 77,341.
Jones,A.L., H.G.Linge and I.R.Wilson, 1972. J.Crystal Growth, 12,201.
Kazmierczak,T.F., M.B.Tomson and G.H.Nancollas. 1982. J.Phys.Chem., 86, 103.
Kern, D.M. 1960. J.Chem.Education 37, 14.
Kitano,Y. 1965. Bull.Chem.Soc.Japan, 35, 1973.
Koutsoukos,P., and G.H.Nancollas.1981. J.Phys.Chem., 85, 2403.
Koutsoukos,P., Z.Amjad, M.B.Tomson and G.H.Nancollas. 1980. J.Amer.Chem.Soc., 102, 1553.
Koutsoukos,P. 1981. Ph.D. Thesis, State University of New York at Buffalo.
Levich,V.G. 1962. Physicochemical Hydrodynamics, Prentice-Hall, New York.
Liu,S.T., and G.H.Nancollas. 1975. J.Colloid Interface Sci., 52,582.
Nancollas,G.H. and K.Sawada.1982. J.Pet.Tech., 645.
Nancollas,G.H. 1979. Adv.Colloid Interface Sci., 10, 215.
Nancollas,G.H. 1979. J.Dental Research 58B, 861.
Nancollas,G.H., T.F.Kazmierczak and E.Schuttringer. 1981. A Controlled Composition Study of Calcium Carbonate Crystal Growth. Nat.Assoc.Corrosion Eng., Paper 226. Corrosion/80. Chicago.
Nancollas G.H., and M.M.Reddy 1971. J.Colloid Interface Sci., 37, 824.
Nancollas, G.H. and M.S.Mohan. 1970. Arch.Oral Biol., 15, 731.
Packter, A. 1975, Kristall und Technik 10, 111.
Pytkowicz,R.M. 1965. J.Geol. 73, 196.
Shyu.L.S., L.Perez and G.H.Nancollas, to be published.
Taft, W.H. 1967., Development in Sedimentology, G.V.Chilingar, M.J.Bissel and R.W.Fairbridge eds., Elsevier, Vol.9B, pp. 151-167.
Tomson,M.B., J.P.Barone and G.H.Nancollas, 1877. Atomic Absorpt. Newsl. 16, 117.

PROBLEMS IN THE UNDERSTANDING OF BIOMINERALS

S. Mann, S.B. Parker, C.C. Perry, M.D. Ross, A.J. Skarnulis
and R.J.P. Williams
Inorganic Chemistry Laboratory, Oxford University,
South Parks Road, Oxford OX1 3QR

Abstract: We describe some recent work on calcium carbonate and silica in biological systems. The object of the work is to use ultra-high resolution electron microscopy to define solids. Associated polymers are examined by nuclear magnetic resonance spectroscopy. Simulation of biomineralisation is described based upon reactions in vesicular space.

The object of this paper will be to describe our approaches to the problems of biomineralisation. There are three main parts to the paper (i) the use of ultra-high resolution electron microscopy for the study of inorganic chemistry: (ii) the use of high resolution nuclear magnetic resonance in the study of the associated organic polymers: (iii) simulations of biological precipitation conditions by carrying out reactions in synthetic-vesicle systems.

ULTRA-HIGH RESOLUTION ELECTRON SPECTROSCOPY

The use of the electron microscope (EM) usually after staining an object is not new in the study of biological materials. Both direct microscopy and electron diffraction have been employed regularly. However ultra-high resolution electron microscopy at a magnification of $>10^6$ using unstained objects has not been employed since the high energies involved destroy organic materials. On the other hand such very high resolution EM studies are frequently made of non-biological minerals with the result that lattice images can be produced. We illustrate this point first directly with a study of calcium carbonates in minerals and in biology.

Calcium Carbonate of the Inner Ear

During a visit by Professor Muriel Ross (Michigan University, Ann Arbor) we decided to make a compartive study of calcium carbonates from a variety of sources - pure minerals such as calcite and aragonite, manufactured calcium carbonates, which are closely related to minerals,

P. Westbroek and E. W. de Jong (eds.), Biomineralization and Biological Metal Accumulation, 171–183.

Figure 1. (a) Electron diffraction pattern from calcite from the rat inner ear showing a (421) face.

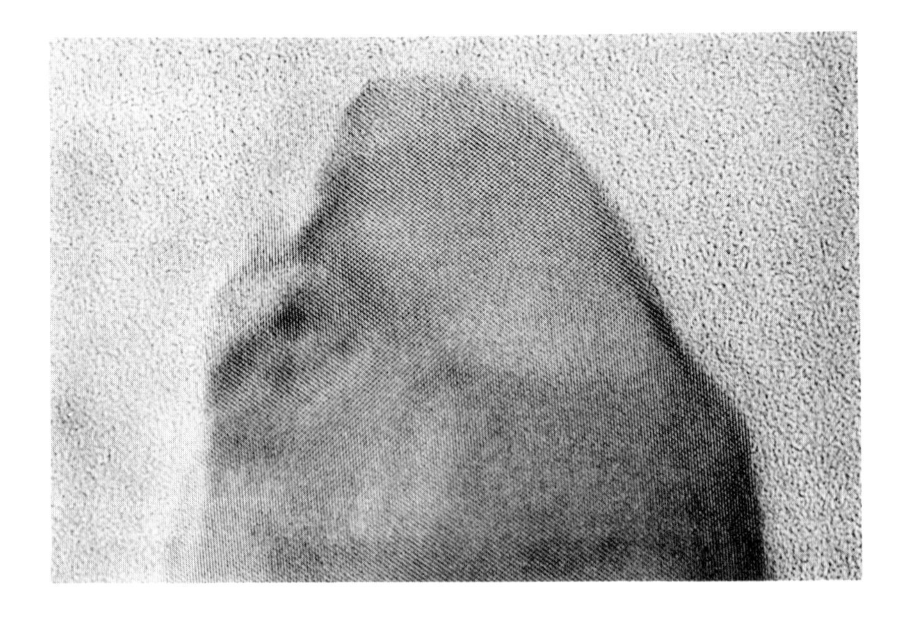

Figure 2. (b) Lattice image of the same crystal (x 4,500,000). Note the curved edges of this fragment.

and calcium carbonates from the inner ear of a variety of animals, Ref. 1.

We illustrate the nature of our observations with Fig. 1 and 2. Fig. 1 shows an electron diffraction pattern from a thin section of calcium carbonate (calcite) from the rat inner ear crystallites. The electron diffraction data allow us to index the face of the crystal which we are observing. It is both the distribution and the intensity of the diffraction pattern which are important and they are checked by simulated diffraction patterns using a programme developed by Dr. A.J. Skarnulis, Ref. 2. The lattice image can be calculated for such a face and this calculated image can be compared with the lattice image seen in Fig. 2 down to the atomic level. Especially significant are the lattice image spacings. These two observations then prove that we are examining a specific face. Detailed examination of the image (and of the diffraction data) may lead to the observation of defects in the lattice, unusual features at the edges of crystals and so on. By making a large number of observations on the biological calcium carbonates and comparing them with mineral or man-made calcium carbonates, which has now been done on some hundred specimens, we hope to be able to describe the manner in which the deposition of the biological crystallites has been achieved.

Table 1 gives a comparative description of mineral calcite, man-made calcite, and calcite from the rat inner ear. The table shows that the calcites differ only in that the major planes observed by indexing are not the same. Man-made calcium carbonates are formed by a rapid progression of events

$$Ca^{2+} + HCO_3 \rightarrow H^+ + \text{"}CaCO_3\text{"} \text{ (hydrated ion-pairs)}$$

$$\text{"}CaCO_3\text{"} + \text{"}CaCO_3\text{"} + Ca^{2+} + CO_3^{2-} \rightarrow CaCO_3 \cdot 6H_2O(a)$$

$$CaCO_3 \cdot 6H_2O(a) \rightarrow \text{crystalline carbonates}$$

where (a) is an amorphous precipitate. This may be the way in which some minerals are seeded as well but it is almost certainly not the route by which biology lays down carbonates. Here the probable route is via association of calcium and/or carbonate with an organic polymer. It is not surprising then that we see different preferred faces of growth of calcite in mineral or man-made and biological carbonates, Table 1.

Now this difference in the nature of crystallites is but one control over the finally observed mineral. There can be at least two others. The first is the way in which crystallites are packed together to make the final mineral material. In the rat otoconia the final mineral phase is an object that looks in some ways like a single crystal of calcite grown on certain preferred faces but curiously this "single crystal" has curved edges, Fig. 3. It is also clearly limited in size so that a mass of "single" crystals of similar dimensions are produced spread over a surface. Clearly the "single crystals" are deposited in a controlled way. Many of our measurements are consistent with the idea that the "single crystals" themselves are in fact ordered assemblies of

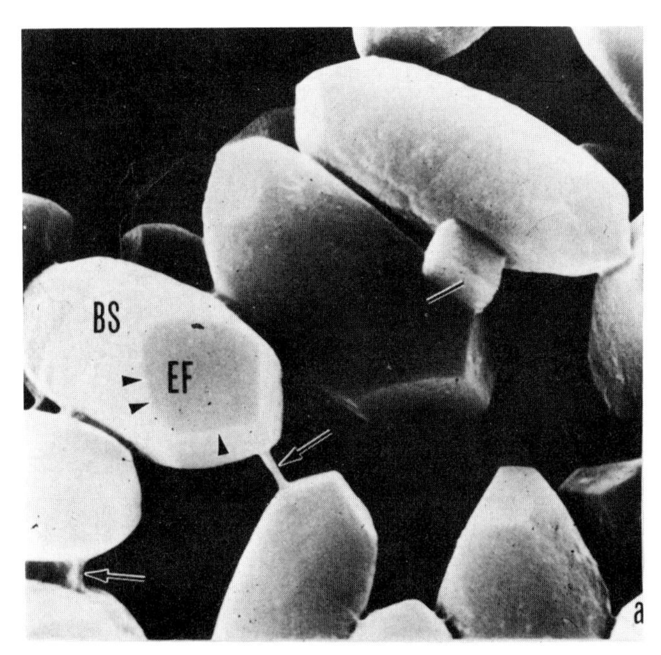

Figure 3. A representation specimen illustrating the curvature of otolith edges (x 90). The question arises: are the round edged fragments of dimension around 1,000Å related to the rounded edges of the larger crystals. If bio-minerals can be made by putting together small units then it is easy to see how shape can be controlled. (See Fig.2).

Table I. Incidence of Crystal Faces of Various Calcium Carbonates

	001	010	011	100	111	211	421	$42\bar{1}$
Aragonite		3						
Fish	3	4						
Frog	4	4	3	1				
Rat	10				3		1	
Calcite	2					4	5	2

Note: 3(201) on rat but only in ion-thinned sample.

micro-crystals and the curved faces result from the control, by organic material, over these growth patterns. The second way in which crystallisation can be controlled is at the level of allotropy. We discuss the allotropes of calcium carbonate below but before doing so note that further differences from in vitro precipitations are that deposition and resolution can be controlled in both space and time. Spatial organisation is necessary, for example, if a permanent magnet is to be used successfully, and the timing of deposition is essential for both endo- and exo-skeletons in obvious ways. The timing and spatial organisation are also essential parts of the gravity devices of the inner ear. The problems which we have not solved greatly outnumber the problems solved so far.

Turning to the deposition of the second allotrope, aragonite, the differences between mineral aragonite and the aragonite from fish otoliths is striking. Even by ordinary analysis the fish aragonite is less than 37% calcium instead of 40%. The initial electron diffraction patterns of fish otolith aragonite give rise to a pattern not too dissimilar from that of aragonite and the 010 face is frequently prominent. Exposure to the electron beam causes with time the fish aragonite but not aragonite itself to change its electron diffraction pattern to a primitive form while the lattice image alters but it is not destroyed. Thus there is a change in the unit cell rather than a destruction in the beam. We can not be looking at a simple aragonite. Analysis in fact gives 0.21% strontium, 0.4% sodium, virtually no magnesium and almost no organic matter. The lattice images remain very good. We are not yet in a position to define the nature of this carbonate more closely, but we believe that some bicarbonate may be trapped in the aragonite. A more detailed study is essential.

We summarise as follows:

(1) Biomineralisation occurs only in prescribed parts of space.
(2) Deposition occurs at specified times.
(3) In the case of calcium carbonate it is based on small crystals. The domains can only be identified with confidence by ultra-high resolution electron microscopy.
(4) The allotropic form is prescribed within domains and crystallites.
(5) The preferred growth planes, and the extent of their growth, are prescribed within domains and crystallites and are not necessarily the same as those observed in vitro.
(6) The domains and crystallites are brought together in a prescribed way but there may be over growth on them. The resultant solid can look like a single crystal.
(7) The resultant solid is itself oriented and may alter shape under stress.
(8) We note that the choice of calcium carbonate as opposed to another insoluble salt or silica is for positive advantage which we do not fully appreciate as yet.
(9) Most of the above could only have been brought about by the interaction between the mineral phase and organic materials.

Organic Polymers Extracted from Biological Carbonates

The preparation of proteins from calcium carbonate shells has been frequently reported but the physical properties of these proteins has not received much attention. The outstanding feature of the proteins is their composition, with a high content of Glx and Asx where no distinction is made between the acids and the amides, and of small amino-acids such as Gly, Ala, Ser, Ref. 3. We have extracted and purified proteins from Atrina vexillum and examined their proton nuclear magnetic resonance, NMR spectra at different pH values. The proteins do not show any gross deviation from the spectrum of a random coil peptide, Fig. 4. It is very likely then that there is only a minimum of tertiary fold although there may well be considerable helical sections. Previously we have examined a protein from dentine and shown that it too has little or no fold though it may have secondary structure. It too was a highly acidic (and phosphorylated) protein, Ref. 4.

These observations lead us to suppose that the purpose of the proteins or of polysaccharides, in defining the growth of calcium carbonate or silica, lies not so much in the formation of nuclei as in the controlled lines along which the mineral is deposited. The restriction could be physical as much as chemical. There are of course many known organic inhibitors of crystal growth, some of which inhibit particular faces preferentially. We know little or nothing about organic polymers which assist growth of crystals, Ref.4.

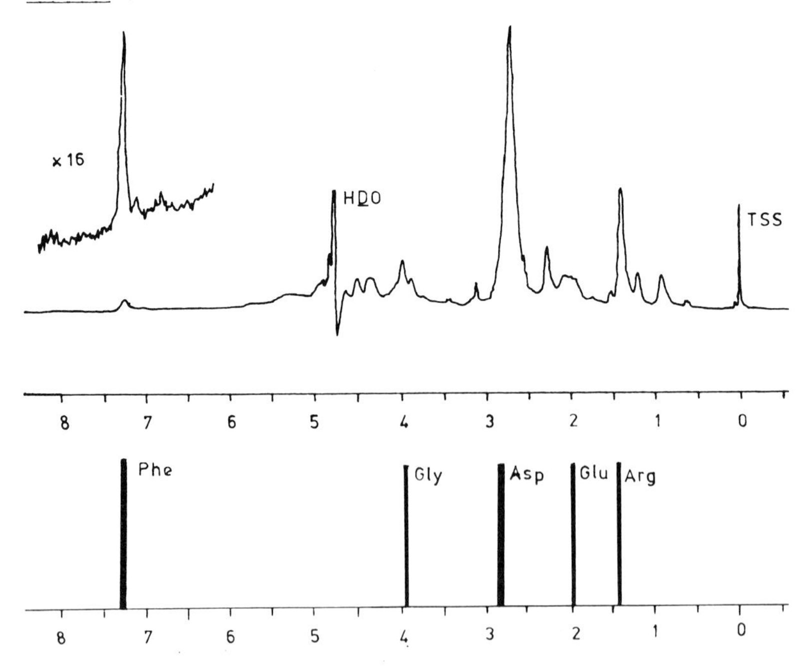

Figure 4. A comparison of the n.m.r. spectrum of a random coil protein and of an extracted protein

Now while bone and shell in animals appear to be related mainly to protein chemistry and structure both silica deposits and calcium carbonate deposits in plants and algae seem to be connected with polysaccharide chemistry. What if any are the specificities involved in these polymer/mineral interactions? We turn to our studies of silica biominerals.

SILICA - AN AMORPHOUS MATERIAL

For many years biologists have pointed out that silica in biological systems is "amorphous". Now "amorphous" is not a description which has any structural value unless the resolution of the examining technique is defined. "Amorphous" judged by X-ray diffraction can be truly amorphous, meaning no molecular repeats down to 10Å resolution, or poly microcrystalline which may mean "amorphous" by methods which resolve to only the 100 to 1,000Å of the microcrystals. Such objects are ordered within but not between crystallites. Electron microscopy at ultra-high resolution distinguishes these possibilities, Ref. 5. We have shown that SiO_2 is amorphous in biology both in plants and protozoa to the level of the absence of repeating [SiO_4] units over as few as three such units in any direction. This type of study requires the use of very high resolution electron microscopy. Once such a definition of amorphous is appreciated a number of new problems can be approached.

(a) Is SiO_2 stoichiometric - i.e. what is the content of hydrogen bound as hydroxyl groups and its relationship to packing?
(b) How is its surface structure to be defined?
(c) How is any morphology maintained?
(d) What is its density? It should not be forgotten that amorphous SiO_2 is not a description such as "calcite" is for $CaCO_3$. The density of amorphous SiO_2 can vary from about 2.0 to 2.6 according to the manner of polymerisation of SiO_2 units. Very high resolution EM pictures or neutron diffraction studies may be able to resolve this problem.
(e) Is SiO_2 a true solid? Can it flow?

We illustrate the nature of the difficulties and the fascination of the problems by the examination of two silica objects from biological sources.

The Lorica of Stephanocca diplocostata Ellis

With much assistance from Dr. B. Leadbeater (Birmingham University) we have been examining the curved SiO_2 rods of this protozoa, Ref. 5. The rods and their overall structural function in the organism are well described by Leadbeater, Ref. 6. Here we concentrate on the nature of the SiO_2. First very high resolution EM images show no lattice or any evidence of regular structure down to the level of 10Å. The SiO_2 is truly amorphous, Ref. 5. Second the rod is not a uniform material. It dissolves preferentially from the centre to produce hollow tubes with opaque (electron dense) walls. There is only silica in these walls with some retaining polymers. Third at T-junctions between rods the link between the rods is made partly of electron dense strands (there is only silicon present as a heavy element - no stain). Presumably these strands are associated with organic material as well as SiO_2.

We make the following conclusions. The silica in the centre of rods is less stable than toward the outside. Now silica dissolves more readily the more hydrated it is. The picture we have of the rod is then that the outside is of quite dense SiO_2 patterned on an organic matrix but the inside contains more hydrated $SiO_2(H_2O)_x$ which flows readily and redissolves in distilled water. When T-joins are made we believe that some of this silica can flow into the join made by organic material initially so that the join is stiffened and in the end appears as quite electron dense material.

Silica Hairs of Phalaris canariensis

The hairs surrounding the seeds of Phalaris canariensis were given to us by Dr. C. O'Neill (Imperial Cancer Institute) after Dr. D. Wynne-Parry (Bangor) had pointed out their interest. Fig. 5(a) shows a micrograph of hairs at low magnification and Fig. 5(b) shows a hair end-on at higher magnification. The hair is made from concentric rings of material. It is overwhelmingly SiO_2 but there is a polysaccharide present as shown by proton nuclear magnetic resonance studies of extracts and chemical studies. The silica is amorphous down to the resolution of about 10Å. However it is not laid down in a single pattern. Fragments of the hair examined at high resolution (100,000x) show that there are at least three distinct ways in which small (perhaps opaline) particles of about 100Å diameter are organised in the hairs. In the first instance the spherical particles are in disorganised arrays though quite densely packed. The second packing is in sheet-like arrangements of spherical particles with the particles still quite discrete. The third arrangement is of lines of particles given a fibriller-like image. The different objects are shown in Fig. 6.

Clearly these SiO_2 deposits are organised by the polysaccharides which we have started to examine by solid state NMR. Efforts must now be devoted to a description of the organic polymers around these deposits and to the reasons for the different organisations.

Amorphous Materials

Truly amorphous materials are well-known to organic polymer chemists; however amorphous silica is rather different from most organic polymers. In the first instance the Si-O-Si bond angle can have wide variations between 109° and 180°. Second the $Si(OH)_4$ monomer has four functional polymerising groups per monomer which can be compared with the two (or three) of most organic monomers. This means that within a polymerised matrix there can be many residual functional groups. The material is $[SiO_{n/2}(OH)_{4-n}]_m$, where n goes from 0 to 4, which is quite unlike a glass

Figure 5(a). Hairs, largely SiO_2, from Phalaris canariensis seeds (x 180).

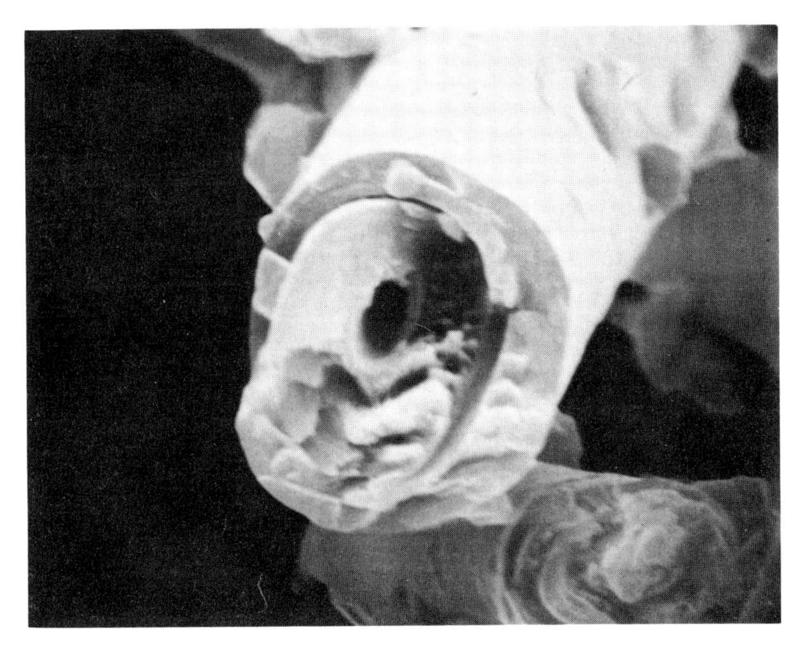

Figure 5(b). SEM picture of cross-section of hairs from (a). Note concentric structure and hollow centre (x 5,850).

Figure 6(a). A detailed view of a hair perpendicular to its length showing concentric structure (x 18,000).

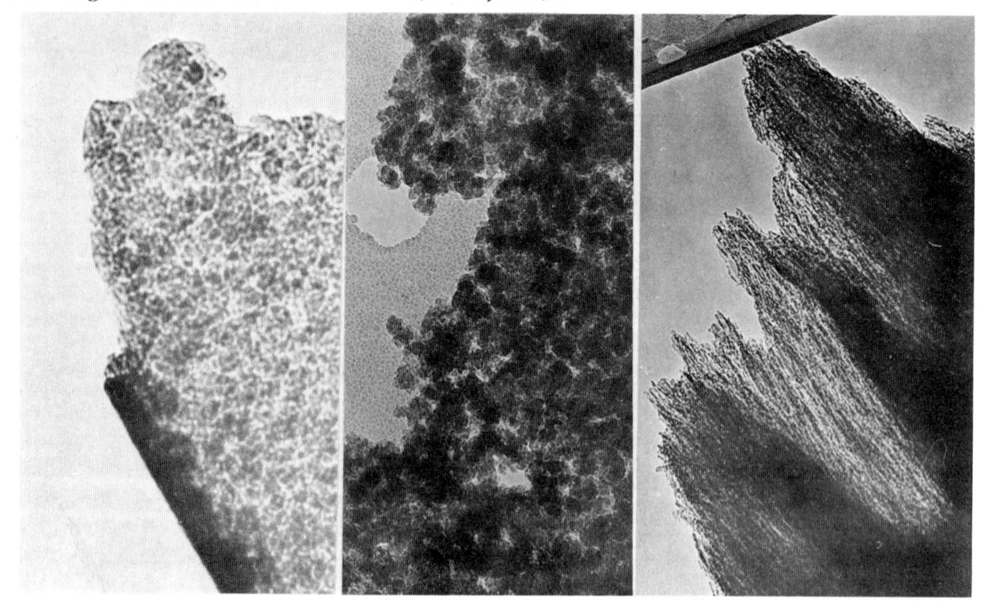

Figure 6(b). Blow-ups (approx. 90,000) of (i) tip of left hand corner; (ii) region nearer the centre; (iii) middle fibrous structure of Fig.6a.

made by melting quartz, SiO_2. The residual Si-OH groups are reactive and can condense or can act in donor H-bonds to a great variety of molecules. It is for this reason that the polymerisation of $Si(OH)_4$ can yield any shape that is demanded by an organic matrix or mould. We may ask do biological systems actually control the degree of hydration? Judging from our work on the silica rods this must be so. Biological SiO_2 is not a single simple material but may vary as much as rubbers, glasses etc., having controlled cross-linking and therefore controlled density, hardness, elasticity and so on. Such is the nature of an amorphous material. It is quite unlike $CaCO_3$ or quartz even in a combination of microcrystals and organic polymers.

FLOW IN MATRICES

It must be made clear that the discontinuity between solid and liquid states is not sharp. Especially amorphous or disordered solids allow motion even translational motion of groups so that within materials such as bone for example, protons migrate readily. This causes it to be piezo-electric. The "aragonite" of fish may have some similar properties. In silica, $[SiO_{n/2}(OH)_{4-n}]_m$ as it is in biology, the totally amorphous material of low density can certainly allow migration of protons but also probably of much larger units according to the water content. Silica gel flows. Thinking along these lines it is possible to understand the dissolution of silica from the centre of a rod or the formation of joints between rods using the silica of the centre as a potential glue. The form of the final object can be decided by moulds just as with cements. The objects can be prestressed too. While calcium phosphates approach well-known crystal forms in some parts of bone in milk the calcium phosphate in casein micelles is quite different. It is amorphous to at least 100Å. The control over the flow of atoms and larger units in such "solids" can be extremely important. It has been stated that the distinction between an assembly of micro-crystals >100Å and truly amorphous materials is of little consequence. This must be incorrect. The variation of both thermodynamic and kinetic (flow) properties with such factors as temperature, pressure and chemical composition are quite different and provide materials of quite different functional value.

MEMBRANE CONTROLS

A further consideration in particle precipitation in vesicles is the control which the membrane exerts over the solution volume and the thermodynamics activities of various ions in it. The membrane acts as a selective barrier to ions since biological membranes have channels, pumps, and prescribed leaks. For example ions such as calcium, phosphate, and bicarbonate can be readily pumped into vesicle spaces such as the the vacuoles of plant cells. The control of pH is also readily demonstrated by the fact that the pH of a vacuole is usually around 5.8 while the cell cytoplasm is at 7.0. Such vacuoles are unfavourable to the precipitation of carbonates and phosphates but permit the precipitation of oxalate, as observed in plants.

We have been studying membrane control in model sytems by incorporating diffusable anions and diffusion channels for cations into phosphatidylcholine vesicles. One procedure is as follows. Sodium phosphate is incorporated in the vesicles. The pH of the external solution is raised and the internal vesicle pH followed using ^{31}P NMR. The pH (outside) can be raised to 13.0 without affecting pH (inside) which remains at 7.0. If the vesicle interior contains sulphate similar observations are made but if nitrate is present hydroxide outside exchanges for nitrate inside and raises the internal pH. A balance is thus struck between the membrane electrochemical potential and the ion flows. This shows that because of the impermeability of membranes to cations the concentration of anions as well is totally controlled by the membrane so that precipitation is under absolute control. We have shown that the introduction of cation selective ion phases into the artificial membranes destroys the permeability barrier and can be used in our (*in vitro*) model systems to produce almost any selected solid (work with Professor E. Grover). In order to understand biological minerals we shall be forced to consider the energised transfer of a multitude of ions. Now it is obvious that the availability of ions and energy are also controlled in biology. This introduces the possibility of the controlled timing of precipitations within the specially designated and organised spaces.

In this paper we shall not describe further our simulation of biological precipitations using vesicles since the work has been described (Ref.5). We draw attention to the demonstration that precipitation of silver oxide in vesicles could be controlled so that (1) the rate of precipitation was limited by pH control in the vesicle (2) the product was a simple micro-crystal. On the other hand precipitation of amorphous materials e.g. CoS and $Fe(OH)_3$ was also possible.

SUMMARY

In this paper we have demonstrated the use of very high resolution electron microscopy, of nuclear magnetic resonance and of simulation. We hope that by bringing together these relatively new methods we shall increase the understanding of biological precipitation.

REFERENCES

1. D. Carlström, "A Crystallographic Study of Vertebrate Otoliths", Biol. Bull. 1963, 125, pp. 441-463.
2. A.J. Skarnulis, "A System for Interactive Electron Image Calculations", J. Applied Crystallography, 1979, 12, pp. 636-638.
3. V.R. Meenakshi, P.E. Hare and K.M. Wilbur, "Amino Acids of the Organic Matrix of Neogastropod Shells", Comp. Biochem. Physiol., 1971, 40B, pp. 1037-1043.

4. D.J. Cookson, B.A. Levine, R.J.P. Williams, M. Jontell, A. Linde and B. de Bernard, "Cation Binding by the Rat-Inner-Dentine Phosphoprotein", 1980, European Journal of Biochemistry, 110, pp. 273-278.
5. S. Mann, J. Skarnulis and R.J.P. Williams, "Inorganic and Bio-inorganic Chemistry in Vesicles", Israel J. of Chemistry, 1981, 21, pp. 3-7.
6. B.S.C. Leadbeater, "Studies of Stephanoeca Ellis", Protoplasma, 1979, 98, pp. 241-262.

A HIGH RESOLUTION ELECTRON MICROSCOPY STUDY OF THE BALANCE ORGANS OF THE INNER EAR

S.B. Parker, S. Mann, M.D. Ross*, A.J. Skarnulis† and R.J.P. Williams
Department of Inorganic Chemistry, University of Oxford, Oxford OX1 3QR U.K. †Department of Chemical Crystallography, University of Oxford, Oxford OX1 3QR U.K. and *Department of Anatomy and Cell Biology, University of Michigan, Ann Arbor, Michigan, U.S.A.

INTRODUCTION

The balance organs, Otoconia and Otoliths, of the inner ear are of interest as the only carbonitic calcified tissue of vertebrates (1). The otoconia are small and may be aragonitic or calcitic; the larger otoliths are aragonitic and grow as the fish ages (2).

MATERIALS AND METHODS

Otoconia from Sprague-Dawley rats and also frogs were dissected as previously described (3). The otoliths of plaice were supplied by Dr. R.W. Blacker, Fisheries Laboratory, Lowestoft. All samples were prepared for observation in the Jeol 200CX ultra-high resolution electron microscope by crushing - a technique chosen as producing the least artifacts in the sample.

The crystal faces of the sample were assigned and the lattice distances calculated after reference to computer simulations obtained by a program of Dr. A.J. Skarnulis (4).

RESULTS

Table 1 (shown in Mann, Parker, Perry, Ross, Skarnulis and Williams, this volume) shows the incidence of identification of various crystal faces from their electron diffraction patterns. In both calcitic and aragonitic mineral types the biominerals showed different results to the geological samples. This change is apparent in aragonite, where the (010) face, which is the cleavage plane, is reduced in importance in the biological samples, but spectacular in the calcite samples, where the emphasis shifts from low faces to high. From this we conclude that the pattern of crystallisation in geological and biological calcium carbonate structures are not identical.

P. Westbroek and E. W. de Jong (eds.), Biomineralization and Biological Metal Accumulation, 185–189.

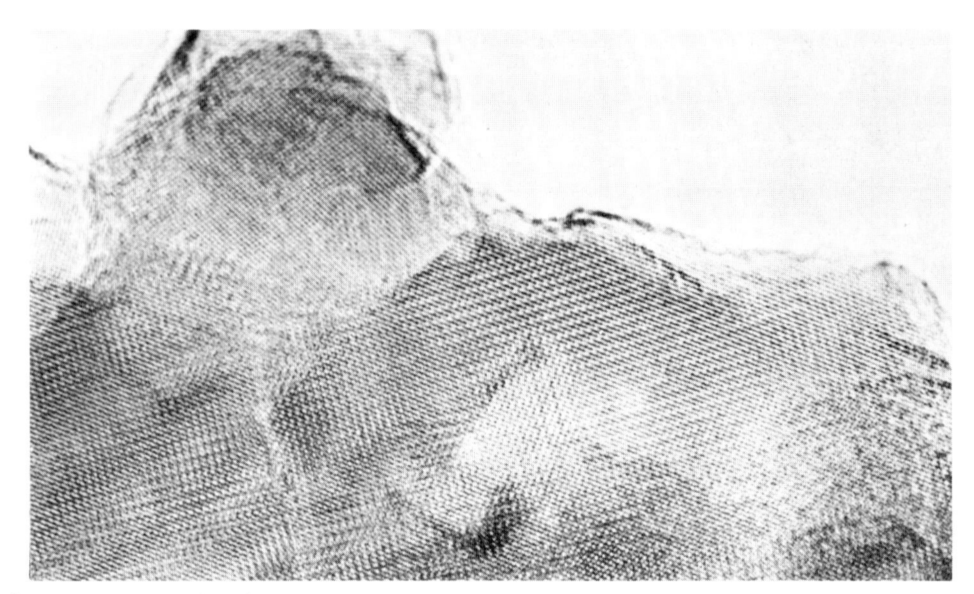

Fig.1 TEM of (010) free of aragonite.

Fig.2 (right) Diffraction pattern of Figure 1.

Examples of the geological minerals will be discussed as they formed the standards for the work. Aragonite (figure 1) can be seen to retain its single-crystal nature, although the example shown has slip-planes running through it, with the lattice spaces extending through the crystal. The face shown was identified as an (010) from the diffraction pattern (figure 2). The lattice planes are due to the (200) reflection, which has a d-spacing of 3.50Å and the nominally forbidden (100) reflection with a value of 4.96Å. This actually appears due to double diffraction. The sample was quite stable under the electron beam showed no signs of decomposing.

The calcite sample also appeared to be a single crystal which is quite stable under the electron beam to even prolonged exposure.

The otoconia of rat, which are made of calcite, cannot be described as a single crystal in the traditional crystallographic sense. The transmission electron micrograph (TEM) (figure 3) clearly shows a

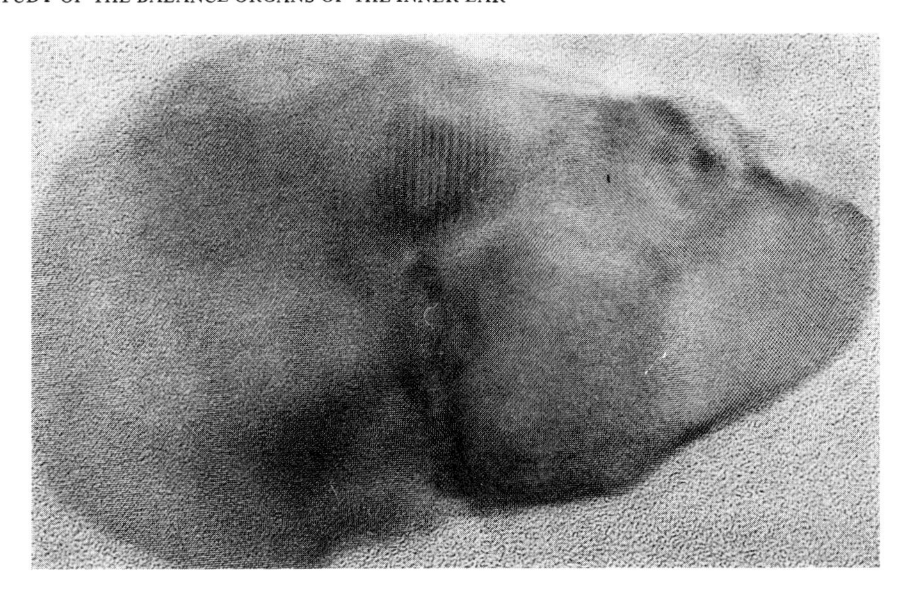

Fig.3 TEM of (421) free of rat otoconium.

Fig.4 (right) Diffraction pattern of Figure 3.

structure of at least four micro crystals which join together. This is established by the presence of Moire fringes, brought about by the actual intersection of different crystal lattices, rather than the fortuitous superposition of the segments on the sample grid. The zone is identified as a (421) zone from the electron diffraction pattern (figure 4) and the TEM shows clearly that the lattice images alter in direction, although all due to the $(0\bar{1}2)$, $(01\bar{2})$, $(\bar{1}12)$ and $(1\bar{1}2)$ reflections with a magnitude of 3.85Å. The 'single crystal' diffraction pattern may be accounted for by the ordering of the various crystallites within the sample, so that the (421) zone is exposed throughout although distinct boundaries exist within the fragment. It should be noted that the calcite sample was quite stable under the electron beam and showed no sign of damage.

The aragonite polymorphs form a distinct group in the discussion of the biominerals in that they are highly beam-sensitive. Consider the sample of plaice (figure 5). This appears to show a single crystal

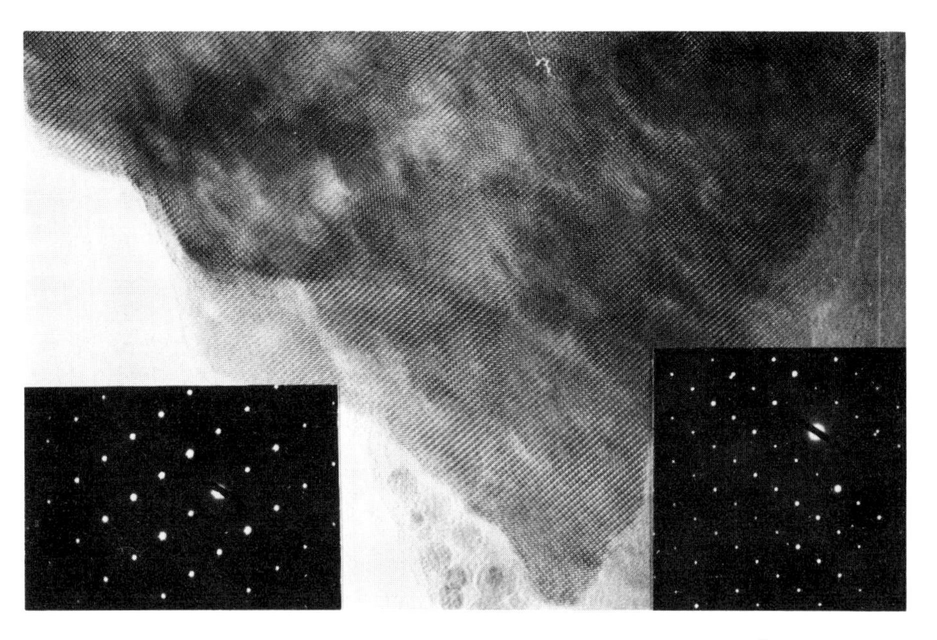

Fig.5 TEM of (010) free of Plaice otolith. Fig.6 (left) Diffraction pattern of Figure 5 before prolonged exposure. Fig.7 (right) Diffraction pattern of Figure 5 showing primitive lattice.

of aragonite which is highly pock-marked and damaged by the electron beam, but consideration of the two electron diffraction patterns (figures 6 & 7) reveal that the beam is causing profound changes to the structure of the sample. The diffraction pattern taken before the sample had been exposed extensively to the beam shows an (010) face very similar to that of the mineral (figure 2). After prolonged exposure to the beam, however, the second pattern reveals that the high symmetry of the structure had been destroyed and a primitive lattice left. This is of great interest in that the change has been brought about by the disappearance of small atoms or molecules within the structure and has left the electron-dense calcium ions intact in their unit cell, deduced from the fact that the diffraction pattern has gained hitherto forbidden reflections, but retained the allowed reflections in their original positions. It has long been known that fish otoliths are non-stoichiometric structures (5) with a small proportion of sodium in their composition. Our analyses have revealed however, (table 2) that there are many more 'rogue' ions in the structure, and work is still in progress to complete the natural composition and determine the reaction being caused by the beam.

The traditional view of the fish otolith as a polycrystalline structure could not be proved or discounted by this study, but the instability of the sample to the electron beam contrasts markedly with the behaviour of the geological mineral after identical preparation and is the subject of continued investigation.

Table 2: Estimated Percentage Composition of Plaice Otolith

Ca	C	N	H	Na	Sr	Mg	P	O
36.35	12.15	0.077	0.128	0.375	0.215	0.185	0.283*	49†

* Result estimated from secondary X-ray emission spectrum

† Result estimated from stoichiometry of $CaCO_3$

Total identified composition = 98.64%

The sample of frog otoconia is of interest in that it appears to mimic the behaviour of the fish otolith in many ways, despite being a much smaller, and ostensibly differently formed, structure. Typical samples showed a crystalline mass which appeared to be of a single crystal interrupted by small (about 50Å) areas of amorphous material. This indicated that the structure is also unstable in the electron beam and either organic material or small ions of some form are being removed from the native structure. The elemental analysis of the frog otoconia has not been made, but the similarity of the two samples lead us to predict that the frog is also likely to show a non-stoichiometric structure.

The nature of the instability of the aragonitic biominerals lead us to suggest that any organic matrix playing a role in the biomineralisation process is secreted throughout the structure and is different to that involved in annular growth and is also influencing the structure to be weak in certain regions so that crushing exposes regions which are not seen in geological samples.

Work is currently underway to try to establish the nature and role of this organic matrix in these structures.

REFERENCES

1) Carlström, D.G.: 1963, Biol. Bull. 125, 441-463.
2) Irie, T.: 1960, J. Fac. Fish. Anim. Husb. Hiroshima Univ. 3, 203-221.
3) Ross, M.D. et al: 1981, Ann. N.Y. Acad. Sci. 374, 808-819.
4) Skarnulis, A.J.: 1979, J. Appl. Cryst. 12, 636-638.
5) Morris, R.W., and Kittleman, L.R.: 1967, Science, 158, 368-370.

MINERALIZATION BY ORGANISMS AND THE EVOLUTION OF BIOMINERALIZATION

H.A. Lowenstam[†] and S. Weiner[‡]
† Division of Geological and Planetary Sciences, California Institute of Technology, Pasadena, California
‡ Incumbent of the Graham and Rhona Beck Career Development Chair, Isotope Department, Weizmann Institute of Science, Rehovot, Israel.

ABSTRACT

An updated compilation of biogenic mineral diversity and distribution is presented and a comparison is made between pathologically- and normally-formed minerals. The known associations of different minerals in individual organisms are tabulated and the various mineral forming processes involved in their deposition, are discussed. In particular, it is noted that "organic matrix-mediated" mineralization seems to be utilized by certain monerans for forming mineral deposits. This observation may have important implications with regard to the evolution of "matrix-mediated" mineralization.

INTRODUCTION

Mineral deposits are being discovered in an ever-increasing number of organisms at both intracellular sites, primarily as storage deposits, and at extracellular sites where they perform a host of different functions in addition to the long-recognized structural roles. Almost all organisms are now known to form mineral deposits and thus the traditional distinction between hard- and soft-bodied organisms is, in essence, obsolete. Furthermore, as many of the newly-discovered minerals are not calcium-bearing minerals, the term "calcification" applies to just one of the many mineral-forming processes embodied by the term "biomineralization". Many of the recent discoveries of biogenic minerals have been made in organisms without conspicuous hard parts such as fungi, protoctists and prokaryotes. Judging from the rate at which new discoveries of biogenic minerals are being made, it is obvious that we are still quite ignorant of the true diversity and distribution of biologically formed mineral deposits.

In this paper we discuss some of the implications of the updated information on mineral types and their distribution among the phyla, and in particular, we compare pathologically- and normally-formed minerals. We draw attention to certain recently reported mineralization

P. Westbroek and E. W. de Jong (eds.), Biomineralization and Biological Metal Accumulation, 191–203.

process in monerans which we think have been formed by "organic matrix-mediated" mineralization processes. We discuss the significance of these findings with respect to the evolution of "matrix-mediated" mineralization, which is commonly thought to be confined to the eukaryotes. A companion paper by Weiner, Traub and Lowenstam (this volume) on the biochemistry of organic matrices, elaborates on some of the topics in this paper.

MINERALS FORMED BY ORGANISMS

Table 1 represents a new compilation of the known diversity of biogenic minerals. Some 40 different minerals are now known to be formed by organisms (compared to 31 reported by Lowenstam, 1981) and the distribution among the phyla has increased considerably. The overall trends outlined by Lowenstam (1981) are still applicable, although the relative proportions of calcium minerals and H_2O and OH-containing minerals are reduced to about half of the total. It should be noted, that the manganese oxide minerals are not confined to monerans (as shown in Table 1). Manganese oxide minerals have been recognized in protoctists and fungi (Krumbein, 1971; Schulz-Baldes and Lewin, 1975; Ivarson and Heringa, 1972) but are not listed in Table 1 as the precise mineral forms are not known.

Table 2 lists minerals which are formed pathologically. They are usually in the form of concretions. About half are phosphate minerals, which probably reflects the fact that most of the occurrences are from vertebrates and in particular, man. Eight of the minerals are unique to pathologic processes. Although the calcium minerals still predominate, the presence of 6 magnesium containing minerals, none of which are found non-pathologically, may result from a breakdown in some aspect of magnesium metabolism.

Figure 1 represents the currently known associations of different minerals at one depositional site. Fourteen of the 22 known associations occur between different calcium minerals. All but two of the remaining associations are between calcium and some other non-calcium mineral. It should be noted that the associated minerals are not generally co-precipitates, but are segregated into discrete microarchitectural units. An important exception is the association of amorphous silica (opal) with amorphous ferric phosphate found in the tooth denticles of certain chiton species (Lowenstam and Rossman, 1975, unpublished data), the skin granules of mature Molpadiidae (holothurians) (Lowenstam and Rossman, 1975) and the sternal shields of the annelids, *Sternaspis* (Lowenstam, 1972, and unpublished data). Almost all the known associations occur in tissues formed by "matrix-mediated" processes. Possible exceptions are the glushinskite-whewhellite association in a lichen (Wilson et al., 1980) (which may also be a co-precipitate) and the amorphous fluorite-amorphous monohydrocalcite association in nudibranch spicules (Lowenstam, 1972).

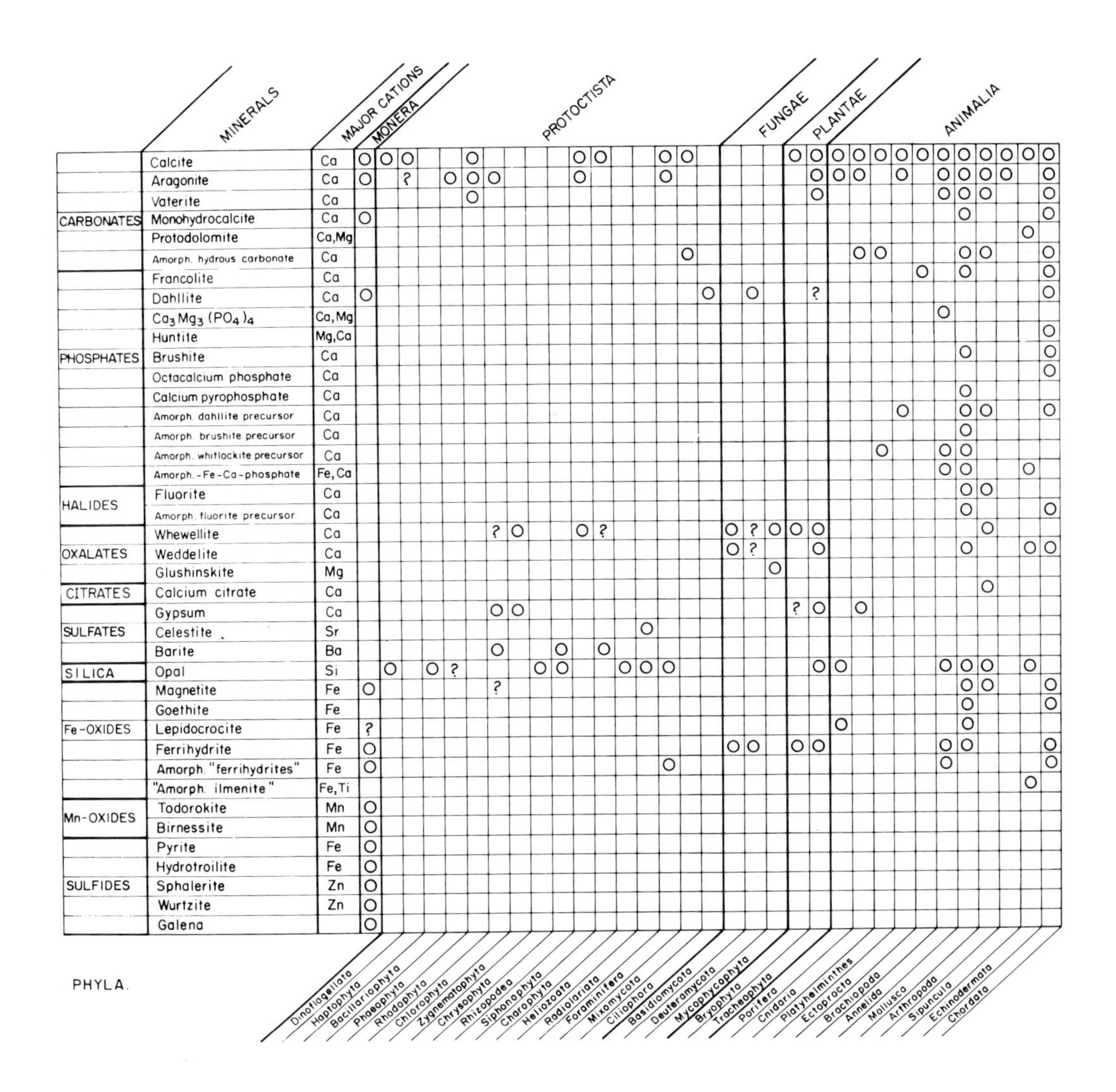

	MINERALS	MAJOR CATIONS	MONERA	PROTOCTISTA															
				Dinoflagellata	Haptophyta	Bacillariophyta	Phaeophyta	Rhodophyta	Chlorophyta	Zygnematophyta	Chrysophyta	Rhizopodea	Siphonophyta	Charophyta	Heliozoata	Radiolariata	Foraminifera	Mixomycota	Ciliophora
	Calcite	Ca	O	O	O			O					O	O			O	O	
	Aragonite	Ca	O		?		O	O	O				O				O		
	Vaterite	Ca						O											
CARBONATES	Monohydrocalcite	Ca	O																
	Protodolomite	Ca,Mg																	
	Amorph. hydrous carbonate	Ca																O	
	Francolite	Ca																	
	Dahllite	Ca	O																O
	$Ca_3Mg_3(PO_4)_4$	Ca,Mg																	
	Huntite	Mg,Ca																	
PHOSPHATES	Brushite	Ca																	
	Octacalcium phosphate	Ca																	
	Calcium pyrophosphate	Ca																	
	Amorph. dahllite precursor	Ca																	
	Amorph. brushite precursor	Ca																	
	Amorph. whitlockite precursor	Ca																	
	Amorph.-Fe-Ca-phosphate	Fe,Ca																	
HALIDES	Fluorite	Ca																	
	Amorph. fluorite precursor	Ca																	
	Whewellite	Ca							?	O			O	?					
OXALATES	Weddelite	Ca																	
	Glushinskite	Mg																	
CITRATES	Calcium citrate	Ca																	
	Gypsum	Ca							O	O									
SULFATES	Celestite	Sr														O			
	Barite	Ba							O			O		O					
SILICA	Opal	Si		O		O	?				O	O			O	O	O		
	Magnetite	Fe	O						?										
	Goethite	Fe																	
Fe-OXIDES	Lepidocrocite	Fe	?																
	Ferrihydrite	Fe	O																
	Amorph. "ferrihydrites"	Fe	O														O		
	"Amorph. ilmenite"	Fe,Ti																	
Mn-OXIDES	Todorokite	Mn	O																
	Birnessite	Mn	O																
	Pyrite	Fe	O																
	Hydrotroilite	Fe	O																
SULFIDES	Sphalerite	Zn	O																
	Wurtzite	Zn	O																
	Galena		O																

MINERALS	FUNGAE			PLANTAE		ANIMALIA										
	Basidiomycota	Deuteromycota	Mycophycophyta	Bryophyta	Tracheophyta	Porifera	Cnidaria	Platyhelminthes	Ectoprocta	Brachiopoda	Annelida	Mollusca	Arthropoda	Sipuncula	Echinodermata	Chordata
Calcite				O	O	O	O	O	O	O	O	O	O	O	O	O
Aragonite					O	O	O		O		O	O	O	O		O
Vaterite					O						O	O	O			O
Monohydrocalcite												O				O
Protodolomite															O	
Amorph. hydrous carbonate							O	O				O	O			O
Francolite										O		O				O
Dahllite		O			?											O
$Ca_3Mg_3(PO_4)_4$											O					
Huntite																O
Brushite												O				O
Octacalcium phosphate																O
Calcium pyrophosphate												O				
Amorph. dahllite precursor									O			O	O			O
Amorph. brushite precursor												O				
Amorph. whitlockite precursor								O			O	O				
Amorph.-Fe-Ca-phosphate											O	O			O	
Fluorite												O	O			
Amorph. fluorite precursor												O				O
Whewellite	O	?	O	O	O								O			
Weddelite	O	?			O							O			O	O
Glushinskite			O													
Calcium citrate													O			
Gypsum				?	O		O									
Celestite																
Barite																
Opal					O	O					O	O	O		O	
Magnetite												O	O			O
Goethite												O				O
Lepidocrocite						O						O				
Ferrihydrite	O	O		O	O						O	O				O
Amorph. "ferrihydrites"											O					O
"Amorph. ilmenite"															O	
Todorokite																
Birnessite																
Pyrite																
Hydrotroilite																
Sphalerite																
Wurtzite																
Galena																

PHYLA.

Table 1. The diversity and distribution of biogenic minerals in extant organisms.
The term "precursor" refers to an amorphous phase which upon heating to 500°C converted to the designated crystalline form.

Organisms	Location	Mineral species	Unique to pathology	Reference
Bivalve	Nephridia	ACP		Doyle et al. (1978)
Teleost fishes	Otoliths	$CaCO_3$: calcite and/or vaterite		Cited in Lowenstam and Abbott (1975)
Dalmatian dog	Bladder	$CaMg(CO_3)_2$: dolomite	+	Mansfield (1980)
Man	Gall bladder	$CaCO_3$: calcite, aragonite		Gibson (1974)
	Kidney, thyroid gland, myocardium	Whewellite: $CaC_2O_4.H_2O$		Gibson (1974)
	Kidney, thyroid gland	Weddellite: $CaC_2O_4.2H_2O$		Lewis et al. (1973)
	Articular crystal deposits	Calcium pyrophosphate: $CaP_2O_7.4H_2O$		Pritzker et al. (1980)
	Kidney	Dahllite: $Ca_{10}(PO_4,CO_3OH)_6OH_2$		Lonsdale (1968)
	"	Hydroxyapatite: $Ca_{10}(PO_4)_6OH_2$		"
	"	Brushite: $CaHPO_4.2H_2O$		"
	"	Monetite: $CaHPO_4$	+	Gibson (1974)
	"	Whitlockite: $Ca_8Mg(PO_4)_6$	+	"
	"	Octacalciumphosphate: $Ca_8H_2(PO_4)_6.5H_2O$		Lonsdale (1968)
	"	Newberyite: $MgHPO_4.3H_2O$	+	"
	"	Struvite: $MgNH_4PO_4.6H_2O$	+	Gibson (1974)
	"	Hannayite: $Mg(NH_4)_2H_4(PO_4)_4.8H_2O$	+	"
	"	Halite: NaCl	+	"
	"	Gypsum: $CaSO_4.2H_2O$		"
	"	Hexahydrite: $MgSO_4.6H_2O$	+	"

Table 2. Minerals formed pathologically in animals.

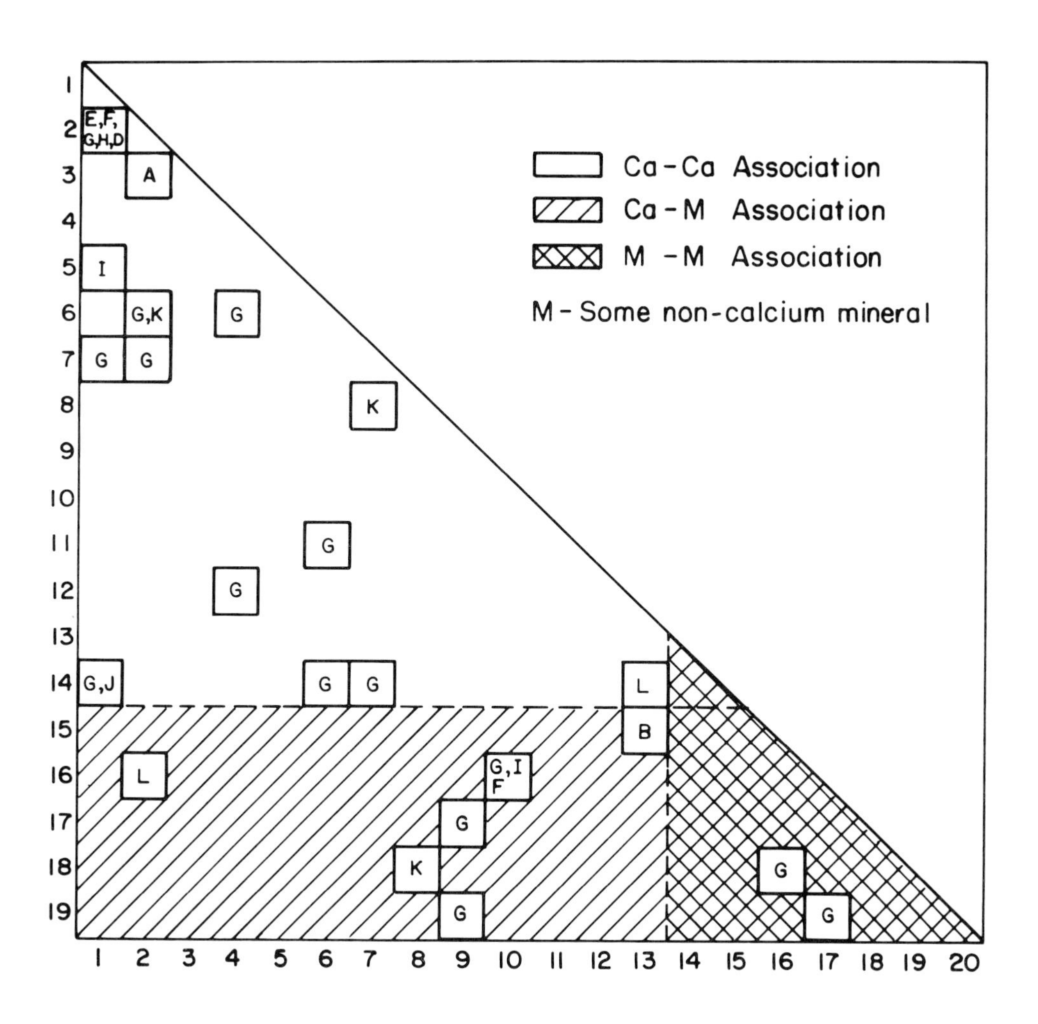

Fig. 1. Biogenic minerals associated at a single depositional site. 1 - calcite; 2 - aragonite; 3 - vaterite; 4 - amorphous monohydrocalcite; 5 - protodolomite; 6 - amorphous calcium phosphate; 7 - brushite; 8 - dahllite; 9 - francolite; 10 - amorphous iron phosphate; 11 - fluorite; 12 - amorphous fluorite; 13 - whewellite; 14 - weddellite; 15 - glushinskite; 16 - opal; 17 - magnetite; 18 - goethite; 19 - lepidocrocite. The phylum in which the mineralized tissue occurs is designated as follows:- A - Chlorophyta; B - Mycophycophyta (lichens); C - Porifera; D - Coelenterata; E - Ectoprocta; F - Annelida; G - Mollusca; H - Arthropoda; I - Echinodermata; J - Chordata-Asidiacea; K - Chordata-Vertebrata; L - Tracheophyta.

Biogenic minerals are utilized for many different functions, some of which are well known viz. skeletal supporting and/or protective structures and as storage sites or reservoirs for mineral metabolites. Some intracellular depositional sites are thought to function as waste disposal sites (Simkiss, 1980). In many cases where minerals are used for mechanical abrasion there is a selection for the harder, more resistant mineral types, for example, magnetite in chiton teeth (Lowenstam, 1962), opal and goethite in limpet teeth (Lowenstam, 1971), and protodolomite in sea urchin teeth (Pilkington, 1969). Biogenic minerals such as fluorite in the statoliths of mysids are used for orientation (Lowenstam and McConnell, 1968). Gypsum statoliths in scyphozoa serve as gravity receptors (Spangenberg and Beck, 1968), as do aragonite statoliths in cephalopods (Lowenstam, Traub and Weiner, in preparation). The otoliths in teleost fishes are thought to function as sound receptors, as well as gravity receptors (Morris and Kittleman, 1967). Magnetite is used by a variety of organisms ranging from bacteria to mammals, for navigational purposes by detection of the magnetic field direction and the changes in the magnetic field strength (Blakemore, 1975; Kirschvink and Gould, 1981).

MINERAL FORMING PROCESSES

Two biomineralization processes have been recognized: one in which mineral formation is induced by organisms as a result of interaction between biologically produced metabolite end products and cations present in the external environment (termed "biologically-induced" mineralization), and the other, a process in which minerals are grown in a pre-formed organic structural framework (termed "organic matrix-mediated" mineralization) (Lowenstam, 1981). The two processes represent end members of an intergrading spectrum, in which the organism exercises increasing control over the mineral species to be deposited as well as crystal growth. At the one end of the spectrum, in the "biologically-induced" type mineralization process, the mineral precipitates adopt crystal habits similar to those formed by inorganic processes and the orientation of the crystallites, in the case of aggregates, is essentially random. In contrast, in the "matrix-mediated" mineralization process, minerals adopt unique crystal habits, their size distribution falls within a narrow range and the crystallites have a well-defined orientation. Furthermore, in some cases the mineral deposited is one that cannot be formed in the biosphere by inorganic processes alone. For example the ferric-ferrous oxide mineral, magnetite (summarized in Lowenstam, 1981).

The "biologically-induced"-type mineralization process is widespread among the monerans at both intra- and extra-cellular locations. Among the Protoctists (in particular certain green and red alge) aragonite crystals are precipitated in this manner between cells (intercellularly) (Lowenstam, 1981). The bioinorganic constituents at intracellular mineral storage sites, commonly found in eukaryotes, are probably also formed by a "biologically-induced"-type mineralization process as the crystal habits usually resemble those formed by inorganic processes.

The "matrix-mediated" mineralization process is widespread among the eukaryotes where it occurs at extracellular locations. Recent observations of mineralized deposits in the prokaryotes, indicate clearly that some monerans may also form minerals by a "matrix-mediated" process. The following is a brief description of the 3 reported cases upon which this observation is based.

The magnetotactic bacteria discovered by Blakemore (Blakemore, 1975), have within their cells a single chain of magnetite crystals. Each crystal is enclosed by an organic sheath. The crystal habits are distinct from their inorganic counterparts and they generally adopt a parallelepiped shape with a hexagonal cross-section (Towe and Moench, 1981). In one case, tear-drop shaped crystals are known (Blakemore et al., 1980). The combination of the unique crystal habits and the presence of the organic sheath, suggest that the magnetite is formed by a "matrix-mediated" process.

Another moneran, an unidentified species of Leptotryx, precipitates a ferric-oxide mineral (also as yet, undefined), on the surface of it's cell walls (Caldwell and Caldwell, 1980). The minerals form a regular hexagonal network. No organic framework has yet been identified, but the unique crystal fabric suggests that a "matrix-mediated" process is involved in it's formation.

Cyanobacterium species of the genus Geitleria (Friedman, 1979) form a thick mineral deposit on their surface, which is composed of successive layers of elongated needle-shaped calcite crystals. In each layer the crystals are aligned. They are also separated from each other by a regular distance. The crystal orientations of alternate layers are mutually perpendicular (Friedman, 1979). After decalcification in dilute HCl, an organic framework was observed (H.A. Lowenstam and S. Golubic, personal observation). These observations suggest that the crystals may have been formed within an organic matrix.

Very few cases of intra-cellular "organic matrix-mediated" mineralization in eukaryotes have been documented. One possible example is that of coccolith formation (De Jong et al., 1976) which usually occurs in the Golgi cisternae of the Coccolithophoridae (Outka and Williams, 1971). The coccoliths, once formed, are moved to the cell surface (Isenberg et al., 1966).

"Biologically-induced" and "matrix-mediated" mineralization processes may be utilized by one organism at different tissue sites or even at the same tissue site. The Pennatulacea form radiating calcitic aggregates in their axial skeletal rods. These are found in association with collagen and are thought to form by a "biologically-induced" type mineralization process. In the cortex, polycrystalline calcite spicules are formed by means of a "matrix-mediated" process (Dunkelberger and

Watabe, 1974; Ledger and Franc, 1978). In Nautilus, the two processes appear to occur at the same site (Fig. 2). In the upper beak regular "matrix-mediated" deposition of calcite layers occurs. Between some layers, crystal aggregates of brushite and weddelite crystals are present. The crystal habits of these minerals are those of their inorganic counterparts and hence are formed, presumably, by a "biologically-induced" type process (Lowenstam, Traub and Weiner, in preparation).

"Organic matrix-mediated" mineralization is generally associated with crystalline minerals and commonly ascribed roles for matrix constituents are the determination of the orientation of the mineral crystallographic axes (Weiner and Traub, 1980) and the shapes of the individual crystal components. The skin granules of the holothurian, Molpadia, are composed of an amorphous hydrous ferric phosphate and opal which contains an organic matrix (see Lowenstam and Rossman, 1975; also Weiner, Traub and Lowenstam, this volume). The mineral phases are composed of tightly packed spherical bodies which are very similar to those observed in inorganically formed amorphous silica (opal). In Molpadia, granules are arranged into concentric shell layers, each of which is separated by an organic matrix envelope. The matrix does not appear to control the shape of the mineral particles. Hence the functions of the matrix associated with this amorphous mineral phase, remain to be determined. This is an important question, considering the fact that about one fifth of the known biogenic minerals, are amorphous (Table 1).

EVOLUTION OF BIOMINERALIZATION

In almost every phylogenetic group, including probably Monera, there are examples of mineralization processes which cover the spectrum between "biologically-induced" and "matrix-mediated" mineralization. In contrast to this view of biomineralization obtained from present-day life, the fossil record apparently presents a very different perspective of the history of biomineralization. The Precambrian appears to represent a period in which organisms produced minerals by the "biologically-induced" type mineralization process, and only towards the end of the Precambrian is more control exercised over the crystal growth processes and "matrix-mediated" mineralization evolved.

One reason for this discrepancy is that most of the information obtained, to date, from the fossil record, which pertains to the evolution of biomineralization, is limited to the preservational record of body fossils. The products of biomineralization may also be recognized when separated from the original organism. Biogenic minerals often have unique shapes and mineralogies, as well as an overprint of disequilibrium (with respect to the environment) chemical signatures. These properties may be exploited to provide a more complete picture of the evolution of biomineralization based on the fossil record. A pioneering paper in this regard, is that of Thode et al. (1951), who used $^{34}S/^{32}S$ ratios in sedimentary pyrites as a tracer of biologic

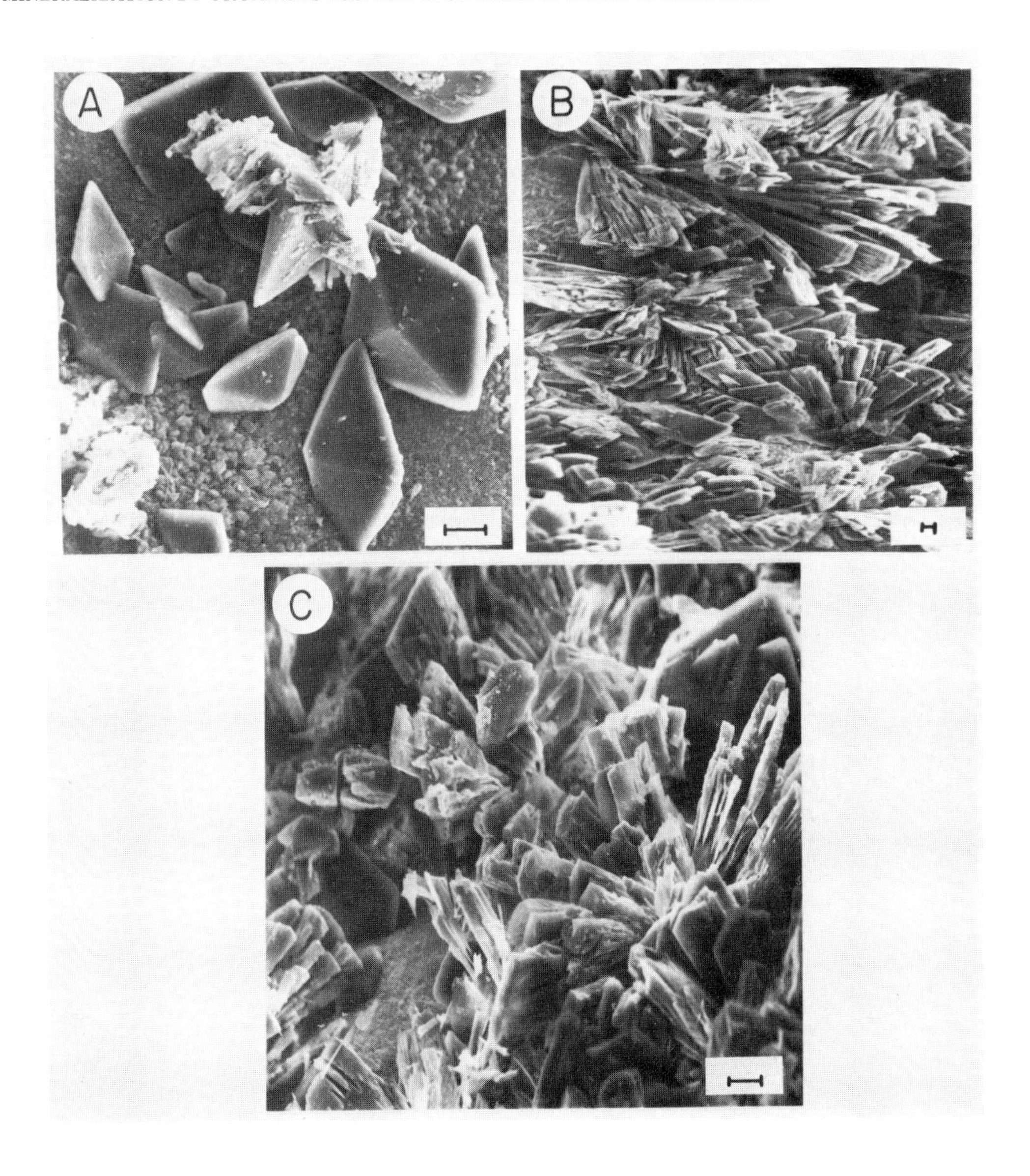

Figure 2. Crystals of A - weddelite, B - brushite and C - brushite and weddelite produced by a "biologically-induced" mineralization process on the surfaces between two calcite layers, in the upper beak of <u>Nautilus belauensis</u>. The calcite is produced by an "organic matrix-mediated" process.

activity in the Precambrian.

In the Precambrian manganese-encrusting bacteria are known from deposits about 1.6 Byr (Muir, 1978). These fossils demonstrate that "biologically-induced" type mineralization was already in existence at that time. The $^{34}S/^{32}S$ ratios in sedimentary pyrites, possibly as old as 3 Byr, show dissimilatory fractionation values similar to those produced by extant sulfate-reducing bacteria (Monster et al., 1979). Thus it appears that this type of mineralization process, may have evolved in the early Precambrian. The fact that "biologically-induced type mineralization is very common in living monerans, is consistent with this view of the fossil record.

As noted above, however, it has recently been shown that certain monerans appear to be capable of producing minerals with the aid of an organic matrix. Significantly, 2 of the 3 examples known, are iron minerals. One is a ferric-ferrous mineral and the other a ferric mineral. This mineral composition may still reflect the reducing conditions and change over from reducing to oxidizing conditions in the biosphere during the early Precambrian. It should, however, be emphasized that nothing is yet known about the biochemistry of the organic matrices of these moneran mineralized hard parts and it still remains to be proved that they are analagous to eukaryotic organic matrices. If indeed the 3 observations are indicative of the fact that "organic matrix-mediated" mineralization did evolve early during the Precambrian, then the widespread exploitation of this process for skeleton building towards the end of the Precambrian, may represent the culmination of a long history of organic matrix evolution and not the initiation of the "matrix-mediated" mineralization strategy.

Close to the Precambrian-Cambrian boundary, the fossil record clearly shows that "matrix-mediated" mineralization was widely used by the eukaryotes for the purpose of building their mineralized skeletons and supporting structures (Lowenstam and Margulis, 1980). Aspects of the biochemistry of organic matrices from various eukaryote phyla show similar properties (Weiner, Traub and Lowenstam, this volume); an observation which is consistent with the notion that "matrix-mediated" mineralization know-how was inherited from some common Precambrian ancestral stock. On the other hand, if "matrix-mediated" mineralization only evolved at the end of the Precambrian, then it would appear that each phylum independently evolved this ability, as hard parts appear to have evolved after the divergence of the eukaryotes into individual phyla.

During the Phanerozoic, biomineralization processes exhibited certain trends both with respect to mineral types utilized and complexity of organization. Just prior to the Cambrian period, all the known biogenic minerals were calcium-minerals and two thirds were composed of calcium phosphate (Lowenstam and Margulis, 1980; Lowenstam, 1980). Within 40 to 50 Myr, calcium carbonate became the dominant mineralization product and in the course of the Phanerozoic, many groups substituted aragonite for calcite in part or all of their mineralized hard parts. Another major development during the Phanerozoic, was the increasing utilization of amorphous silica by members of various eukaryotic phyla (Lowenstam, 1974; Tappan, 1980).

Trends in time with respect to the degree of organization of mineralized deposits also occurred. The labyrinth of the vestibulary apparatus of fishes contains mineralized hard parts which function as part of the balancing apparatus and apparently also as sound detectors. Primitive shark labyrinths have a sand-like slurry of isolated crystals (otoconia) which all have specific crystal habits. More advanced sharks also have a otaconia, but the constituents are crystal aggregates and not single crystals (see Fig. 3 in Lowenstam, 1981). A more advanced group of fishes, the primitive teleosts have organized crystal aggregates (otoliths) with well-defined morphology and fabric. Grooves within the aggregate are still, however, filled with single crystals. Finally the advanced teleosts have structurally well organized otoliths, but without the single crystals in the grooves (Lowenstam, 1981; Lowenstam and Fitch, in preparation). Evolution of mineralized skeletal deposits within the mollusks, demonstrates a similar trend. The most primitive mollusks contain only isolated mineralized spicules. Intermediate groups have both spicules and shell plates, whereas the common, generally more advanced mollusks, have structurally well organized shells, often composed of both calcite and aragonite (Lowenstam, 1981). A late evolutionary development in mollusk mineralization occurred in the cephalaspidian gastropods where the gizzard plates are composed of minerals such as amorphous calcium phosphate and fluorite, whereas their shells consist of aragonite (Lowenstam and McConnel, 1968; Lowenstam, 1972). It should also be pointed out that one class of mollusks exhibits an opposite trend in mineralization, namely towards a reduction of the mineral component. Primitive tetrabranchian cephalopods have 4 mineralized sites (shell, beak, statoliths and uroliths). Evolution in one group of cephalopods, the Coleoidea, has resulted in the reduction in the number of mineralization sites and in the size of the shell, to the point where in extant squids and octopi, the hard parts are composed entirely of an insoluble structural organic material (protein-chitin complex), except for the statoliths (Lowenstam, Traub and Weiner, in preparation).

Another observed trend in biomineralization during the Phanerozoic is demonstrated by the Foraminifera, where the earliest forms constructed their shells entirely out of an insoluble structural organic material. Later evolving forms reenforced the organic substrate initially with detrital particles and thereafter they adopted the organic matrix strategy for constructing their mineralized tests (data from Loeblich and Tappan, 1964).

CONCLUDING REMARKS

Despite the impressive body of information that already exists on the subject, it appears that we are just beginning to appreciate the great diversity of biogenic mineral types formed, the fact that in some biomineralization processes tens and tens of macromolecules may be involved in nucleation and crystal growth and that enormously complicated regulatory, supply and biosynthetic pathways are needed for the controlled provision of raw materials to the mineralization site.

Since biologically formed minerals, and in particular those formed by "organic matrix-mediated" processes, are unique in crystal habit, trace element and isotopic composition and occassionally even in mineral species, they may still be recognized even after their dispersal among the sediments. This could provide a means of more fully evaluating the impact of biomineralization products on the biosphere, improve our understanding of the evolution of biomineralization during the Phanerozoic and, in particular, aid in resolving some very fundamental questions relating to the evolution of "organic matrix-mediated" mineralization during the Precambrian.

ACKNOWLEDGEMENTS

We are indebted to S. Golubic and D.E. Caldwell for samples. Part of the research was funded by NSF Grant No. EAR79-05730 to H.A.L. and the U.S.-Israel Binational Science Foundation (BSF) Grant to S.W. and H.A.L. The manuscript was prepared at the Weizmann Institute of Science during the tenure of a Jakob and Erna Michael Visiting Professorship by H.A.L. This is Contribution No. 3783 from the Division of Geological and Planetary Sciences, California Institute of Technology, California.

REFERENCES

Blakemore, R.P.: 1975, Science, 190, pp. 377-379.
Blakemore, R.P., Frankel, R.B., and Kalmijn, A.J.: 1980, Nature, 286, pp. 384-385.
Caldwell, D.E., and Caldwell, S.J.: 1980, Geomicrobiol. Jour., 2, pp. 39-53.
De Jong, E.W., Bosch, L., and Westbroek, P.: 1976, Eur. J. Biochem., 70, pp. 611-621.
Doyle, L.J., Blake, N.J., Woo, C.C. and Yevich, P.: 1978, Science, 199, pp. 1431-1433.
Dunkelberger, D.G., and Watabe, N.: 1974, Tissue Cell, 6, pp. 573-586.
Friedman, I.: 1979, Plant Syst. and Evolution, 131, pp. 169-178.
Gibson, R.I.: 1974, Amer. Min., 59, pp. 1177-1182.
Isenberg, H.D., Douglas, S.D., Lavine, L.S., Spicer, S.S., and Weissfellner, H.: 1966, Am. New York Acad. Sci., 136, pp. 191-210.
Ivarson, K.C., and Heringa, P.K.: 1972, Canadian Jour. Soil Science, 52, pp. 401-416.
Kirschvink, J.L., and Gould, J.L.: 1981, Biosystems, 13, pp. 181-201.
Krumbein, W.E.: 1971, Naturwissenschaften, 58, pp. 56-57.
Ledger, P.W., and Franc, S.: 1978, Cell. Tiss. Res., 192, pp. 249-266.

Lewis, R.D., Lowenstam, H.A., and Rossman, G.R.: 1974, Arch. Pathol., 98, pp. 149-155.
Loeblich, A.R. Jr., and Tappan, H.: 1964, in Treatise of Invertebrate Paleontology (ed. Moore, R.C.) Univ. of Kansas Press, Lawrence, Kansas, Part C, Protista, pp. 900.
Londsdale, K.: 1968, Science, 159, pp. 1199-1207.
Lowenstam, H.A.: 1962, Bull. Geol. Soc. Amer., 73, pp. 435-438.
Lowenstam, H.A., and McConnell, D.: 1968, Science, 162, pp. 1496-1498.
Lowenstam, H.A.: 1971, Science, 171, pp. 487-490.
Lowenstam, H.A.: 1972, Chem. Geol., 9, pp. 153-166.
Lowenstam, H.A.: 1974, in The Sea, Vol. 5, Marine Chemistry (ed. Goldberg, E.D.) John Wiley and Sons, New York, pp. 715-796.
Lowenstam, H.A., and Abbott,D.P.: 1975, Science, 188, pp. 363-365.
Lowenstam, H.A., and Rossman, G.R.: 1975, Chem. Geol., 15, pp. 15-51.
Lowenstam, H.A.: 1980, in Biogeochemistry of Amino Acids, (eds. Hare, P.E., Hoering, T.C., and King, K.) John Wiley and Sons, New York, pp. 3-16.
Lowenstam, H.A., and Margulis, L.: 1980, Biosystems, 12, pp. 27-41.
Lowenstam, H.A.: 1981, Science, 211, pp. 1126-2231.
Mansfield, C.F.: 1980, Geochim. Cosmochim. Acta, 44, pp. 829-839.
Monster, J., Appel, P.W.U., Thode, H.G., and Schidlowski, M.: 1979, Geochim. Cosmochim. Acta, 43, pp. 405-413.
Morris, K.W., and Kittleman, L.R.: 1967, Science, 158, pp. 368-370.
Muir, M.D.: 1978, in Environmental Biogeochemistry and Geomicrobiology (ed. Krumbein, W.E.) Ann Arbor Science, Ann Arbor, Michigan, vol. 3, pp. 937-944.
Outka, D.E., and Williams, D.C.: 1971, J. Protozool., 18, pp. 288.
Pilkington, J.B.: 1969, J. mar. biol. Ass. U.K., 49, pp. 857-877.
Pritzker, K.P.H.: 1980, J. Amer. Geriatrics. Soc., 28, pp. 439-445.
Schulz-Baldes, M., and Lewin, R.A.: 1975, Science, 188, pp. 1119-1120.
Simkiss, K.: 1980, in The Mechanisms of Biomineralization in Animals and Plants (eds. Omori, M., and Watabe, N.) Tokai University Press, pp. 13-18.
Spangenberg, D.B., and Beck, C.W.: 1968, Trans. Amer. Microsc. Soc., 87, pp. 329-335.
Tappan, H.: 1980, in The Paleobiology of Plant Protists, W.H. Freeman, San Francisco, pp. 1028.
Thode, H.G., Kleerekoper, H., and McElcheran, D.: 1951, Research, 4, p. 581.
Towe, K.M., and Moench, T.T.: 1981, Earth Planet. Sci. Lett., 52, pp. 213-220.
Weiner, S., and Traub, W.: 1980, FEBS Lett., 111, pp. 311-316.
Wilson, M.J., Jones, D., and Russel, J.D.: 1980, Miner. Mag., 43, pp. 837-840.

ORGANIC MATRIX IN CALCIFIED EXOSKELETONS

S. Weiner*, W. Traub† and H.A. Lowenstam‡
* Incumbent of the Graham and Rhona Beck Career Development Chair, Isotope Dept., Weizmann Institute of Science, Rehovot, Israel
† Structural Chemistry Dept., Weizmann Institute of Science, Rehovot, Israel
‡ Division of Geological and Planetary Science, California Institute of Technology, Pasadena, California, U.S.A.

ABSTRACT

A speculative model of organic matrix structure and function, based primarily on aspects of mollusk shell matrix biochemistry, is presented. A brief review of the biochemistry of organic matrices from various other phyla, suggests that at least the essential elements of the mollusk matrix model are more generally applicable viz. the widespread presence of soluble acidic proteins, as well as insoluble "framework" macromolecules. The latter show considerable variation between the various phylogenetic groups. The observed common matrix properties provide an encouraging basis for more effectively utilizing the comparative biochemical approach for studying organic matrix functions in biomineralization.

INTRODUCTION

The degree of control exercised by organisms over the minerals they form varies greatly. At one end of the spectrum, crystals are grown in a pseudo-inorganic manner and the involvement of the organism is minimal. This mineral forming process has been termed "biologically-induced" mineralization (Lowenstam, 1981). At the other end of the spectrum, the organism synthesizes a structural framework composed of organic macromolecules into which the appropriate ions are introduced, and then induced to crystallize and grow. The structural framework determines, at least in part, the mineral phase to be formed, the orientation of its crystallographic axes and the final morphology of the individual microarchitectural units. This mineral-forming process has been termed "organic-matrix-mediated" mineralization (Lowenstam, 1981).

Organic matrices are composed of many different macromolecules, each of which presumably performs a specific function. Even in the

P. Westbroek and E. W. de Jong (eds.), Biomineralization and Biological Metal Accumulation, 205–224.

best studied mineralized tissues, a complete inventory of matrix constituents is not available, let alone information on the functions they perform prior to, during or after mineralization. For most phyla, our knowledge of matrix biochemistry is extremely limited. One outcome of this situation is that little insight has been gained into matrix functions by exploitation of the powerful technique of comparative biochemical studies.

Recent advances in our understanding of vertebrate matrix biochemistry (see Veis, this volume) and mollusk matrix biochemistry (reviewed in this paper), have resulted in the presentation of somewhat speculative, but similar models of matrix organization and function. In this paper we evaluate the applicability of, at least, the essential concepts of the model derived from mollusks to other phyla in order to assess whether or not there are common facets, which could provide a more meaningful basis for future comparative studies.

MOLLUSK SHELL ORGANIC MATRICES

Biochemistry

Biochemical analyses of mollusk shell organic matrices were initiated more than a century ago. An exhaustive list of publications and abstracts of results (up to 1970) has been compiled by Grégoire (1972). These studies in essence report that the organic matrix generally comprises between 0.01 to almost 10% by weight of the shell (Hare and Abelson, 1965) and is composed primarily of proteins and polysaccharides. The matrix constituents can be fractionated according to their relative solubilities in various solvents and each fraction has a different protein amino acid composition (e.g. Voss-Foucart, 1968). In addition protein amino acid compositions differ between shell layers of the same species (Hare, 1963) and between shells obtained from different species (Gregoire et al., 1955; Hare and Abelson, 1965; Degens et al., 1967). The polysaccharide fraction includes neutral (Meenakashi and Scheer, 1970) and amino sugar-containing polymers (Hare and Abelson, 1965; Crenshaw, 1972), sulfated mucopolysaccharides (Wada, 1964; Simkiss, 1965) and a few taxa contain chitin as a minor matrix component (Jeuniaux, 1963: Peters, 1972). Small amounts of chitin are reported to be present in all mollusk shells (Poulicet et al., 1982). A lipid fraction has also been detected histochemically (Beedham, 1958; Crenshaw and Hedley, 1967). More detailed biochemical studies of matrix constituents are dependent upon the use of a mild decalcification procedure for removing the mineral component, which ensures that the macromolecules remain intact. The calcium chelating agent ethylenediaminotetraacetic acid (EDTA) effectively fulfils these requirements. During the decalcification process, however, a significant portion of the matrix dissolves (Meenakshi et al., 1971). Crenshaw (1972) first exploited this technique to examine various aspects of the biochemistry of soluble matrix components of the clam *Mercenaria mercenaria* and thus initiated a new "era" of organic matrix biochemical studies. The proportions of soluble and insoluble organic matrix after EDTA decalcification[1] of the shell vary considerably (Weiner and Hood, 1975). For example, the matrix of Nautilus pompilius is approximately 90% insoluble (Weiner and Hood, 1975) whereas the matrices of Tridacna squamosa and Strombus gigas are almost entirely soluble (Weiner, 1979 and unpublished observations). These two fractions are quite distinct biochemically and their properties will, therefore, be summarized separately.

The insoluble fraction is primarily composed of proteins rich in the amino acids glycine alanine, phenylalanine and tyrosine (Meenakshi et al., 1971). The insolubility of this fraction is probably due, in part, to cross-linking of proteins by phenoloxidase, an enzyme which has been isolated from mollusk shells (Samata et al., 1980; Gordon and Carriker, 1980). The insoluble matrix (after EDTA decalcification) may be partially solubilized in some solvents. N-terminus analyses and gel electrophoresis of the solubilized material suggest that this fraction may be composed of only a few different proteins (Tanaka et al., 1960; Voss-Foucart, 1968; Jope, 1979). X-ray diffraction analyses of insoluble matrices from individual shell layers of different mollusks indicate that the predominant conformation adopted by the protein fraction is the anti-parallel β-sheet (Weiner and Traub, 1980). The diffraction patterns are very similar to those obtained from insect silk-fibroin. Infra-red spectra of insoluble matrices also reportedly indicate the presence of proteins with β-sheet conformations as well as some with helical conformations (Hotta, 1969). The X-ray diffraction study did not, however, detect helical conformation. In some mollusks, chitin is also a major constituent of the insoluble fraction (Jeuniaux, 1963; Goffinet and Jeuniaux, 1969; Peters, 1972). The chitin in Nautilus and Trochus shells is present in the β-form (Weiner and Traub, 1980). β-Chitin was, in fact, initially identified in the pen of the squid, Loligo (Lotmar and Picken, 1950). X-ray diffraction of Nautilus repertus nacreous layer indicates that the chitin fibrils are oriented perpendicular to the protein polypeptide chains (Weiner and Traub, 1980). The chitin and protein are present as two distinct phases, but their relative spatial distributions are not understood.

The soluble matrix fraction is composed of a heterogeneous mixture of macromolecules which can be most effectively separated into subclasses using ion-exchange chromatography, isoelectric focusing (Krampitz et al., 1976; Weiner, 1979) and high performance liquid chromatography (Weiner, 1982); techniques which exploit the charged nature of the molecules. Molecular sieve chromatography is less effective (unpublished observations) and gel electrophoresis may be misleading in that only a small portion of the soluble constituents bind commonly used dyes (Weiner, 1977; Wheeler et al., 1981). The major constituents of the soluble matrix are proteins rich in aspartic acid and to a lesser extent glutamic acid (Weiner, 1979). These residues are predominantly present as their acidic forms, as has been demonstrated (Weiner 1979) by titration of the available carboxyl groups using the method of Hoare and Koshland (1967). Some matrix components which elute from the ion-exchange column in low ionic strength solutions tend to have relatively large amounts of glutamic acid (as compared to aspartic acid) and serine (Krampitz, 1976; Weiner, 1979). Proteoglycans may also be present, as inferred from the identification of substituted seryl and threonyl residues in the soluble matrices of Mercenaria mercenaria (Fig. 1).[2] These observations suggest the presence of O-glycosidic linkages, which are common in proteoglycans (Lindahl and Rodén, 1966).

A quantitatively important amino acid sequence present in the soluble protein fraction is one in which aspartic acid is present every second residue (Weiner and Hood, 1975). The sequence may constitute up to about 40% of the protein of certain soluble organic matrices. Some of the soluble matrix constitutents are capable of binding calcium in vitro (Crenshaw, 1972; Krampitz et al., 1976; Wheeler et al., 1981). Because of the stoichiometry of calcium binding observed in Mercenaria (one mole of calcium bound for each two moles of ester sulfate) Crenshaw (1972) suggested that one mode of calcium binding may be chelation by ester sulfate groups from two adjacent polysaccharide chains. A second mode of calcium binding, which does not exclude the first and for which there is, as yet, no direct evidence, is that the aspartic and/or glutamic acid residues of the acidic proteins also bind calcium (Hare, 1963; Matheja and Degens, 1968; Weiner and Hood, 1975). In general, this type of calcium binding is common in globular calcium-binding proteins and requires cooperativity between carboxyl groups (Kretsinger, 1976) It would thus be difficult to measure in vitro.

Organic Matrix Organisation and Relationship to the Mineral Phase

Electron microscopy of mollusk shells clearly show that all or part of the organic matrix constituents envelope or delineate individual mineral units such as prisms or lamellae (e.g. Grégoire, 1962; Towe and Hamilton, 1968; Mutvei, 1970). In some cases these units are further subdivided with each sub-division separated by a matrix layer (Mutvei, 1980). It is still not clear whether some matrix constituents are located totally within the mineral phase itself. Observations showing that the soluble matrix components are not extractable without prior demineralization (Meenakshi et al., 1971) or are relatively protected from destruction by oxidizing agents (Crenshaw, 1972), may indicate the presence of organic molecules that are totally enclosed within the mineral phase. Alternatively, these molecules may be in intimate contact with the mineral surfaces and hence only relatively protected from the oxidizing agents as compared to the insoluble material.

In Nautilus repertus nacreous shell layer, X-ray diffraction of the interlamellar layers shows that, at the molecular level, the chitin and protein molecules of the insoluble matrix are well aligned with the a and b axes of the adjacent aragonite (Weiner and Traub, 1980). Thus these insoluble matrix constituents appear somehow to direct the mineral crystal formation. It is not yet known whether this is a more general phenomenon, as X-ray diffraction patterns from other matrices are very variable in quality.[3]

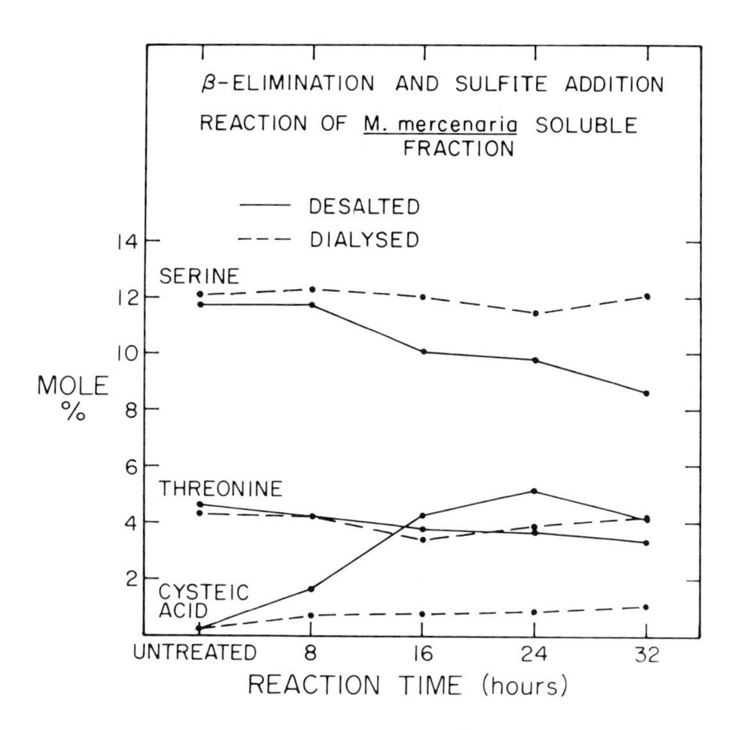

Fig. 1. β-Elimination and sulfite addition of reaction of M. mercenaria soluble fraction, according to the conditions of Simpson et al. (1972). An aliquot was removed every 8 h. Half was dialyzed exhaustively against water and half was dialyzed exhaustively against water and half was desalted on Sephadex G-25. The desalted samples decreased with increasing hydrolysis time in serine and threonine and increased in cysteic acid, indicating that about 25% of the hydroxyamino acids are substituted. Phosphate concentrations are very low (6 nanomoles/mg) and thus could not account for the observed differences. The dialysed samples showed no obvious trend and presumably the protein moiety to which the carbohydrate(?) was linked, passed through the dialysis bag.

Iwata (1975) reported that in nacreous interlamellar layers the soluble matrix constituents coat the surfaces or fill in the pores of the insoluble matrix. The conclusion was based on comparisons of fixed and unfixed matrices viewed in the transmission electron microscope (TEM). Stained sections of these matrices observed at high magnification in the TEM reveal a "sandwich-like" substructure in which the core of an individual matrix layer is electron luscent and the surfaces are electron dense (Goffinet et al., 1977; Bevelander and Nakahara, 1975; Nakahara, 1979). Other shell layers such as the crossed-lamellar layer of Strombus gigas have only electron dense matrix layers (Bevelander and Nakahara, 1980). Significantly, Strombus gigas contains almost no insoluble fraction after EDTA decalcification (Weiner, 1979) and hence it is reasonable to assume that the electron dense layers observed in TEM are composed, at least in part, of soluble matrix constituents.

We therefore envisage an individual matrix layer to be generally composed of a core of silk-fibroin-like protein covered on both surfaces by layers of soluble matrix constituents (Fig. 2) (Weiner and Traub, 1981). In certain cases such as the Strombus gigas crossed-lamellar

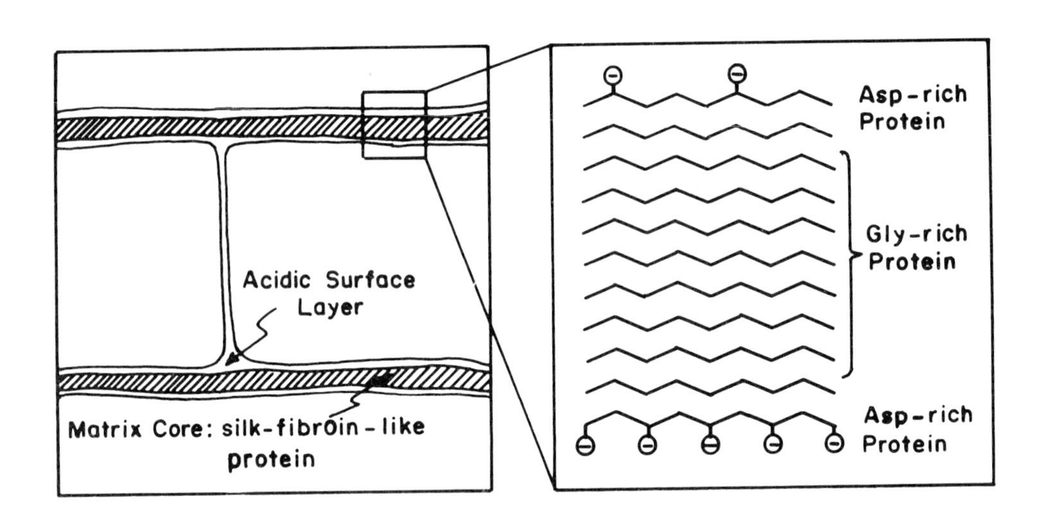

Figure 2. A schematic representation of a section through a mollusk organic matrix layer showing the "framework" core overlain by the surface acidic proteins.

layer or the "intercrystalline matrix" of Turbo (see "IC" Fig. 8 in Nakahara, 1979), the core is apparently absent.

The functions performed by different matrix constituents are still poorly understood. As a working hypothesis, Weiner and Traub (1981) have proposed that the silk-fibroin-like protein core, where present, acts primarily as a non-mineralizing structural framework. "On its surface the soluble matrix proteins are aligned with the silk-fibroin-like polypeptide chains of the core. These presumably also adopt a β-sheet conformation and, as the major constituents of the soluble fraction are proteins rich in aspartic acid (Weiner, 1979), they may form a two-dimensional sheet with regularly spaced negatively-charged groups on the surface. The precise spatial arrangement of these carboxyl groups determines the function of that area of the matrix surface with regard to its role in crystal formation" (Weiner and Traub, 1981, pp. 476-477). Degens (1976) has previously presented a model of the matrix structure which incorporates some of the same elements.

In a first attempt to obtain some evidence to test this hypothesis, aspects of the amino acid sequences of soluble mollusk shell proteins were characterized by analysis of the peptides after cleavage of the proteins on both sides of the aspartic acid residues (Weiner, 1981). This approach provides a means of determining repeating amino acid sequences which contain aspartic acid residues and hence potential calcium binding sites. When the calcitic and aragonitic layers of *Pinctada margaratifera* were compared, each layer contained a few unique peptides and hence unique aspartic acid-containing sequences; an observation consistent with an epitaxial growth model, in which specific template proteins determine which polymorph of $CaCO_3$ is to be precipitated.

The major conclusions of this survey of mollusk shell organic matrix biochemistry are: 1) The soluble matrix constituents (primarily acidic proteins and possibly proteoglycans) fulfill important, but as yet poorly understood, functions in crystal growth. 2) The electron dense layers observed in TEM (termed "envelopes" by Bevelander and Nakahara, 1969) correspond to part or all of the soluble matrix constituents. 3) Insoluble "framework" macromolecules are not present in all matrices. When present, however, they are presumed to act as a substrate for soluble matrix constituents and in addition contribute to the mechanical properties of the formed shell.

Organic Matrices from the Mineralized Tissues of Various Other Phyla.

The matrix model presented for mollusks envisages a well-ordered structure in which the core or framework of an individual matrix layer is composed primarily of insoluble matrix macromolecules and the surfaces, of various soluble acidic proteins. To effectively evaluate the applicability of such a model to the matrices of mineralized hard parts of other phyla, proteins having common properties and fulfilling similar functions should be recognized in each case. A promising start has been made within the Chordata (Fincham et al., 1982). Among the invertebrates, however, the data available is so limited that at this stage we confine ourselves to evaluating whether or not candidates for "framework" macromolecules and "surface" proteins, are indeed present.[4] The results are summarized in Table 1.

Table 1. The nature of the insoluble and soluble constituents of matrices from different phyla.

Phylum	Skeletal Hard Part† (Mineral Type)	Soluble Acidic Proteins	Insoluble "Framework" Macromolecules (Predominant Amino Acids)
Chordata	Bone (HA)	*Isol. (Asp, Glu)	Collagen (Gly, Pro, Hypro)
	Cartilage (HA)	Isol. (Asp, Glu)	Collagen, proteoglycan
	Tooth dentine (HA)	Isol. (Asp, Glu)	Collagen
	Tooth enamel (HA)	Isol. (Asp, Glu)	Amelogenin (Pro, Gln)
Echinodermata	Sea urchin tooth (C, protodomite)	**Inf. (Asp)	Protein (Gly)
	Skin granules (Ca-Fe-phosphate)	Inf. (Asp, Glu)	Protein (Asp, Glu, Gly)
Mollusca	Shell (C, A)	Isol. (Asp)	Silk-fibroin-like protein (Gly±Ala) ± B-chitin
Annelida			
Serpulida	Tubes (C, A)	Inf. (Asp, Glu)	Protein (Gly, Ser)
Arthropoda			
Crustacea	Cuticle ($CaCO_3$, ACP)	Inf.	Protein (Gly, Ala), α-chitin
Brachiopoda			
Articulata	Shell (C)	No data	Protein (Gly)
Inarticulata	Shell (HA)	No data	Protein (Gly, Ala, Hypro), chitin
Craniacea	Shell (C)	No data	Protein (Gly, Ser, Asp), chitin
Ectoprocta	Zoecium (C)	No data	Protein, chitin
Cnidaria	Coral skeleton (A)	Isol. (Asp)	Absent ? Few have chitin
Protozoa			
Foraminifera	Shell (C)	Inf. (Asp, Glu)	Sulfated acid mucopolysaccharides

† HA-hydroxyapatite; C-calcite; A-aragonite; ACP-amorphous calcium phosphate.
* Isol.-isolated from other soluble constituents. The predominant acidic amino acids are designated in parentheses.
** Inf.-inferred to be present based on amino acid composition analysis of total soluble fraction.

The Chordata have been extensively studied and reviewed (see A. Veis, this volume). Unusually acidic proteins have been isolated from among the soluble components of tooth dentine (Veis et al., 1972) where they were first identified by A. Veis and his coworkers in 1963 (Schleuter and Veis, 1963). Acidic matrix proteins have also been identified in bone (Shuttleworth and Veis, 1972; Cohen-Solal et al., 1979; Termine et al., 1981), in cartilage (Choi et al., 1981) and in developing tooth enamel (Fincham et al., 1981). The associated "framework" macromolecules vary in these tissues. Bone and dentin are both composed predominantly of type 1 collagen, whereas cartilage contains type II collagen (reviewed by Piez, 1981) as well as large amounts of proteoglycan. Tooth enamel on the other hand contains proline- and glutamine-rich proteins called amelogenins (Eastoe, 1971; Fincham et al., 1981). The importance of the acidic proteins in mineralization has been highlighted by radioautographic experiments which show that in dentine these matrix proteins rapidly migrate to the mineralization front (Weinstock and Leblond, 1973; Dimuzio and Veis, 1978). Recently proposed structural models of dentine, enam 1 and bone matrix organization also envisage the insoluble "framework" acting as the substrate for acidic proteins, some of which in turn bind cations and/or anions in such a way as to initiate apatite formation (Dimuzio and Veis, 1978; Glimcher, 1979; Termine et al., 1981; Veis, 1981).

Echinodermata form diverse mineralized skeletons, spicules and granules. The spines and test plates of the echinoid (sea urchin) skeleton are remarkably free of protein (Klein and Currey, 1970) although some organic material (in particular pigments) is present (Pilkington, 1969; Pearse and Pearse, 1975). The sea urchin teeth do, however, have an organic matrix (Markel, 1969; Kniprath, 1974) which comprises about 0.3% by weight of the tissue. After EDTA decalcification, about 90% of the matrix protein is soluble and aspartic acid and glutamic acid together account for about 20 mole % of the protein (see appendix). The soluble and insoluble matrix protein compositions resemble each other except for a few significant differences (see glycine and lysine contents in appendix).

Another echinoderm matrix that has been analyzed with respect to its protein amino acid composition is that of the skin granules in the holothurian Molpadia. The granules are composed of an amorphous, hydrous, ferric phosphatic mineral (Lowenstam and Rossman, 1975) and about 0.13% by weight organic matrix. Despite the fact that the associated mineral is not a calcium mineral, the matrix protein compositions are rich in aspartic and glutamic acids, together with relatively large proportions of glycine and alanine. The soluble and insoluble protein compositions are somewhat similar, except for the appreciably higher glycine and alanine and lower arginine contents of the soluble fraction.

The apparent absence of a protein containing matrix in sea urchin tests and spines is in marked contrast to the demonstrated presence of matrix in sea urchin teeth, holothurin skin granules and of course, in almost all known mineralized skeletons. Presumably the "absence" of the matrix protein is due to some maturation process after mineral deposition has occurred (Pilkington, 1969).

The Annelida. Polychaete worms which form both calcified tubes (Serpulidae) and non-calcified tubes (Sabellariidae) have been studied (Mitterer, 1971). The serpulid tubes were decalcified in HCl and only total amino acid compositions are reported for both tube types. Comparisons of the calcified and non-calcified tubes show that the calcified tubes have an additional protein fraction rich in aspartic and glutamic acids. Presumably, the insoluble "framework" proteins of the calcified serpulid tube are similar in composition to the non-calcified tube viz. rich in glycine and serine.

The Arthropoda. The most conspicuous mineralizers in this phylum are the Crustacea. The crustacean cuticle contains between 15 and 45% by weight organic matrix, almost all of which is an insoluble protein-α-chitin complex (Welinder, 1974). The chitin accounts for between 52 and 72% by weight and the remainder is protein which contains about 20 mole % aspartic and glutamic acid and 14 mole % glycine (Welinder, 1974). The soluble protein of a decapod was analyzed after EDTA decalcification (Hackman, 1974) and its amino acid composition is similar to that of the insoluble fractions reported by Welinder (1974). The soluble fraction of crustaceans contains numerous proteins which have isoelectric points in the range pH 3.5 to pH 5.7 (Hackman, 1974; Vranckx and Durliat, 1980). Some of the components are therefore, fairly acidic.

The Brachiopoda are divided into two classes. The Articulata shells are composed of calcite and contain about 0.5% by weight protein, whereas those of the Inarticulata are composed of hydroxyapatite, contain up to 25% by weight protein together with chitin. One group within the Inarticulata, the Craniacea, have calcitic shells (Jope, 1977). The organic matrices of both Articulata and Inarticulata are predominantly insoluble. The soluble fraction has not been studied. The insoluble matrix can be solubilized in denaturing solvents and N-terminus analyses as well as gel electrophoresis of this material indicates that it may be composed of only a few different proteins (Jope, 1979). The proteins of the Articulata matrix are rich in Gly, whereas those of the Inarticulata (excluding Crania) contain relatively large amounts of hydroxyproline in addition to Gly and Ala. No characteristic collagen 640Å banding pattern was observed (Jope, 1967).

Ectoprocta (Bryozoa) exoskeletons have organic matrices which comprise between 5 and 20% by weight of the skeleton (Schopf and Manheim, 1967). They are composed of protein and chitin in about 10:1 proportions (Schneider, 1963; Hunt, 1972). Hunt (1972) determined the total protein amino acid composition after removing the soft tissues and found the proteins rich in glycine, aspartic acid and glutamic acid. Many bryozoan skeletons have appreciable amounts of phosphate (reviewed in Schopf and Manheim, 1972) which is removable by sodium hypochlorite oxidation.[5] It is thus presumably part of the organic matrix (Lowenstam, unpublished observations).

The Cnidaria (Coelenterata). A major reef-building group within this phylum, the scleractinian corals, have aragonitic skeletons which contain about 0.03% by weight protein (Wilfert and Peters, 1969; Young 1971a; Mitterer, 1978). Three species of corals are known to contain chitin in their skeletons (Wainwright, 1963; Wilfert and Peters, 1969). A lipid fraction composed largely of cetyl palmitate, is also present in the skeleton (Young et al., 1971b). As extended extraction with warm 1N NaOH (Mitterer, 1978) or 2-3 min. boiling in 5N NaOH (Young, 1971a) was required to remove coral tissues, the possibility exists that the matrix constituents themselves were altered. After HCl decalcification, the matrices are soluble (except for those that contain chitin) and their protein fractions are particularly rich in aspartic acid (Young,1971a; Mitterer, 1978). Similar results were obtained from the decalcified spicules of alcyonarian octacorals (Mitterer, 1978). The latter contained almost ten times more protein than the scleractinian coral skeletons and in some cases aspartic acid constituted up to 75 mole % of the protein (Mitterer, 1978). The interpretation of these results may be further complicated, as Johnston (1980) has reported that most of the organic matrices of Pocillopora damicornis (except presumably for the chitin) may have degraded and disappeared after skeletal mineralization was complete. Significantly, in mammalian tooth enamel, where matrix degradation and removal does occur, the last remaining remnants are also primarily soluble acidic proteins (Glimcher et al., 1977). If matrix degradation in corals occurs generally and as the preparation procedures employed may have affected the matrix constituents, the impression that in newly formed coral skeletons, the matrices are composed primarily of very small amounts of soluble acidic proteins, may well be misleading.

The Gorgoniacea, Pennatulacea and Zoantharia all have skeletons constructed almost entirely out of an insoluble structural organic material. This material may be analogous to the insoluble "framework" constituents of mineralized matrices. In the Pennatulacea and Gorgoniacea the material is composed primarily of collagen (Ledger and Franc, 1978; Goldberg, 1976) whereas in the Zoantharia, the organic skeleton is primarily a complex of protein rich in glycine and alanine (Goldberg, 1976) together with β-chitin (Ellis et al., 1980).The protein comprises about 97% by weight of the skeleton and may be in the β-sheet conformation (Ellis et al., 1980). The insoluble organic matrix within the spicules of one species of Gorgoniacea is reported to contain collagen-like proteins (Silberberg et al., 1972).

Forminifera shell proteins constitute only 0.02 to 0.06% by weight of the calcitic shell (King and Hare, 1972; appendix, this paper). Part of the matrix is insoluble after decalcification in HCl and histochemical tests demonstrate the presence of protein, lipid and a polymerized sulfated acid mucopolysaccharide (Hedley, 1960; Moss in Bé and Ericson, 1963). The benthic foramifer, Heterostigina depressa (Gulf of Elat) contains about 0.16% by weight organic matrix. About two thirds is insoluble, of which only 2.2% by weight is protein. The rest is presumably the polysaccharide identified histochemically. More than

95% of the total protein is in the soluble fraction and it contains proteins rich in serine, glycine and lysine and to a lesser extent aspartic and glutamic acids (see appendix). Protein comprises about 0.06% by weight of the shell. The planktonic foraminifera analyzed by King and Hare (1972) which were about 1000-2000 years old, have total amino acid compositions rich in glycine, aspartic and glutamic acids. Based only on these few analyses, it appears that foraminifera have an insoluble matrix which contains primarily polysaccharide and a soluble fraction, rich in acidic proteins.

The major conclusions of this survey are: 1) Soluble acidic proteins are present in all organic matrices for which data are available. 2) The insoluble matrix fractions show considerable variation with respect to the nature of their protein and polysaccharide constituents. 3) The proportions of insoluble and soluble matrix constituents vary greatly. In some cases, the insoluble fraction is apparently absent.

DISCUSSION

The presence of soluble acidic matrix constituents (proteins and proteoglycans) in all matrices analysed to date emphasizes their importance in "organic matrix-mediated" mineralization. Only a few of these proteins have been characterized. Neither the conformations adopted *in vivo* by these soluble proteins nor their precise locations within the organic matrix, are known. In contrast, the conformations of several insoluble "framework" macromolecules including collagen, mollusk silk-fibroin-like protein and chitin, have been determined. Some roles envisaged for these soluble matrix macromolecules, according to the proposed model (Fig. 2), include crystal nucleation and crystal growth inhibition. Presumably the precise functions performed are related to their primary amino acid sequences, which in turn determine the spatial arrangement of calcium binding sites on the matrix surface. One corollary of the model is that matrices associated with a particular mineral type, should have some common constituents, on the assumption that crystal nucleation occurs on the same template. This attribute is currently being tested.

The role of proteoglycans in biomineralization is enigmatic. They are major components of mineralized cartilage (Shatton and Schubert, 1954) and are present in small amounts in bone and dentine (Termine, 1981; Firez et al., 1981). They are subjected to selective biodegradation and removal with the onset of mineralization (Toole and Linsenmayer, 1977). They may be present in mollusk shell organic matrices (see Fig. 1 and accompanying text) and possibly in other invertebrates.

The insoluble "framework" macromolecules are thought to perform two primary functions: they act as a substrate on which the acidic proteins are aligned and secondly, contribute to the overall mechanical properties of the formed tissues. As insoluble "framework" macromolecules *appear* to be absent in some skeletons, both these roles may not be essential for all mineralized hard parts. The striking variability in the nature of the insoluble constituents in the mineralized hard parts of the phyla surveyed (Table 1) is consistent with the notion that these matrix constituents, where necessary, contribute certain mechanical properties to the formed product according to the specific requirements of the organism. There is also no correlation between the nature of the insoluble constituents and the associated mineral type. In the model proposed, the insoluble components form a "core" or "framework" upon which the acidic proteins are located, and thus there is no required role for the insoluble constituents in directly determining mineral type.

It is interesting to note that the "framework" proteins resemble each other in that their amino acid compositions are essentially all characterized by varying proportions of glycine, alanine, proline and/or hydroxyproline. Collagen is rich in Gly, Pro and Hypro, amelogenin in Pro and mollusk shell silk-fibroin-like protein in Gly ± Ala. Rudall (1968) has previously noted similarities between collagen and insect silk-fibroin which he, in part, ascribed to a possible evolutionary relationship.

The presence of organic matrices in mineralized tissues containing only amorphous minerals, such as the holothurian skin granules (described above in the paragraph on Echinodermata) suggests that matrices do have a role to play in mineralization processes without the necessity of structural complementary between matrix and mineral. On the other hand, in cases such as *Nautilus* nacreous layer, a specific epitaxial structural relationship, can exist. It should be noted that hardly any studies of matrices from non-calcium minerals have been made. Comparisons between the matrices from calcium and non-calcium mineral deposits could be most informative.

The inter-phylum survey of organic matrix properties shows that matrices of many different organisms are composed of "framework" macromolecules and acidic proteins. If indeed these observed similarities extend to the functions of specific macromolecules and this in turn reflects the presence of common underlying mechanisms, then "matrix-mediated" mineralization may well have evolved from some common ancestor early in the Precambrian (see discussion in Lowenstam and Weiner, this volume). The widespread utilization of "organic matrix-mediated" mineralization for building exoskeletons which occurred at the end of the Precambrian (Stanley, 1976; Lowenstam and Margulis, 1980) may then represent the end product of a long evolutionary history and not the inception of this type of mineralization process. The alternative that each phylum independently "invented" the matrix strategy would have to be qualified by invoking an explanation for the observed common inter-phylum biochemical properties of organic matrices. One possible explanation is that some pre-existing appropriate organ or tissue was adapted for the purpose of inducing minerals to form under the controlled conditions of an organic framework.

It is interesting to note that if our speculations are correct and crystal nucleation takes place on a β-sheet surface, then the simplest repeating distance of alternating amino acids along the polypeptide chain (about 6.9Å) and between adjacent polypeptide chains (about 4.7Å) are very similar to the Ca-Ca distances along the c and a axes of hydroxyapatite, respectively. In aragonite, however, where the b axis is parallel to the matrix polypeptides (of the insoluble fraction) (Weiner and Traub, 1980), the compatibility between the mineral Ca-Ca distances and the protein repeat distances, would require a more complicated amino acid repeat sequence. The same situation probably applies to calcite. The fossil record shows that about 75% of the skeletal deposits formed by animals at the end of the Precambrian were composed of calcium phosphate and only later did calcite and then aragonite become the common skeletal-forming minerals (Lowenstam and Margulis, 1980). It is intriguing to speculate that the sequence of events may in part reflect, an increasing complexity of the matrix template!

CONCLUDING REMARK

The importance of evaluating the common facets of organic matrices from the mineralized hard parts of different phyla, goes well beyond the historical interest. The resolution of questions relating to a divergent as opposed to a convergent evolutionary pathway of "matrix-mediated" mineralization, are essential, before a solid foundation for future comparative studies in this field, can be laid.

ACKNOWLEDGEMENTS

We thank Yonatan Erez, Marine Laboratory, Hebrew University, Elat, for supplying the Foraminifera samples. Part of this research was supported by a U.S.-Israel Binational Science Foundation (BSF) grant and a U.S. National Science Foundation (NSF) grant (EAR 79-05730) to HAL.

NOTES

[1] In this paper the terms "soluble" and "insoluble" will refer to those matrix constituents which are soluble and insoluble after EDTA decalcification at or close to neutral pH. The precise conditions for decalcification do, however, have a minor effect on the end products (Weiner, 1976). Note also that not all the EDTA can be removed by dialysis and hence quantitation of the soluble fraction yields is subject to error.

[2] Crenshaw (1980) using a different technique did not find any evidence for substituted seryl and/or threonyl residues in *Mercenaria*. The details of the experiment, however, have not been published.

[3] See discussion in Weiner and Traub (1981) of X-ray diffraction studies of other mollusk matrix-mineral relationships.

[4] In almost all the studies discussed below, no determinations of aspartic acid and glutamic acid versus asparagine and glutamine have been made. We will assume that the acidic forms are the predominant ones.

[5] The skeleton of Flustra contains approximately 6% by weight P_2O_5. The infra-red spectrum of a sample treated with 5.2% sodium hypochlorite was compared with a untreated sample and the phosphate absorption peak, showing disorded ACP, at about 1100 cm^{-1} was absent in the treated material.

REFERENCES

Beedham, G.E.: 1958, J. Microsc. Sci., 99, pp. 341-347.

Bevelander, G., and Nakahara, H.: 1969, Calc. Tiss. Res., 3, pp. 84-92.

Bevelander, G., and Nakahara, H.: 1975, Earth Sci., 28, pp. 87-91.

Bevelander, G., and Nakahara, H.: 1980, in The Mechanisms of Biomineralization in Animals and Plants (eds. Omori, M., and Watabe, N.) Tokai University Press, pp. 19-27.

Choi, H., Johnson, T., Pal, S., Tang, L., Rosenburg, L., Reiner, A., and Poole, A.R.: 1981, in the Chemistry and Biology of Mineralized Connective Tissues (ed. Veis, A.) Elsevier North Holland Inc., pp. 343-348.

Cohen-Solal, L., Lian, J.B., Kossiva, D., and Glimcher, M.J.: 1979, Biochem. J., 177, pp. 81-98.

Crenshaw, M.A.: 1972, Biomineralization, 6, pp. 6-11.

Crenshaw, M.A.: 1980, in Skeletal Growth of Aquatic Organisms (eds. Rhoads, D.C., and Lutz, R.A.) Plenum Press, New York and London, pp. 115-132.

Crenshaw, M.A. and Hedley, J.D.: 1967, J. Dent. Res., 46, p. 65.

Degens, E.T.: 1976, Topics in Current Chemistry, 64, 1-112.

Degens, E.T., Spencer, D.W., and Parker, R.H.: 1967, Comp. Biochem. Physiol., 20, pp. 553-579.

Dimuzio, M.T., and Veis, A.: 1978, Calcif. Tiss. Res., 25, pp. 169-178.

Eastoe, J.E.: 1971, in Comprehensive Biochemistry (eds. Florkin, M. and Stotz, E.) Elsevier North Holland Inc., 26C, pp. 785-834.

Ellis, L.C., Chandross, R.J. and Bear, R.S.: 1980, Comp. Biochem. Physiol., 60B, pp. 163-165.

Fincham, A.G., Belcourt, A.B., Termine, J.D.: 1981, in The Chemistry and Biology of Mineralized Connective Tissues (ed. Veis, A.) Elsevier North Holland Inc., pp. 523-529.

Fincham, A.G., Belcourt, A.B., Lyaruu, D.M., and Termine, J.D. : 1982, Calcif. Tiss. Int., 34, pp. 182-189.

Firoz, R., Prince, C.W., Caterson, B., Christner, J.E., Baker, J.R., and Butler, W.T.: 1981, in The Chemistry and Biology of Mineralized Connective Tissues (ed. Veis, A.) Elsevier North Holland Inc., pp. 389-393.

Glimcher, M.J.: 1979, J. Dent. Res., 58B, pp. 790-806.

Glimcher, M.J., Brickley-Parsons, D., and Levine, P.T.: 1977, Calcif. Tiss. Res., 24, pp. 259-270.

Goffinet, G., and Jeuniaux, C.: 1968, Comp. Biochem. Physiol., 29,

pp. 277-282.
Goffinet, G., Gregoire, C., and Voss-Foucart, M.G.: 1977, Arch. Internat. Physiol. Biochem., 85, pp. 849-863.
Goldberg, W.M.: 1976, Mar. Biol. 35, pp. 253-267.
Gordon, J. and Carriker, M.R.: 1980, Mar. Biol., 57, pp. 251-260.
Grégoire, C.: 1962, Bull. Inst. roy. Sci. natur. Belg., 38, pp. 1-71.
Grégoire, C.: 1972, in Chemical Zoology VII (eds.M. Florkin and B.T. Scheer) Academic Press, New York, pp. 45-102.
Grégoire, C., Duchateau, G., Florkin, M.: 1955, Annls. Inst. oceanogr. Monaco, 31, pp. 1-36.
Hackman, R.H.: 1974, Comp. Biochem. Physiol., 49B, pp. 457-464.
Hare, P.E.: 1963, Science, 139, pp. 216-217.
Hare, P.E. and Abelson, P.H.: 1965, Carnegie Inst. Wash. Year Book, 64, pp. 223-232.
Hedley, R.H.: 1960, Quart. J. icroscop. Sci., 101, pp. 279-293.
Hoare, D.G. and Koshland, D.E.: 1967, J. Biol. Chem., 242, pp. 2447-2453.
Hotta, S.: 1969, Chikyu Kagaku, 23, pp. 133-140.
Hunt, S.: 1972. Comp. Biochem. Physiol., 43B, pp. 571-577.
Iwata, K.:1975, J. Fac. Sci. Hokkaido Univ. Ser. 4, 17, pp. 173-229.
Jeunieux, C.: 1963, Chitine et Chitinolyse (Masson, Paris).
Johnston, I.S.: 1980, Int. Rev. Cytology, 67, pp. 171-214.
Jope, M.F.: 1967, Comp. Biochem. Physiol., 20, pp. 593-600.
Jope, M.F.: 1977, Amer. Zool., 17, pp. 133-140.
Jope, M.F.: 1979, Comp. Biochem. Physiol., 63B, pp. 163-173.
King, K., and Hare, P.E.: 1972, Micropaleontology, 18, pp. 285-293.
Klein, L., Currey, J.D.: 1970, Science, 169, pp. 1209-1210.
Kniprath, E.: 1974, Calc. Tiss. Res., 14, pp. 211-228.
Krampitz, G., Engels, J., Cazaux, C.: 1976, in The Mechanisms of Mineralization in the Invertebrates and Plants (eds. Watabe, N. and Wilbur, K.M.) University of South Carolina Press, Columbia, pp. 155-173.
Kretsinger, R.H.: 1976, Coordination Chemistry Reviews, 18, pp. 29-124.
Ledger, P.W. and Franc, S.: 1978, Cell. Tiss. Res., 192, pp. 249-266.
Lindhal, U., and Rodén, L.: 1966, J. Biol. Chem., 241, pp. 2113-2119.
Lotmar, W. and Picken, L.E.R.: 1950, Experientia, 6, pp. 58-59.
Lowenstam, H.A.: 1981, Science, 211, pp. 1126-1131.
Lowenstam, H.A., and Rossman, G.R.: 1975, Chem. Geol., 15, pp. 15-51.
Lowenstam, H.A. and Margulis, L.: 1980, Biosystems, 12, pp. 27-41.
Markel, K.: 1969, Z. Morph. Tiere, 66, pp.1-50.
Meenakshi, V.R. and Scheer, B.T.: 1970, Comp. Biochem. Physiol., 34, pp. 953-957.
Meenakshi, V.R., Hare, P.E., and Wilbur, K.M. 1971, Comp. Biochem. Physiol., 40B, 1037-1043.
Mitterer, R.M.: 1971, Comp. Biochem. Physiol., 38B, pp. 405-409.
Mitterer, R.M.: 1978, Bull. Mar. Sci., 28, pp. 173-180.
Moss, M.L.: 1963, in Bé, A. and Ericson, D.B., Annals N.Y. Acad. Sci., 109, pp. 65-81.
Mutvei, H.: 1970, Biomineralization, 6, pp. 96-100.
Mutvei, H.: 1980, in The Mechanisms of Biomineralization in Animals and Plants (eds. Omori, M., and Watabe, N.) Tokai University Press, pp. 49-56.

Nakahara, H.: 1979, Jap. J. Malacology, 38, pp. 205-211.
Pearse, J.S., and Pearse, V.B.: 1975, Amer. Zool., pp. 731-753.
Peters, W.: 1972, Comp. Biochem. Physiol., 41B, pp. 541-550.
Piez, K.: 1981, in The Chemistry and Biology of Mineralized Connective Tissues (ed. Veis, A.) Elsevier North Holland, pp. 3-6.
Pilkington, J.B.: 1969. J. mar. biol. Ass. UK, 49, pp. 857-877.
Poulicek, M., Voss-Foucart, M.F., and Jeuniaux, C.: 1982, unpublished results.
Rudall, K.M.: 1968, in Treatise on Collagen (ed. Gould, B.S.), Academic Press, London and New York, Part A, Vol. 2, pp. 83-138.
Samata, T., Sanguansri, P., Cazaux, C., Hamm, M., Engels, J. and Krampitz, G.:1980, in The Mechanisms of Biomineralization in Animals and Plants (eds. Omori, M. and Watabe, N.), Tokai University Press, pp. 37-47.
Schleuter, R.J., and Veis, A.: 1963, Fed. Proc., 22, p. 478.
Schneider, D.: 1963, in The Lower Metazoa, (eds. Dougherty, E.C., Brown, Z.N., Hanson,E.D., and Hartman, W.D.)University of California Press, pp. 357-371.
Schopf, T.J.M., and Manheim, F.T.: 1967, J. Paleontol., 41, pp. 1197-1225.
Shatton, J., and Schubert, M.: 1954, J. Biol. Chem., 211, pp. 565-573.
Shuttleworth, A., and Veis, A.: 1972, Biochim. Biophys. Acta, 257, pp. 414-420.
Silberberg, M.S., Ciereszko, L.S., Jacobson, R.A., and Smith, E.A.: 1972, Comp. Biochem. Physiol., 43B, pp. 323-332.
Simkiss, K.: 1965, Comp. Biochem. Physiol., 16, pp. 427-435.
Simpson, D.L., Hranisavljevic, J., Davidson, E.A.: 1972, Biochemistry, 11, pp. 1849-1856.
Stanley, S.M.: 1976, Am. J. Sci., 276, pp. 56-76.
Tanaka, S., Hatano, H., and Suzue, G.: 1960, J. Biochem., 47, pp. 117-123.
Termine, J.D.: 1981, in Matrix Vesicles, Wichtig Editore, Milano, pp. 155-189.
Termine, J.D., Kleinman, H.K., Whitson, S.W., Conn, K.M., McGarvey, M.L., and Martin, G.R.: 1981, Cell, 26, pp. 99-105.
Termine, J.D., Belcourt, A.B., Conn, K.M., and Kleinman, H.K.: 1981, J. Biol. Chem., 256, pp. 10403-10408.
Toole, B.P., and Linsenmayer, T.F.: 1977, Clin. Orthop. Rel. Res., 129, pp. 258-278.
Towe, K.M., and Hamilton, G.H.: 1968, Calc. Tiss. Res., 1, pp. 306-318.
Veis, A., Spector, A.R., and Zamoscianyk, H.: 1972, Biochim. Biophys. Acta, 257, pp. 404-413.
Veis, A., Stetler-Stevenson, W., Takagi, Y, Sabsay, B., and Fullerton, R.: 1981, in The Chemistry and Biology of Mineralized Connective Tissues (ed. Veis, A.) Elsevier North-Holland Inc., pp. 377-387.
Voss-Foucart, M.F.: 1968, Comp. Biochem. Physiol., 26, pp. 877-886.
Vranckx, R., and Durliat, M.: 1980, Comp. Biochem. Physiol., 65B, pp. 55-64.
Wada, K.: 1964, Bull. Jap. Soc. Fish., 30, p. 998.

Wainwright, S.A.: 1963, Q. Jl. microsc. Sci., 104, pp. 169-183.
Weinstock, M., and Leblond, C.P.: 1973, J. Cell. Biol., 56, pp. 838-845.
Welinder, B.S.: 1974, Comp. Biochem. Physiol., 47A, pp. 779-787
Weiner, S.: 1976, Aspects of the biochemistry of the organic matrix of extant and fossil mollusks, Ph.D. thesis, California Institute of Technology.
Weiner, S.: 1979, Calcif. Tiss. Int., 29, 163-167.
Weiner, S.: 1981, in The Chemistry and Biology of Mineralized Connective Tissues (ed. Veis, A.) Elsevier North Holland Inc., pp. 517-521.
Weiner, S.: 1982, J. Chromatography (in press).
Weiner, S., and Hood, L.: 1975, Science, 190, pp. 987-989.
Weiner, S., Lowenstam, H.A., Hood, L.: 1977, J. exp. Mar. Biol. Ecol., 30, pp. 45-51.
Weiner, S., and Traub, W.: 1980, FEBS Lett., 111, pp. 311-316.
Weiner, S., and Traub, W.: 1981 in Structural Aspects of Recognition and Assembly in Biological Macromolecules (eds. Balaban, M., Sussman, J.L., Traub, W., and Yonath, A.) Balaban ISS, Rehovot and Philadelphia, pp. 467-482.
Wilfert, M., and Peters, W.: 1969, Z. Morph. Tiere, 64, pp. 77-84.
Young, S.D.: 1971a, Comp. Biochem. Physiol. 40B, pp. 113-120.
Young, S.O., O'Connor, J.D., and Muscatine, L.: 1971b, Comp. Biochem. Physiol., 40B, pp. 945-958.

APPENDIX

Amino acid compositions of the organic matrices of various mineralized hard parts.

	Sea Urchin Tooth[1] *Paracentrotus lividans*		Holothurian Skin Granules[2] *Molpadia intermedia*		Benthic Foraminifer Shell[3] *Heterostigina depressa*	
Mole %	Insoluble	Soluble	Insoluble	Soluble	Insoluble	Soluble
ASP+ASN	12.79	12.40	12.40	13.47	7.83	10.10
THR	2.96	3.97	7.04	9.05	0.78	5.00
SER	5.53	9.98	5.76	7.67	0.26	16.79
GLU+GLN	7.37	7.95	10.83	11.58	9.65	12.89
PRO	5.64	8.31	6.00	6.33	11.83	3.40
GLY	33.18	22.48	10.32	19.19	14.61	15.70
ALA	10.22	13.44	7.48	15.47	11.22	8.79
CYS	0.73	0.10	2.32	-	-	-
VAL	3.24	3.32	7.80	8.30	5.13	2.71
MET	2.29	1.06	1.44	0.09	1.57	0.45
ILE	2.13	1.57	4.89	1.53	4.17	1.99
LEU	3.63	2.61	6.24	2.21	24.78	2.87
TYR	1.62	1.69	2.95	0.47	0.43	1.78
PHE	2.18	1.55	3.28	0.37	1.04	1.15
HIS	1.17	1.38	1.91	1.04	0.52	2.84
LYS	2.63	7.01	3.51	3.14	5.13	13.04
ARG	2.46	1.23	5.70	trace	1.04	0.53
Protein Content % (protein/g. dry shell weight)	0.07	0.76	n.d.	n.d.	0.002	0.06
$\frac{\text{ASP+GLU}^4}{\text{ASP+ASN+GLU+GLN}}$ (mole %)		43				51

Notes to Appendix

n.d. Not determined.

1. The mature, consolidated zone of the tooth was decalcified according to the method of Weiner and Hood (1975).

2. Granules were dissolved in 0.1M HCl following the method of Lowenstam and Rossman (1975). EDTA does not dissolve the mineral. The granules used for the soluble fraction analysis were briefly treated with sodium hypochlorite to release them from the skin tissue and the analysis may be misrepresentative.

3. Tests were used from individuals which had undergone schizogeny and were presumably free of protoplasm. The method of Weiner and Hood (1975) was used for decalcification. The analysis of the insoluble fraction may be imprecise, due to the small quantity of material available.

4. Carboxyl groups were titrated according to the method of Hoare and Koshland (1967). As total soluble fractions were titrated, carboxyl groups from sources other than Asp and Glu could also have been titrated.

CALCIFICATION OF GASTROPOD NACRE

Hiroshi Nakahara
Department of Oral Anatomy, Josai Dental College, Sakado, Saitama Pref., 350-02, Japan

Abstract: The growth surface of the nacreous layer of archaeogastropod shells (5 genera) were studied by means of transmission and scanning electron microscopy. Aragonite crytals forming the stacks of nacre grow within pre-formed, multi-layered compartments, partitioned with horizontally-arranged organic sheets. The organic sheets contain an electron-dense central core which is not observed in bivalve nacre and is probably composed mainly of a chitinous substance. Perforations similar to those observed in interlamellar matrices are observed in the organic sheets. Growing crystals within the compartments are always surrounded by a thin organic envelope which is thought to be involved in the acceleration of crystal growth.

The nacreous layer of mollusc shells is composed of the aragonitic form of calcium carbonate crystals in a structure of successive laminae, and between the crystals are filled with a complex organic matrix. In connection with the calcification of bivalve nacre, Bevelander and Nakahara (1969, 1980) observed compartment formation prior to mineral crystal growth. Later Wise (1970 a,b) proposed that the successive compartments in gastropod nacre might also be present prior to crystal growth. Recently Nakahara (1979, 1981) demonstrated that a similar mechanism of compartment formation indeed takes place during calcification of the nacre in some gastropod species.

It has been shown that two distinct organic materials are involved in the formation of nacre. One is a thin, relatively electron-dense "envelope" which intimately encloses the growing crystals, the other is the "sheet-like" substance appearing within the extrapallial fluid which is arranged parallel to the growth surface and forms "compartments" in which crystals grow (Nakahara 1979, 1981; Bevelander and Nakahara 1980; Nakahara et al. 1982). The present paper introduces recent studies dealing with the formation of the nacreous layer of certain archaeogastropod species, with special regard to the fine structure of the growth surface and decalcified mature nacre.

Materials employed for the study are members of the following

P. Westbroek and E. W. de Jong (eds.), Biomineralization and Biological Metal Accumulation, 225–230.

genera: Monodonta, Calliostoma, Tegula, Turbo, and Haliotis. The nacreous layer of these species shows the formation of conical stacks consisting of piled up aragonite tablets (Fig. 1). Because of the difficulty in preserving the normal relationship between the marginal shell and mantle tissue together, both were fixed separately. After fixation, the structures directly related to shell growth (organic compartments and growing stacks) were attached to the shell side without much disturbance (Figs. 2,7). For ordinary transmission electron microscopic observation, specimens were fixed in cacodylate-buffered glutaraldehyde, post-fixed in O_sO_4, then embeded in Araldite. For SEM observation, the shell surface was treated with 6% sodium hypochlorite for 10 minutes to remove surface organic material (Nakahara et al., 1982).

Sections of shell growth surfaces show regularly spaced organic sheets arranged parallel to the surface, usually at intervals of 0.45 to 0.6 μm. These horizontal sheets are connected with the interlamellar matrix located between adjacent aragonite tablets (Fig. 3); thus the sheets apparently partition part of the extrapallial space into compartments in which the crystals grow (Nakahara 1979, 1981; Nakahara et

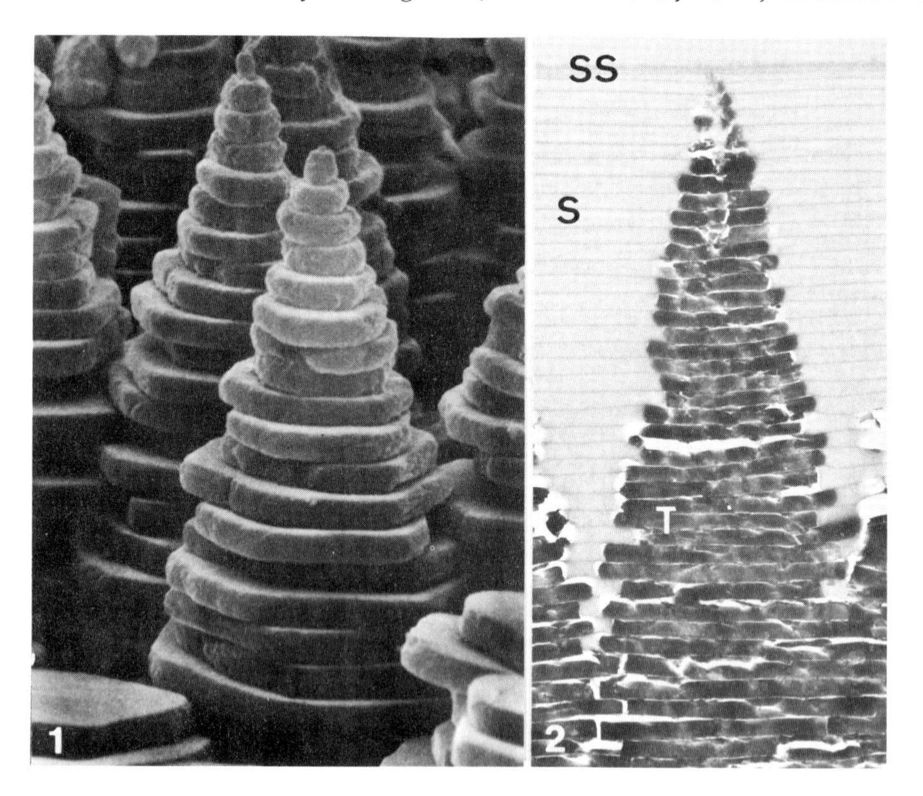

Fig. 1 Growth surface of nacreous layer of Monodonta labio, oblique view. x6,300.

Fig. 2 Unstained section showing growing stack (T) of Monodonta. Note regularly-spaced organic sheets (S) and surface sheets (SS). x3,600.

al. 1982). These sheets appear more prominent than those of bivalves reported by Bevelander and Nakahara (1969, 1980). The sheets in the surface area (near the mantle epithelium) are more narrowly spaced. Two or more sheets, together with an unorganized cluster of dense organic material at the outermost area, form a group of surface sheets which may provide a wall of resistance of disturbances caused by mantle movement (Nakahara, 1979).

One of the well-known features of the interlamellar matrix in calcified nacre is the numerous fenestrations demonstrated in bivalves and Nautilus as well as gastropods (Grégoire et al., 1955). These perforations apparently exist prior to the formation of adjacent aragonite crystals since similar fenestrations are clearly observed in the horizontal sheets of growing surface in sectioned materials cut nearly parallel to the sheets (Fig. 6). The fenestrations seem to provide a passage way for outgoing and incoming substances between the compartments and extrapallial fluid adjacent to the mantle epithelium. Such substances as calcium, carbonate, H_2O, and organic compounds would traverse these openings during the growth of the nacre. These substances must also pass through the surface sheets even though the latter do not clearly show the perforations seen in the regular horizontal sheets. However, higher magnification electron micrographs of surface sheets show that they have a rather spongy appearance (Nakahara, 1979). This type of structure might permit the easier passage of substance and probably only functions as a barrier to large particles.

An electron-dense central core, about 10 nm in thickness, was observed in each organic sheet (about 40 nm in total thickness) in every species examined. This central core apparently persists after treatment in hot alkali solution (unpublished data, Fig. 5). Goffinet (1969) demonstrated that the alkali-resistant fibrillar residue of Nautilus nacre consists of chitin. Further, Peters (1972) observed in several archaeogastropod species a similar fibrous component which appears to be chitin. Consequently, the central core of the organic sheets is probably composed mainly of chitin. Peters also reported that calcified bivalve shells (with the exception of few taxodont species) do not contain detectable amounts of chitin. This fact agrees with the illustrations shown by Bevelander and Nakahara (1969, 1980) that the central core has not been demonstrated in the organic sheets of two bivalve species, Pinctada and Mytilus. Though not quite clearly visible in the section of decalcified mature shell shown in Fig. 4, the central core is evidently retained within the interlammellar matrix and may provide mechanical strength for the mature nacre as well as for the organic structure of the growing surface area.

The number of tablets in each stack varies by species, individual, or position in the shell surface; however, there are usually 15 to 40 (36 tablets were counted in case of Fig. 2). Wise (1970a, b) stated that the mode of stack formation is more advantageous for gastropod nacre in which the growth surface is very limited, because it allows a faster growth rate than that of pelecypod types. In bivalve nacre,

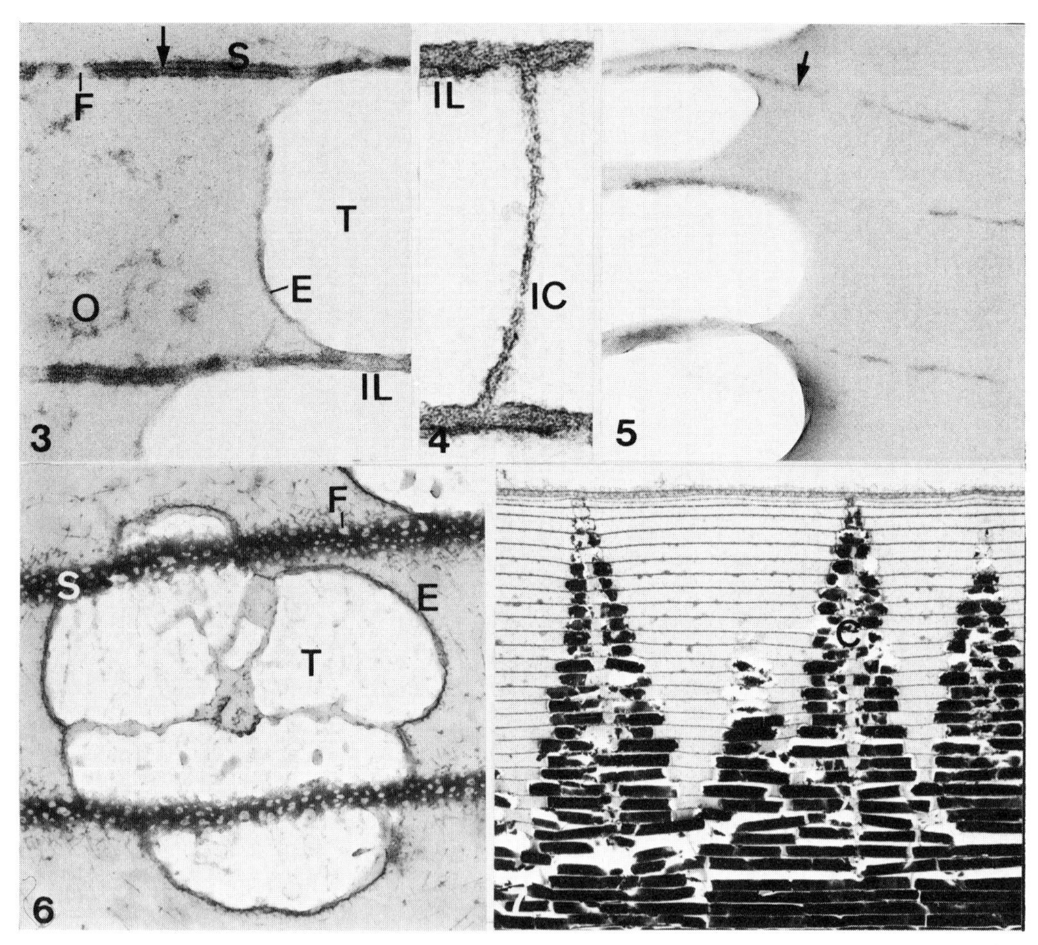

Fig. 3 Part of a growing stack of Monodonta. Calcium carbonate crystals (T) were dissolved due to uranyl acetate-lead citrate double stain. Crystals are covered with envelopes (E). Central cores (arrow) and fenestrations (F) are clearly seen within the organic sheets (S). IL: interlamellar matrix. O: clusters of organic substance. x81,000.

Fig. 4 EDTA decalcified nacre, cut perpendicular to the surface. Double stain. Interlamellar matrix (IL) consists of two electron dense outer layers and an electron lucent inner layer. Turbo cornutus. x117,000. IC: intercrystalline matrix.

Fig. 5 Section of a growing stack of Monodonta treated with 1N sodium hydroxide at 100°C for one hour. Double stain. Note that the central portion of the sheets remain (arrow). x34,000.

Fig. 6 Calliostoma unicum cut nearly parallel to surface. Tablet (T) is divided into sectors. E: envelope, S: Sheets, F: Fenestration. Double stain. x15,000.

Fig. 7 Section of Calliostoma. Note an organic core (C) at the center of a stack. Lead citrate stain. x3,150.

the growth surface construction in most cases has the well-known imbricate or step-like pattern. Since the nacre covers a much wider area in pelecypods, stack formation is unnecessary (Wise, 1970a). The irregular profile of the stacks due to the uneven growth rate of tablets is often observed in scanning electron micrographs (Fig. 1) and sectioned materials (Figs. 2,7). Because of this irregularity, Wise (1970a) stated that crystals make contact in a staggered sequence, assuring a tighter bond between stacks.

According to the previously stated concept (Nakahara, 1979; Bevelander and Nakahara 1980; Nakahara et al. 1980, 1981), the most important organic structure involved in biological mineralization is the "envelope" which is clearly demonstrated in various kinds of mollusc shell structures. The function of the organic envelope is considered to be involved in the induction or accerelation of crystal growth. The envelope is composed mainly of aspartic acid-rich protein which is supposed to be an essential substance for the process of biological mineralization. In the case of gastropod nacre, envelopes always surround the growing crystals within a compartment similar to that observed in bivalve nacre (Fig. 3) and they also have a fine beaded appearance as seen in many other calcified tissues. The compartments contain clusters of scattered organic materials (Fig. 3) which may become part of the envelope as crystal growth progresses. The envelope materials are finally incorporated in mature calcified nacre as three distinct structures: (1) outer layers of the interlamellar matrix (Fig. 4); (2) intracrystalline matrix; and (3) the intercrystalline matrix which exhibits a double layer of beaded material (Fig. 4) similar to that of bivalves and *Nautilus* (Nakahara, 1981).

The tablets have a roughly hexagonal shape in most species, however, in *Calliostoma*, an irregularly curved outline was observed (Fig. 6). Cross sections of stacks in *Calliostoma unicum* show that tablets are subdivided into from two to several sectors (Fig. 6), which is similar to the observation reported by Mutvei (1978). Adjacent sectors are divided by a thin organic wall, and a small mass of organic substance is located at the center of the tablets (Fig. 6,7). A similar central organic structure is also shown in *Monodonta* (Nakahara, 1981) and probably corresponds to what Mutvei (1978) has described as the "central cavity". It is presumed that these organic structures have a important role in the successive nucleation of new tablets within the surface group of sheets; however more investigation is necessary to clarifty this point.

References

Bevelander, G., and Nakahara, H.: 1969. An electron microscope study of the formation of the nacreous layer in the shell of certain bivalve molluscs. Calc. Tiss. Res. 3, pp.84-92.

Bevelander, G., and Nakahara, H.: 1980. Compartment and envelope formation in the process of biological mineralization. In: The Mechanisms of Biomineralization in Animals and Plants. M. OMORI

and N. WATABE (ed.), pp.19-27, Tokai Univ. Press, Tokyo.
Goffinet, G.: 1969. Etude au microscope electronique de structures organisees des constituants de la conchioline de nacre du Nautilus macromphalus Sowerby. Comp. Biochem. Physiol. 29, pp.835-839.
Grégoire, Ch., Dechâteau, Gh., and Florkin, M.: 1955. La Trame protidique des nacre et de perles. Ann. Inst. Océanogr. 31, 1-36.
Mutvei, H.: 1978. Ultrastructural characteristics of the nacre in some gastropods. Zoologica Scripta 7, pp.287-296.
Nakahara, H.: 1979. An electron microscope study of the gorwing surface of nacre in two gastropod species, Turbo cornutus and Tegula pfeifferi. Venus (Jap. J. Malac.), pp.38:205-211.
Nakahara, H.: 1981. The formation and fine structure of the organic phase of the nacreous layer in mollusc shell. in: Study of Molluscan Paleobiology (Prof. Omori Mem. Vol.), pp.21-27. Niigata Univ.,Japan. (in Japanese with English summary).
Nakahara, H., Bevelander, G., and Kakei, M.: 1982. Electron microscopic and amino acid studies on the outer and inner shell layers of Haliotis rufescens. Venus (Jap. J. Malac.), 41, pp.33-46.
Nakahara, H., Kakei, M., and Bevelander, G.: 1980. Fine structure and amino acid composition of the organic "Envelope" in the prismatic layer of some bivalve shells. Venus (Jap. J. Malac.), pp.39:167-177.
Nakahara, H., Kakei, M., and Bevelander, G.: 1981. Studies on the formation of the crossed lamellar structure in the shell of Strombus gigas. Veliger, pp.23:207-211.
Peters, W.: 1972. Occurrence of chitin in Mollusca. Comp. Biochem. Physiol. 41B, pp.541-550.
Wise, S.W.: 1970a. Microarchitecture and mode of formation of nacre (mother-of-pearl) in pelecypods, gastropods and cephalopods. Eclogae geol. Helv. pp.63:775-797.
Wise, S.W.: 1970b. Microarchitecture and deposition of gastropod nacre. Science 167, pp.1486-1488.

ORGANIC MATRICES OF MOLLUSC SHELLS

G. Krampitz, H. Drolshagen, J. Häusle, and
K. Hof-Irmscher

Abteilung für Biochemie, Institut für Anatomie,
Physiologie and Hygiene der Haustiere
Universität Bonn, Katzenburgweg 7-9, D 5300 Bonn 1,
Federal Republic of Germany

Abstract

The organic matrix of mollusc shells includes organic components soluble in aqueous solutions and constituents not soluble in water and organic solvents. The matrix components are regarded to be closely involved in biomineralization processes. The concept of an organic matrix depends on the hypothesis, that macromolecular organic constituents largely control the deposition of solid inorganic matter. The insoluble matrical fraction is composed of varying quantities of macromolecules, and its surface may be hydrophobic. The chemical structure of the insoluble matrix and its biological significance on the molecular level needs more elucidation, although studies on the secondary structure revealed in some cases antiparallel ß-pleated sheet conformation. The soluble matrical fraction from molluscs consists primarily of sulfated, high molecular weight glycoproteins at least some of them selectively bind calcium. Among the soluble biopolymers of the matrix the enzymes polyphenol oxidases have been detected.

1. INTRODUCTION

The organic chemical substances found in mollusc shells as well in calcareous egg shells have generally been described as the organic matrix. This expression does not define the chemical characteristics of the organic molecules being present in shells. However, this word is Latin and describes in this context the function of the uterus. In other words, when this expression was introduced one had in mind that something is produced in or on the organic part namely the crystals forming a shell. Since the early days of biomineralization it has generally been accepted that the organic matrix plays an important role in processes leading to the

P. Westbroek and E. W. de Jong (eds.), Biomineralization and Biological Metal Accumulation, 231–247.

formation of mollusc shells. There is no a priori reason to assume that all components of the matrix have a function in the mineralization process or that mineralization is the only function of the matrix. For example, the organic matrix appears to have a role in determining the mechanical properties of the shell of molluscs (1). The organic substances of mollusc shells can be obtained by dissolution of the mineralized layers of the shell. This procedure yields (47) two fractions, one that is soluble and the other insoluble in aqueous solutions. The content of organic substances of molluscan shells varies with species and is commonly between 0.01 % and 10 % (45). In analyses of shells of 14 species of estuarine bivalves a range from 1.4 % to 21.4 % was found (26). Reported values for a particular species may vary considerably (26). The organic components of mollusc shells may be divided into two regions: a sklerotized, non-mineralized layer covering the outer surface of the shell, termed the periostracum, and the organic material associated with the mineralized layers of the shell, termed the organic matrix.

2. SOLUBLE MATRIX

The soluble matrix has been regarded to be intracrystalline (34, 23, 5), while the insoluble matrix has been thought to be intercrystalline (13). The concentration of soluble matrix components varies widely among species, ranging from 14 % to 64 % (48). The chemical composition of the soluble matrix of molluscan shells seems to be far from being simple with respect of the number of components observed within a single species (18, 38, 40, 21, 30), between species (46, 23, 18, 38, 39, 15) and between different crystal layers of a single bivalve shell (27). The soluble matrices from bivalves and cephalopods are 40 to 80 % proteins. From 60 to 70 % of the amino acids are aspartic acid, serin and glycine. Aspartic acid amounts to about 30 % of the total (34, 36). Aspartic acid appears to be present in the native protein as its amide, asparagine (7, 46). The protein core may be a polymer of $(Asp\text{-}Y)_n$, where Y is predominantely glycine or serine (36).

2.1 Soluble high molecular weight protenaceous substances

The first report on the extraction and at least partial purification of a soluble matrix component came from CRENSHAW (1972). This author isolated a water-soluble highly sulfated glycoprotein of 160.000 daltons (5), which was evenly distributed throughout the shell and which is one primary constituent of the soluble matrix from the clam, Mercenaria mercenaria (aragonite). It has been shown that in this substance are negative by charged groups predominate due to ester sulfate associated with hexosamine residues (46, 47). The predominant soluble compound of the matrix from the oyster Crassostrea virginica (calcite) appears to be similar with that from M. mercenaria having a molecular weight of 170.000 daltons (43). These molecular weights are approximate because the glycoprotein is probably fibrillar. The carbohydrate content also prevents their staining in electrophoretic gels with Coomassie blue, the stain most commonly used to detect proteins.

The amino acid compositions of above mentioned sulfated glycoproteins is very similar to those of insoluble mollusc shell matrices except that the isolated sulfated glycoprotein had a higher aspartic acid content. All the dicarboxylic amino acid residues are present in the protein as their amides. A soluble fraction extracted from Crassostrea virginica shells exhibited an apparent molecular weight of 1 million daltons. Also from Crassostrea irrescens, Mercenaria mercenaria and Nautilus pompilius soluble proteins have been isolated and analyzed for amino acid composition. It has been presumed that a significant proportion of the soluble protein of the organic matrix of molluscan shells is composed of a repeating sequence of aspartic acid by either glycine or serine. This regularly negatively charged aspartic acid molecule may function as a template upon which mineralization occurs (36). However, the amino acid composition of the soluble protein of Mercenaria mercenaria (36) does not correspond with that purified by CRENSHAW (4).

Fossil glycoproteins of the soluble matrix are present in an 80 million year old mollusc shell from the late Cretaceous period. Discrete molecular weight components are preserved. A particular repeating amino acid sequence $(\text{-Asp-Y-})_n$ found in contemporary mollusc shell protein was identified in fossil glycoproteins (37). A more recent study of the amino acid composition of the major subfractions of the soluble matrix of Nautilus repertens, Strombus gigas and Plicatula plicata confirms earlier findings that the proteins of all subfractions are enriched in acidic and polar amino acids. The subfraction which contains the major portion of total protein also contains the highest concentration of aspartic acid (39). Included in this article should be some observations obtained from Crassostrea gigas shells. A very high proportion of C.gigas cultures at the French Atlantic seabord as well as the Bay de Lion (Mediterranean Sea) exhibit symptoms of shell malformations. The shells show depressions at their inner surface which are filled with some kind of a jelly. The jelly contains pro-

teins with an unusual amino acid composition. The predominant amino acid is threonine in contrast to probably all other shell proteins. No calcium-binding of any of the jelly constituents could be obtained. Other constituents of the jelly are carbohydrates as well as phospholipids with polyunsaturated fatty acids.
The jelly is covered with an extreme thin calcium carbonate layer (calcite). The reasons for this phenomenon are still unknown. At some places at the French Atlantic coast it is already sometimes impossible to find shells without jelly formation and chambering. Extracts of the massive shell portion of normal and pathological shells and the calcareous thin layers upon the jelly (paper shells) yielded at least two major fractions. One fraction exhibits an apparent molecular weight of $\hat{=} 5 \times 10^6$ daltons while the major peak has an approximate molecular weight of 100.000 - 500.000 daltons. Probably due to the carbohydrate content no fraction gave a positive reaction with Coomassie blue, but stained positively with PAS reagent. Neither peak could be applied succesfully to PAGE or isoelectrofocussing because of the extreme high molecular weight and the acidic pH of their solution. This material did not penetrate into the gels.

2.2 Smaller Molecular Weight Substances

The soluble fraction may also contain a number of smaller molecular weight components collectively representing up to 10 % of the total protein (38). A comparative study of soluble matrix components with a molecular weight up to 15.000 daltons has been reported earlier (18). Recent studies revealed that e.g. EDTA-shell extracts of Crepidula fornicata also did not contain any soluble high molecular weight components but only those in the $<$ 30 kilodalton range. The amino acid compositions of those substances are very similar to those of the high molecular fractions as well as the insoluble matrices. Amino acid analyses yield usually a limited information on the characteristic of a protein or polypeptide. Another approach has been made for a more detailed characterisation of typical substances of this category. By stepwise degradation as commonly used in amino acid sequence analysis fragments of the most prevailing and homogenous substance from shells of normal Crassostrea gigas exhibited the presence of glutamine and asparagine residues. Similar results were obtained with small polypeptides extracted from fossil oyster shells (Ostrea carrissima, Lopha diluvania, Lopha marshi) which were comparable with those from Crassostrea gigas.

The prevailing amino acid sequence from recent and fossil oyster shells were nearly the same except for the exchange of some amino acid residues probably due to evolutionary process (21, 30). Smaller peptides of the soluble fraction from mollusc shells are much easier to handle rather than those with extreme high molecular weights. However, some small peptides react easily with EDTA which is commonly used for decalcification of the shells. EDTA attached to small peptides cannot be removed by ultrafiltration or dialysis. EDTA-peptides can only be identified by a color reaction with calcium after acid hydrolysis of the complex (24). EDTA-peptides have been excluded from further characterisation in own studies. Some smaller peptides from mollusc shells (EDTA-free), however, also bind Ca^{2+}. Very recently a new aspect of calcium-binding proteins has been revealed (10). In molluscs the shell is produced by the mantle tissue (46). The mantle edge secretes the primary proteinaceous outer layer, the periostracum, resulting in enlargement of the shell at the border (28, 32), whereas the next parts of the mantle are involved in secondary thickening (calcium carbonate deposition and matrix formation). In the fresh water snail Lymnea stagnalis, body and shell growth is controlled by the hormone of the neuroendocrine light green cells (LGC) of the cerebral ganglia (11). In the absence of this molluscan growth hormone, periostracum formation and shell calcification at the edge (the formation of the outer crystalline layer) stop, whereas matrix formation and calcium carbonate deposition an the entire inner surface of the shell continue. Moreover, the calcium concentration in the mantle edge decreases after LGC removal. From the mantle edge of L. stagnalis a calcium-binding protein has been isolated which is specific for this tissue. Removal of the growth hormone-producing cells resulted in a strong decrease of the amount of that calcium-binding protein. It has been concluded that L.stagnalis possesses a hormone dependent calcium-binding protein, probably responsible for the maintenance of a high calcium concentration in that part of the mantle that produces the outer crystalline layer of the shell. Whether the Lymnea calcium-binding protein plays a role as crystallisation nuclei, or acts as an intracellular calcium carrier, or acts as an enzyme modulator, is at present unknown. As far as the known chemical data are reported this specific calcium-binding protein is probably not identical with the calcium-binding proteins referred to in the next section (10).

2.3 Calcium-binding by Soluble Matrix Components

The soluble matrix from Mercenaria mercenaria and Crassostrea virginia selectively bind calcium (5, 43). Binding studies on the protein of M.mercenaria by flow rate dialysis indicate that this material has 0.65 µmoles of high affinity calcium binding sites per mg protein with a dissociation constant of 1 x 10^{-5} M and 1.3 umoles of lower affinity sites per mg protein with a dissociation constant of 3 x 10^{-4} M. The binding parameters were not significantly different over a pH range of 3.2 to 7.4 (8). These values are remarkably similar to those of the soluble, calcium-binding polysaccharide isolated from coccoliths (9). The soluble matrix of C.virginia has a much higher binding capacity of 20 - 25 µmoles calcium/mg protein (43). Recently several soluble high molecular weight fractions from normal and pathological shells as well as "papershells" from C.gigas have been tested for calcium-binding properties. At least two different soluble substances with calcium-binding properties from each extract have been identified. The data of binding studies are comparable with those from M. mercenaria (5). Jelly produced by oysters C.gigas had no affinity to calcium. The calcium-binding properties of the soluble high molecular weight components are specific as demonstrated by CRENSHAW (5) even in the presence of an excess of other cations. The same has been observed with high molecular weight components from C.gigas shells, indicating that even under extreme conditions 100.000 fold excess of other cations) calcium is specifically bound. However, it has recently been reported that the soluble matrix of C.virginia binds calcium, but also prevents or reduces the growth of calcium carbonate crystals (43).
The soluble matrix of nacre may be localized in the pores of the interlamellar matrix (16). When nacre was demineralized with agents that dissolve the soluble matrix, the pore patterns typical of the molluscan class were evident in the interlamellar matrix. The interlamellar matrix was solid, with no pores evident, when the demineralization was done with agents that fixed the soluble matrix as it was exposed (8).

2.4 Possible Roles of Soluble Matrix

Calcium-binding, sulfated, polysaccharide sites were found in the interlamellar matrix when cephalopod nacre was demineralized with low ionic strength solutions containing a quaternary ammonium salt (6). When the quaternary ammonium salt was omitted or solutions of high ionic strength were used, the calcium-binding sites were not detected and the typical cephalopod pore pattern was seen instead. The solubility characteristics of the quaternary ammonium complex, with respect to ionic strength,led to the conclusion that the calcium-binding sites was correlated with the site of initial crystal deposition (8). A recent study of initial deposition of mineral into glass cover slips inserted between the mantle and shell of bivalves yielded results that are consistent with the hypothesis that the soluble matrix is at the site of initial crystal formation (33). Microprobe analysis showed that only those organic granules with a high sulfar content mineralized. Histochemical staining indicated that the specific calcifying granules contained protein very acidic glycosaminoglycans and ester sulfate (33). Several roles for the soluble matrix have been suggested. The only one is that, in solution, it inhibits calcium carbonate precipitation from a solution sufficiently supersaturated to precipitate spontaneously (43). This may be correct for some fractions of the soluble matrix, but not for all (42). The soluble matrix may serve as a template for calcium carbonate nucleation. It has been suggested that the aspartic acid residues bind calcium at distances corresponding to some multiple of the spacing of calcium ions in the crystal plan to provide a nucleation surface. When the interaspartyl distances do not match the crystal, the insoluble matrix may inhibit crystal growth (41). If the matrix protein contains asparagine rather than aspartic acid (7, 19,46) this mechanism is not possible (8). Another proposed function envisages the soluble matrix to act, alternatively, as an inhibitor and as a nucleator in the following manner. The glycoprotein is selectively absorbed into the growing centers of a crystal because the centers have the highest fixed concentration of free calcium, and it inhibits further growth of that crystal. The free portions of the absorbed glycoproteins selectively concentrate calcium and secondarily, carbonate to nucleate a second crystal by the progress of ionotropy (6). Consequently the soluble matrix may have opposite functions, depending upon whether it is in solution or absorbed into a surface.

3. INSOLUBLE MATRIX

Polyphenol oxidases have been found in the soluble matrix of Crassostrea gigas, Strombus aureus (20, 29) and Mercenaria mercenaria (12). Substances for these enzymes as tyrosine and other phenolic substances have been indentified in shell matrices (27, 12). It has been concluded that polyphenol oxidases and phenolic components may play a role in polymerization of the insoluble matrix. In contrast to observations obtained from the soluble fraction of avian egg shells no carbonic anhydrase activity has been detected (22). The insoluble matrix is the classical conchiolin. It is primarily an aggregate of proteinaceous material although a significant polysaccharide content has been detected in many species (13, 23, 41). As much as 70 % of the total nitrogen in this fraction is present as amino nitrogen. It has been reported that glycine and alanine are the most abundant amino acids and amount to 30-60 % of the total amino acid residues (13). However, in comparing of conchiolins of various species and sections of shells other results have been obtained. A large number of the observations made in conchiolin have been emumerations of the amino acids present. Any differences found in the amino acid composition of the conchiolin associated with a particular microstructure or calcium carbonate polymorph are less than the variation in the amino acid composition of that particular conchiolin within a taxonomic group (13). This is not surprising, because conchiolin contains varying amounts of several proteins and polysaccharides. An example of the complexity of conchiolin is furnished by the serial extraction of nacreous conchiolin from cephalopods, bivalves and archeogastropods (13). Nacre is aragonitic and in vertical section appears a brick wall; the crystalline tablets being the bricks and matrix the mortar. When the mineral is dissolved, the vertical intercrystalline matrix collapses upon the horizontal interlamellar matrix to form outlines of the positions of the original polygonal crystals. The interlamellar conchiolin forms lacelike network with pore patterns that are characteristic of the molluscan class (13). When it is extracted with borate, nareous conchiolin retains its lacelike structure. The extract contains about 24 % of the total nitrogen, and 30 % of the amino acids are glycine and alanine. A second extraction with 5 % sodium hydroxide removes 71 % of the total nitrogen. Glycine and alanine amount to approximately 55 % of the total amino acid residues in this extract. The residue is fibrous and contains 6 % of the total nitrogen. As much as 35 % of this residue is chitin. Over 80 % of the amino acids are glycine and alanine. While the significance of the observations in terms of calcium carbonate mineralization is not known,they confirm earlier indications that conchiolin is a refractory mixture and show that total amino acid analysis of conchiolin may be of limited value in the determination of any role conchiolin may have in mineralization. The insoluble matrix has cross-linked proteins which may be quinone-tanned as indicated by the high tyrosine content (3, 23), indicating that the polyphenol oxidases may crosslink proteins after they have been secreted by the mantle.

Examination of the X-ray diffraction patterns given by the the insoluble matrix revealed that either an α or ß-keratin pattern was obtained, depending on the species studied (44). Also the molecular conformation of proteins present in the mollusc shell and those of the decalcified films of the nacreous layer were measured for the regions of 4000-700cm^{-1}. The decalcified films were found to consist primarily of proteins. The structure of proteins constituting the decalcified films of the nacreous layer is coiled and has also a ß-conformation (14). More recently X-ray diffraction patterns of the proteinaceous components of insoluble matrices from nacreous shell of species representing the three major orders of molluscs shows an antiparallel ß-pleated sheet conformation similar to that of silk fibroin (41). The polysaccharide components, where present, exhibits a structure equivalent to that of chitin. Protein of the insoluble matrix of prismatic layers of some species, however, show no well oriented pattern (47).
As coincidence of matrix axis with crystallographic axis was found only in the case of the c-axis of aragonite in one cephalopod (25). Such coincidence was observed earlier an was considerated to be due to the growth direction and movement of the mantle (13). Therefore, the alignment of the two axes may not be directly related, and the orientation of the c-axis of the crystal may not be determined by any orientation in the matrix. However, the X-ray diffraction studies are incomplete, and they may yield important insight into the possible role of conchiolin in calcium carbonate mineralization and crystal orientation.
The formation of initial calcium carbonate crystals on the surface of the organic matrix has been observed in many organisms (35). However, the characteristics of this surface have not been investigated. The striking feature from the amino acid analyses of conchiolin (13) is the preponderance of amino acids having hydrophobic side chains (31). The hydrophobicity of conchiolin may be greater than indicated by the amino acid composition because the hydrophilic effect of the peptide bound may be nullified by the formation of interpeptide bonds formed in its ß-sheet structure (31). Conchiolin stains with Sudan Black, a typical lipid stain (4). However, exhaustive extraction of the matrix with chloroform-methanol, acidified chloroform-methanol or with diethyl ether did not reduce this staining nor could any lipids be detected in the extracts (8). Therefore, the Sudan Black staining, which depends upon partition rather than upon a chemical reaction, appears to provide additional evidence that conchiolin is hydrphobic. The hydrophobic nature of coralline aragonite suspensions (17) may also be due to a hydrophobic matrix. The possible hydrophobic nature of the insoluble matrix is not directly documented; however, this characterization of this surface is

of paramount importance in deducing any possible role that it may have in calcium carbonate mineralization. Recent studies revealed that conchiolin from oyster shells (C. gigas) has a concentration effect on calcium. This concentration effect may not be identical with calcium binding properties as has been reported for soluble fractions of the matrix. However, it is remarkable that conchiolin concentrates calcium in a microenvironment in the range of 150 µg calcium/mg conchiolin. The affinity for calcium is low, since calcium can be removed from conchiolin by washing with water. It is not clear why a concentration effect occurs while the pH of a conchiolin suspension is in the alkaline range (8.3 - 9.4). Perhaps Ca-complexes are formed in the alkaline range which are comparable with mechanism of EDTA. The concentration of calcium attracted by conchiolin is not prevented in the presence of an excess of other cations. Even Sr^{2+} or Ba^{2+} and La^{3+} only decrease slightly the concentration of Ca^{2+} attached to the insoluble matrix using an unrealistic excess of 5 x 10^5. On the other hand in equilibrium dialysis conchiolin also concentrates HCO_3^{1-} probably due to the alkaline pH of the suspension. Other competing anions did not reduce the concentration of bicarbonate drastically, even in a high excess, except PO_4^{3-}, which prevented the enrichment of HCO_3^{1-}, however, in an excess of 5 x 10^5. It is remarkable that conchiolin of the bivalve C. gigas simultaneously and specifically attracs both Ca^{2+} and HCO_3^{1-}. Similar observations have been made using conchiolin from gastropods (Crepidula fornicata, Patella vulgaris), corals (Acropora formosa) as well as from avian egg shells (Gallus domesticus). The significance of these observations is not clear as yet, however, the simultaneous concentration of Ca^{2+} and HCO_3^{1-} in an alkaline milieu is consistent with the conditions under which $CaCO_3$ crystals are formed. When the insoluble matrix (A. formosa) is treated stepwise by various solutions (e.g. borate, urea, NaOH) at least groups of different proteins can be extracted. The individual extracts e.g. borate solutions followed by 8M urea and 5 % NaOH yielded high molecular weight substances as controlled by gel filtration. The average apparent molecular weight of most of the proteins is in the range of 5 x 10^6 daltons, except for a group of proteins obtained by NaOH extraction which have apparent molecular weight in the range of 100.000 to 500.000 daltons. It cannot be ruled out that those fractions are breakdown products caused by NaOH treatment.

3.1 Functions of the Matrix

Three functions of matrix presumed but not definitively are crystal nucleation, limitation of crystal growth and crystal orientation. A fourth function, the binding together of crystals and crystalline layers, is important in the physical behaviour of the shell (47).

3.2 Crystal Nucleation

The weight of available evidence indicates that calcium carbonate mineralization occurs on the surface of the organic matrix and that the matrix may have a dominant role in mineralization. The mechanisms by which the matrix may induce and controll crystallization may not be so specific as epitaxy, which requires more order of the matrix than has been reported. Ionotrophic nucleation is proposed as an alternative mechanism. Postnucleation crystal growth may be controlled by the matrix and the bathing fluid. The following description of the ionotrophic hypothesis is made in terms of molluscan nacre, although it is geenerally applicable. The insoluble matrix has hydrophobic surfaces cannot nucleate mineral, but they may exert some control over post nucleation growth. The soluble, calcium-binding matrix is fitted into the pores of the soluble matrix. The total assembled matrix presents a hydrophobic field interrupted by specifically placed hydrophilic, calcium-binding sites. These sites bind calcium and concentrate additional calcium from the parent fluid by the Donnan effect (2). A secondary layer, specifically enriched in carbonate, is formed around the calcium-enriched sites. A local concentration change, such as an increase in pH or a decrease in carbon dioxide occurs, and a crystal seed is formed at the site. Some of the seeds grow and others dissolve. The survival of a seed may be determined by local transport processes or by the original, local concentration of the calcium-binding matrix. The initial, rapid growth of the crystal is along the vertical c-axis. Growth along the c-axis is stopped by the next episodic deposition of a complete matrix assembly. The calcium-binding sites of this matrix assembly register on the growth centers of the crystal, such as slip or screw dislocations, because these growth centers have the highest localized concentration of free and loosely bound calcium. The horizontal growth of the crystal continues, albeit at a slow rate, until the space between adjacent growing crystals is filled.

3.4 Limitation

The organic matrix has been hypothesized to limit crystal growth through periodic deposition of a uniform thickness of matrix in which a crystal layer growth (46); by uncontrolled sclerotization of the surface of the matrix layer (46), or by adsorption or binding of newly deposited matrix on the growing crystal lattice and so preventing the addition of further Ca^{++} (6). It has been suggested that if the distance between negatively charged groups of the matrix is different from the Ca^{2+} - Ca^{2+} distance in the crystalline lattice of $CaCo_3$, then calcium-binding by the negative groups could inhibit further crystal growth. This would presumably require an alteration of the matrix structure by a change within the microenvironment of shell formation or secretion of altered matrix.

3.5 Crystal Orientation

WEINER and TRAUB (40, 41) have demonstrated a well-defined spatial relationship exists between the protein, chitin and aragonitic crystals of the shell of Nautilus repertus. The a-axis of the crystals is well lined up with the chitin, b-axis and the b- and c-axis lie along the b- and c-axis of the protein present in the ß-sheet conformation.

4. CONCLUDING REMARKS

To summarize, recent investigations have demonstrated the complexity of the organic matrix, particularly the soluble matrix. It may be that certain components which bind calcium will also be found to be involved in crystal nucleation and the inhibition of crystal growth. These or other components may polymerize to form the insoluble matrix. The temporal patterns of secretion and deposition of the soluble and insoluble matrix components may be important to follow in elucidating the mechanisms of shell formation.

Acknowledgement

We wish to express our gratitude to the Deutsche Forschungsgemeinschaft for financial support (Kr 119/26).

References

1) ABELL, A.K., CRENSHAW, M.A., and TURNER, D.T. 1981. Limiting hardness of polymer/ceramic composites.In: Biomedical and Dental Application of Polymers,ed. F.K.KOBLITZ, pp.347-355. Oxford: Pergamon Press

2) ADAMSON, A.W.,1976. Physical Chemistry of Surfaces,New York: John Willy and Sons

3) BEEDHAM, G.E.,and OWEN, G. 1965. The mantle and shell of Solemya parkinson. Proc.Zool.Soc.London 145:405-430

4) CRENSHAW, M.A.,and HEELY, J.D. 1967 Sudanophilia at sites of mineralization in molluscs. J.Dent.Res.49B:65

5) CRENSHAW, M.A. 1972. The soluble matrix from Mercenaria mercenaria shell. Biomineralisation 6: 6-11

6) CRENSHAW, M.A.,and RISTEDT, H., 1976. The histochemical localization of reactive groups in the septal nacre from Nautilus pompilius L. In: Mechanisms of Mineralization in the Invertebrates and Plants; eds. N. WATABE and K.M. WILBUR, pp.335-367. Columbia: University of South Carolina Press

7) CRENSHAW, M.A. 1980. Mechanisms of shell formation and dissolution. In: Skeletal Growth of Aquatic Organisms; eds. D.C.RHOADS and R.A.LUTZ, pp.115-132. New York: Plenum Press

8) CRENSHAW, M.A. 1982. Mechanisms of normal biological mineralization of calcium carbonates. In: Biological Mineralization and Demineralization; ed. G.H. NANCOLLAS,pp. 243-257; Dahlem Konferenzen, Springer,Berlin-Heidelberg-New York

9) DE JONG, L.W., DAM, W., WESTBROEK, P., and CRENSHAW,M.A. 1976. Aspects of calcification in Emiliania huxley. In: Mechanisms of Mineralization in the Invertebrates and Plants; eds. N. WATABE and K.M. WILBUR, pp.135-153. Columbia: University of South Carolina

10) DOGTEROM, A.A.,and DODERER, A.1981. A hormone dependent calcium-binding protein in the mantle edge of the freshwater snail Lymnea stagnalis. Calcif.Tissue Int. 33: 505-508

11) GERAERTS, W.P.M. 1976. Control of growth by the neurosecretory hormone of the light green cells in the freshwater snail Limnea stagnalis. Gen.Comp.Endocrinol. 29: 69-71

12) GORDON, J., and CARRIKER, M.R. 1980. Sclerotized protein in the shell matrix of a bivalve mollusc. Mar.Biol. 57: 251-260

13) GRÉGOIRE, C. 1972. Structure of the molluscan shell. In: Chemical zoology. VII. Mollusca; ed. FLORKIN, M., and SHEER, B.T., pp. 45-102. New York: Academic Press

14) HOTTA, S. 1969. Infra-red spectra and conformation of protein constituting the nacreous layer of molluscan shell. Earth Sci. Tokyo 23: 133-136

15) HUNT, S. 1970. Polysaccharide-protein complexes in invertebrates; 329 pp. London: Academic Press

16) IWATA, K. 1975. Ultrastructure of the conchiolin matrices in molluscan nacreous layer. J. Fac.Sci.Hokkaido Univ.Ser. 4 17: 173-229

17) KITANO,Y., KANAMORI, N., and YOSHIOKA, S. 1980. Aragonite to calcite transformation in corals in aquatic environment. In: The Mechanisms of Biomineralization in Animals and Plants; eds. OMORI, M., and WATABE, N., pp. 269-278. Tokyo:University Press

18) KRAMPITZ, G., ENGELS, J., and CAZAUX, C. 1976. Biochemical studies on water-soluble proteins and related components of gastropod shells. In: The Mechanisms of Mineralization in the Invertebrates and Plants; ed. WATABE,N., and WILBUR, K.M., pp. 355-367. Columbia, South Carolina: University of South Carolina Press

19) KRAMPITZ, G., ENGELS, J., HAMM, M., KRIESTEN, K., and CAZAUX, C. 1977. On the molecular mechanism of the biological calcification. 1. Ca-ligands from gastropod shells, egg shells and uterine fluid of hens. Biomineralization Res.Rep. 9: 59-72

20) KRAMPITZ, G., and WITT, W. 1979. Biochemical aspects of biomineralization. Topics in Current Chemistry 78, 57-144.

21) KRAMPITZ, G. 1980. Calcium-binding proteins in mollusc shells. Haliotis 10: 82

22) KRAMPITZ, G. 1982. Structure of the Organic Matrix in Mollusc Shells and Avian Egg Shells. In: Biological Mineralization and Demineralization; ed. G.H. NANCOLLAS, pp. 219-232, Dahlem Konferenzen, Springer: Berlin-Heidelberg - New York

23) MEENAKSHI, V.R., HARE, P.E., and WILBUR, K.M. 1971. Amino Acids of the organic matrix of neogastropod shells. Comp.Biochem.Physiol. 40B: 1037-1043

24) MEISEL, H. 1982: Molekularmechanismen der Biomineralisation: Biochemische Untersuchungen an Ovocalcin. Dissertation, University of Bonn

25) PLUMMER, L.N., PARKHURST, D.L., and WIGLEY, T.M.L. 1979. Critical review of the kinetics of calcite dissolution and precipitation. In: Chemical Modelling in Aqueous Solutions; ed. E.A. JENNE, pp. 537-573. Washington: American Chemical Society

26) PRICE, T.J., THAYER, G.W., LA CROIX, M.W., and MONTGOMERY, G.P. 1976. The organic content of shells and soft tissues of selected estuarine gastropods and pelecypods. Proc.Nat.Shellfish Assoc. 65: 26-31

27) RAVINDRANATZ, M.H., and RAVINDRANATZ, M.H.R. 1974. The chemical nature of the shell of molluscs: I. Prismatic and nacreous layers of a bivalve Lamellidans marginalis (Unionidae). Acta histochem. 48: 26-41

28) SALEUDDIN, A.S.M. 1976. Ultrastructural studies on the structure and formation of the Periostracum in Helisoma (Mollusca). In: eds. N. WATABE, and K.M. WILBUR. The Mechanism of Mineralization of the Invertebrates and Plynts; pp. 309-337, University of South Carolina Press, Columbia

29) SAMATA, SANGUANSRI, P., CAZAUX, C., HAMM, M., ENGELS, J., and KRAMPITZ, G. 1980 Biochemical studies on components of mollusc shell. In: The Mechanisms of Biomineralization in Animals and Plants; eds. OMORI, M., and WATABE, N.; pp. 37-47. Tokyo: Tokai University Press

30) SAMATA, T., and KRAMPITZ, G. 1982. Ca^{2+}-binding polypeptides in oyster shells. Haliotis (in press)

31) TANFORD, C. 1980. The Hydrophobic effect: Function of Micelles and Biological Membranes. New York: John Wiley and Sons

32) TIMMERMANS, L.P.M. 1969. Studies on shell formation in molluscs. Neth. J. Zool. 19: 417-523

33) Wada, K. 1980. Initiation of mineralization in bivalve molluscs. In: The Mechanisms of Biomineralization in Animals and Plants; eds. M. OMORI, and N. WATABE; pp.79-92. Tokyo: Tokai University Press

34) WATABE, N 1965. Studies on shell formation. XI Crystal-matrix relationships in the inner layers of mollusc shells. J. Ultrastruct.Res. 12: 351-370

35) WATABE, N.1981. Crystal growth of calcium carbonate in invertebrates. In: Progress in Crystal Growth and Characterization. Vol.4; ed. B. RAMPLIN (in press) Oxford: Pergamon Press

36) WEINER, S., and HOOD, L. 1975. Soluble protein of the organic matrix of mollusc shells: A potential template für shell formation. Science 190: 987-989

37) WEINER, S., LOWENSTAM, H.A., and HOOD,L. 1976. Characterization of 80 million year old mollusc shell proteins. Proc.Nat.Acad.Sci. (Wash.) 73: 2541-2545

38) WEINER, S., LOWENSTAM, H.A., and HOOD, L. 1977. Discrete molecular weight components of the organic matrices of mollusc shells. J.Exp.Mar.Biol.Ecol. 30: 45-31

39) WEINER, S. 1979. Aspartic acid rich proteins: Major components of the soluble organic matrix of mollusc shells. Calc.Tiss.Int. 29: 163-167

40) WEINER, S., and TRAUB, W. 1980. X-ray diffraction study of the insoluble organic matrix of mollusc shells. FEBS Letters 111: 311-316

41) WEINER, S., and TRAUB, W 1981. Organic matrix-mineral relation-ship in mollusc-shell nacreous layers. In: Structural aspects of Recognition and Assembly in Biological Macromolecules; eds. M. BALABAN, J. SUSSMAN, A. YONAT, and W. TRAUB, pp. 467-482. Glenside: International Sciences Services

42) WESTBROEK, P. 1982. Personal communication.

43) WHEELER, A.P., GEORGE, J.W., and EVANS, C.A. 1981. Control of calcium carbonate nucleation and crystal growth by soluble matrix of oyster shell. Science 212: 1397-1398

44) WILBUR, K.M., and WATABE, N. 1963. Experimental studies on calcification in molluscs and the algae Coccolithus huxleyi. Ann. New York Acad.Sci. 109: 82-112

45) WILBUR, K.M., and SIMKISS, K. 1968. Calcified shells. In: Comprehensive Biochemistry 26 A; ed. FLORKIN, M., and STOTZ, E.H.; pp. 229-295, New York: Elsevier

46) WILBUR, K.M. 1976. Recent studies of invertebrate mineralization. In: Mechanisms of Mineralization in the Invertebrates and Plants; eds. N. WATABE, and K.M. WILBUR; pp. 79-108. Columbia: University of South Carolina Press

47) WILBUR, K.M., and MANYAK, D.M. 1981. Biochemical aspects of shell mineralization. ONR Symposium on Marine Biodeterioration (in press)

IMMUNOLOGICAL STUDIES ON MACROMOLECULES FROM INVERTEBRATE SHELLS
- RECENT AND FOSSIL -

P. Westbroek, J. Tanke-Visser, J.P.M. De Vrind, R. Spuy,
W. van der Pol and E.W. De Jong
Department of Biochemistry, State University of Leiden,
Wassenaarseweg 64, 2333 AL Leiden, The Netherlands

The study of macromolecules that are contained in biominerals is relevant for our understanding of the biomineralization process. Moreover, comparative investigations are likely to give valuable systematic information. In our laboratory it was found on two occasions that the antigenic properties of soluble macromolecules from 70 million year old cephalopod shells are still preserved (1,2). Also biochemical analyses carried out elsewhere have revealed that macromolecules from calcified tissues may be preserved over considerable periods of geological time (3-5). Because they are protected from degradation by the mineral phase, these macromolecules are an attractive, albeit hardly explored, source of paleontological and geological information. Both our previous findings and the results of the present study support the view that immunology may become an important tool for the retrieval of the information contained in the macromolecules from biominerals, both fossil and recent. The present investigation concentrates on recent shells of invertebrates and is mainly concerned with the systematic specificity of such immunological reactions. Evidence is presented that certain antigenic determinants are only recognized in a limited group of closely related taxa, while others appear to be very conservative from an evolutionary point of view: they were detected in different classes, or even phyla. Significant reactions were also obtained with fossil materials, so that our earlier observations were further substantiated.

EDTA-soluble macromolecules were extracted from shells of the heterodont bivalve mollusc *Ensis ensis*, as described elsewhere (6). The extract was fractionated by gel filtration on an Ultrogel AcA 44 (LKB) column (for more details see 6). Three fractions were collected: E1, E2 and E3, two of which, E1 and E3 were well separated. Antisera prepared in rabbits against the latter two fractions were used in this study. The IgG-antibodies of the antisera (aE1 and aE3) and pre-immune sera of the same rabbits were purified by ammonium sulphate precipitation, followed by passage through a column of DE22 cellulose (Whatman Ltd.) (7).

The Enzyme Linked Immuno Sorbent Assay (ELISA) as described by Clark and Adams (7) was slightly adapted. Small, mechanically cleaned

P. Westbroek and E. W. de Jong (eds.), Biomineralization and Biological Metal Accumulation, 249–253.

fragments of shell material (about 3 mm in diameter) were etched in 10% EDTA (pH 8.0) for 15 min, washed and placed in the wells of microtiter plates (flat bottom microelisa plates, M 129 A, Dynatech). They were incubated during 1 h at 37°C with a solution of antibodies (0.01 mg/ml) and thoroughly washed. Then the fragments were incubated in the presence of goat anti-rabbit IgG alkaline phosphatase conjugate (Sigma) (1 h, 37°C; dilution 1: 1,000) and subsequently rinsed. The preparations were then incubated with a p-nitrophenyl phosphate solution as a substrate (7). In order to arrest the phosphatase reaction 3 N NaOH was added; the shell fragments were removed from the wells and the adsorption of the remaining stained solution was measured automatically at 405 nm with a Titertek Multiskan photometer (Flow Laboratories). Each of these experiments was carried out in duplicate and blank reactions using antibodies from the pre-immune sera were performed in the same run. In one day more than one hundred reactions could be carried out. Most of the (duplicate) reactions were repeated several times.

We assume that the intensity of the reactions in these experiments depended on the number of determinants exposed on the surface of the etched shell fragments. This is likely to depend not only on the phyletic distance of the investigated species from *Ensis ensis* but also on the size of the exposed etched shell surface and on shell structure. Minor variations in the experimental conditions were found in some cases to have a considerable effect on the outcome of the experiments (evidence not shown). All these factors together may account for the considerable standard deviations that were obtained in a number of cases (Figure 1). At this stage the method can only be considered as semi-quantitative.

The EDTA-soluble macromolecular extract of *Ensis ensis* was found by Krampitz and his collegues to contain at least 20 different molecular species (Krampitz, personal communication). Elsewhere we report evidence showing that both E1 and E3 are very complex macromolecular mixtures, and that aE1 and aE3 are polyvalent antibody preparations (6).

The results of the immuno assays are shown in Figure 1. In most cases the blank reactions were very weak. From this it can be concluded that significantly stronger reactions, obtained with aE1 and aE3, resulted from specific immunological interactions.

It is evident from the figure that, in so far as the reactions with recent shells are concerned, aE3 is much more specific than aE1. The former antibody preparation, directed against the relatively small-sized macromolecules (6), gave only significant reactions with a number of representatives of the Heterodonta, in particular the Solenacea and the Tellinacea. Both reagents gave by far the strongest reactions with *Ensis ensis* itself, probably owing to the presence of antibodies directed against antigens with a very narrow systematic range. It is likely that many of the antigenic determinants that are responsible for the specific reactions represent relatively novel acquisitions of evolution. The possibility cannot be ruled out, however, that these sites also occur in more distant taxa, but that they were not sufficiently exposed during the etching procedure in order to be recognized by the antibodies.

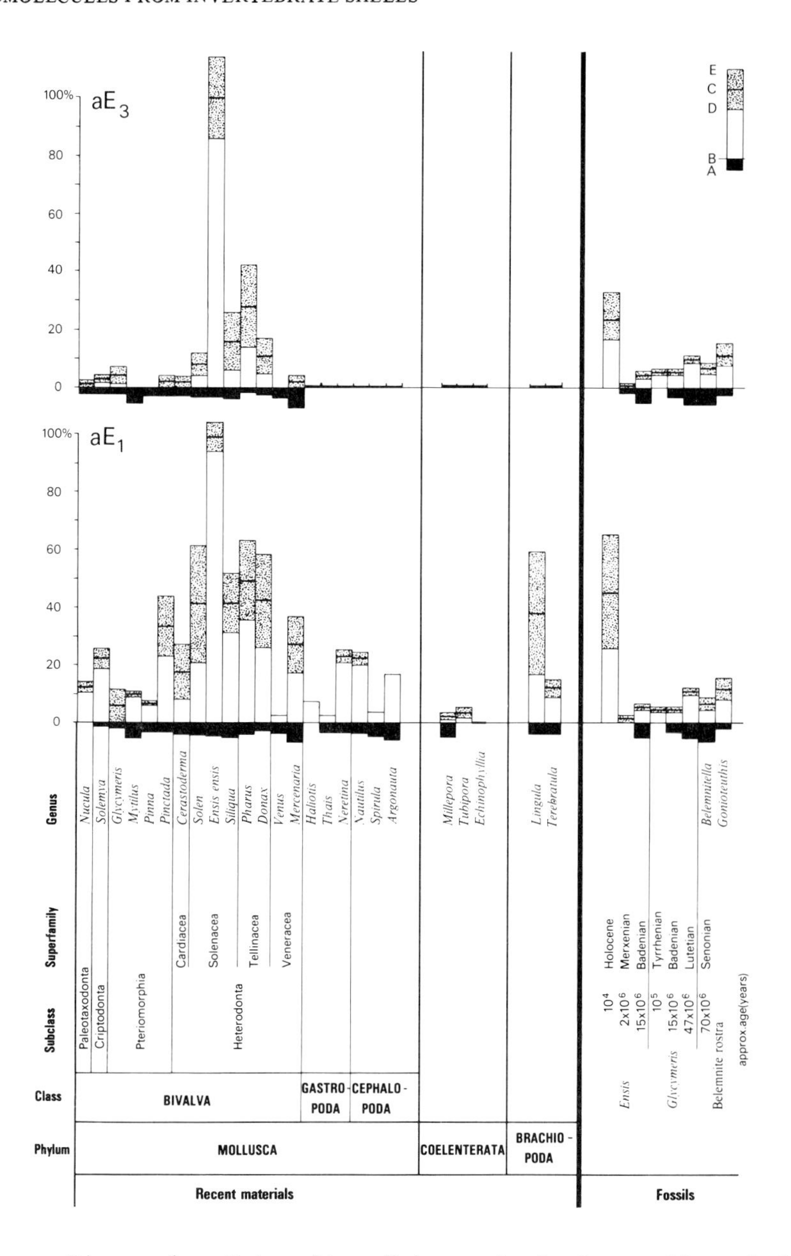

Figure 1. Intensity of immunological reactions between antibody preparations (aE1 and aE3) directed against different mixtures of soluble macromolecules extracted from shells of the bivalve *Ensis ensis* (E1 and E3, respectively), and etched shell fragments of recent and fossil invertebrates. For experimental details see text. Insert: AB, blank reactions with antibodies from pre-immune serum; BC, mean intensity of reaction with aE3 (top) or aE1 (bottom), minus blank reaction, expressed as percentage of mean reaction with *Ensis ensis*; CD and CE: standard deviation.

Positive reactions with a very broad variety of taxa were obtained with aE1 (directed against relatively large-sized macromolecules). Even with representatives of the gastropods and the cephalopods significant reactions occurred. The affinity of the antibodies with skeletal macromolecules of coelenterates was found to be very weak, or even absent. Very striking is a strong reaction that was found between aE1 and *Lingula* (Brachiopoda). The determinants that are responsible for these broad-range, but definitely immunological reactions must be very conservative structures that have undergone little change since the beginning of the Phanerozoic or even the late Precambrian. Possibly, they play an essential role in calcification.

Significant but not very strong reactions were obtained with fossil *Ensis* and *Glycymeris* and with belemnite rostra (Figure 1). Both *Ensis* and *Glycymeris* shells consist of aragonite and their fine structure is easily destroyed by diagenesis. Many invertebrates have shells of calcite, the most stable form of calcium carbonate, and these materials are likely to give a better preservation of the macromolecules that are contained in them. Systematically, the Belemnoidea are remote from *Ensis*; they represent the oldest materials used in this study (70 myr). It is very striking, however, that both aE1 and aE3 gave significant reactions with *Gonioteuthis* sp. In the past, excellent results were obtained with the same material using antisera directed against macromolecules from *Nautilus* shells (2). It is remarkable that aE3 reacts with a fossil cephalopod, while experiments with recent shells suggest that its affinity is limited to some heterodont bivalves. Possibly, certain relevant determinants have a broad systematic distribution, but are not accessible for the antibodies in fresh shells; they may become exposed upon diagenesis.

In future research, antibody preparations directed against conservative determinants may be used for the study of systematic relations between distant taxa, *e.g.* at the class, or even the phylum level. On the other hand, reagents with a small systematic range might be suited for more refined phylogenetic analysis. Our previous finding that antigenic properties of macromolecules from calcified tissues may be preserved over prolonged periods of geological time is further substantiated by the present data. It seems to be worth investigating whether an immunological approach can be successfully applied in the study of geological and paleontological problems, such as the systematic position of fossil taxa *incertae sedis*, and the recognition *in situ* of fossil mecroorganisms.

As a first step towards such applications of immunology the production is advocated of reagents that only interact with one or a few determinants. A particularly promising avenue seems the use of monoclonal antibodies. Although the immuno assay used in this work is superior in many respects to the more cumbersome Ouchterlony technique that was applied in earlier investigations (1,2), improvements with respect to sensitivity and reproducibility are needed.

The shells that were used for these studies were obtained from various sources. Fresh shells of *Ensis ensis* were collected at the sea shore of Nieuwpoort (Belgium). *Glycymeris*, *Mytilus*, *Cerastoderma* and *Mercenaria* were purchased alive at a wholesale firm and *Belemnitella* was collected for us by P.J. Felder, Museum of Natural History, Maastricht (The Netherlands). The other materials came from museum collections and were kindly provided by Drs. E. Gittenberger, J.C. Den Hartog and M.R.R.B. Wijsman (State Museum of Natural History, Leiden, The Netherlands), Dr. A.W. Janssen, State Museum of Geology and Mineralogy, Leiden, and Dr. M.K. Howarth, British Museum of Natural History (London). We thank Dr. J.W. Bruning and Dr. J. Beijersbergen for valuable suggestions. The biotechnical work for this research was carried out by F. Leupe and his staff (Department of Medical Microbiology, State University of Leiden). This Research was supported by the Netherlands Foundation for Earth Science Research (AWON) with Financial aid from the Netherlands Organization for the Advancement of Pure Research (ZWO).

REFERENCES

1. De Jong, E.W., Westbroek, P., Westbroek, J.F., and Bruning, J.W.: 1974, Nature 252, p. 63.
2. Westbroek, P., Van der Meide, P.H., Van der Wey-Kloppers, J.S., Van der Sluis, R.J., De Leeuw, J.H., and De Jong, E.W.: 1979, Paleobiology 5, pp. 151-167.
3. Weiner, S., Lowenstam, H.A., and Hood, L.: 1976, Proc. Natl. Acad. Sci. USA 72, pp. 2541-2545.
4. Krampitz, G., Weise, K., Potz, A., Engels, J., Samata, T., Becker, K., Hedding, M., and Flagis, G.: 1977, Naturwissenschaften 64, p. 583.
5. Samata, T.: 1979, Ph.D. Thesis, Bonn, West Germany.
6. Westbroek, P., Spuy, R., Tanke-Visser, J., De Vrind, J.P.M., and De Jong, E.W., in preparation.
7. Clark, M.F., and Adams, A.N.: 1977, J. Gen. Virol. 34, pp. 475-483.

METHODOLOGICAL ASPECTS OF THE ULTRASTRUCTURAL ANALYSIS OF THE ORGANIC AND MINERAL COMPONENTS IN MOLLUSC SHELLS

J.P. Keller and Y. Dauphin
Laboratoire de Pétrologie sédimentaire et Paléontologie,
Université d'Orsay, 91405 Orsay Cedex, France
(ERA 765 du CNRS).

1. RESEARCH ORIENTATION

The purpose of this study is to specify the characteristics of the mineral phase in carbonate tests of several modern and fossil invertebrate taxa. This requires an amelioration of present knowledge on topographic and biochemical relationships between organic and mineral components. There are double implications to the present study:

- to specify the value of the approximations presented by various authors (1, 2) based on the microstructural aspects of the tests, and to propose the homologies established on the identity of biomineralization processes;
- from a paleontological and sedimentological point of view, this research is indispensable for specifying the modalities of fossilization, that is to say an analysis of the diagenetic evolution of organic and mineral phases in various categories of burial environments.

2. METHODS AND EXAMPLES OF APPLICATION

Among the various techniques used together, two are illustrated here.

2.1. Enzymatic Proteolysis

In the qualitative approach towards the organic matrix-crystal elements relationships, the demonstration of these relationships, without affecting their structure, depends upon a carefully controlled destruction of the matrix with a technique that does not cause carbonate dissolution. One of the most effective ways is enzymatic proteolysis (3).

Thus *Pinna* prisms which appear homogeneous and compact in a simple break (Fig. 1), show in longitudinal section after pronase treatment, certain fine-grained mineral elements, presenting at times oblique lamella in relation to the growth axis (Figs. 2-5). Furthermore, the

P. Westbroek and E. W. de Jong (eds.), Biomineralization and Biological Metal Accumulation, 255–260.

importance of the volume of the organic matrix is demonstrated (4, 5).

In the Trochidae shell, four layers were defined (6, 7): the distal (outer part), outer prismatic, nacreous and internal or oblique prismatic. Among these layers, the outer prismatic layer of *Monodonta* is presented here for comparative biomineralization purposes. Between the distal layer and the nacreous layer, the units vertically oriented and made of successive spherulite sectors constitute the "prismatic" layer. By means of the enzymatic proteolysis (pronase), the topographic relationships of aragonitic microcrystals and organic matrix are clearly shown (Fig. 6) in the spherulite sectors. From the top down, acicular microcrystals are disposed in a feathery fan manner. Towards the bottom of the sector, their size changes by increasing width and decreasing length. The gaps between the microcrystals resulting from proteolysis suggest that each microcrystal is surrounded by a protinaceous sheath. On the other hand, the vertical limits between contiguous sectors ("vertikalen Trennfugen", 7) are not homogeneous but remain visible when specimens are decalcified (Figs. 7-8) or treated with pronase (Figs. 9-10): some of them are filled by a membrane that is very resistant to protease action.

The descriptions of the external wall of the *Spirula* test mention a variable number of layers with diverse structures: two prismatic (8, 9, 10, 11), or three: outer prismatic, median lamellar (nacreous) and inner prismatic (12). Sometimes the mural part of a septum is also visible. Simple breaks do not permit detailed observation of the test structure (Fig. 11), which explains in part the divergence. A considerable improvement is obtained by treating the specimens: the successive action of NaOH, pronase and chitinase, by eliminating the organic matter that was very abundant initially, favours the determination of the structure. Therefore, the outer wall of the *Spirula* appears to represent two prismatic layers (Figs. 12-13) whose limits are not always very clear.

The treatment of the tests by the enzymes brings positive results at the various structural levels considered in these examples. Furthermore, it allows us to question the definition of terms frequently used (in this case: prism) based upon the morphology as opposed to the processes of their formation.

2.2. Ionic Microanalysis

The distribution images of Ca, Na and K on the outer wall of *Spirula*, obtained by ionic microanalysis, confirm the preceding results on the number and the structure of the layers. In effect, one observes a discontinuity between the tubercles and their basal layer on the one hand (outer prismatic) and the rest of the wall on the other (inner prismatic), a previously recognized topography (11) (Figs. 14-16). Furthermore, the distribution of Na and especially K, demonstrates as well the perpendicular aspect of the outer wall from elements of the inner layer.

The large radial prisms in the Belemnitidae are generally considered as a primary structure, and have served notably as a standard for paleotemperature curves. The study of the rostrum in Triassic forms (Turkey - Austria), conserved under exceptional conditions (13), places once again this postulate in question. Fundamentally, the rostrum of the three genera studied (*Ausseites*, *Prographularia* and *Aulacoceras*) is made of concentric prismatic layer alternating with more thin, supposedly organic strata (Fig. 17) (14). Colorations and X-ray diffraction have shown the aragonitic nature of the prismatic layers. Often reduced to the membraneous state, the "organic" layers present locally the typical aspect of a microcrystalline calcite due to the recrystallization of an initial structure. An examination of *Ausseites* by electronic microprobe reveals the presence of wide zones rich in Sr, corresponding to the aragonitic prismatic layers and alternating with more narrow zones lacking Sr. The distribution images of Sr obtained by ionic microanalysis have confirmed this correlation by showing a constant association of Sr, Na and K, equivalent to that observed in a number of other organisms. The comparison of the SEM micrographs (Fig. 16) and ionic microanalyser images (Figs. 17-20) corroborate the correspondence between the "organic" layer and the localization of Mg. Thus it seems that from an initial internal basal zonation of the prisms, which is not visible habitually but indicating real periodic

Figs. 1-5: *Pinna* (Bivalvia)
1. homogeneous and compact appearance of prisms in simple break; x 90
2. continuous growth lines in contiguous prisms; x 450
3. prisms etched with pronase; x 2700
4. growth lines (pronase); x 630
5. granular mineral elements (pronase); x 4500

Figs. 6-10: *Monodonta* (Gastropoda)
6. prismatic layer with growth lines; x 450
7. limit between spherulite sectors (RDO); x 450
8. organic membranes between spherulite sectors (EDTA); x 315
9. acicular microcrystals disposed in a feathery fan manner (pronase); x 450
10. detail; x 4500

Figs. 11-16: *Spirula* (Cephalopoda)
11. simple fracture in external wall; el: external layer; il: internal layer; x 270
12. external wall etched with NaOH, pronase and chitinase; x 405
13. general aspect with external tubercles; x 160
14-16. distribution images of Ca, Na and K, obtained by ionic microanalysis; x 190

Figs. 16-20 *Ausseites* (Cephalopoda)
16. transverse section in rostrum, with concentric layers; x 90
17-20. distribution images of Mg, Na and Sr obtained by ionic microanalysis; x 160

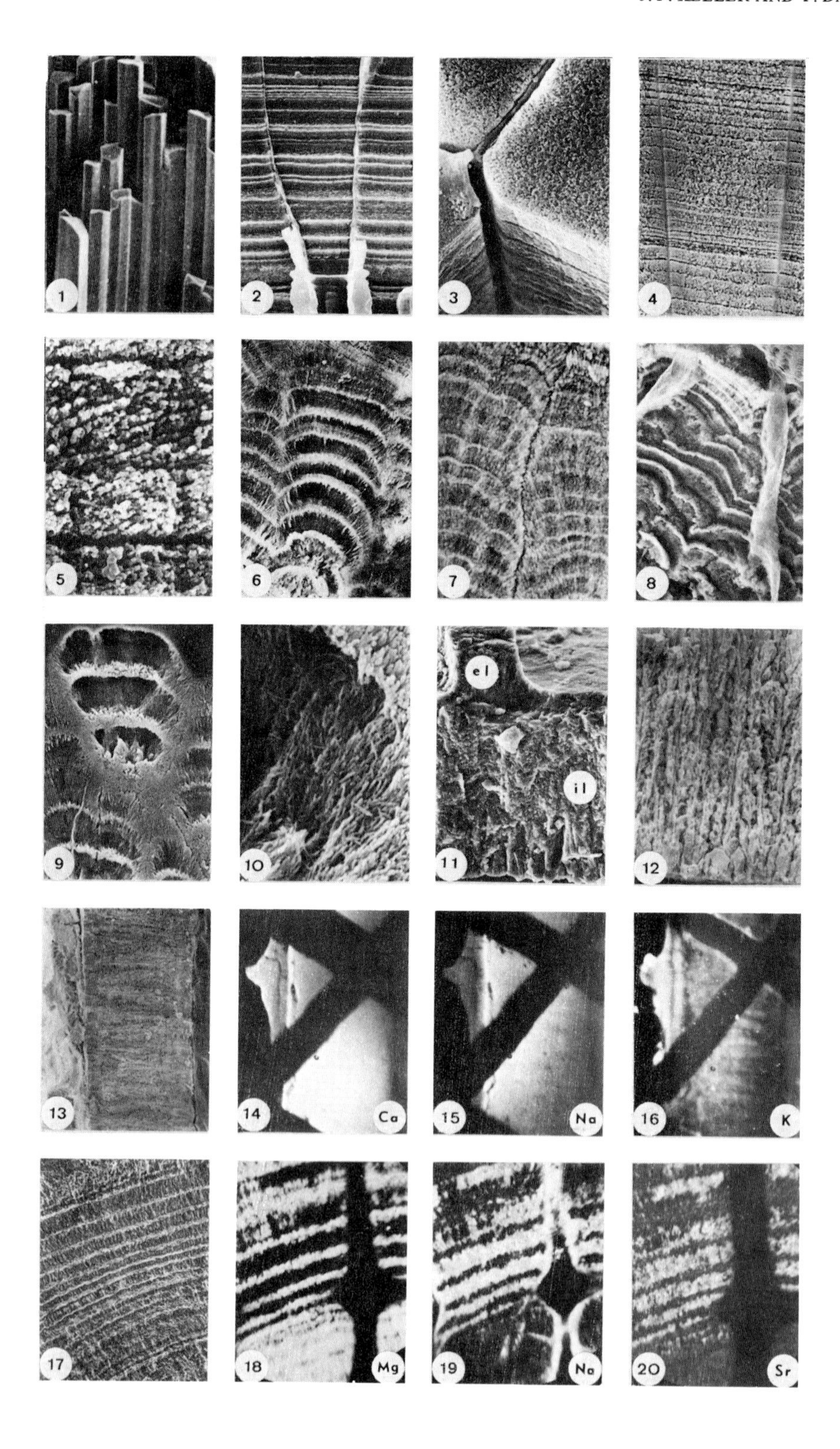
1
2
3
4
5
6
7
8
9
10
11
el
il
12
13
14
Ca
15
Na
16
K
17
18
Mg
19
Na
20
Sr

growth phenomena, diagenesis starts out in a privileged fashion with "organic layers" which are precocious depositional zones of calcite. The remobilization of the prismatic layers occurs eventually at varying degrees.

Thus, the association of results obtained by enzymatic proteolysis and localized microanalysis supplies a good basis to structural studies on invertebrate tests.

3. REFERENCES

1. Taylor, J.D., Kennedy, W.J., and Hall, A. (1973). The shell structure and mineralogy of the Bivalvia. II. Lucinacea, Clavagellacea, conclusions. Bull. Brit. Mus. (nat. hist.), Zool. 22, pp. 255-294.
2. Suzuki, S., and Uozumi, S. (1981). Organic components of prismatic layers in Molluscan shells. Jour. Fac. Sci. Hokkaido Univ., ser. IV, 20, pp. 7-20.
3. Keller, J.P. (1981). Le dégagement du matériel minéral des tests d'Invertébrés (Bivalves) par protéolyse enzymatique de la trame organique. Geobios. 14, pp. 269-273.
4. Cuif, J.P., Dauphin, Y., Denis, A., Gaspard, D., and Keller, J.P. (1980). Continuité et périodicité du réseau organique intraprismatique dans le test de *Pinna muricata* L. (Lamellibranche). C. R. Acad. Sci., Paris, t. 290, 12, sér. D., pp. 759-762.
5. Cuif, J.P., Denis, A., and Gaspard, D. (1981). Recherche d'une méthode d'analyse ultrastructurale des tests carbonatés d'Invertébrés. Bull. Soc. Géol. France, (9), XXIII, pp. 525-534.
6. Wise, S.W. Jr., and Hay, W.W. (1968). Scanning electron microscopy of molluscan shell ultrastructures. II. Observations of growth surfaces. Trans. Amer. Micros. Soc. 87, p. 419.
7. Erben, H.K. (1971). Anorganische und organische Schalenkomponenten bei *Cittarium pica* L. (Archaeogastropoda). Biomineralization 3, pp. 51-64.
8. Appellof, A. (1893). Die Schalen von *Sepia*, *Spirula* und *Nautilus* studien über den Bau und das Wachstum. K. svenska Vetensk Akad. Handl., Stockholm, 25, pp. 1-106.
9. Mutvei, H. (1964). On the shells of *Nautilus* and *Spirula* with notes on the shell secretion in non-cephalopod Mulluscs. Arkiv för Zoologi 16, pp. 221-278.
10. Barskov, I.S. (1973). Microstructure of the skeletal layers of *Sepia* and *Spirula* compared with the shell layers of other Mollusks. Paleont. J. 3, pp. 285-294.
11. Dauphin, Y. (1976). Microstructure des coquilles de Céphalopodes I. *Spirula spirula* L. (Dibranchiata, Decapoda). Bull. Mus. Nat. hist. nat., 3è sér., Sci. Terre 54, 382, pp. 197-238.
12. Bandel, K., and van Boletzky, S. (1979). A comparative study of the structure, development and morphological relationships of chambered cephalopod shells. Veliger 21, pp. 313-354.

13. Cuif, J.P. (1972). Note sur des Madréporaires triasiques à fibres aragonitiques conservées. C. R. Acad. Sci., Paris. sér. D, t. 274, pp. 1272-1275.
14. Dauphin, Y., and Cuif, J.P. (1980). Implications systématiques de l'analyse microstructurale des rostres de trois genres d'Aulacocéridés triasiques (Cephalopoda, Coleoidea). Palaeontographica A, 169, pp. 28-50.

BIOMINERAL FORMATION OF GASTROPODS, IN COMPARISON WITH THAT OF PELECYPODS

Iwao Kobayashi
Department of Geology and Mineralogy, Niigate University, Japan,
Katsutomo Mano, Fumio Isogai
Institute of School Education, University of Tsukuba, Japan,
Masae Omori
Faculty of General Education, Azabu University, Japan.

Abstract

In this report, the growing plane of a shell in gastropods is examined by a scanning electron microscope. Next, the specific distribution of the shell structure is described. On the basis of these data, we discuss the comparison of biomineral formation between these taxa.

1. INTRODUCTION

Studies on the structure and biomineral formation of gastrophod shells are scarse in comparison with studies on pelecypods. The mechanisms of calcification, biomineral formation and the shell structure of gastropods are similar to those of pelecypods, since the phylogenetical relationships between the two taxa are very close. There are, however, significant differences between these groups.

Bøggild (1930) described the shell structures of twelve families of gastropods. MacClintock (1967) examined the specific distribution of shell structures in patelloid and bellerophontoid gastropods. Gainey and Wise (1976) observed the growing surface of shell of archaeogastropoda under a scanning electron microscope. They pointed out the relation between the shell and mantle surfaces. Erben (1971) examined the shell structure of *Cittarium*, the ontogenetic change and organic components of shell matrix. Togo (1977) proposed the differentiation of the shell structures, while Iwata (1980) examined the mineralization of a larval shell of *Haliotis*. Nakahara (1979) showed that crystal tablets grow within preformed multi-layered compartments in the shell of *Turbo* and *Tegula*. Mutvei (1978) discussed the ultrastructural characteristics of the gastropod nacre as compared with the pelecypod nacre. In addition there have been many studies on shell formation of land snails.

P. Westbroek and E. W. de Jong (eds.), Biomineralization and Biological Metal Accumulation, 261–266.

2. GROWING PLANE OF SHELLS OF SEVERAL ARCHAEOGASTROPODS

Morphological patterns of the growing plane of the shell surface of four species of archaeogastropods were examined by the use of scanning electron microscope. *Batillus cornutus* (Lightfoot), a Turbinid, is one of the most popular species for food in Japan. Living specimens of *Batillus* were collected at Sado Island, Niigata. Three or four morphological types of shell structure were detected in the shell of this species by using thin sections, namely, complex prismatic (partly homogeneous grained), composite prismatic and nacreous structures. Openings of a canal structure were found on the inner surface near the apex (Figure 8). The morphological pattern of the growing planes near the aperture can be classified into the following types from the outer lip inwards: (a) Patterns of subrectangular blocks (Figure 1) with spherolite spectors (Erben, 1971) which increase inwards (Figure 2); (b) A gradual change to the irregular granule pattern (Figure 3) that appears smoother; (c) The pattern of columnar nacre (Figure 5). Figure 4 shows a transition region by the nacre; (d) This gradually changes to the fish-scaled pattern (Figures 6 and 7); and (e) The patterning becomes obscure.

Specimens of *Pomaulax japonicus* (Dunker) collected from Sado Island show a different pattern except for the nacre. The peripheral region forms spherolite sectors and hexagonal prisms with a dendritic pattern. The specimens of *Monodonta labio* (Linné) and *Chlorostoma argyrostoma turbinatum* (A. Adams) were collected from the beach of Niigata City. The growing plane of the shell of the former specimen shows an intricate change from the outer lip inwards as follows: (1) The aggregation of small hexagonal and rectangular plates in an irregular direction; (2) columnar nacre; (3) small granular pattern; (4) the pattern of spinose crystal; and (5) fish-scaled pattern (inner prismatic layer). The growing plane of a shell of *Chlorostoma* is also different from the above species except for the nacre.

3. SPECIFIC DISTRIBUTION

Gastropod shells fit into two groups on the basis of their shell structure namely those with and those without the nacre shell layer. The first group includes the archaeogastropoda. The second group are generally made of 2, 3 or 4 crossed lamellar layers often oriented at 90 degrees to one another (Gainey and Wise, 1976).

The shells of archaeogastropods consist of a combination of various shell structures, namely homogeneously grained, complex prismatic, composite prismatic, columnar nacreous, foliated, crossed lamellar and complex crossed lamellar structures. The complex prismatic layer may be made of calcite crystals. Shells of some genera of the Haliotiidae and Trochidae are of this structure. As pointed out by Gainey and Wise (1976) the distinctive feature of the archaeogastropod is that the inner calcareous layer is composed of the columnar nacreous structure. "The inner

Table 1. Specific distribution of shell structure
Hog gra:homogeneously grained st. Cpo pri:composite prismatic st. Cro lam:crossed lamellar st. Obs cro lam:obscure crossed lamellar st. Cpl cro:complex crossed lamellar st. Tra-pri:trans-prismatic st. Nac:nacreous st. Cpl pri:complex prismatic st. Fol:foliated st. A:aragonite C:calcite

Species	Mineral	Shell structure
Archaeogastropoda		
Sulculus supertexta		Hog gra, Cpl pri, Nac
Umbonium moniliferum		Cpo pri, Nac
U. giganteum	A	Cpo pri, Nac
Chlorostoma argyrostoma lischkei	A,C	Cpl pri, Cpo pri, Nac
Ch. argyrostoma turbinatum		Cpl pri, Cpo pri, Nac
Omphalius pfeifferi		Cpl pri, Nac
O. rusticus	A,C	Cpl pri, Cpo pri, Nac
Monodonta labio	A	Cpo pri, Nac
M. perplexa	A	Cpo pri, Nac
Cantharidus japonicus		Cpo pri, Nac
Batillus cornutus	A,C	Hog gra, Cpl pri, Cpo pri, Nac
Pomaulax japonicus	A,C?	Cpo pri, Nac
Marmarostoma stenogyrum		Hog gra, Cpo pri, Nac
Lunella coronata		Cpo pri, Nac
Astralium haematragum		Cpo pri, Nac
Theliostyla albicilla		Fol, Cro lam, Cpl cro
Purperita japonica	A,C	Fol, Cro lam, Cpl cro
Mesogastropoda		
Littorina brevicula	A,C	Cpl pri, Cro lam, Tra-pri
Cerithidea cingulata		Obs cro lam
Batillaria multiformis		Obs cro lam
Crepidula grandis		Cro lam
Neverita didyma		Cpo pri, Cro lam
Erosaria helvola		Cro lam
Ravitrona caputserpentis		Cro lam
Palmadusta artuffeli		Cro lam
Casmaria cernica		Cro lam
Semicassis persimilis		Cro lam, Tra-pri
Phalium flammiferum	A	Cro lam, Tra-pri
Monoplex echo		Cro lam
Canarium mutabilis		Cro lam
Tonna luteostoma		Cro lam
Neogastropoda		
Thais clavigera	A	Cro lam
Ergalatax constrictus		Cro lam
Rapana thomasiana		Cpl pri, Cro lam
Neptunea polycostata		Cpl pri, Cro lam
N. intersculpta	A,C	Cpl pri, Cro lam
Japeuthria ferrea	A	Cro lam
Babylonia japonica	A	Cro lam
Pleuroploca trapezium audouini		Cro lam
Hemefusus ternatanus	A	Cro lam

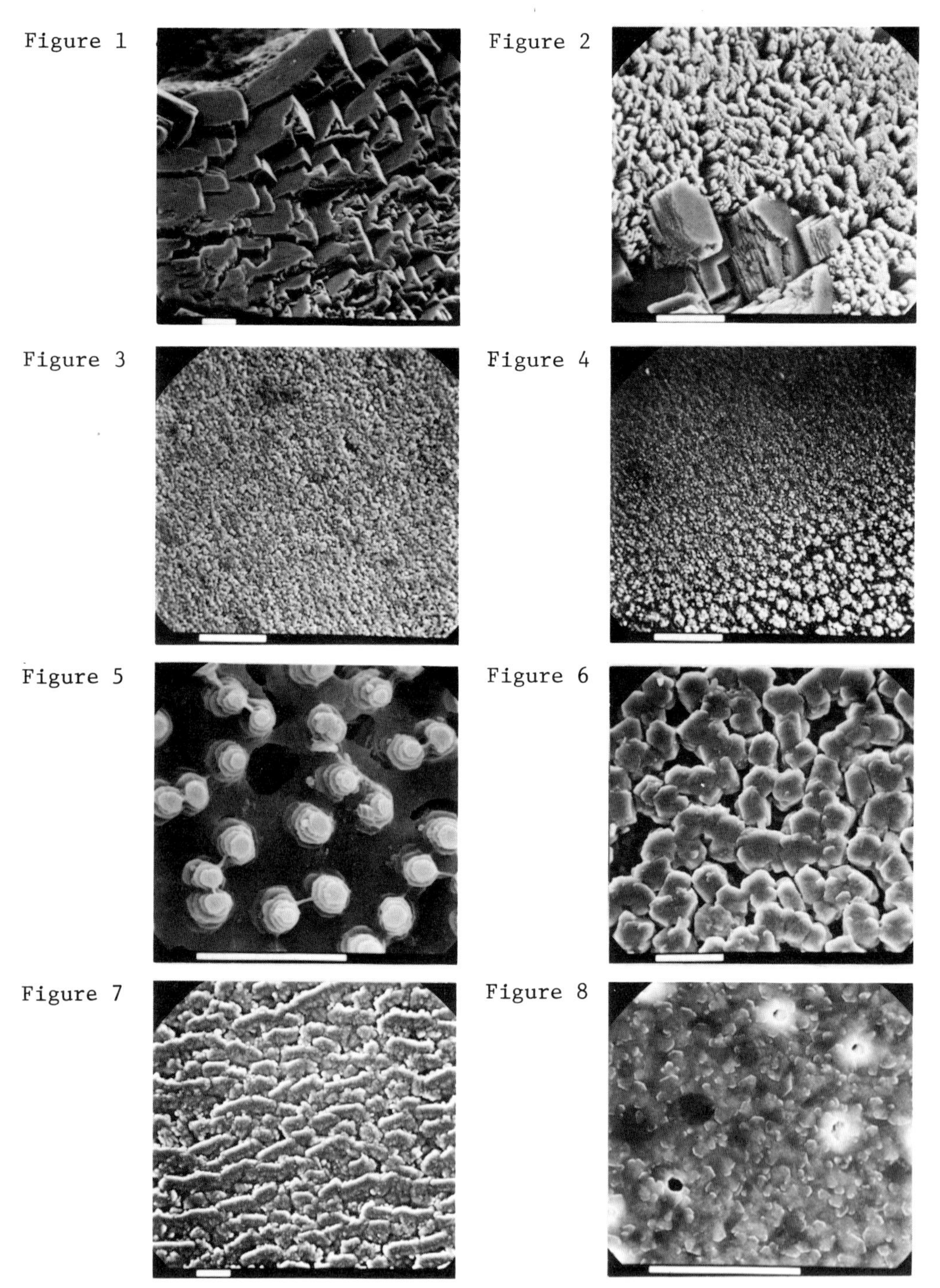

Figures 1-8. Scanning electron micrographs of the growing plane of *Batillus cornutus* (Lightfoot).
All scales: 10 μ.

prismatic layer" cannot easily be discriminated from the nacreous layer. The growing plane of this layer displays the fish-scaled pattern as seen in scanning electron micrographs. In the shell of mesogastropods and neogastropods, the crossed lamellar structure is widely distributed. The uniformity of shell structure in them contrasts markedly with the variety of shell structure in archaeogastropods. There are, however, some exceptions, for example, the shell of *Littorina* consists of the complex prismatic layer and the shell of *Neverita* contains a composite prismatic layer. The shells of *Neptunea* and *Rapana* are also composed of the complex prismatic structure.

4. COMPARISON WITH SHELL STRUCTURES OF PELECYPODS

The formation of crossed lamellar and composite prismatic structures in the gastropods is similar to that of the pelecypods. Sheet nacre is well developed in pelecypods but is replaced by columnar nacre in gastropods. The mineralization of the outer prismatic layer in the shell of archaeogastropods differs from that of the normal prismatic layer of pelecypods. Generally, in the mineralization of the gastropods, the pattern of the growing plane is more intricate than in those of pelecypods. Sometimes the crystal growth shows quite different patterns even between closely related species.

Gastropods have the same ability as the pelecypods in forming aragonite and calcite crystals. Aragonite occurs more commonly in gastropods than in pelecypods and the distribution of calcite layers in gastropods raises interesting questions on the control of calcite and aragonite formation.

Compared with pelecypods gastropods show the following differences. The sheet nacre is well developed in pelecypods, whereas columnar nacre is common in gastropods. The crossed lamellar structure is commonly observed in both taxa, showing some differences in the orientation of the first order lamellae; parallel with the shell margin in pelecypods, but parallel and vertical to the aperture margin in gastropod shell. It may be mentioned that palaeotaxodonta and pteriomorphia of pelecypods and archaeogastropods which appeared and flourished in the early geologic age have various shell structures, such as prismatic, nacreous, foliated and crossed lamellar structures etc. On the other hand, heterodonta of pelecypods, mesogastropods and neogastropods which flourished in the later age have the crossed lamellar group, including homogeneous structure. The features mentioned seem to suggest that there are some trends of parallel evolution between these two taxa.

REFERENCES

Bøggild, O.B.: 1980. K. Danske Vidensk. Selsk. Skre., ser. 9,2, pp. 231-325.
Erben, H.K.: 1971. Biomineralization 3, pp. 51-64.

Gainey, L.F., and Wise, S.W.: 1976. Proc. Workshop Zool. Appl. SEM. IIT Res. Inst., pp. 585-590.
Iwata, K.: 1980. Jour. Fac. Sci., Hokkaido Univ., ser. IV, 19, pp. 305-320.
MacClintock, C.: 1967. Peabody Mus. Nat. His., Yale Univ., B.22, pp. 1-140.
Mutvei, H.: 1978. Zoologica Scripta 7, pp. 287-296.
Nakahara, H.: 1979. Jap. Jour. Malac. 38, pp. 205-211.
Togo, Y.: 1977. Jour. Geol. Soc. Japan 83, pp. 567-573.

ULTRASTRUCTURAL EVOLUTION OF MOLLUSCAN NACRE

H.Mutvei
Swedish Museum of Natural History, Stockholm, and
Paleontological Institute, Uppsala University, Uppsala,
Sweden.

Ultrastructurally well preserved nacre is described in two Ordovician orthoconic cephalopods (Isorthoceras and Catoraphiceras), a gastropod (Murchisonia) and a bivalve (Palaeoconcha). Aragonite in the nacre has been pre-diagenetically substituted by calcium phosphate. In the orthoconic cephalopods and in the gastropod, the nacre is generally similar to Recent Nautilus and gastropods except that it is porous. Due to the porosity, the nacre was less strong but more flexible than in corresponding Recent molluscs. In the Ordovician bivalve, the nacre was practically solid, but in other respects differs considerably from Recent bivalves.

INTRODUCTION

Molluscan nacre is composed of aragonite. In early Paleozoic rocks the aragonite is not preserved but is usually substituted by calcite. In consequence, the nacreous structure is completely lost. However, under certain conditions, pre-diagenetic substitution of aragonite by calcium phosphate occurred. In such cases, as described here, the nacreous structure is sufficiently well preserved for detailed ultrastructural studies.

The material available comprised secondarily phosphatized, Ordovician molluscs, as follows:
(1) Orthoconic cephalopod: Isorthoceras sociale (Hall), bivalve: Palaeoconcha sp., and gastropod: Murchisonia sp., from the late Upper Ordovician Maquoketa shale, Graf, Iowa, USA (about 450 million years old). (2) Orthoconic cephalopod: Catoraphiceras sp., from the early mid Ordovician Aseri Stage, Harku quarry, Tallinn, Estonian SSR.

The nacreous layer was investigated using a scanning electron microscope on untreated vertical and horizontal fracture planes.

This study was financed by Grants 287-105 and 107 from the Swedish Natural Science Council, whose support is gratefully acknowledged.

P. Westbroek and E. W. de Jong (eds.), Biomineralization and Biological Metal Accumulation, 267–271.

RECENT NACRE

Except in the dibranchiate cephalopod Spirula, the nacre in Recent molluscs is composed of small aragonite tablets, commonly with pseudohexagonal outlines at early growth stages. These tablets form thin mineral lamellae, separated by inter-lamellar organic membranes. The structure and arrangement of the nacreous tablets show considerable differences at class level.

1. Nautilus-gastropods

The nacreous tablets in the shell wall of the cephalopod Nautilus and in gastropods consist of a variable number (2 to 50) of crystalline sectors which radiate from the centre of the tablet to the periphery (Fig. 2B). These sectors represent cyclic and interpenetrant crystal twins. The surface of each sector shows parallel crystalline laths which indicate lamellar (polysynthetic) twinning of the sector. Most characteristic is the occurrence of a calcified organic accumulation in the centre of each tablet. The nucleation of a new tablet is probably determined by this organic accumulation. As a result, the tablets in the consecutive mineral lamellae form vertical columns (Mutvei, 1978, 1980).

2. Bivalves

Two structurally different layers can be distinguished in each nacreous tablet. The thin outer layer is composed of two pairs of crystalline sectors which are probably cyclic crystal twins. Of these sectors, one pair has much higher solubility than the other. Nucleation of a new tablet takes place on the less soluble pair of sectors, and the tablets in the consecutive mineral lamellae acquire therefore a "brick wall" arrangement. The surface of the inner layer shows parallel crystalline laths (Fig. 3B) which indicate lamellar (polysynthetic) twinning of this layer (Mutvei, 1977, 1979, 1980).

ORDOVICIAN NACRE

1. Cephalopods-gastropods

The nacre in the shell wall of the two orthoconic cephalopods Isorthoceras sociale and Catoraphiceras sp. and the gastropod Murchisonia sp. is closely similar to that in Recent Nautilus and gastropods. The nacreous tablets show: (1) vertical stacking (Figs. 1A,B), and (2) a central accumulation of calcified organic matrix (indicated by an arrow in Fig. 1A). The crystalline laths radiate from the centre of the tablet to the peripheri (Figs. 1A.B, 2A). In several tablets, the arrangement of the laths is not fully radial but each tablet is subdivided into numerous, narrow sectors, and in each sector the laths are parallel to one another. The latter structure is found occasionally in the nacre of Recent Nautilus and in the gastropod Haliotis (Fig. 2B; see also Mutvei, 1978).

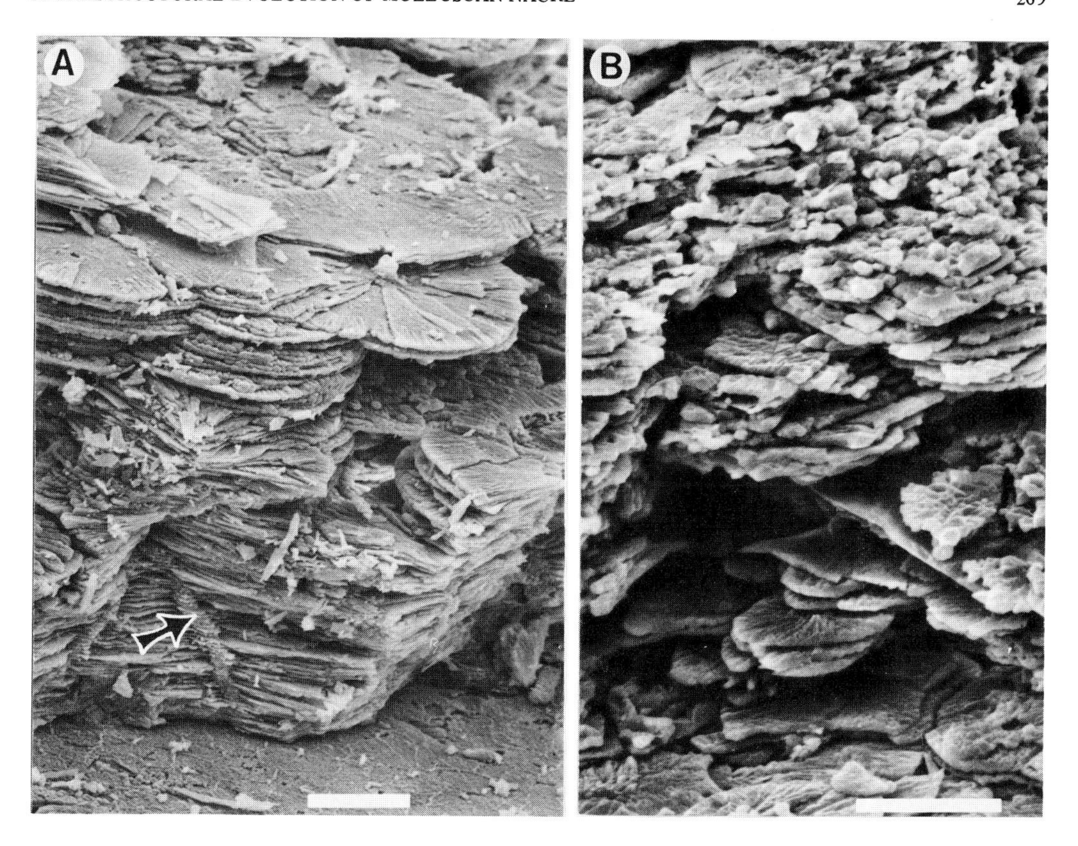

Figure 1. Nacre in Ordovician orthoconic cephalopods. A. Isorthoceras sociale and B. Catoraphiceras sp. Calcified organic accumulation in the centre of a tablet is indicated by an arrow. Bar=10 μm.

Figure 2. Gastropod nacre. A. Ordovician Murchisonia sp., and B. Recent Haliotis sp. Bar=10 μm.

The most striking difference between the Ordovician cephalopod-gastropod nacre and the corresponding Recent nacre is the porous structure of the former. The porosity is due to numerous vertical interspaces left between adjacent stacks of tablets (Figs. 1A, 2A). These interspaces must originally have been filled by an uncalcified organic matrix. In consequence, the nacre in these Ordovician cephalopods and gastropods was less strong but more flexible than in Recent Nautilus and gastropods.

2. Bivalves

The nacre in Palaeoconcha sp. differs in the following features from Recent bivalves: (1) the tablets have highly variable size and arrangement (Fig. 3A), instead of the regular size and "brick wall" arrangement of Recent bivalves, and (2) the crystalline lahts in the tablets commonly have a radial orientation, less usually a parallel or sub-parallel orientation (Fig. 3A), instead of the strictly parallel orientation of Recent bivalves (Fig. 3B). In addition, in Recent bivalves the tablets have an outer, thin layer which could not be observed in Palaeoconcha sp.

By comparison with the Ordovician cephalopod-gastropod nacre, there are the following differences: (1) the nacre is not porous but is practically solid, (2) the tablets usually do not form vertical stacks, (3) they vary much more in size, and (4) they lack a central organic accumulation.

DISCUSSION

This study suggests that major evolutionary changes in the structure of the molluscan nacre took place in the early Paleozoic. It is premature to draw general conclusions about the evolution, because there is considerable ultrastructural variation in Ordovician nacre. The variation has not yet been studied in detail.

The Ordovician cephalopods and gastropod studied has porous nacre. This nacre was therefore more flexible than in Recent Nautilus and gastropods. The cosequences of this condition on the mode of life in early Paleozoic orthoconic cephalopods will be discussed in detail in a forthcoming paper (Mutvei, 1983).

REFERENCES

Mutvei, H.: "The nacreous layer in Mytilus, Nucula and Unio (Bivalvia). Crystalline composition and nucleation of nacreous tablets". 1977, Calcif. Tiss. Res. 24, pp.11-18.

Mutvei, H.: "Ultrastructural characteristics of the nacre in some gastropods". 1978, Zool. Scr. 7, pp.287-296.

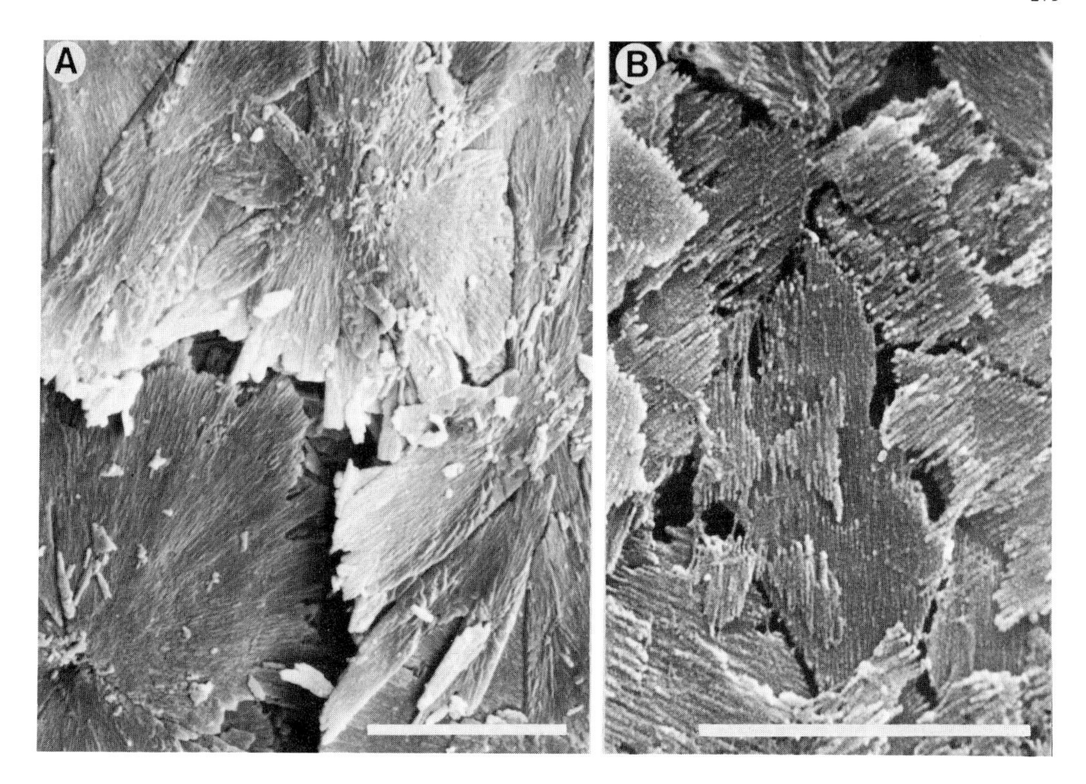

Figure 3. Bivalve nacre: A. Palaeoconcha sp. from the Ordovician, and B. Mytilus edulis, Recent (etched with a glutaraldehyde-acetic acid solution). Note differences in orientation of the crystalline laths, and in size and arrangement of the tablets. Bar=10 μm.

Mutvei, H.: "On the internal structures of the nacreous tablets in molluscan shells". Scanning Electron Microscopy/1979/II, pp.457-462.

Mutvei, H.: "The nacreous layer in molluscan shells". In M.Omori & N.Watabe (eds.): Proc. 3rd Intern. Biomineral. Symp. 1980, pp.49-56.

Mutvei, H.: "Flexible nacre in Isorthoceras (Cephalopoda:Nautiloidea), with remarks on the evolution of cephalopod nacre". Lethaia 1983 (in press).

BONE AND TOOTH FORMATION. INSIGHTS INTO MINERALIZATION STRATEGIES.

Arthur Veis and Boris Sabsay
Northwestern University Dental School, Chicago,
Illinois 60611, USA

Mineralized tissues take either of two possible forms, distinguished on the basis of the relationship between mineral and matrix and the organization of the mineral phase. In organisms with shells, carapaces and exoskeletons the mineral crystals are large, form the major continuous phase, and determine the principal physical properties of the tissue. Bone, on the other hand, is a composite-like tissue in which the organic matrix is the continuous phase while the mineral is dispersed as discontinuous microcrystals regularly throughout the matrix. In both types of tissues the organic phase is the cellular product deposited before mineral accretion begins. In both situations it has been postulated that the organic matrix directs the deposition of the mineral phase, from nucleation through ultimate limitation of crystal growth. The central question under investigation here is that of the mechanisms whereby the matrix control of mineralization is expressed. Is there a common strategy used in all systems, in spite of the evident diversity?

The mammalian tooth is an excellent model for study of these questions. The enamel mineral phase accounts for more than 90% of the enamel by dry weight, is comprised of large, well organized calcium hydroxyapatite rods. but shows distinct relationships between mineral and matrix during formation. Dentin is only about 65% mineral by weight on a dry basis. The bulk of the matrix is a network of interwoven collagen fibers. The dense apatite microcrystals take up less than half the space of the dentin and are embedded within the fiber matrix. Enamel and dentin are less cellular than bone and since neither remodels, biosynthetic analyses are not complicated by concomitant resorptive reactions as in bone.

In our laboratory we have emphasized studies of dentin and most of the following discussion will focus on the dentin system. However, we begin with a brief review of tooth formation and enamel mineralization so that the enamel and dentin systems may be fruitfully compared and contrasted .

P. Westbroek and E. W. de Jong (eds.), Biomineralization and Biological Metal Accumulation, 273–284.

I. TOOTH DEVELOPMENT

Relatively late in embryonic development a layer of cranial neural crest cells lying between the epithelium and mesenchyme invaginates into the mesenchyme to form the multi-layered dental lamina. At the earliest stage a basement membrane separates the epithelial and mesenchymal tissues. The dental lamina folds into the mesenchyme partially enclosing that portion destined to form the tooth. The cells on the exterior form the enamel organ epithelium, the inner cells, the preodontoblasts. Some, as yet undefined, epigenetic interaction across the basement membrane leads to differentiation of the ectomesenchymal cells into odontoblasts, and shortly thereafter, the epithelial ameloblasts form. Both cell types become columnar and polarized so that the cell nuclei are distal to their cell-cell junction. They secrete extracellular matrix at the region proximal to the dentino-enamel (DE) junction. As the enamel and dentin form, the two pallisades of cells recede from the DE-junction on the bases of matrix they have produced.

II. ENAMEL MATRIX AND MINERALIZATION

Ameloblasts secrete a complex mixture of two classes of protein, amelogenins and enamelins. The amelogenins are present in larger amount but, in fetal bovine teeth may be extracted with denaturants without demineralization. The enamelins may also be extracted in 4.0 M Guanidine HCl (Gd. HCl) but only during or after demineralization (1). The amelogenins are Glu and Pro rich. In the highest molecular weight fraction these two amino acids comprise more than 50 residue % of the molecule. The Asp and Gly contents of enamelin are higher while Glu and Pro levels are lower than in the amelogenins. Both types of protein are secreted at the apical end of the ameloblast and mineralization follows rapidly with the initial crystal long axis formation perpendicular to the DE-junction. Cells retreat leaving the Tomes processes in contact with matrix but tilted with respect to the cell body. Mineral deposition follows in regions vacated by the Tomes process and enamel prisims are formed, with their long axes parallel to the process direction, producing the twisting of the enamel rods (2). The enamelins may be incorporated into the enamel crystal phase at this time but the amelogenins remain on the enamel tissue surface. At a later stage in enamel maturation, ameloblasts exhibit resorptive character, and this may be the period of amelogenin removal. Mature post-eruptive enamel contains only enamelins.

Although the mechanism of mineralization is not evident enamel formation is clearly an example of matrix directed mineralization.

III. DENTIN MATRIX AND MINERALIZATION

The odontoblasts anchor themselves to the DE-junction by means of the odontoblastic processes. As the cells secrete collagen and other matrix proteins they recede from the DE junction and the odontoblastic processes elongate. Viable cell contents remain in the odontoblastic processes. Thus, although the main bodies of the odontoblasts may be distant from the most mature dentin, the cells remain in communication with much of the dentin at all times. As in bone, the matrix adjacent to the cell membrane is non-mineralized. Mineralization takes place in a highly regulated fashion at the predentin-dentin junction several µm distant from the cell body.

In our attempt to understand the mineralization regulatory mechanisms we have proceeded from the perspective that the extracellular matrix proteins are the crucial participants. In the following sections, we shall consider two aspects in particular, the nature and composition of the matrix components, and, the interaction between matrix components as they might influence mineral depositon.

A. Dentin Collagen

Dentin collagen is entirely Type I (3). It is, in rat, bovine and human dentin essentially completely insoluble, presumably as a result of a unique set of intermolecular cross-linkages (4) involving the collagen α2 chain. Dentin does have a higher degree of hydroxylation of lysine than the soft tissue collagens but glycosylation is primarily as monosaccharides rather than dissaccharides (3). Dentin collagen does contain a few residues of covalently bound phosphate moieties (3,5). It is difficult to relate these small differences between mineralized and soft tissue Type I collagen to the mineralization problem.

B. Non-collagenous Proteins

Concomitant with demineralization, 0.5M EDTA (ethylene dinitrilo acetic acid) at pH 7.4 brings almost all of the non-collagenous proteins (NCP) of dentin into solution. There are proteolytic enzymes in the matrix and the presence of a mixture of protease inhibitors is an absolute requirement during demineralization if one wishes to study the matrix NCP. The NCP are a very heterogeneous mixture. Figure 1 A shows the gel-filtration chromatography of the unfractionated mixture of EDTA soluble proteins from rat incisor under denaturing conditions, while 1 B shows the gel-electrophoresis of the same material. These proteins are largely acidic in character but differ markedly in properties and composition. Chromatography on anion exchange columns, Figure 2, permits isolation of a set of very anionic proteins unique to dentin. These proteins, rechromatographed in the 4.0M quanidine HCl system on BIOSIL TSK 400 are resolved into three components with apparent molecular weights of 62000, 37,000 and 16,000 (line B Figure 1). Their compositions are shown in Table 1. These three proteins are

A.

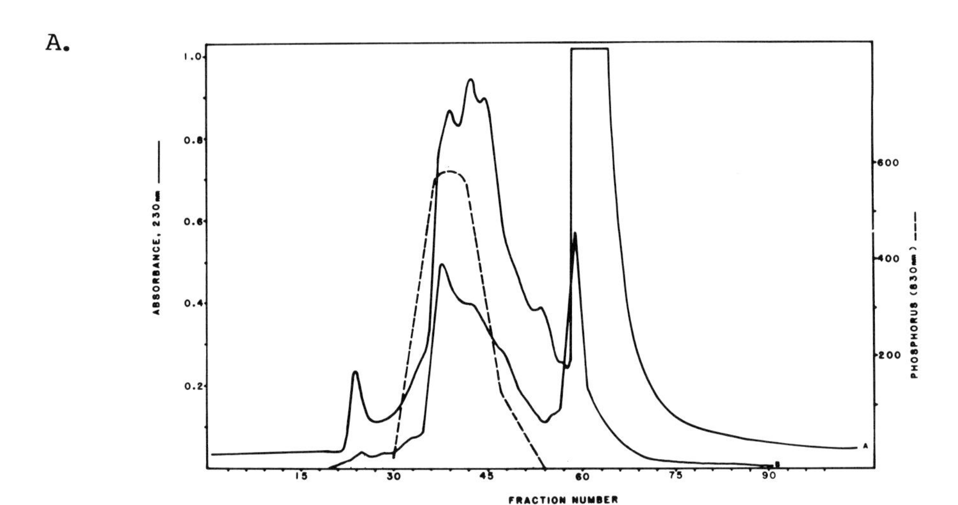

B.

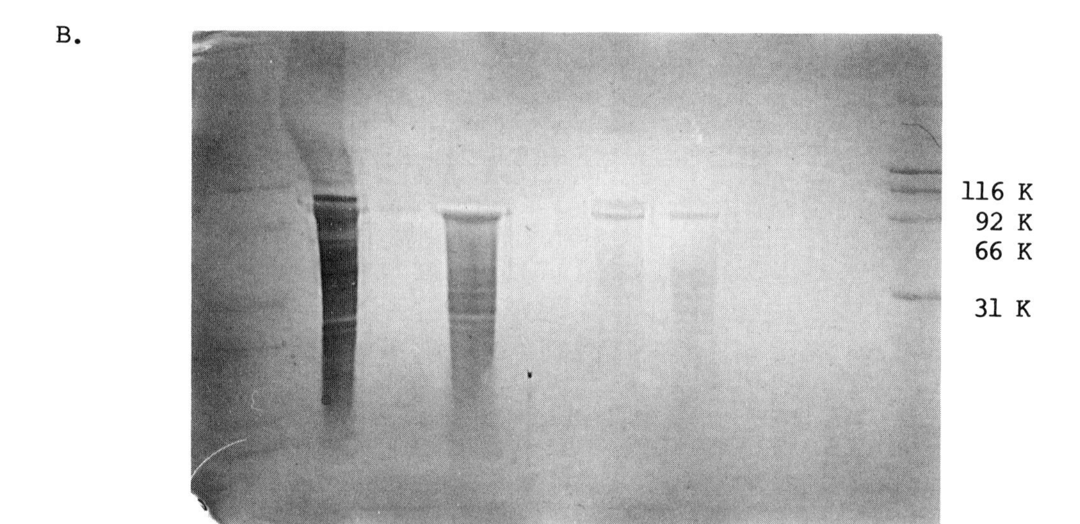

Figure 1 A. Gel filtration of EDTA extracts from rat incisors on Biosil TSK-400 in 4.0M Guanidine HCl. Upper solid line (A), absorbance of crude extract. Lower solid line (B), absorbance of 0.25-0.29M NaCl elution peak from DEAE chromatography of crude extract. Dashed line, phosphorus content of components from DEAE chromatography.

Figure 1 B. Gel electrophoresis of: Lane 1, Crude EDTA Extract; Lane 2, DEAE 0.25-0.29M NaCl peak; Lanes 3,4,5 phosphoryn fractions 1,2,3 from Biosil TSK gel filtration rechromatography of DEAE 0.25-0.29M NaCl eluate. Double stained with Stains-All and Coomassie Blue, arrows show positions of molecular weight standards.

TABLE 1

Amino Acid Compositions of Phosphophoryns Isolated by Gel Filtration Chromatography in 4.0M Guanidine. HCl From the 0.25-0.29M NaCl Eluate From DEAE-Chromatography of EDTA Extracted Protein of Rat Incisor. Data Expressed as Residues per 1000 Amino Acid Residues.

	Apparent Molecular Weight		
	62K(93)[a]	37K(55)[a]	16K(24)[a]
Asp	353	296	282
Thr	10	20	27
Ser + Ser(P)	512	423	372
Glu	36	91	118
Pro	7	16	26
Gly	29	51	65
Ala	10	25	29
Half-Cys	4	-	-
Val	6	8	14
Met	-	2	-
Ile	2	3	2
Leu	3	10	12
Tyr	2	5	6
Phe	2	4	5
Lys	11	15	18
His	8	8	6
Arg	3	14	17

a. Stetler-Stevenson and Veis (8) have shown that the molecular weight estimated from 4.0M Gdn HCl chromatography on Biosil TSK 400 columns yields a value, for phosphophoryns, which is approximately 67% of the true weight due to some intramolecular association. The numbers in parentheses indicate the probable correct molecular weight. The column headings are the direct operational values.

obviously related, all are of the class we call phosphophoryns-phosphate carrier proteins (6). These are rich (>75%) in Asp and Ser residues and the Ser residues are almost all phosphorylated. Similar phosphophoryns have been isolated from all dentin species examined. However, in bovine molars there is only a single phosphophoryn component (7,8). Like the enamelins of enamel, the phosphophoryns of dentin can be extracted only upon dissolution of the mineral phase (12). Studies of very carefully prepared predentin showed that the phosphophoryns were not present (13) whereas proteoglycans are present in greater amount in predentin than in dentin (14). The phosphophoryns bind calcium ions with high affinity (7,9) and adhere selectively to calcium hydroxyapatite crystal surfaces (10). They can be precipitated from solution upon the addition of Ca^{2+} ion under conditions where the other dentin NCP remain in solution (7,11). Every phosphophoryn phosphate group and carboxylate bind Ca^{2+} ion with high affinity. A single molecule is capable of localizing hundreds of calcium ions. Upon binding Ca^{2+} ions phosphophoryns in solution in vitro change from a random to an ordered form (15). These properties along with the localization of the phosphophoryns to the mineralized dentin, have led us to suggest that the phosphophoryns might be the interactive components involved with regulating dentin matrix mineralization.

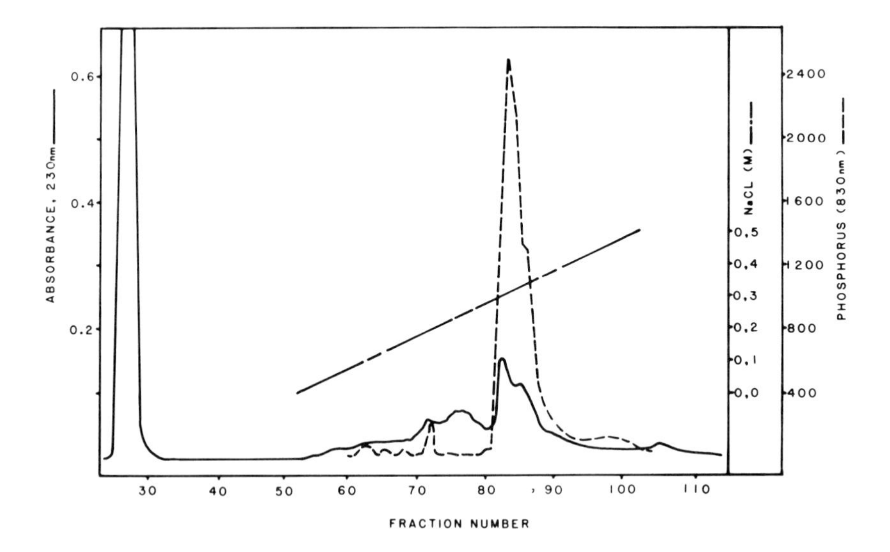

Figure 2. DEAE-cellulose ion exchange chromatography of rat incisor EDTA soluble NCP. ________ Absorbance 230 nm; - - - phosphorus content, arbitrary units; - - - - ionic strength gradient. The high absorbance of the initial fraction is due to the proteinase inhibitors present. The material in fractions 80-87 was taken for gel filtration chromatography, Line B, Fig. 1.

C. Collagen-Phosphophoryn Interactions

In early studies of bovine dentin collagen we noted (3) that CNBr digestion of the matrix left about 5% of material that was not soluble in dilute acetic acid. This residue proved to have a high content of organic phosphorus but also had a well-defined amount of hydroxyproline, indicating residual collagen. A major portion of this material could be solubilized at pH 7.4 to 8.3 and that soluble fraction yielded, after chromotography on DEAE-cellulose or hydroxyapatite columns, a set of high molecular weight protein components having both collagen and phosphophoryn moieties (10,15).

A.

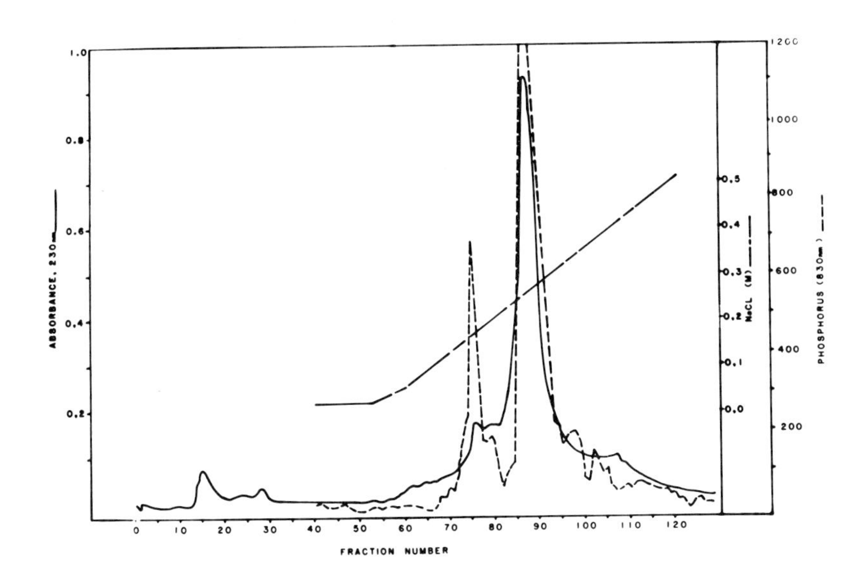

B.

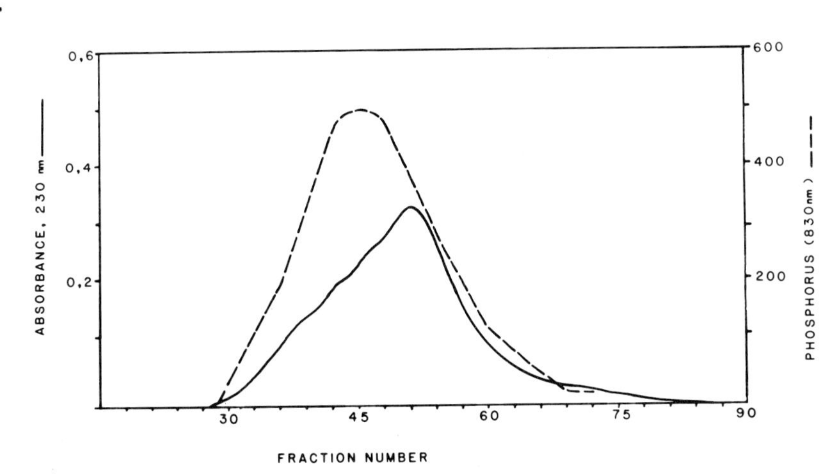

Figure 3. A. DEAE-cellulose chromatography of CNBr digest of rat incisor EDTA insoluble matrix, acid-insoluble residue.

Figure 3. B. Rechromatography of DEAE-cellulose 0.25-0.29M NaCl peak on BIOSIL TSK 400 gel filtration chromatography. In each case solid line is absorbance at 230 nm; dashed line, phosphorus content in arbitrary units.

TABLE 2

The Amino Acid Composition of Rat Incisor Phosphophoryn Collagen Conjugate Purified by DEAE-Cellulose Chromatography. Data Expressed as Residues Per 1000 Amino Acid Residues.

4-OH Pro	16
Asp	257
Thr	28
Ser + Ser (P)	322
Glu	84
Pro	44
Gly	78
Ala	40
Half-Cys	9
Val	12
Met	-
Ile	9
Leu	29
Tyr	4
Phe	2
OH Lys	1
Lys	25
His	10
Arg	23
Homo Ser	3

Figure 3 shows the DEAE-cellulose and gel-filtration chromatography of the comparable rat incisor component. Its composition is listed in Table 2. This component appears to be a covalent conjugate of a portion of the collagen molecule and a phosphophoryn. The exact nature of covalent bonding is a subject of current study but there is little doubt of the fact that the interaction is covalent, and that it is an *in vivo* produced compound (15,16).

D. Dynamic Aspects Of Mineralized Tissue Formation. Collagen And Phosphophoryn Synthesis As Related To The Mineralization Process.

Collagen is deposited in extracellular compartment at the cell proximal edge of the predentin within 30-45 minutes after synthesis begins (17,18). Phosphophoryn secretion can be detected within 15 minutes in protein labeling studies (18,19). The phosphophoryn follows a different secretory pathway than the collagen, it bypasses the predentin and appears directly at the mineralization front where it is rapidly incorporated into the mineralized dentin. It is likely that the phosphophoryns are secreted via the odontoblastic processes.

In a double labeling study, published as yet only in preliminary form, we (16) have demonstrated that the phosphophoryn-collagen conjugate is formed from newly synthesized phosphophoryn and collagen which was secreted at least 10 hours earlier. Dimuzio and Veis (18) had reported that the phosphophoryns turned over to some extent, that is, a part of the secreted protein was degraded. This was confirmed by Maier et al (16), with the further recognition that one of the two major phosphophoryn components turned over much more rapidly than the than the other. It has been known for a long time that alkaline phosphatase activity was highest at loci of active calcification (20 and the mineralization front may also be a locus for the other degraditive enzymes of the extracellular matrix.

E. A Model for Dentin Mineralization

We are now in position to construct an hypothetical mechanism for the ordered deposition and stabilization of the mineral phase in dentin.

The facts presented are:

1. An essentially unremarkable Type I collagen is secreted at the apical end of the odontoblast layer, forming the predentin.
2. The predentin is rich in proteoglycan but poor in phosphophoryn, relative to the mineralized dentin.
3. Newly formed phosphophoryns are secreted directly in the region of the mineralization front.
4. Some of the phosphophoryns are degraded while the remainder become incorporated into the dentin. A small portion of the latter becomes covalently conjugated to the collagen.
5. The phosphophoryn has a very high affinity for calcium ion and can bind hundreds of ions per molecule.
6. Phosphophoryns bind avidly to the surfaces of calcium hydroxyapatite crystals.
7. Apatite crystals first appear in dentin in association with the collagen fibers and are ordered by them.

All of these facts can be accommodated in the dynamic scheme depicted in Figure 4 and representing a continuously erupting rodent incisor system. The predentin can be considered as a zone of mineralization inhibition in which the matrix components are first assembled. Collagen fibers, various non-collagenous proteins and cell-mediated calcium are all secreted in the apical region proximal to the main body of the odontoblast. At the same time phosphophoryns, and perhaps other matrix constituents such as alkaline phosphatase, arrive via the odontoblastic processes and are secreted directly at the mineralization front, the zone of active mineral deposition. We postulate that in this region several synthetic and degradative events take place. On the synthetic side, some phosphophoryn binds specifically to the surfaces of the already formed collagen fibers and a part of this binding becomes covalent. Another portion of the

phosphophoryn is degraded locally, first by alkaline phosphatase creating locally high concentrations of phosphate ion; secondly, the dephosphorylated core protein is degraded and removed. At the same time, the proteoglycan component is degraded and the inhibition to calcium phosphate precipitation is released, permitting crystallization on the high calcium affinity matrix bound phosphophoryn. This action specifically locates the early nuclei on the collagen surface. Crystal growth is supported by the concomitant local release of phosphate from degraded phosphophoryn and calcium from calcium-proteoglycan complexes. Crystal size varies, but the size range is limited by the excess, free phosphophoryns in the region which bind to the crystal surface. Once bound, the crystal surface phosphophoryns are not accessible to proteases and create the dentin zone of crystal stabilization.
We presume that the same types of reactions occur in bone, except that in place of the unique dentin components (phosphophoryns) one can substitute the unique bone extracellular calcium binding protein, osteonectin (21).

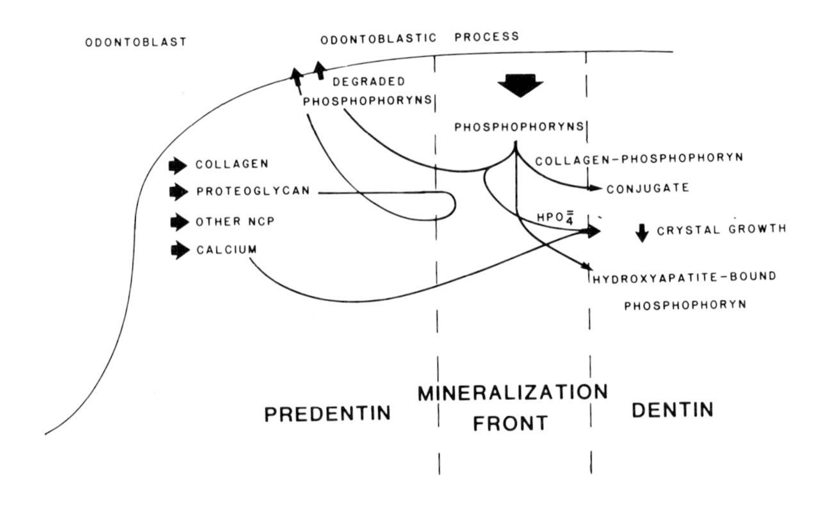

Figure 4. Molecular events related to mineralization of dentin, an hypothetical scheme.

IV. GENERAL CONCEPTS OF BIOMINERALIZATION

Are there any generalizations that can be made from these two examples of coupled but distinctly different modes of mineralization? The most salient feature is that the matrix is a two component system. There is a structural bulk phase which sets the architecture of the system. In enamel formation this may be the role of the amelogenin component. It is obviously the function of the collagen of dentin. The second class of component is interactive and strongly associated with the mineral phase. In enamel, the enamelin reacts intensely with the apatite. It is present in fetal enamel at all stages of development (1) but its relative content increases in mature enamel as a result of resorptive removal of the amelogenins. As postulated above, the phosphophoryns are the mineral interactive components of dentin and they serve two roles: the localization of apatite nucleation sites on the structural matrix; and, the limitation of crystal size.

Other components of the extracellular matrix must be important in inhibition of crystallization from the supersaturated calcium phosphate mileu of the extracellular spaces, and in the varied enzymatic processes required. How all these components operate to nucleate, direct crystal placement, limit crystal growth and stabilize the mineral phase requires detailed study in every case. However, it is likely that all mineralizing systems follow the same plan or general strategy and that in spite of differences in both matrix and mineral phases components of corresponding function will be found.

We are pleased to acknowledge that this work has been supported by grant DE 01732 from the National Institute of Dental Research.

REFERENCES

1. Termine, J.D., Belcourt, A.B., Christner, P.J., Conn, K.M. and Nylen, M.U.: 1980, J. Biol. Chem. 255, pp. 9760-9768.

2. Simmelink, J.W.: 1981, in "Chemistry and Biology of Mineralized Connective Tissues" A. Veis, ed. Elsevier North-Holland, Inc. N.Y., pp. 299-302.

3. Volpin, D. and Veis, A.: 1973, Biochemistry 12, pp. 1452-1464.

4. Scott, P.G.: 1980, Biochemistry 19, pp. 6118-6124.

5. Li, S.T., and Katz, E.P., 1978, J. Dent. Res. 57A, pp. 279.

6. Dimuzio, M.T. and Veis, A.: 1978, Calcif. Tiss. Res. 25 pp. 169-178.

7. Lee, S.L., Veis, A. and Glonek, T.: 1977, Biochemistry 16 pp. 2971-2979.

8. Stetler-Stevenson, W.G. and Veis, A: 1981, Fed. Proc. 5, pp. 1438.

9. Zanetti, M. de Bernard, B., Jontell, M. and Linde, A.: 1981, Eur. J. Biochem. 113 pp. 541-545.

10. Curley-Joseph, J. and Veis, A.: 1979, J. Dent. Res. 58 pp. 1625-1633.

11. Kuboki, Y., Fujisawa, R. Aoyama, K. and Sasaki, S.: 1979, J. Dent. Res. 58, pp. 1926-1932.

12. Termine, J.D., Belcourt, A.B., Miyamoto, M.S. and Conn, K.M.: 1980, J. Biol. Chem. 255 pp. 9769-9772.

13. Carmichael, D.J., Chovelon, A., and Pearson, C.H.: 1975, Calcif. Tiss. Res., pp. 263-271.

14. Hjerpe, A. and Engfeldt, B.: 1976, Res. 22 pp. 173-182.

15. Lee, S.L. and Veis, A.: 1980, Calcif. Tiss. Res. 31 pp. 123-134.

16. Maier, G.D., Lechner, J.H. and Veis, A: 1981, in "Chemistry and Biology of Mineralized Connective Tissues" A. Veis, Ed. Elsevier North-Holland, Inc. pp. 477-487.

17. Weinstock M, and Leblond, C.P.: 1974, J. Cell Biol. 60 pp. 92-127.

18. Dimuzio, M.T. and Veis, A.: 1978, J. Biol. Chem. 253 pp. 6845-6852.

19. Weinstock, M. and Leblond, C.P.: 1973, J. Cell Biol. 56, pp. 838-845.

20. Robison, R.: 1923, J. Biochem 17, pp. 286-293.

21. Termine, J.D., Kleinman, H.K., Whitson, S.W., Conn, K.M., McGarvey, M.L. and Martin, G.R.: 1981, Cell 26, pp. 99-105.

A SYSTEMATIC APPROACH TO SOME FUNDAMENTAL QUESTIONS OF CARBONATE CALCIFICATION

C. Steven Sikes and A. P. Wheeler
University of South Alabama, Mobile, Alabama 36688, USA
Clemson University, Clemson, South Carolina 29631, USA

Our studies of the physiology of carbonate calcification have been organized to address five general questions about the process: 1)what are the sources and forms of dissolved inorganic carbon, 2)what is the role of active transport, 3)what is the role of carbonic anhydrase, 4)how is pH regulated, and 5)what is the role of organic matrix? The studies feature several organisms including coccolithophorids, oysters, freshwater clams, and sea urchin embryos. The results are summarized here and current interpretations presented.

DISSOLVED INORGANIC CARBON (DIC)

In coccolithophorids, the evidence suggests that HCO_3^- from the sea enters the cells as a source of inorganic carbon for both photosynthesis and coccolith formation. In photosynthesis, CO_2 appears to be the actual substrate of carbon fixation, with HCO_3^- influx acting to supplement internal CO_2 supplied directly as CO_2 from the sea. In coccolith formation, CO_3^{2-} clearly is the form of carbon fixation but it originates as HCO_3^- in the medium rather than as CO_2 or CO_3^{2-} produced internally or provided externally. Several lines of evidence in support of these conclusions are available (Sikes et al. 1980, 1981, Sikes and Wilbur 1982, Sikes and Wheeler 1982).

A HCO_3^- flux also is implicated in oyster calcification. For example, when pieces (50-150 mg wet weight) of isolated mantles of "Crassostrea virginica" were exposed to $DI^{14}C$ at pH 8.5 (DIC only 0.2% as CO_2), ^{14}C was incorporated into both an acid volatile pool (DIC) and an acid soluble pool (amino and carboxylic acids). The DIC pool had a turnover rate with $t_{1/2}$ less than 15 min and the acid soluble pool turned over with $t_{1/2}$ more than 60 min (Wheeler, 1980). Because of its rapid turnover rate, the mantle DIC pool which is supplied externally is likely the most significant source of shell carbonate. However, the acid soluble pool was as large or larger than the DIC pool and is released primarily as metabolic CO_2 from the mantle. This source of DIC could contribute substantially to shell formation. Such a situation has been demonstrated in spicule formation by sea urchin embryos (Sikes et al., 1981).

P. Westbroek and E. W. de Jong (eds.), Biomineralization and Biological Metal Accumulation, 285–289.

ACTIVE TRANSPORT

Kinetics of calcium uptake, light dependence, effects of inhibitors, and even presence of Ca ATPases all argue in favor of active transport of calcium during algal calcification (refs. in Sikes and Wilbur, 1982). However, there is a strong inwardly-directed electrochemical force for passive uptake of calcium in coccolithophorids (Sikes and Wilbur, 1982). In addition, calcium influx in these cells is nearly as high as rate of photosynthetic carbon fixation itself, calling for high levels of energy consumption if standard mechanisms of calcium active transport are invoked. Consequently, the mechanism of calcium transport in coccolithophorids is an open question, as it is in other systems as well.

In oyster mantle, values for DIC fluxes may indicate active transport of HCO_3^- toward the shell. Bidirectional flux studies were carried out on paired pieces of isolated mantle mounted in Ussing chambers and exposed to $DI^{14}C$ in artificial seawater at pH 7.8-7.9. The flux rates were determined before and after treatment of the mantle with 10^{-5} M acetazolamide (carbonic anhydrase inhibitor). As seen in Figure 1 (Wheeler, 1975), the ratio of the flux into the experimental equivalent of the extrapallial cavity to the flux out of the equivalent of the cavity was near unity (0.91-1.28, mean=1.02, n=10) in all paired experiments before acetazolamide treatment but was always significantly greater than unity after such treatment (1.67-10.4, mean=5.31, n=10). With the poten-

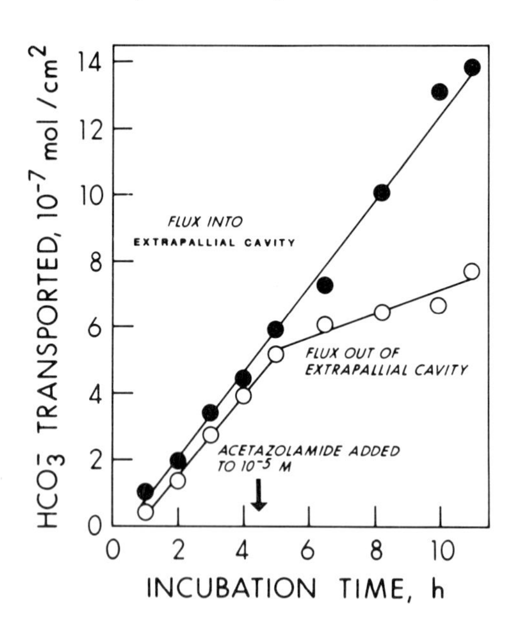

Figure 1. Bidirectional flux of dissolved inorganic ^{14}C across whole oyster mantles. Each data point is the mean of replicate samples in paired experiments in which fluxes were measured in either direction using separate pieces of mantle from one valve of an oyster.

tial difference across the mantle near zero (Wheeler, unpubl.), application of the Ussing criterion for active transport suggests that the presence of acetazolamide exposed some active transport of DIC toward the extrapallial fluid. In addition, the presence of a HCO_3^- ATPase isolated from the mantle of the freshwater clam "Anodonta cataracta" also suggests that active transport of the ion occurs. Furthermore, cell fractionation studies show that some of the HCO_3^- ATPase activity is associated with plasma membrane (Wheeler and Harrison, 1982), as might be expected for an enzyme involved in HCO_3^- transport.

In coccolithophorids, as mentioned, a HCO_3^- influx against an electrochemical gradient is thought to occur (op cit.). Although this may be indicative of operation of an active mechanism of uptake, studies of activity and distribution of HCO_3^- ATPase are not available.

CARBONIC ANHYDRASE (CA) and pH REGULATION

Surprisingly, our studies of coccolithophorids indicate only a minor role for CA in this important system. In "Coccolithus (Emiliania) huxleyi", both calcified and non-calcified cells seemed to lack the enzyme altogether. Low levels of activity were observed in "Hymenomonas carterae" (Sikes and Wheeler, 1982). In oyster mantle, the enzyme is much more prominent and cell fractionation studies showed that about 70% of CA activity is particulate, with part of this having a peak isopycnic density identical with that for plasma membrane markers (Wheeler, 1975, 1979).

In either system, the data suggest that CA functions to facilitate movements of DIC by the mechanism of elimination of CO_2 depletion in unstirred layers. In the coccolithophorids, such facilitated diffusion was observed when CA was added to the medium and the rate of photosynthesis increased by about 30% (Sikes and Wheeler, 1982). In oyster mantles, as shown in Figure 1, when CA was inhibited, the outward facilitated movement of DIC away from the shell was inhibited. This result also has a bearing on the problem of pH regulation. Facilitated removal of a weak acid (dissolved CO_2) from the mineralization site would promote carbonate formation through the maintenance of sufficiently-high pH. In the case of coccolith formation, in fact, the operation of such a mechanism alone can counteract the otherwise acidifying effects of $CaCO_3$ deposition (Sikes et al., 1980).

ORGANIC MATRIX

Oyster soluble matrix (SM), extracted from shell with EDTA, is a very potent inhibitor of $CaCO_3$ precipitation. As shown in Figure 2, at concentrations as low as 10^{-8} M, SM extended the time required for the initiation of crystal growth in solutions highly supersaturated with respect to $CaCO_3$. In addition, when SM was added to solutions in which precipitation already had begun, further precipitation was inhibited (Wheeler et al., 1981). SM also dramatically affected the ultrastructure of $CaCO_3$ crystals grown in vitro at levels of SM sufficient to slow but

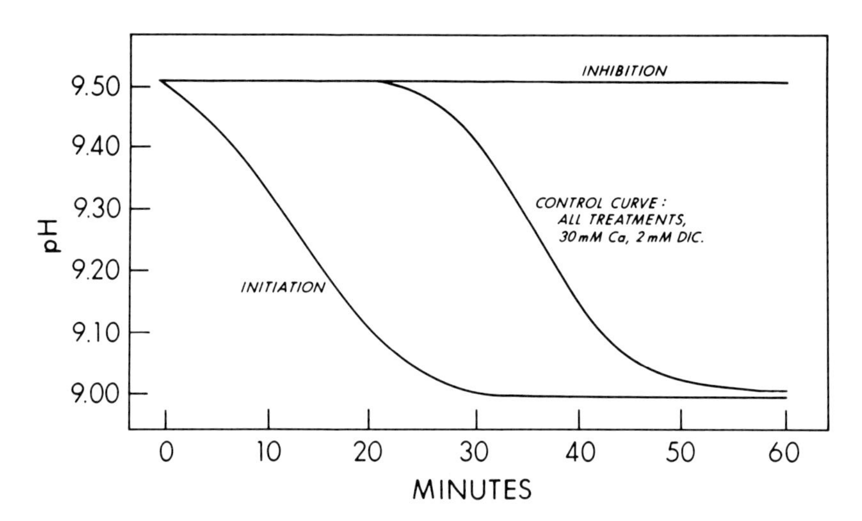

Figure 2. Recorder tracings of $CaCO_3$ nucleation and crystal growth in vitro in the presence and absence of inhibitory or initiating substances. Precipitation of $CaCO_3$ is measured as a function of the pH decrease in artificial seawater accompanying removal of CO_3^{2-} from supersaturated solutions (Wheeler et al. 1981, Sikes and Wheeler 1982). Inhibitory compounds increase the time until onset of pH decrease: initiating compounds hasten fall in pH. We have observed inhibitory effects using oyster SM at 10^{-8} M, and polyaspartate and polyglutamate (Sigma) at 10^{-7} M. Curves essentially the same as those for control treatments (no organic additives) were observed when neutral or positively charged polyamino acids, or when free amino acids of all types were used at concentrations equivalent to those used in inhibitory treatments (i.e., the same number of residues per volume). Initiation of crystal growth has been observed using decalcified crab carapace (Sikes and Wheeler, 1982) and also using non-specific substances such as glass beads or even cheese cloth.

not to stop crystal growth (Sikes et al., unpubl.). These results demonstrate that SM has the capacity to regulate both nucleation and crystal growth, and crystal morphology as well. Although SM binds calcium, less than 2% of the calcium in these experiments was bound to the matrix. Thus the effects of SM were not due simply to removal of free calcium.

As seen in Figure 2, when synthetic polyamino acids were added to supersaturated solutions, inhibition of crystal growth occurred only when negatively charged plymers of aspartate or glutamate were used. These residues are known to comprise a significant portion of molluscan soluble matrix and the possibility of their functioning as sites of inhibition of $CaCO_3$ crystal growth has been hypothesized (Weiner and Traub, 1981). Our results support the concept that it is the negatively charged residues of proteinaceous matrix that act as the functional sites of inhibition. Furthermore, the residues must exist with the discrete spacing of the polymeric form in that free amino acids of aspartate and glutamate had no effect on precipitation. It may be that other anionic groups also participate in regulation of $CaCO_3$ formation, as the matrix ranges from 7-11% phosphate and 4-6% sulfate by protein weight (Wheeler and George, unpubl.).

All of the anionic sites presumably bind calcium and in fact the matrix can bind calcium in excess of the number of total possible binding sites (Wheeler and George, 1981). This is suggestive of a non-specific binding or ionotropy that can result from an initial specific binding of calcium. This interpretation is supported by the finding that SM has two kinds of affinity for calcium as determined by equilibrium dialysis. There is a high-affinity binding (K_D less than 10^{-3} M), perhaps representing specific calcium binding sites. There also is a low affinity binding (K_D more than 10^{-3} M) that may represent ionotropy (Wheeler and George, unpubl.). Calcium binding of this type often has been viewed as having a function in promoting rather than inhibiting nucleation of $CaCO_3$ crystals. Although we have observed initiation of nucleation using preserved, decalcified crab carapace (Sikes and Wheeler, 1982), our studies to date have not revealed, nor do they rule out, this function for oyster SM.

REFERENCES

Sikes, C.S., Roer, R.D., and Wilbur, K.M. 1980, Limnol. Oceanogr. 25, pp. 248-261.
Sikes, C.S., Okazaki, K., and Fink, R.D. 1981, Comp. Biochem. Physiol. 70A, pp. 285-291.
Sikes, C.S. and Wilbur, K.M. 1982, Limnol. Oceanogr. 27, pp. 18-26.
Sikes, C.S. and Wheeler, A.P. 1982, J. Phycol., in press.
Weiner, S. and Traub, W. 1981, pp. 467-482 in M. Balaban et al., eds. Structural aspects of recognition and assembly in biological macromolecules. Balaban ISS, Rehovat and Philadelphia.
Wheeler, A.P. 1975, Dissertation, Duke University, Durham NC, USA.
Wheeler, A.P. 1979, Am. Zool. 19, pp. 995.
Wheeler, A.P. 1980, Am. Zool. 20, pp. 801.
Wheeler, A.P., George, J.W., and Evans, C.A. 1981, Science 212, pp. 1397-1398.
Wheeler, A.P. and George, J.W. 1981, Am. Zool. 21, pp. 941.
Wheeler, A.P. and Harrison, E.W. 1982, Comp. Biochem. Physiol., in press.

CALCIFICATION IN COCCOLITHOPHORIDS

E.W. de Jong, P. van der Wal, A.H. Borman, J.P.M. de Vrind,
P. van Emburg, P. Westbroek and L. Bosch
Department of Biochemistry, State University of Leiden
Wassenaarseweg 64, 2333 AL Leiden
The Netherlands

INTRODUCTION

Coccolithophorids, a group of monocellular, photosynthetic marine algae belonging to the class of Haptophyceae Parke and Dixon, are characterized by their ability to form calcified bodyscales, so-called coccoliths. These coccoliths are fine round or oval structures which, on analysis, are composed of $CaCO_3$. The morphology of coccoliths is species-specific and enormous variations in sizes and shapes has been described (cf. Black, 1965; Okada and McIntyre, 1977). Fossil coccoliths are important biostratigraphical tools. Geographically, nowadays, these coccolithophorids have a wide area of distribution: they have been found in tropical, subtropical as well as in arctic waters (Klaveness and Paasche, 1979) and contribute to a considerable extent to the formation of extensive lime ores in the deepsea. Also in the geological past the coccolithophorids have been partly responsible for mass accumulations of $CaCO_3$, especially in the Cretaceous.

The physiology and biochemistry of the process of coccolith formation has been studied in various laboratories (*e.g.* Wilbur and Watabe, 1963; Isenberg *et al.*, 1966; Outka and Williams, 1971; Pienaar, 1971; Klaveness and Paasche, 1979). Here a cursory review will be given of the investigations we performed on calcification in *Emiliania huxleyi* and *Hymenomonas carterae*.

CALCIFICATION IN *EMILIANIA HUXLEYI*

Morphology of Cells and Coccoliths

The coccoliths formed by this species consist of 20-40 radially arranged entities inside an oval ring (Fig. 1). Each entity is composed of a flat lower part and a T-shaped upper part connected by a vertical stem. Watabe (1967) demonstrated that each radial spoke yields an electron diffraction pattern characteristic of a single crystal. These crystallites differ greatly from the rhombohedral calcite crystals generally formed in inorganic precipitation. This difference in crystal

P. Westbroek and E. W. de Jong (eds.), Biomineralization and Biological Metal Accumulation, 291–301.

shape can be explained by assuming that the living cells are able to regulate $CaCO_3$ crystallization. *Emiliania huxleyi* has a specialized intracellular machinery for coccolith formation (Wilbur and Watabe, 1963; Klaveness, 1972a).

Morphologycally, this synthetic process is localized in rather specialized areas, characterized by special organelles (Fig. 2). Close to the nucleus a large long vesicle can be observed which is connected with a system of anastomizing tubules called the reticular body (Wilbur and Watabe, 1963). The coccolith (or its remnants) can be observed to be present inside the vesicle. The presence of Golgi zones suggests that at least part of the synthetic processes take place here. In an early stage of coccolith formation small rhombohedral crystals can be observed in the periphery of the coccolith vesicle (Fig. 3, see also Klaveness, 1979). Usually the coccolith obtains its final shape within one hour and is then excreted. The continuity of the lumen of the coccolith vesicle with that of the reticular body suggests a function of the latter in calcification. After completion of the coccolith the reticular body gradually disappears (Fig. 4). Moreover, the reticular body degenerates when coccolith formation is stopped (by placing the cells in the dark, see below). It is regenerated when the algae are subsequently incubated in the light (van der Wal, unpublished results). The exact function of the reticular body is unknown. Wilbur and Watabe (1963) suggested that it may be the site of synthesis of the organic constituents of coccoliths. At least, part of these constituents are polysaccharides (see below). Special staining techniques indicate the presence of polysaccharides in the reticular body as well as in the coccolith vesicle (van der Wal, in preparation; cf. Klaveness, 1976). The reticular body may be the site of synthesis and/or transport of this material.

Composition of Coccoliths

Coccoliths are associated with organic material. Part of this material can be visualized with electron microscopy. In sections of embedded cells the mineral usually is dissolved. Klaveness (1972) demonstrated the presence in such sections of an organic matrix with the outline of a coccolith. Cytochemical reactivity suggested that the matrix material consists of glycoprotein.

We cultivated the algae in batches of 300-1,000 l. The coccoliths can be isolated from the cells by ultrasonication and density centrifugation (de Jong *et al.*, 1976). When the isolated coccoliths are dissolved in EDTA (ethylenediamine tetraacetate, 10%, pH 8) two macromolecular fractions are obtained: a soluble and an insoluble material. Preliminary investigations suggest that the insoluble material contains some glycoprotein (Westbroek *et al.*, 1982). Whether it is identical to the matrix material reported by Klaveness remains to be investigated. We have demonstrated that the soluble material is a polysaccharide. Presumably the soluble and insoluble materials cooperate in regulating $CaCO_3$ crystallization. So far, most of our attention has been focussed on the EDTA-soluble polysaccharide.

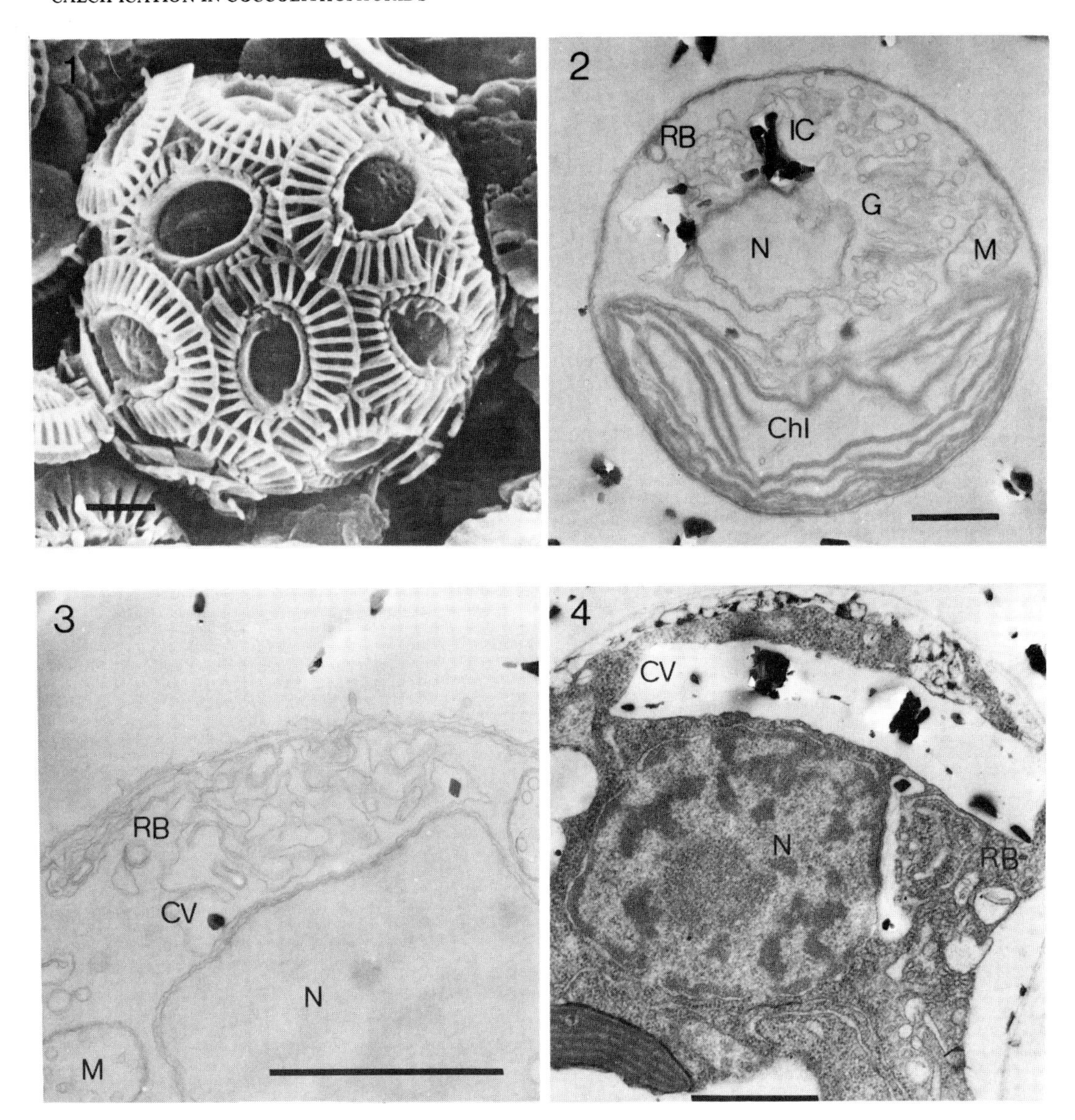

Fig. 1. Scanning electron micrograph of coccolith cover and loose coccoliths of *Emiliania huxleyi*. Bar = 1 μ.

Figs. 2-4. Electron micrographs of ultrathin sections through cells of *E. huxleyi*. Figs. 2 and 3: fixed in glutaraldehyde and OsO_4 solutions to which aminotriazole had been added. Fig. 4: fixed in glutaraldehyde and OsO_4, Pb-citrate stained. Chl: chloroplast; CV: coccolith vesicle; G: Golgi complex; M: mitochondrion; N: nucleus; IC: intracellular coccolith; RB: reticular body. Part of the $CaCO_3$ was dissolved during preparation leaving holes in the sections. Bar = 1 μ.

The Coccolith Polysaccharide, Properties and Biosynthesis

The primary structure of the coccolith polysaccharide has been investigated by Fichtinger-Schepman, Kamerling and Vliegenthart (Fichtinger-Schepman *et al.*, 1981). The molecule is of a remarkable complexity. The partial structure of the mean repeating unit is depicted in Fig. 5. The polysaccharide contains at least 14 different monosaccharides. Several monosaccharides contain one or two methyl instead of hydroxyl groups. They may represent relatively hydrophobic regions in the molecule. Some of these monosaccharides have never been found before to be constituents a naturally occurring polysaccharide.

The polysaccharide is acidic: it harbours sulphate esters and uronic acids. Generally, the primary structure of a polysaccharide is determined by the specificity of the glycosyltransferases. The biosynthesis of the complex coccolith polysaccharide would require a large investment of enzymes and energy. We therefore assume that this complicated molecule must fulfil an important role in coccolith formation. Chemically its association with the calcite is rather strong: it cannot be solubilized by treating coccoliths with alkaline solutions not containing in the meantime calcium-chelating agents too. We found that the polysaccharide is able to bind Ca^{2+} considerably stronger than Mg^{2+} and Na^{+} (de Jong, 1976). Moreover, it affects $CaCO_3$ precipitation *in vitro* (Borman *et al.*, this volume, and Borman *et al.*, submitted for publication). The soluble matrices from *Mercenaria mercenaria* and oyster shells are reported to have similar properties (Crenshaw, 1972; Wheeler and Evans, 1981, Sikes and Wheeler, this volume). Although our findings suggest a regulatory function of the polysaccharide in coccolith formation, they do not give a decisive answer about the activity of the molecule *in vivo*. The effect of the polysaccharide on $CaCO_3$ crystallization may depend on its conformation (Borman *et al.*, submitted for publication) and on its interaction with other components in the coccolith vesicle.

We found that from complete cells that were labelled with radioactive bicarbonate two macromolecular fractions could be isolated with gel electrophoresis, carrying several of the monosaccharides characteristic of the coccolith polysaccharide. These fractions presumably represent polysaccharide precursors (Westbroek *et al.*, 1982). At this stage it cannot be decided whether these fractions are different in composition or that they represent aggregates of a single molecule. The precursors probably are associated with protein. The polysaccharide obtained from isolated coccoliths is protein-free. Possibly the polysaccharide is anchored to other components in the calcifying apparatus by means of this protein fraction at a certain stage of its biosynthesis. It is not known whether the material is functional in calcification while it is in this associated form.

NON-CALCIFYING CELLS

Cells of *E. huxleyi* can loose the ability to form coccoliths (*e.g.* Crenshaw, 1964; Blankley, 1971; de Jong *et al.*, 1979). The cause

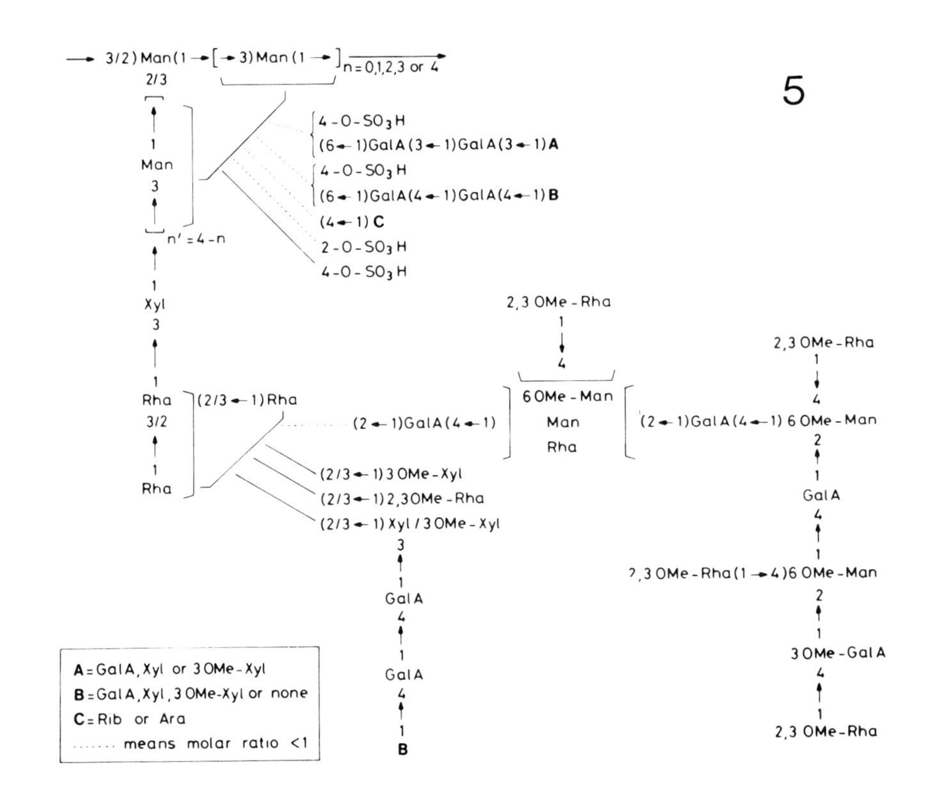

Fig. 5. Preliminary structure of the mean repeating unit of the polysaccharide associated with coccoliths of *E. huxleyi*.

of this phenomenon is unknown. Although some authors have reported that calcification can be restored by modification of the culturing conditions (*e.g.* Klaveness, 1979) this phenomenon could never be reproduced in our laboratory. The morphology of the non-calcifying cells is very similar to that of the calcifying ones (Klaveness, 1971; van der Wal, in preparation). The morphological entities related to calcification, coccolith vesicle plus reticular body, is fully developped but no visual evidence for calcification is found.

When non-calcifying cells are grown in seawater for several days an acid polysaccharide can be isolated from the culture medium (de Jong *et al.*, 1979). The excreted polysaccharide is very similar to that obtained from coccoliths. However, contrary to the coccolith polysaccharide it contains about 20% (w/w) protein. The protein component probably present in the coccolith polysaccharide precursors (see above) is split off during coccolith formation. We are presently investigating the possibility that the protein moiety of the polysaccharide excreted into the medium by non-calcifying cells is identical with that of the coccolith polysaccharide precursor(s).

COCCOLITH FORMATION IN *HYMENOMONAS CARTERAE*

Morphology of the Cells

The sequence of events leading to coccolith formation is different in *E. huxleyi* and in *H. carterae*. Coccoliths of *H. carterae* consist of body scales decorated with calcite elements (Manton and Leedale, 1969; Pienaar, 1971). The calcite elements are associated with an organic matrix (*e.g.* Outka and Williams, 1971). Non-calcified scales and coccoliths are formed in cisternae of the Golgi apparatus (Fig. 6). The cis-

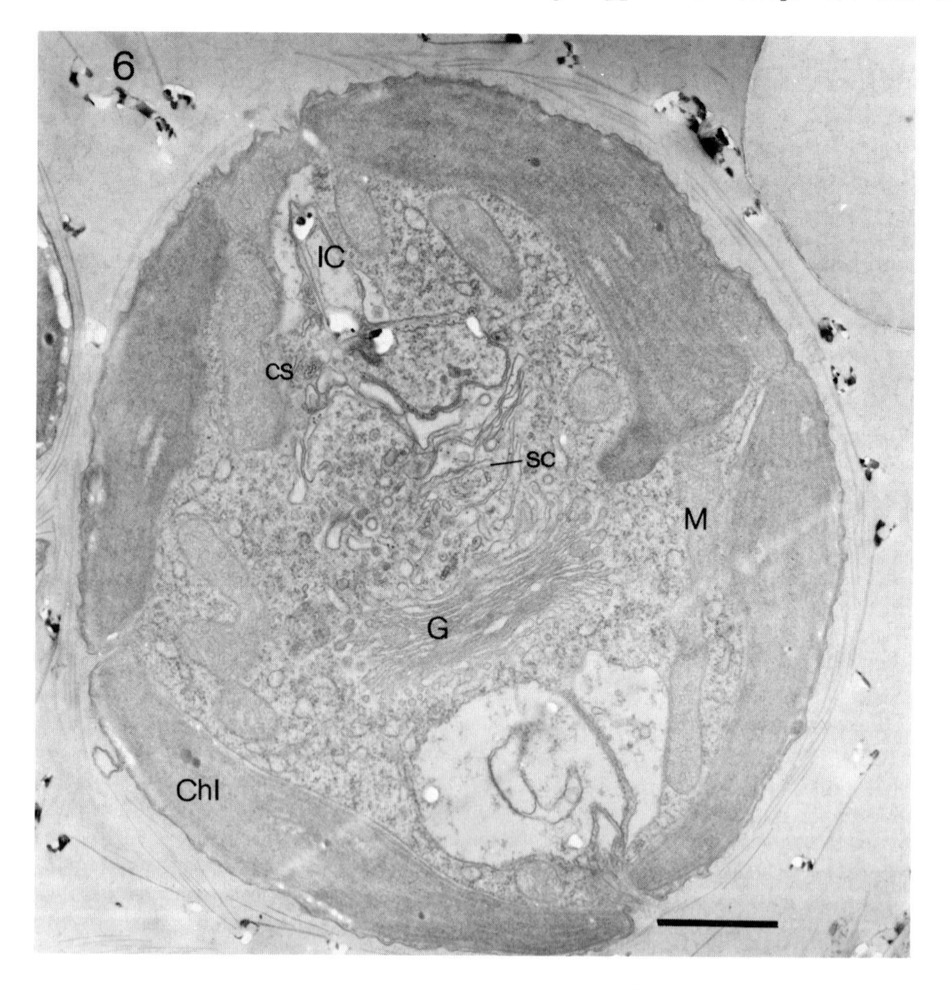

Fig. 6. Electron micrograph of ultrathin section through a cell of *Hymenomonas carterae*. cs: coccolithosomes; sc: scale. For other abbreviations see legend to Figs. 2-4. Fixed in glutaraldehyde and OsO_4. Pb-citrate stained. Bar = 1 μ.

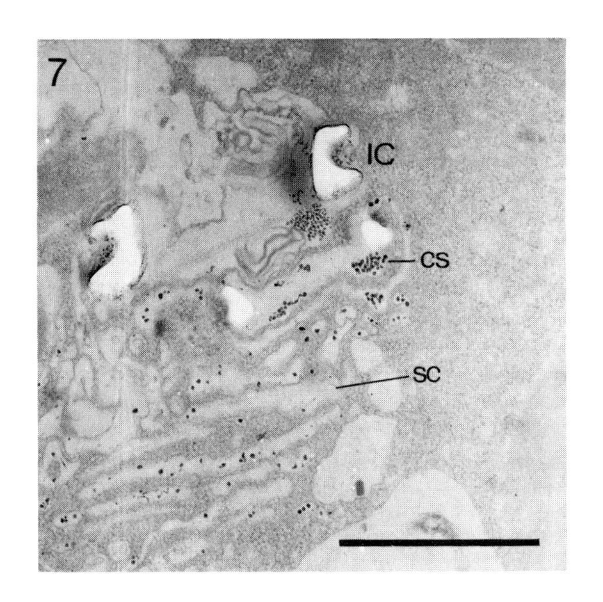

Fig. 7. Electronmicrograph of ultrathin section through part of the Golgi complex of *H. carterae*. Vesicles containing coccoliths (c) scales (sc) and coccolithosomes (cs) are present. White areas are holes left by dissolution of $CaCo_3$. Fixed in glutaraldehyde and OsO_4. Pb-citrate stained. Bar = 1 µ.

ternae containing the organic coccolith base plates also harbour small granular bodies called coccolithosomes (Outka and Williams, 1971). Vesicles merely containing coccolithosomes are also present (fig. 7). This suggests that coccolithosomes are synthesized in separate vesicles which eventually fuse with those containing the base plates. Outka and Williams (1971) suggested that the coccolithosomes are the precursors of the organic matrix associated with the calcite elements. We demonstrated that both coccolithosomes and matrix are heavily stained after treatment with CPC (cetyl pyridinium chloride) and silverproteinate, indicative for the presence of acidic polysaccharides (v.d. Wal, in preparation). X-ray microanalysis of such coccolithosomes showed that they contain Ca, S and Mg in a ratio of 25:3:2 (v.d. Wal *et al.*, 1982). It can be suggested that they may be the carriers of both the calcium and the organic coating necessary to form the calcite pinnacles on the organic base plates. Coccolithosomes are absent in *Emiliania huxleyi*.

Chemical Composition of Coccoliths

The chemical analysis of coccoliths of *H. carterae* is seriously hampered by the difficulty of obtaining them in a pure form (de Jong, 1975). A preparation of coccoliths usually contains uncalcified scales as well. When the $CaCO_3$ of such a preparation is dissolved in EDTA, two macromolecular fractions are solubilized which can be separated by electrophoresis (de Jong, 1975) or by ion exchange chromatography. Both fractions appear to consist mainly of highly acidic polysaccharides containing over 50% uronic acids. Both differ greatly in composition from the polysaccharide associated with coccoliths of *Emiliania huxleyi*. Their localization is uncertain: they may be constituents of the acidic coating associated with the calcite crystals or of the organic base plate of the coccoliths (cf. Brown *et al.*, 1973).

LIGHT DEPENDENCE OF COCCOLITH FORMATION IN *EMILIANIA HUXLEYI* AND *HYMENOMONAS CARTERAE*

Under normal culture conditions calcification in both species stops when cells are placed in the dark. Non-illuminated cells do not form coccoliths and show reduced Ca-accumulation (e.g. Crenshaw 1964, Blankley 1971, de Jong 1979). Sikes *et al.* (1980) suggested that calcification and photosynthesis may be mutually dependent according to the reaction:

$$2HCO_3^- \rightleftarrows \underset{\substack{\downarrow \\ \text{calcification}}}{CO_3^{2-}} + \underset{\substack{\downarrow \\ \text{photosynthesis}}}{CO_2} + H_2O$$

Although there is some evidence to support this hypothesis, the situation *in vivo* may be more complicated (cf. Klaveness, 1979, and Erez, this volume). Ariovich and Pienaar (1979) reported coccolith formation by decalcified cells of *H. carterae* in the dark. They assumed that the Ca^{2+} of the coccoliths was derived from an intermediate reservoir formed by the cells during illumination. We showed, however, that decalcification *per se* induces Ca^{2+} accumulation by cells of *H. carterae* in the dark (v.d. Wal *et al.*, 1982). No evidence for an intermediate Ca-reservoir was found. We suggest that calcification in *H. carterae* is dependent on energy stored by the cells during the light period. This picture is complicated by the fact that decalcified cells of *E. huxleyi* hardly accumulate Ca^{2+} or form coccoliths in the dark.

CONCLUDING REMARKS

The title of this paper "Calcification in Coccolithophorids" suggests a more or less uniform mechanism for coccolith formation in different species. Although there may be a common principle underlying coccolith synthesis as we will discuss below, the variety in mineralizing systems is considerable. Not only may coccoliths be formed in different areas of the cell (e.g. in the coccolith vesicle in *E. huxleyi*, or in the Golgi zone in *H. carterae*), the composition of their organic constituents may also differ substantially. Moreover, the calcifying system may vary within one species, the morphology or production of coccoliths depending on the life cycle of the algae. For instance, apart from coccolith bearing cells at least two non-calcifying cell types of *E. huxleyi* exist (Klaveness and Paasche, 1971): naked cells, also described above, and scale-bearing cells. The latter do not calcify but have an internal vacuole containing a crystalline Ca-salt (v.d. Wal, unpublished results). Whether the interconversion of these different cell types responds to variations in the environment is unknown.

Descriptive work on fossil and recent coccoliths show that, although coccolith formation in one species must be submitted to a strict biochemical regime, the system is very plastic in an evolutionary sense. An enormous morphological variety of coccoliths is known, and those of *E. huxleyi* and *H. carterae* appear to belong to the more simple types.

In spite of the apparent variability of the calcifying systems in coccolithophorids one still may point to common features. Coccoliths of both *E. huxleyi* and *H. carterae* are formed in intracellular vesicles; they are associated with EDTA-soluble acidic polysaccharides. Coccoliths of *E. huxley* also contain an EDTA-insoluble fraction, presumably glycoprotein, and in *H. carterae* a similar situation exists.

Consideration of various systems of biomineralization, particularly in molluscs and dentin, suggest that similar principles are operating. Virtually all mollusc shells contain soluble acidic proteins thought to play a role in the regulation of $CaCO_3$ crystallization by stimulating and directing nucleation and growth and/or by termination of crystal growth. They are presumably anchored onto insoluble carrier proteins (Weiner *et al.*; Krampitz *et al.*; Nakahara, this volume). In dentin collagen functions as a carrier for the soluble and highly acidic phosphophorins which probably regulate hydroxyapatite growth (Veis and Sabsay, this volume). Both the soluble acidic proteins from several mollusc shells and from dentin are able to bind Ca^{2+} (Crenshaw 1972; Weiner *et al.*; Krampitz *et al.*; Veis and Sabsay, this volume). Phosphophorins have a high affinity for hydroxyapatite crystals (Veis and Sabsay, this volume). Oyster shell matrix was shown to inhibit $CaCO_3$ precipitation *in vitro* (Wheeler *et al.*, 1981; Sikes and Wheeler, this volume) and this is presumably also due to binding to crystal nuclei.

The acid polysaccharide from coccoliths of *E. huxleyi* also binds Ca^{2+} and inhibits $CaCO_3$ precipitation *in vitro*. The possibility that in the early stages of coccolith formation the polysaccharide is anchored onto the insoluble glycoprotein component is presently under investigation. Preliminary experiments suggest that the polysaccharide is bound to slowly growing $CaCO_3$ crystals. It is therefore conceivable that calcification in coccolithophorids has basic features in common with calcification in invertebrates and vertebrates.

We must realize, however, that our knowledge is still far from complete. Although it is likely that the organic components of coccoliths are functional in calcification, the evidence is circumstancial.

A special feature of the studied coccolithophorids is that calcification occurs in intracellular vesicles. As Mann *et al.* (this volume) pointed out the vesicle membrane will play an important role in regulating mineralization, e.g. by controlling the intravesicular pH and ion concentrations.

We are only beginning to understand the mechanisms of biomineralization. Coccolithophorids seem a suitable model for studying calcification. Here all the instruments needed for the production of biominerals of a highly specific morphology are combined in a single cell.

ACKNOWLEDGEMENTS

The morphological work was performed in collaboration with Dr. W.C. de Bruijn and Prof.Dr. W.Th. Daems, Laboratory of Electron Microscopy, University of Leiden, Leiden, The Netherlands. Part of the investigations was supported by the Netherland Organization for the Advancement of Pure Research (Z.W.O.).

REFERENCES

Ariovich, D. and Pienaar, R.N., 1979. The role of light in the incorporation and utilization of Ca^{2+} ions by *Hymenomonas carterae* (Braarud et Fagerl.), Braarud (Prymnesiophyceae). Br. Phycol. J. 14, 17-24.

Black, M., 1965. Coccoliths. Endeavour 24, 131-137.

Blankley, W.F., 1971. Auxotrophic and heterotrophic growth and calcification in coccolithophorids. Ph.D. thesis, Univ. California, San Diego.

Brown, R.M., Herth, W., Franke, W.W. and Romanovics, D., 1973. The role of the golgi apparatus in the biosynthesis and secretion of a cellulosic glycoprotein in *Pleurochrysis*: a model system for the synthesis of structural polysaccharides. In: Biogenesis of plant cell wall polysaccharides (Loewus, F., ed.) pp. 207-257.

Crenshaw, M.A., 1964. Coccolith formation by two marine coccolithophorids, *Coccolithus huxleyi* and *Hymenomonas sp.*, Ph.D. thesis. Duke University, Dept. of Zoology, Durham, N.C.

Crenshaw, M.A., 1972. The soluble matrix from *Mercenaria mercenaria* shell. Biomineral. Res. 6, 6-11.

Fichtinger-Schepman, A.M.J., Kamerling, J.P., Versluis, C. and Vliegenthart, J.F.G., 1981. Structural studies of the methylated, acidic polysaccharide associated with coccoliths of *Emiliania huxleyi* (Lohmann) Kamptner, Carbohydrate Res. 93, 105-123.

Isenberg, H.D., Douglas, S.D., Lavine, L.S., Spicer, S.S. and Weisfellner, H., 1966. A protozoan model of hard tissue formation. Ann. N.Y. Acad. of Sci. 136, 155-190.

de Jong, E.W., 1975. Isolation and characterization of polysaccharides associated with coccoliths. Ph.D. thesis, University of Leiden, the Netherlands.

de Jong, E.W., Bosch, L., and Westbroek, P., 1976. Isolation and characterization of a Ca^{2+}-binding polysaccharide associated with coccoliths of *Emiliania huxleyi* (Lohmann) Kamptner. Eur. J. Biochem. 70, 611-621.

de Jong, E.W., van Rens, L., Westbroek, P. and Bosch, L., 1979. Biocalcification by the marine alga *Emiliania huxleyi* (Lohmann) Kamptner Eur. J. Biochem. 99, 559-567.

Klaveness, D. and Paasche, E., 1971. Two different *Coccolithus huxleyi* cell types incapable of coccolith formation. Arch. Mikrobiol. 75, 382-385.

Klaveness, D., 1972. *Coccolithus huxleyi* (Lohmann) Kamptner. I. Morphological investigations on the vegetative cell and the process of coccolith formation. Protistologica. T. VIII, fasc. 3, p. 335-346.

Klaveness, D., 1976. *Emiliania huxleyi* (Lohmann) Hay & Mohler. III. Mineral deposition and the origin of the matrix during coccolith formation. Protistologica. T. XII, fasc. 2, p. 217-224.

Klaveness, D. and Paasche, E., 1979. Physiology of Coccolithophorids. In: Biochemistry and physiology of Protozoa: 191-213.

Manton, I. and Leedale, G.F., 1969. Observations of the microanatomy of *Coccolithus pelagicus* and *Cricosphaera carterae*, with special reference to the origin and nature of coccoliths and scales. J. Marine Biol. Ass. U.K. 49, 1-16.

Okada, H. and McIntire, A., 1977. Modern coccolithophores of the Pacific and North Atlantic Oceans. Micropaleontology 23, 1-55.

Outka, O.E. and Williams, D.C., 1971. Sequential coccolith morphogenesis in *Hymenomonas carterae*. J. Protozool. 18, 285-297.

Pienaar, R.N., 1971. Coccolith production in *Hymenomonas carterae*.Protoplasma 73, 217-224.

Sikes, C.S., Roer, R.D. and Wilbur, K.M., 1980. Photosynthesis and coccolith formation: In organic carbon sources and net inorganic reaction of deposition. Limnol. Oceanogr. 25, 248-260.

van de Wal, P., de Jong, E.W. and Westbroek, P., 1982. Calcification in the coccolithophorid alga *Hymenomonas carterae*. Ecol. Bull. in press.

Watabe, N., 1967. Crystallographic analysis of the coccolith of *Coccolithus huxleyi*. Calc. Tiss. Res. 1, 114-121.

Westbroek, P., de Jong, E.W., v.d. Wal, P., Borman, A.H., de Vrind, J.P.M., van Emburg, P., and Bosch, L., 1982. Coccolith formation, wasteful or functional? Ecol. Bull. in press.

Wheeler, A.P., George, J.W. and Evans, C.A., 1981. Control of Calcium Carbonate nucleation and crystal growth by soluble matrix of oyster shell. Science 212, 1397-1398.

Wilbur, K.M. and Watabe, N., 1963. Experimental studies on calcification in molluscs and the alga *Coccolithus huxleyi*. Ann. N.Y. Acad. Sci. 109, 82-112.

INHIBITION OF $CaCO_3$ PRECIPITATION BY A POLYSACCHARIDE ASSOCIATED WITH COCCOLITHS OF *EMILIANIA HUXLEYI*

A.H. Borman, E.W. De Jong, M. Huizinga and P. Westbroek
Department of Biochemistry, State University of Leiden,
Wassenaarseweg 64, 2333 AL Leiden, The Netherlands

The coccolithophoridae are characterized by their ability to form a calcified cell cover consisting of calcite platelets called coccoliths. In *Emiliania huxleyi* the coccoliths are formed intracellularly under strict cellular control (De Jong *et al.*, this volume). If the $CaCO_3$ of isolated coccoliths is dissolved in EDTA-NaOH pH 8.0, two macromolecular fractions can be obtained: an insoluble and a soluble fraction. When the insoluble fraction is submitted to acid hydrolysis (6 N HCl, 18 h, 100°C) monosaccharides and amino acids are released, indicating that the insoluble material contains polysaccharide and protein (1). Most of the insoluble fraction resists hydrolysis, however. The nature of this material remains to be investigated. The soluble fraction consists of a Ca^{2+}-binding polysaccharide containing uronic acid and ester sulphate groups (2). Presumably part of the cellular control over calcification is mediated by this polysaccharide. Much of the work in our laboratory is focused on the question whether this polysaccharide plays a regulatory role in the nucleation, growth and/or termination of calcite crystallization.

In this paper we describe the effect of the coccolith polysaccharide on $CaCO_3$ precipitation *in vitro*. As reported by Wheeler *et al.* (3) the precipitation of $CaCO_3$ can be studied by recording the decline in pH resulting from the reaction:

$$Ca^{2+} + HCO_3^- \rightleftharpoons CaCO_3 + H^+$$

Upon mixing of $CaCl_2$ and $NaHCO_3$ the pH drops rapidly and then remains stable for a few minutes. Subsequently a precipitate is formed causing a further decrease in pH (Figure 1, ◘). Thus precipitation proceeds in two distinct steps. First crystal precursors are formed (probably amorphous $CaCO_3$ hydrates, cf. Nancollas, this volume) followed by crystal growth resulting in a visible precipitate. Generally a precipitate is formed within five minutes after the mixing of $CaCl_2$ and $NaHCO_3$ (Figure 1, ◘). The precipitation can be delayed for about 10 minutes when 20 μg of polysaccharide is added to the reaction mixture (Figure 1, Δ). Addition of 50 μg polysaccharide delays precipitation for at least 6 hours (Figure 1, O). These results can be explained by assuming

P. Westbroek and E. W. de Jong (eds.), Biomineralization and Biological Metal Accumulation, 303–305.

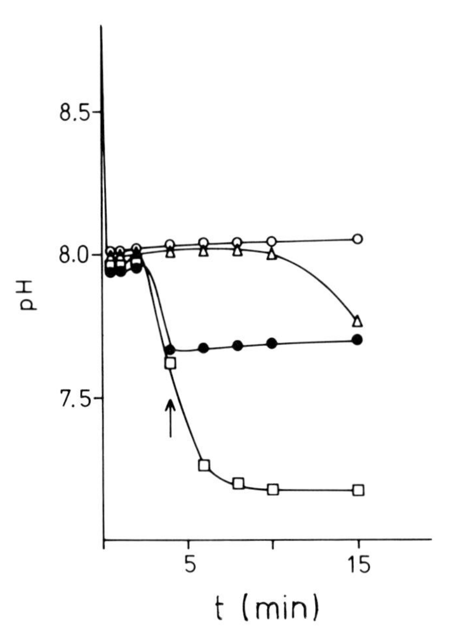

Figure 1. Effect of coccolith polysaccharide on $CaCO_3$ precipitation. The rate of $CaCO_3$ precipitation was determined by recording the decrease in pH. Three ml of 20 mM $CaCl_2$ (pH approximately 8.7) was added to a solution containing 20 mM $NaHCO_3$ (pH approximately 8.7) and 0.3 ml of H_2O (□) or 0.3 ml of a solution containing 20 µg (Δ) or 50 µg (O) of coccolith polysaccharide. In one experiment 50 µg polysaccharide was added 4 minutes after onset of precipitation (arrow, ●).

an interaction of the polysaccharide with crystal nuclei, thus blocking further growth of the latter. Addition of 50 µg polysaccharide after onset of precipitation immediately stops the decrease in pH (Figure 1, ●), probably through binding to crystal nuclei and to the growth sites of the crystals.

Subsequent experiments revealed the following:
(1) The effect of the polysaccharide on $CaCO_3$ precipitation depends upon the nature of the cations which are bound to the acid groups. The polysaccharide used in the experiments of Figure 1 contained Na^+ ions. The Ca^{2+} salt of the polysaccharide is much less effective in inhibiting $CaCO_3$ precipitation than the Na^+ salt.
(2) Only the carboxyl groups of the uronic acids play a role in the inhibition of $CaCO_3$ crystallization. (3) However, the presence of uronic acids is not the only factor involved in this inhibition. We found that the conformation of the molecule may play a role in this process (more detailed information is presented in Borman *et al.* (1982) submitted for publication).

REFERENCES

1. Westbroek, P., De Jong, E.W., Van der Wal, P., Borman, A.H., De Vrind, J.P.M. , Van Emburg, P., and Bosch, L.: 1982, Ecological Bull., in press.
2. De Jong, E.W., Bosch, L., and Westbroek, P.: 1976, Eur. J. Biochem. 70, pp. 611-621.
3. Wheeler, A.P., George, J.W., and Evans, C.A.: 1981, Science 212, pp. 1397-1398.

CALCIFICATION RATES, PHOTOSYNTHESIS AND LIGHT IN PLANKTONIC FORAMINIFERA

Jonathan Erez
The Hebrew University of Jerusalem
The H. Steinitz Marine Biology Laboraotry
P.O.B. 469, Eilat
Israel

Introduction

Two processes that are of major importance in the global carbon cycle, photosynthesis and calcification, are widely associated in the marine environment. This association can be direct, as in coccolithophores and calcareous algae, or in the form of symbiosis, as found in hermatypic corals, planktonic and benthonic foraminifera. These groups of organisms are the major calcifiers in the ocean and precipitate the bulk of $CaCO_3$ in the pelagic and the neritic environments.

It is generally suggested that photosynthesis enhances calcification by uptake of CO_2 which modifies the carbonate system to yield higher supersaturation conditions and therefore faster precipitation rate of calcium carbonate. This mechanism has been suggested for hermatypic corals and calcareous algae (Goreau, 1960; 1961; 1963). It was similarly suggested that coccolithophores enhance photosynthesis by the formation of their calcitic coccoliths (Sikes et al., 1980; Sikes & Wilbur, 1982). Therefore photosynthesis and calcification may be mutually enhanced. Generally, the chemical reaction involved in this process is presented as follows:

$$Ca^{++} + 2HCO_3^- \rightleftarrows \downarrow CaCO_3 + CO_2\uparrow + H_2O$$

(Photosynthesis above the arrow; Calcification below)

The major evidence for this model comes from consistent observations on light enhanced calcification. Goreau (1963) has shown that both hermatypic corals and calcareous algae calcify much faster in the light compared to dark controls. This has been shown also in benthonic foraminifera (Lee & Bock, 1976; Erez, 1978). In these light-dark experiments the effects of photosynthesis and light were not separated. Vandermulen et al. (1972) tried to separate these two factors using a specific photosynthetic inhibitor (DCMU). They concluded that photosynthesis is responsible for the light enhanced calcification in hermatypic corals.

P. Westbroek and E. W. de Jong (eds.), Biomineralization and Biological Metal Accumulation, 307–312.

Goreau's (1960) model was also adopted to describe carbon isotope fractionation between the skeleton and the symbionts in hermatypic corals (Weber & Woodhead, 1970; 1972; Goreau, 1977a; 1977b). Sikes & Wilbur (1982) showed evidence that this model may indeed hold for carbon isotope fractionation in coccolithophores. But a direct attempt to test this model for benthonic foraminifera and hermatypic corals (Erez, 1978) showed that when photosynthesis increases, the isotopic composition of the skeleton becomes lighter rather than heavier as expected from the model (Weber & Woodhead 1970; 1972). Land et al. (1977) measured the isotopic composition of hermatypic corals and their symbionts at different depths. Their data does not support the above-mentioned model either.

The present study was carried out in order to measure rates of calcification and photosynthesis in planktonic foraminifera and to find if calcification is enchanced by photosynthesis and/or by light. Despite their major importance in $CaCO_3$ production in the ocean, very little is known about the processes of calcification in planktonic foraminifera. In part this is because, until recently, it was not possible to culture them in the laboratory. The pioneer work of Bé, Spindler and Hemleben (Bé et al. 1977) demonstrated that it is possible to collect these organisms from the oceanic water and culture them in the laboratory until they reach reproduction. This can be done only in areas where oceanic water is found close to land. In the Gulf of Eilat we have ideal conditions for culturing planktonic foraminifera: Deep water of oceanic nature is found at a distance of only 2 km from the H. Steinitz Marine Biology Laboratory of the Hebrew University.

Methods

Small planktonic foraminifera were collected using a 200 µm plankton net from surface water off the H. Steinitz Marine Biology Laboratory, Eilat, Israel. After two days of recovery and feeding they were labelled with ^{45}Ca and ^{14}C and kept for 5 hrs (1200-1700 hrs) near a window. After incubation they were transferred to non-radioactive sea water for 15 min., then rinsed 10 times with sea water, using an 0.45 µm glass fiber filter, with a final quick rinse with distilled water. After drying in a dessicator for 48 hours, they were weighed using a Cahn microbalance, and analyzed for radioactivity. The method is described in detail by Erez (1978). Briefly, the skeletons were dissolved in 2 ml of 8.5% H_3PO_4 and the $^{14}CO_2$ from the skeleton was trapped - in 2 ml of Carbosorb. The acid fraction was filtered on an 0.45 millipore filter to collect the insoluble organic matter. Thus each sample yielded 3 fractions:

1. The Carbosorb fraction which represents the ^{14}C in the skeleton.

2. The millipore filter which contains the ^{14}C in the insoluble organic matter.

3. The filtered acid which contains the ^{45}Ca of the skeleton and some ^{14}C from the soluble organic matter.

Each fraction was counted using a Packard scintillation counter using Instagel scintillation liquid. The last fraction (No.. 3) was recounted 3 months later and the exact initial ^{45}Ca activity was calculated according to the known decay constant. The remaining ^{14}C activity from this fraction was added to the activity of fraction No. 2. Fraction No. 2 represents the photosynthetically fixed ^{14}C and fraction No. 3, after correction, represents the ^{45}Ca incorporated into the skeleton.

Two separate experiments were carried out on different days with different treatments: light, with varying concentrations of DCMU (10^{-5} to 10^{-7}M),and a dark control. The details are shown in Table 1.

Results and Discussion

The results are shown in Table 1. Photosynthesis is presented in μgC/mg total weight. Calcification is presented in % weight of skeleton added for the duration of the experiment and this was calculated separately for the ^{45}Ca and for the ^{14}C incorporation. Experiments E8-1 and E8-2 show that calcification in the light is 4 times higher than in the dark. In this respect planktonic foraminifera are similar to benthonic foraminifera, hermatypic corals and calcareous algae. The calcification rates in the light were very high, 4.4% and 4.8% for 5 hours. If we assume a 12 hour light/dark cycle, the daily calcification rates reach values of 13.3% to 14.4%. These are indeed very high rates compared to any other calcifying organism known (compare to 2.5%/day for benthonic foraminifera (Erez, 1978) and roughly 3%/day for corals (Goreau, 1963)). These rates of calcification explain the high $CaCO_3$ production by planktonic foraminifera that is observed in sediment traps in the ocean (Honjo, 1980, 1978). The rates of photosynthesis are also very high. They are ∿ 5 times higher than rates measured in benthonic foraminifera (Erez, 1978). If one assumes a population density of 500 foraminifers/m^3 (Bé, 1977), a photic zone of 100m and an average weight/individual of 25 μg, then the daily photosynthetic rate of planktonic foraminifera can reach a few percent of the total productivity in open oceanic water (Ryther, 1962).

The most important part of this study is the DCMU experiments. At high DCMU concentration (10^{-5}) both photosynthesis and calcification are inhibited. But when DCMU concentration is lowered the calcification rate increases while photosynthesis is still completely inhibited. At a DCMU concentration of 5×10^{-6} calcification reaches 50% of its normal value and at DCMU concentrations of 10^{-6} to 5×10^{-7} we get 90% of the normal calcification rates while photosynthesis is completely inhibited. Only when DCMU concentration is 10^{-7}, photosynthesis starts to increase to a rate which is still 10 times lower than normal light value; calcification then reaches the normal light values. These results clearly demonstrate that light, and not photosynthesis, is responsible for the fourfold difference between light and dark calcification. It is possible that photosynthesis may enhance calcification indirectly, but this effect seems secondary compared to the effect of light. Vandermuelen et al. (1972) who conducted similar DCMU experiments on hermatypic corals concluded that photosynthesis is the main factor which influences calcification. They, however, used very high concentrations of DCMU (10^{-4}). It is possible that such a high concentration of DCMU inhibited calcification in corals in the same way that a concentration of 10^{-5} inhibited both photosynthesis and calcification in the planktonic foraminifera (Table 1, experiment E8-3). Only at DCMU concentrations of 10^{-6} to 10^{-7}, that are 2-3 orders of magnitude lower than those used by Vandermuelen et al. (1972), was it possible to show that light, and not photosynthesis, is the main factor in foraminifera calcification. Direct effect of light on calcification was also observed in coccolithophores (Paasche, 1964; 1968). Ariovich and Pienaar (1979) suggested that light influences the Ca^{++} uptake in the coccolithopore _Hymenomonas carterae_. The role of light in calcification and Ca^{++} uptake in foraminifera and corals is presently being studied in our laboratory.

Comparison of the calcification rates measured by ^{45}Ca and ^{14}C incorporation is shown in Table 1 as a $^{14}C/^{45}Ca$ calcification ratio. This ratio should have been unity if both isotopes were ideal tracers for this process. In most of the experiments ^{14}C lags behind ^{45}Ca and gives lower calcification rates. Similar phenomenon was found by Goreau (1963) for hermatypic corals and by Erez (1978) for hermatypic corals and benthonic foraminifera. In the latter study the $^{14}C/^{45}Ca$ calcification ratio showed linear correlation with the skeletal stable isotope composition (carbon and oxygen) for the coral _Stylophora pistillata_. These observations were interpreted to represent dilution of the ^{14}C tracer in an internal metabolic carbon pool. When the metabolic pool is larger, the $^{14}C/^{45}Ca$ calcification ratio is lower and the stable isotope composition is lighter because a larger fraction of the skeletal carbonate is of metabolic origin. In experiment E8 it is clearly shown that photosynthesis may increase the size of the metabolic pool. As a result, the $^{14}C/^{45}Ca$ ratio is lowest (0.13) in the light without DCMU. At 10^{-6} DCMU calcification approaches the normal value; there is no photosynthesis and the $^{14}C/^{45}Ca$ ratio is 0.26. This may indicate that the carbon pool is smaller because of lack of photosynthesis. The ^{14}C calcification values in experiment E9 are much higher than those in E8. These experiments were carried out on different days, at different light levels. In addition, the feeding was not equally timed. It is possible that the individuals of E8 were fed on the morning of the experimental day while those of E9 were fed one day before the experiment, and therefore their carbon pool was smaller. Despite this difference, the same trend in the $^{14}C/^{45}Ca$ calcification ratio is seen in experiments E8 and E9. When photosynthesis is inhibited (e9-3) this ratio is higher than when photosynthesis is not inhibited (E9-1). These results are preliminary, and more experiments are needed before this problem is clarified. Nevertheless it seems that photosynthesis contributes to the increase in the size of the carbon pool and this supports the conclusions of Erez (1978) with respect to hermatypic corals.

In conclusion, I have shown that planktonic foraminifera calcify at a very high rate, and their symbionts may contribute to oceanic primary productivity. Calcification in these organisms is enhanced by light, and probably not by photosynthesis of their symbiotic algae.

Treatment	No. of Individuals	Average weight/individual (μg)	Photosynthesis (μg C/mg $CaCO_3$)	Calcification % weight ^{14}C	^{45}Ca	$^{14}C/^{45}Ca$ Calcification ratio
E8-1 Light	16	12	2.05	0.57	4.38	0.13
E8-2 Dark	17	14	0.04	0.24	1.13	0.21
E8-3 Light 10^{-5} DCMU	14	8	0.08	0.28	0.20	1.4
E8-4 Light 5×10^{-6} DCMU	18	13	0.02	0.45	2.16	0.21
E8-5 Light 1×10^{-6} DCMU	9	20	0.03	1.03	4.05	0.26
E9-1 Light	20	9.8	2.81	2.63	4.80	0.55
E9-2 Light 5×10^{-6} DCMU	21	9.9	0.03	1.16	2.07	0.56
E9-3 Light 5×10^{-7} DCMU	18	12.6	0.04	3.08	4.31	0.71
E9-4 Light 1×10^{-7} DCMU	21	11.4	0.26	2.82	4.77	0.59

Duration of experiments 5 hrs. (1200-1700)

Table 1. Calcification and photosynthesis in the planktonic foraminifer *Globigerinoides sacculifer*. Experiments were carried out in natural sunlight in the laboratory, and lasted 5 hours between 1200 and 1700. Rates of photosynthesis and calcification are for the duration of the experiments. DCMU concentrations are given in mole / liter. Note difference between light and dark calcification (E8-1, E8-2) and the high rates of calcification (E8-5, E9-3, E9-4) when photosynthesis is inhibited by DCMU. This suggests that calcification in these organisms is enhanced by light and not by photosynthesis of their symbionts.

References

Be, A.W.H., 1977. Oceanic Micropaleontology, ed. Ramsay, A.T.S., p.1-100, Acad. Press, Lond.
Be, A.W.H., Hemleben, C., Anderson, O.R., Spindler, M., Hacunda, J. & Tuntivate-Choy, S., 1977. Micropaleontology. 23:155-179.
Erez, J., 1978. Nature. 273(5659):199-202.
Goreau, T.F., 1960. In: The Biology of Hydra p. 269-285, H.N. Lenhoff & W.F. Loomis eds. Univ. of Miami press, Florida.
Goreau, T.F., 1961. Endeavour. 20:32-39.
Goreau, T.F., 1963. Ann. N.Y. Acad. Sci. 109:127-167.
Goreau, T.J., 1977a. Proc. R. Soc. London. B., 196:291-315.
Goreau, T.J., 1977b. Proc. 3rd Int. Coral Reef Symp. p. 396-401.
Honjo, S., 1978. Jour. Mar. Res. 36(3):469-492.
Honjo, S., 1980. Jour. of Mar. Res. 38(1):53-97.
Land, L.S., J.C. Lang & B.N. Smith, 1977. Lim. & Ocean.20(2):283-287.
Lee, J.J. & Bock, W.D., 1976. Bull. Mar. Science. 26(4):530-537.
Ryther, J.H., 1962. In: The Sea, M.N. Hill ed., V. 2, p. 347-380. Interscience Publishers.
Sikes, C.S., Roer, R.D., Wilbur K.M., 1980. Lim. & Ocean. 25:248-261.
Sikes, S.C. & Wilbur, K.M., 1982. Limnol. Oceanogr. 27(1):18-26.
Vandermuelen, J.H., Davis, N.D. & Muscative, L., 1972. Mar. Biol.16:185-191
Weber, J.N. & P.M.J. Woodhead, 1970. Chem. Geol. 6:93-117.
Weber, J.N. & P.M.J. Woodhead, 1972. Mar. Biol. 15:293-297.
Ariovich, D. & R.N. Pienaar, 1979. Br. Phycol. J. 14:17-24.
Paasche, E., 1964. Physiol. Plant. 3(suppl):1-82.
Paasche, E., 1968. Annu. Rev. Microbiol. 22:77-86.

STROMATOLITES, FOSSIL AND RECENT: A CASE HISTORY

Stjepko Golubic
Department of Biology
Boston University
Boston, MA, U.S.A.

ABSTRACT

Stromatolites built by the coccoid cyanophyte *Entophysalis major* in Shark Bay, W. Australia undergo seasonal lithification by carbonate precipitation within the polysaccharide envelopes of the organism. Mineral incorporation obliterates the biological structures. The lithified stromatolite surface is then colonized by destructive, carbonate penetrating microbial endoliths. Stromatolite growth resumes when *Entophysalis* recolonizes the surface. *Entophysalis* stromatolites serve as a direct interpretational model for domal stromatolites built by *Eoentophysalis* (a silicified microfossil) in Precambrian strata. *Eoentophysalis* stromatolites occurred worldwide over a period of 1 Ga in the Precambrian (1.9-0.9 Ga). *Entophysalis* may be its direct descendant.

INTRODUCTION

Stromatolites are defined by most workers as bio-sedimentary structures built by microbial trapping and binding or/and precipitating of sediments (Awramik *et al.*, 1976). This and other definitions are presented and discussed by Semikhatov *et al.* (1979). While stromatolites were widespread in the geological past, particularly in the Precambrian, their occurrence today is limited to only a few special environmental settings. Comparisons between modern and ancient stromatolites are therefore burdened by serious limitations. One of these stems from the fact that only few ancient stromatolites contain recognizable preserved microfossils of the organisms that may have built them. Another limitation refers to the fact that most modern stromatolites in marine environments trap and bind allochthonous sediments. They are not hardened by autochthonous precipitation of carbonate, and thus are fragile and ephemeral formations with a low preservation potential. The present paper deals with stromatolites of high preservation potential which are built by *Entophysalis major* (Cyanophyta, Cyanobacteria), a microorganism with a long paleontological history.

P. Westbroek and E. W. de Jong (eds.), Biomineralization and Biological Metal Accumulation, 313–326.

THE STROMATOLITES

Domal stromatolites and stratiform algal mats are growing today on the sublittoral platform and on tidal flats along the coasts of Hamelin Pool, Shark Bay, W. Australia (Playford and Cockbain, 1976). Several (11) different types of mats have been distinguished on the basis of their microbial composition, internal structure and external morphology (Golubic, 1976; and unpublished). Each mat is a microbial community dominated by one or more microbial species. A mat is also a differentiated microenvironment within which microbial populations are arranged along often sharp boundaries and steep gradients of critical environmental determinants (e.g. oxygen supply, Eh, pH, light). An algal mat functions as an ecosystem with photosynthetic production and fermentative and respirative degradation of organic matter. It may completely recycle nutrients (Doemel and Brock, 1977), and has a high degree of homeostatic regulation.

A mat can also be viewed as a structure which is composed of supportive organic and inorganic elements. Organic elements include microorganisms as well as their products (e.g. polysaccharide cell envelopes and sheaths, or skeletons). Inorganic elements include trapped sediment particles as well as autochthonous mineral precipitates. The architecture of an algal mat provides the framework within which metabolic functions take place. Therefore, structural properties of a mat not only determine its external morphology but also its physical, and indirectly, its chemical properties. For example, water retention and drainage, which depend on structural properties of the mat, influence the exchange rates of gases and solutes and thus determine the red-ox state within the microenvironments of the mat. Structural properties of stromatolites also determine their resistance to environmental energies. Stromatolitic reefs form effective wave breakers and modify the pattern of currents and sediment transport. Finally, structural properties determine the durability and preservation potential of a stromatolite.

In Hamelin Pool lithified intertidal stromatolites formed by the coccoid cyanophyte *Entophysalis major* are the most conspicuous feature which dominates the landscape and serves as an important coast stabilizing agent (Fig. 1A). These structures are comparable to some Precambrian domal stromatolites. The shaping of these stromatolites progresses through the processes of colonization, sediment stabilization, selective erosion, lithification and isolation of the structure through erosional removal of surrounding sediment.

THE MICROORGANISM

Entophysalis major Ercegović is a coccoid cyanophyte with spherical cells (3-)5-9 µm in diameter, that divide in one to three planes of division and produce copious amounts of extracellular polysaccharide which holds them together in a mat-like colony. The polysaccharide envelopes have a loose fibrous ultrastructure which forms an elastic fabric.

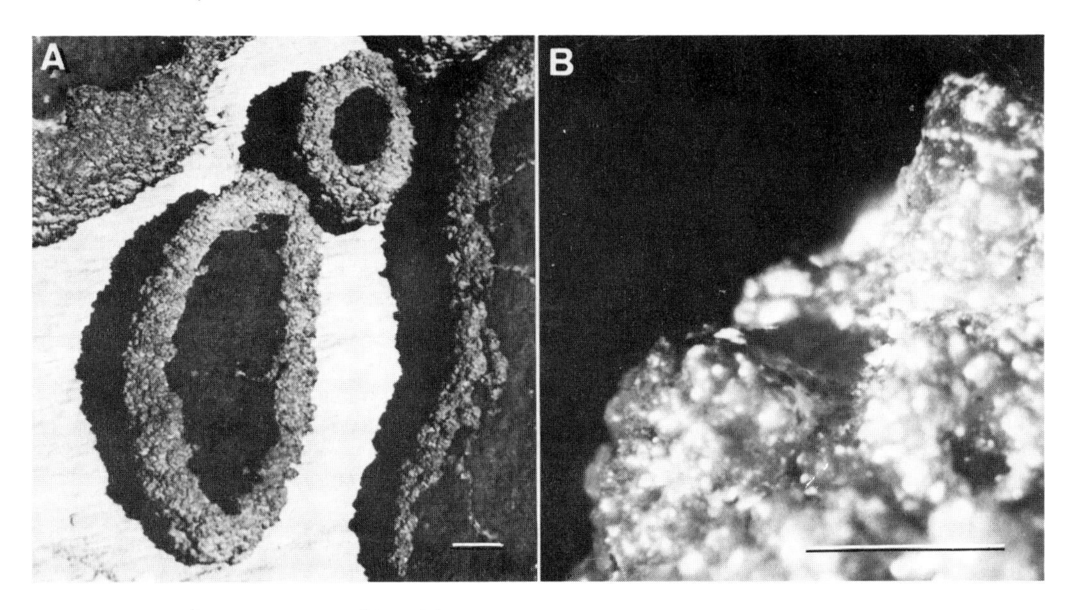

Fig. 1. *Entophysalis* stromatolites, intertidal zone, Hamelin Pool, Shark Bay, W. Australia. A: two elongated stromatolite domes with lithified tops and margins overgrown by soft mamillate mat. Scale bar is 100 mm. B: mamillate mat, close-up. Scale bar is 1 mm.

This fabric expands as cells divide, and new polysaccharide layers are laid down inside of the old ones (Golubic, 1980a; Golubic and Hofmann, 1976). The envelopes are stained brown by the photoreactive pigment scytonemine, which is particularly dense in layers close to the mat surface. The envelopes are also highly hydrated which gives the mat a translucent gelatinous appearance and consistency, and provides for a convenient water storage during the time of emersion and air-exposure of the mat. Directed (polar) excretion of gel and/or predominant divisions of cells in one plane can be responsible for the arrangement of cells in vertical rows that is often evident in interior layers of the mat. The cells closer to the surface divide predominantly in three perpendicular planes and form isodiametric packages which swell and protrude through the surface forming billowy and warty outlines of the mat (Fig. 1B). Because of this warty appearance of the mat's surface, *Entophysalis* mats are also called mamillate mat (Golubic, 1973; is "cynder mat" of Kendall and Skipwith, 1968 and "pustular mat" of Logan *et al.*, 1974). The mat gradually disintegrates in the lowermost parts which leaves parts of the mat hollow.

The microorganism spreads by detachment of entire cell groups from the colony (Fig. 2A). After storms, when the entire mat is covered by sediment, individual cells are released by dissolution of the gel. These cells, surrounded by a single layer of polysaccharide, are then distributed by the tide and start new colonies.

GEOGRAPHIC DISTRIBUTION AND ENVIRONMENTAL SETTING

Entophysalis major was first described by Ercegović (1932) from organically enriched habitats in the Adriatic Sea. There it occurs today in protected embayments and along the edges of evaporation pans in salt works, and forms small, inconspicuous colonies (Golubic, 1980b). Small colonies of *Entophysalis major* are also common on intertidal flats of Andros Island, Bahamas, where it occurs together with other coccoid cyanophytes. Formation of contiguous mats by *Entophysalis* has been reported from the lagoons of Baja California, Mexico (Horodyski and Vonder Haar, 1975). Extensive mats of *Entophysalis* cover vast surfaces in the lower intertidal zone of the lagoons of the Arabian (Persian) Gulf at Abu Dhabi, U.A.E., mostly as colonizers of loose sediments (Golubic, 1973; 1976; Kinsman and Park, 1976). Such mats are widespread in protected embayments of Shark Bay, notably Nilemah flats, Hutchison and Gladstone embayments. On wave-exposed coasts of the Hamelin Pool, Shark Bay, mamillate mat stabilizes large platforms which are subsequently carved by waves and lithified to form domal and head-shaped stromatolites (Golubic, 1976).

From the observations detailed above, we may conclude that *Entophysalis major* is a cosmopolitan species which favours intertidal (lower) settings in warm and moderately hypersaline (2-3x sea water salinity) waters. It requires frequent periodic alternation of emersion and flooding and good drainage. It grows in protected embayments as well as on coasts moderately exposed to waves. It does not grow subtidally, nor in intertidal pools having stagnant conditions. Although the organism occurs abundantly in a wide variety of environments, it forms lithified stromatolites under particular conditions which have, to date, been encountered only along the exposed coasts of the hypersaline Hamelin Pool, Shark Bay, Western Australia.

Entophysalis mat in Hamelin Pool colonizes and stabilizes loose sediment or, more often, it colonizes surfaces already stabilized by the smooth *Microcoleus* mat. Extensive platforms with mildly rounded outlines and gentle slopes are covered by this mat. The mat surface is broken during storms at regularly spaced intervals following the patterns of coastal wave interference. Once the protective cover of *Entophysalis* is damaged, the receding waves mobilize sediment, and the erosion process carves channels across the platform that are roughly perpendicular to the coast. This process subdivides the platform in elongated strips. The strips of mat are undercut and droop over the edges. Overgrowth of the surfacial mat along the vertical sides soon stabilizes the edges of the structure which now has a more prominent profile. Lithification occurs in the mat at seasonal intervals. This protects the tops of stromatolites. Continuing erosional episodes deepen the depressions around these hardened structures isolating them. Erosion of relatively softer stromatolite slopes steepens and undercuts the structure leaving even a narrow neck as a base. Stromatolites accrete by recolonization of lithified stromatolites with mamillate mat. Reoccurring events of mat colinization and lithification are recorded in the internal structure of stromatolites as laminae.

THE GEOLOGICAL HISTORY

Mid-Precambrian stromatolites found in the ca. 1.9 Ga old Kasegalik and McLeary Formations, Belcher Islands, Canada, are similar in size and shape to the modern lithified stromatolites of Shark Bay. They contain microfossils preserved in silica (Hofmann and Jackson, 1969). The dominant microfissil was described by Hofmann (1976) as *Eoentophysalis belcherensis*. This microfossil, which has been subsequently found throughout the Precambrian, exhibits marked similarities to the modern *Entophysalis* (Fig. 2B) with respect to: (1) the shape and size of the stromatolites they form; (2) the form of the mat and the billowry outlines of its surface; (3) the multiple layered extracellular envelopes which are pigmented on the upper surfaces of the colonies; (4) the frequency and spatial positioning of dividing cells, following a division pattern in three planes; (5) the specific mode of cell shrinkage and post-mortem degradation which in the case of the fossil, has been arrested in various stages; and (6) the environment which is an evaporitic, probably intertidal setting in both cases. The number of correlated properties shared by these microorganisms which are separated in time by the enormous span of almost 2 billion years is significantly higher than: (a) the properties shared by contemporaneous microfossils within the Belcher Island assemblage (see Hofmann, 1976), and (b) the properties shared by modern microorganisms which are morphologically and taxonomically closest to *Entophysalis*. *Eoentophysalis* microfossils with similar properties, but younger in age have been found throughout the world. In the strata of the mid Proterozoic McArthur Group which has been dated between 1.57 and 1.39 Ga (see Oehler, 1978) microfossils of this type have been found in its oldest carbonate unit, the Amelia Dolomite (Muir, 1976), as well as in its youngest unit, the Balbirini Dolomite (Oehler, 1978). Although microfossils of the Amelia Dolomite have been reported under different

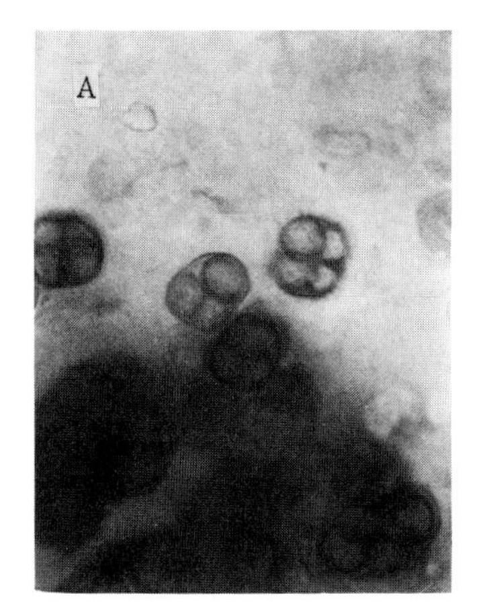

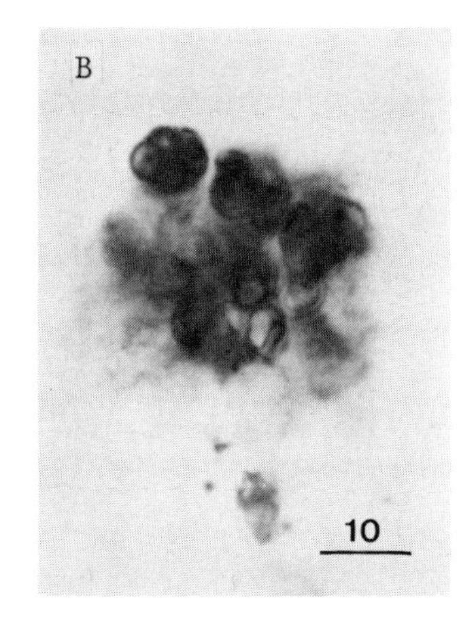

Fig. 2. Modern *Entophysalis* (A) and its Precambrian counterpart *Eoentophysalis* (B) (Gaoyuzhuang Formation, China, 1.5 Ga, courtesy Zhang Yun). Scale bar is 10 μm.

names, many of them are clearly recognizable as variants of *Eoentophysalis* (see also Oehler, 1978, p. 280; and Zhang, 1981, p. 490). Well preserved stromatolitic mats of *Eoentophysalis* within the same time span, 1.5-1.4 Ga, have been described by Zhang (1981) in the Gaoyuzhuang Formation, North China. *Eoentophysalis* has been identified as the principal mat-building fossil microorganism in stromatolitic cherts of the Elery Creek outcrop of the Late Proterozoic Bitter Springs Formation, Amadeus Basin, Australia, with an age of 1.0-0.7 Ga (Knoll and Golubic, 1979).

Morphological and paleontological correspondance between various fossil findings of *Eoentophysalis* (over a time span of over a billion, 10^9, years), and the modern *Entophysalis* together with their respective environmental settings (in an interval less than one billion years later) is striking. This suggests that *Entophysalis* is a direct descendant of *Eoentophysalis*. The study of the modern form may therefore have direct relevance for the understanding of processes which lead to formation of ancient stromatolites and the conditions under which they fossilized, and may provide a key organism for study of cell-level mineralization reactions.

LITHIFICATION OF STROMATOLITES

The irregular mamillate surface of *Entophysalis* mat traps sediment particles which are overgrown and incorporated in the mat (Golubic, 1973). These particles concentrate in isolated islands within the mat, because they are excluded from the extracellular polysaccharide envelopes as they are excreted from the cells. These envelopes remain largely free from sediment and are soft during most of the year. The actual hardening of the structure starts with lithification in the summer when the saturation of the ambient water with respect to $CaCO_3$ is at its highest level.

Lithification ensues as precipitation of calcium carbonate <u>within</u> the polysaccharide envelopes of *Entophysalis*. There is no evidence of exclusion of precipitated carbonate from polysaccharide organic matter, as there is no evidence of exclusion of organic matter by the growing mineral fabric. Rather, the precipitate which starts as granular-nodular concretion soon permeates the entire gel. The process is rapid and it is difficult to encounter transitional stages. However, examining fractured or sectioned lithified stromatolites it is possible to find microenvironments which have been exposed to different degrees of lithification. These sections expose a yellow to olive green nucleus and a white peripheral rim of a hardening stromatolite. Under scanning electron microscope (SEM) this effect is inverted, so that the rim appears darker than the nucleus (Fig. 3A). When etched briefly with dilute HCl, the carbonate recedes leaving behind an insoluble yellow-green gelatinous residue, particularly in the area of the nucleus. Such specimens have been critical-point-dried and studied under SEM. More or less shrivelled cells of *Entophysalis* are found only in pockets which have been spared

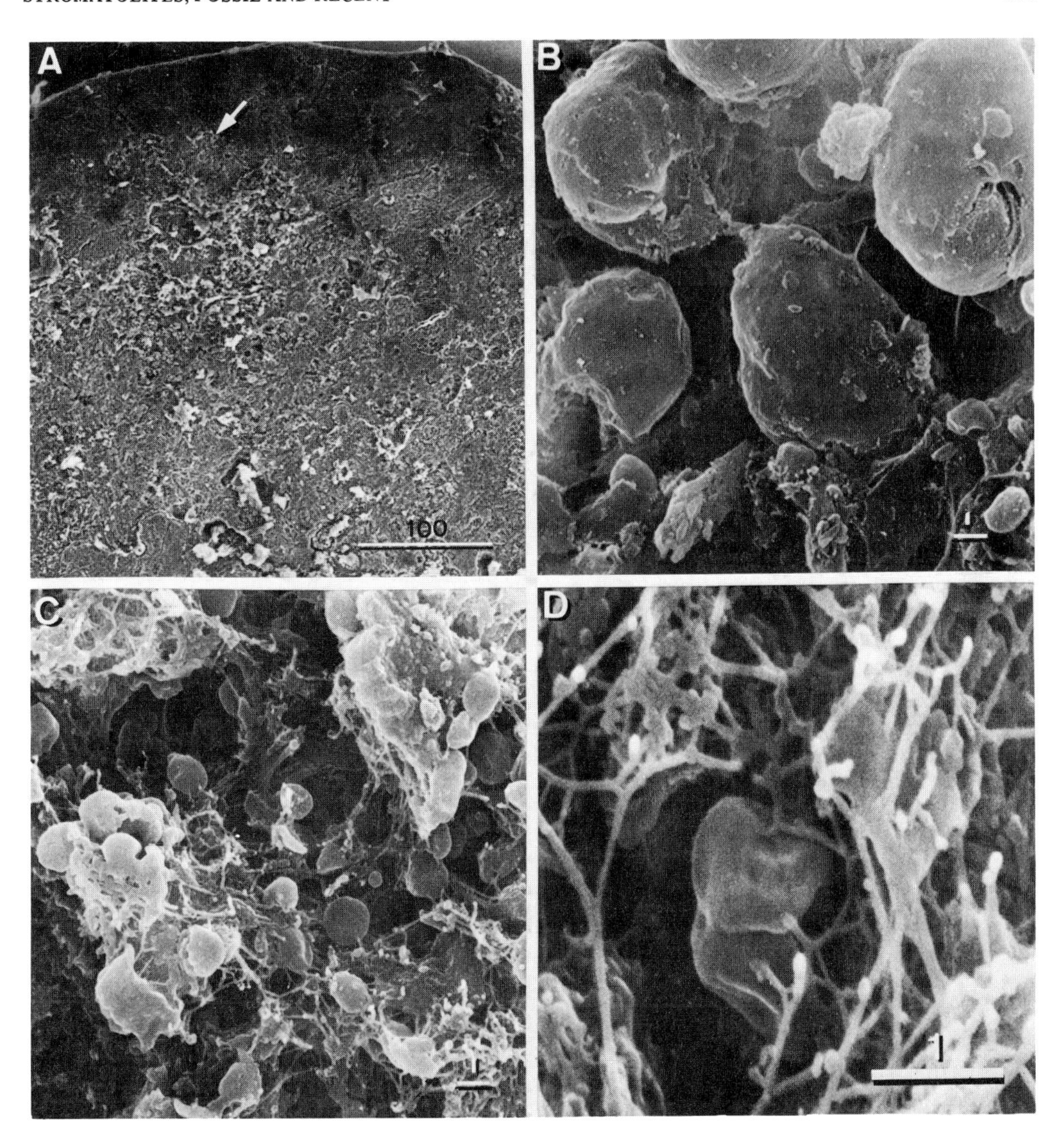

Fig. 3. Organic component of a stromatolite in the process of lithification. A: Polished, etched and critical-point-dried section through a hardened *Entophysalis* stromatolite. Arrow marks the boundary between the yellow-green nucleus and a white (here dark) rim; note the billowy outline of this boundary: organic residue is present mostly in the nucleus. B: Partially shrivelled cells of *Entophysalis* preserved within non-lithified prockets of the nucleus. C: Cluster of coccoid bacteria in the insoluble organic residue. D: Closer view of the three-dimensional network of polysaccharide fibers with a shrivelled dividing bacterial cell. Scale bars in µm.

from intensive lithification (Fig. 3B). In the bulk of the organic residue such cells have been completely obliterated by precipitating carbonate. Only the much smaller bacterial cells have occasionally escaped obliteration (Fig. 3C). The organic residue is composed mostly of polysaccharide fibers which are seen under higher magnification as a three-dimensional network (Fig. 3D).

The nature of the mineral precipitate has been examined on specimens in which the organic matter has been destroyed by sodium hypochlorite (clorox). Calcium carbonate precipitates within *Entophysalis* colonies, following precisely their external morphology (Fig. 4A., compare with Fig. 1B). The fine structure of this precipitate is finely granular-nodular with no crystallographic characterization. Its surface is covered with minute bumps (granules) averaging 0.08-0.1 µm in diameter. It is almost identical with the amorphous calcium carbonate described by Golubic and Campbell (1980) for the concretions in the thalli of marine *Rivularia* (Cyanophyta). On the clorox-treated section through a stromatolite, amorphous carbonate is located in the nucleus correlating with the presence of incorporated gel (Figs. 4C, n; 4D). At the periphery of the nucleus and in the transition to the external rim, the precipitate asumes fibrous crystalline morphology. The rim consists of aragonite needles without preferred orientation (Fig. 4C, r). It is concluded that the rim represents surface precipitation on, rather than within *Entophysalis* colony. This view is supported by the evidence that the boundary between the nucleus and the rim shows billowy outlines corresponding the mamillate surface of the *Entophysalis* mat (Fig. 3A, arrow), while the outer surface of a lithified stromatolites appears smoother.

Lithified stromatolites are porous. The cavities represent gaps that are bridged over by the growing *Entophysalis* mat, or they may originate from decompositional cavities of the aging mat. They are initially void with bumpy irregular walls (Fig. 4E) which are lined by several layers of uniform size aragonite crystals, 1.5-2.5 x 0.2-0.5 µm (Fig. 4F). In the advanced stages of lithification the cavities within stromatolites are filled by large grains of sparry calcite (Figs. 4G, 4H). It is noteworthy that this type of void filling appears restricted to stromatolites of the intertidal zone. It is different from void filling within beachrock and calcareous crusts of Shark Bay outside of the stromatolite areas, in which all cavities are filled with a typically fibrous radiating aragonite (Figs. 5A-C).

Two stages in the process of lithification are readily distinguishable. In the first stage, the mat hardens and loses its translucent properties, becoming opaque and dark brown on the surface. It is still dominated by *Entophysalis* which is rapidly desiccating and shrinking. The interior is hard, but has the consistency of plaster indicating that a carbonate mineral fabric is porous. At this stage, all transitions between non-lithified and lithified mat are found in islands of softer mat surrounded by carbonate encrustation. In the second, more advanced stage of lithification the hardness of the structure increases due to

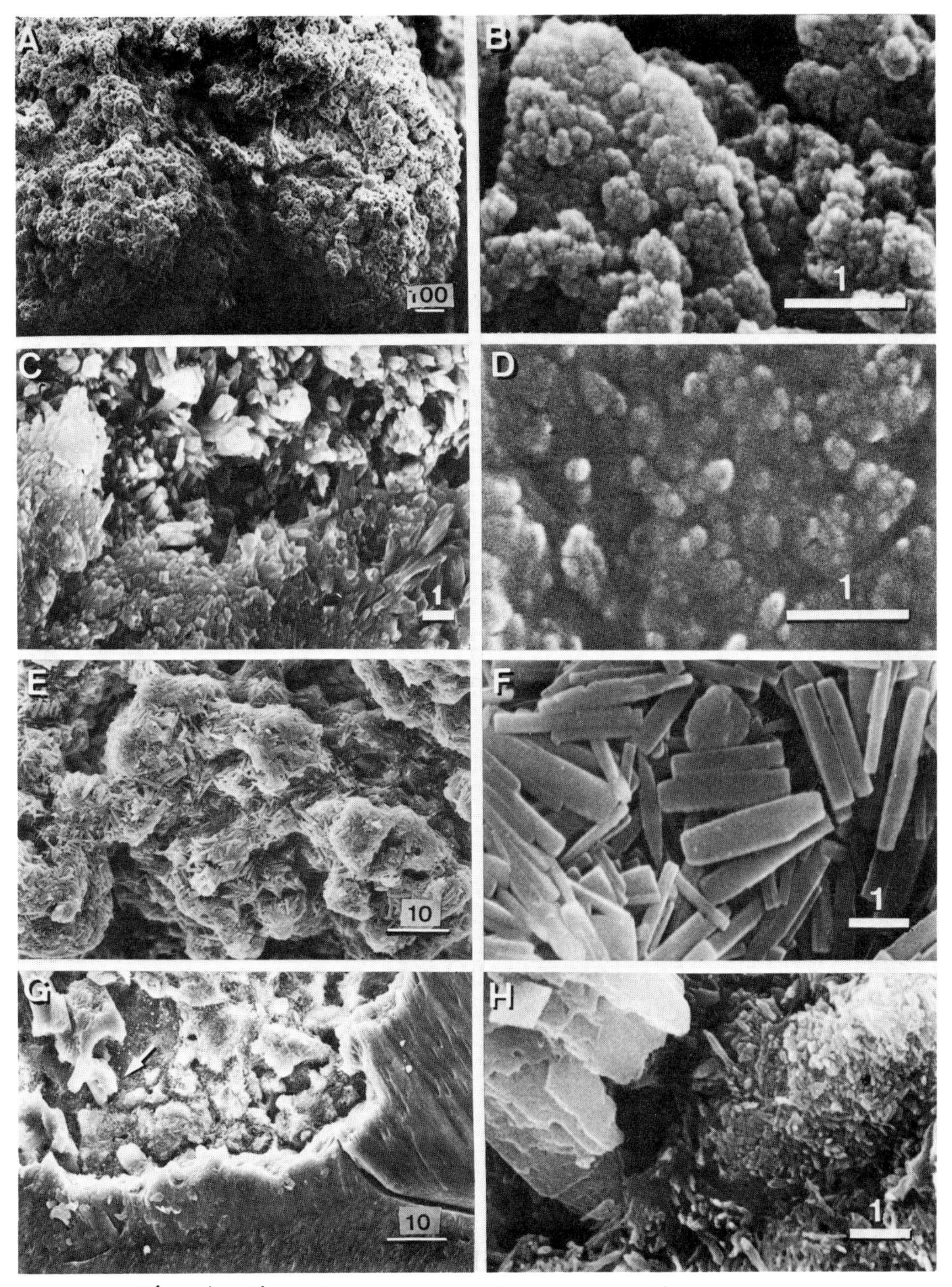

Fig. 4. Mineral component of a stromatolite. A, B: external surface; C, D: fractured surface; E, F: lining of internal cavities; G, H: mineral cavity filling. See text for explanations. Scale bars in µm.

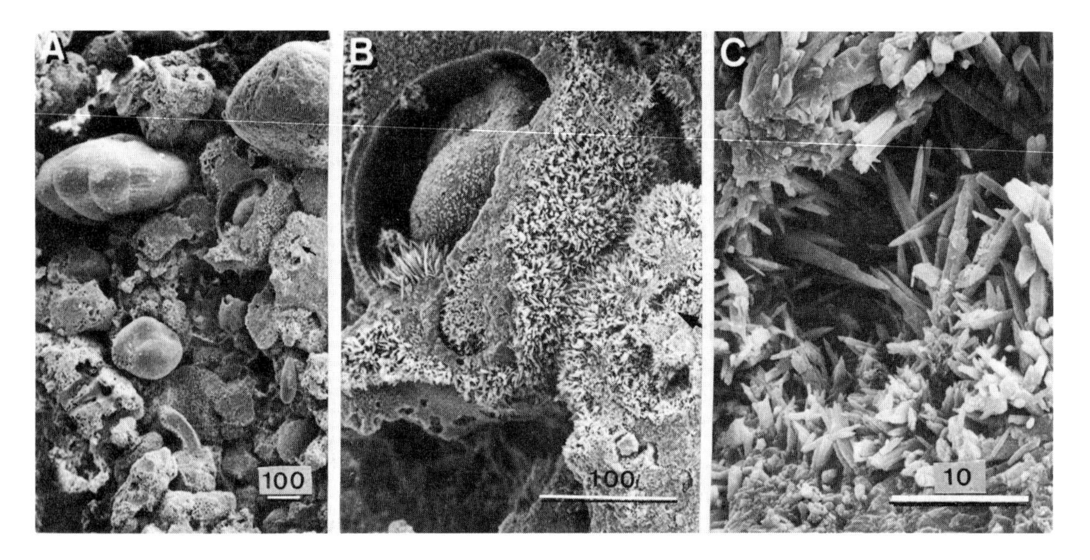

Fig. 5. Cements and mineral fill of cavities in a non-stromatolitic calcareous crust. A: Foraminiferal tests and other skeletal fragments composing the crust. B and C: Details from Fig. 5A marked by arrows. Scale bars in μm.

progressive cementation of the intercrystalline spaces and pores, resulting in the consistency of hard rock. This difference in texture of the two lithification stages is reflected in thudding vs. pinging sounds when the structures are struck with a geological hammer. The surface coloration changes from dark brown to bluish-gray as a consequence of deterioration of *Entophysalis* and colonization of the now lithified stromatolite surface by the endolithic cyanophyte *Hormathonema violaceo-nigrum*.

THE COLONIZATION OF STROMATOLITES BY ENDOLITHS

The indurated surface of a lithified stromatolite is colonized by endoliths, which form a thin bluish-black veneer that tightly adheres to the carbonate surface. This cover is slippery when wet and has been termed "film mat" (Logan *et al.*, 1974). Golubic (1976) determined that film mat consists of stromatolite destroying rather than stromatolite building microflora. As compared with endolithic vegetation of carbonate coasts in normal marine environments (see LeCampion-Alsumard, 1978) the film mat of the hypersaline Hamelin Pool shows a lower species diversity, a higher proportion of epilithic vs. endolithic cover, and a lower depth of penetration. The film mat is dominated by predominantly epilithic thalli of *Hormathonema violaceo-nigrum* Ercegović which send short endolithic filaments into the substrate. They can remove as much as 40-60% of carbonate from a thin, ca. 50 μm deep surface layer of the carbonate crust (Fig. 6A). In order to assess the depth of penetration, the

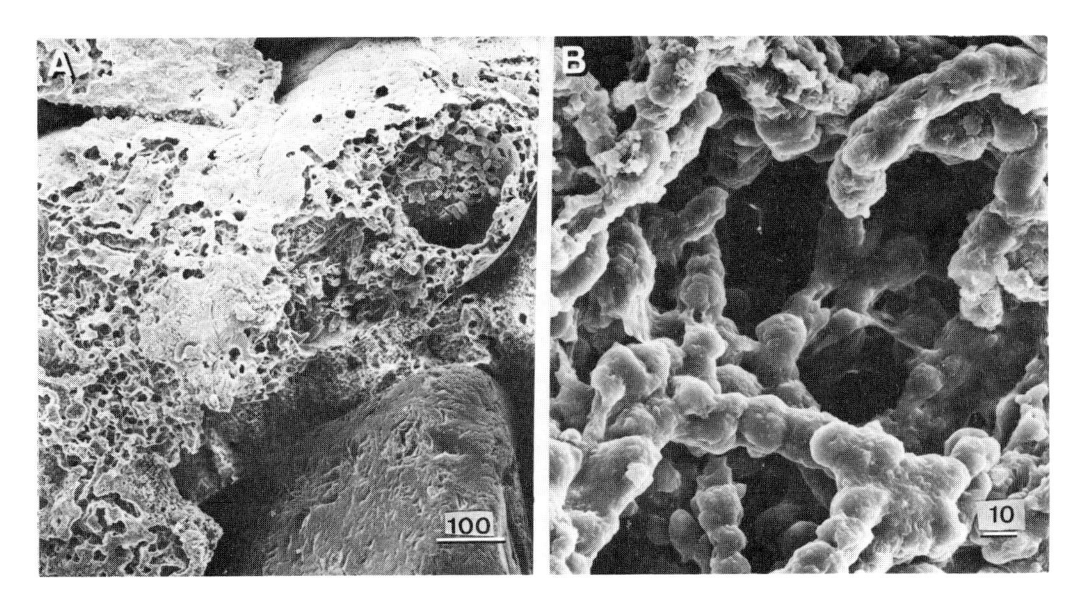

Fig. 6. Endoliths on lithified crusts and stromatolites. A: Clorox-treated surface of an intertidal carbonate crust showing severe surface damage by endoliths. B: Resin-cast endolith tunnels of a lithified stromatolite dome. Scale bars in µm.

tunnels are filled with polymerizing resin and the carbonate removed by dilute HCl (Golubic *et al.*, 1970). The exposed resin casts of borings have been examined by SEM (Fig. 6B). Their density is estimated to be on the order of 2×10^4 per mm^2. The endolith attack in the hypersaline Hamelin Pool is not followed by an intensive snail grazing as in normal marine environments (see Torunski, 1979). Therefore it seems to be limited to the surface and does not cause progressive degradation of the crust.

ACCRETION OF STROMATOLITES

Soft, non-lithified *Entophysalis* mats are easily destroyed by storms. Particularly the older mats which become hollow inside peel off regularly and are deposited by currents in quiet pools and depressions as detrital fragments. In such habitats there is no accretion of the mat. Accretion is only possible where the mat becomes indurated by lithification. Lithification, on the other hand, terminates the growth of *Entophysalis*, and the subsequent colonization by endoliths introduces a non-growing, destructive phase. Thus, phenomena described above represent but one cycle in the growth of *Entophysalis* stromatolites of Shark Bay.

Growth (accretion) resumes when lithified stromatolites become recolonized by the mamillate *Entophysalis* mat. The recolonization process can be observed everywhere in the zone where the mamillate mat grows. It starts on the slopes and margins of lithified stromatolite domes and encroaches upward. Sometimes it is incomplete leaving the tops of stromatolites "bald" (Fig. 1A). The growth of soft mamillate mat (with some sediment entrapment) then proceeds gradually through the austral winter and spring, when evaporation rates and carbonate saturation levels are lower, while the lithification sets in abruptly the following summer. The overgrowth of mamillate mat over a lithified stromatolite causes a destruction of endoliths and, to some degree, a softening of the underlying carbonate crust. Both recolonization and lithification are patchy in distribution, so that large surfaces of lithified and non-lithified mat always exist side by side. Multiple, but incomplete recolonization cycles remain "frozen" by lithification as hardened mat fronts. Direct measurements of stromatolite accretion rates (Playford, 1979) have shown that only a small proportion of stromatolites are actively accreting, while the average algal growth in balanced by erosion. Therefore, in a longitudinal section through a stromatolite dome, the laminae which record the cycles of stromatolite accretion are not always clear and complete, nor can they be counted as yearly increments to determine the age of the structure.

Extensive fields of lithified stromatolites exist in the upper tidal range along the coast of Hamelin Pool. They are outside of the zone of growth of mamillate mat but preserve the same morphology. These stromatolites are interpreted as "fossil" residues of once higher water levels. Immediately above the zone of active *Entophysalis* (and stromatolite) growth, the stromatolite domes are permanently covered by the film mat. Higher up, they become bleached retaining the dark endolith cover only along cracks, where sufficient wetting is assured by capillary transport from the interior of the still porous structure. Bleached stromatolites are either beige or rust-red in coloration. These colors are mineral and not biological in origin. Cracked stromatolites in this zone often collapse and are destroyed and hollowed out by erosion.

In conclusion, the accretion of intertidal Shark Bay stromatolites is not a continuous process that follows the biological growth of the constructing microorganisms, or their upward movement as in cases shown by Golubic and Focke (1978). Rather, it is a complex cyclic process which involves the following stages: (1) colonization and growth of *Entophysalis* mat with some sediment entrapment; (2) seasonal lithification of this mat which terminates its growth; (3) colonization of the hardened crust by endolithic microorganisms (a biological succession); and (4) recolonization and overgrowth of stromatolites by a new generation of *Entophysalis* mat, which starts a new accretion cycle.

FOSSILIZATION

The ecology and life cycle of *Entophysalis major* as a stromatolite building organism raises several interesting questions that concern the interpretation of its ancient counterparts. In order to assess the relevance of such recent-to-fossil comparisons, we need to determine (a) the preservation potential, and (b) diagenetic changes which accompany (and follow) the fossilization process. We have determined that stromatolites have the best preservation chance when they are lithified by carbonate precipitation within the stromatolite-forming mat. However, the same process which preserves the stromatolite as a structure is destructive for the organisms that have built it. This may explain why Precambrian (and Phanerozoic) stromatolites which are preserved in carbonates do not contain preserved cellular structures (see Hofmann, 1975). They may have lost them within a single season in the process of primary lithification, rather than post-depositionally during diagenetic recrystallization and dolomitization processes.

Microfossils are often preserved within cryptocrystalline silica. When such fossiliferous chert lenses occur *in situ* within carbonate stromatolites, it means that they represent primary silica deposit which formed synsedimentary or prior to the (seasonal) lithification in carbonate. Preservation of microfossils through later (diagenetic) silicification, however, can be postulated when there is evidence of burial of non-lithified mat under anaerobic conditions. Stratiform *Entophysalis* mats in protected lagoons and embayments of Shark Bay are frequently exposed to this type of burial. Similar conditions are met when dislocated mat fragments are transported by currents and then deposited in pools as secondary burial sites. *Entophysalis* remains preserved within anaerobic organic-rich layers with shrivelled cells, but with intact and pigmented multiple envelopes (Golubic and Hofmann, 1976). If anaerobic conditions persist, organic structures may remain partially preserved for thousands of years (Golubic, 1973), while under aerobic conditions their degradation is fast and almost always complete.

ACKNOWLEDGEMENTS

Research was supported by NSF Grants: GA43391, OCE12999-A02, EAR7911200, EAR8107686, NASA Grant NSG7588, NSG141 and grants by Boston University. A Travel support was received from the Bureau of Mineral Resources and Baas Becking Institute, Canberra, Australia. Scanning electron microscopy was carried out by Ed Selling, MCZ, Harvard University. Dr. S.E. Campbell critically read the manuscript and provided support and valuable discussions.

REFERENCES

Awramik, S.M., Margulis, L., and Barghoorn, E.S. (1976). In: "Stromatolites, Development in Sedimentology" (M.R. Walter, ed.), Vol. 20, Elsevier, Amsterdam-Oxford-New York, pp. 149-162.

Doemel, W.N., and Brock, T.D. (1977). Appl. Environ. Microbiol. 34, pp. 433-452.

Ercegović, A. (1932). Rad, J.A.Z.U. 244, pp. 129-220.
Bull. Internat. Yougosl. Acad. Sci. Art. (1932), pp. 33-56.
Golubic, S. (1973). In: "The Biology of Blue Green Algae" (N. Carr and B.A. Whitton, eds.), Blackwell Sci. Publ., Oxford, pp. 434-472.
Golubic, S. (1976). In: "Stromatolites, Development in Sedimentology" (M.R. Walter, ed.), Vol. 20, Elsevier, Amsterdam-Oxford-New York, pp. 113-126.
Golubic, S. (1980a). In: "Life Sciences and Space Research (COSPAR)", (R. Holmquist, ed.), Vol. 18, Pergamon Press, Oxford-New York, pp. 101-107.
Golubic, S. (1980b). Origins of Life 10, pp. 169-183.
Golubic, S., Brent, G., and LeCampion, T. (1970). Lethaia 3, pp. 203-209.
Golubic, S., and Campbell, S.E. (1980). In: "Phanerozoic Stromatolites" (C.L.V. Monty, ed.), Springer, New York-Heidelberg-Berlin, pp. 209-229.
Golubic, S., and Focke, J.W. (1978). J. Sedim. Petrol. 48, pp. 751-764.
Golubic, S., and Hofmann, H.J. (1976). J. Paleont. 50, pp. 1074-1082.
Hofmann, H.J. (1975). Am. J. Sci. 275, pp. 1121-1132.
Hofmann, H.J. (1976). J. Sedim. Petrol. 50, pp. 1040-1073.
Hofmann, H.J., and Jackson, G.D. (1969). Can. J. Earth Sci. 6, pp. 1137-1144.
Horodyski, R.J., and Vonder Haar, S.P. (1975). J. Sedim. Petrol. 45, pp. 894-906.
Kendall, C.G.S., and Skipwith, P.A. (1968). J. Sedim. Petrol. 38, pp. 1040-1058.
Kinsman, D.J.J., and Park, R.K. (1976). In: "Stromatolites, Development in Sedimentology" (M.R. Walter, ed.), Vol. 20, Elsevier, Amsterdam-Oxford-New York, pp. 421-433.
Knoll, A.H., and Golubic, S. (1979). Precambrian Research 10, pp. 115-151.
LeCampion-Alsumard, T. (1978). Les cyanophycées endolithes marines. Systematique, ultrastructure, ecologie et biodestruction, Thèse II, Université d'Aix-Marseille, pp. 1-197.
Logan, B.W., Hoffman, P., and Gebelein, C.D. (1974). Amer. Assoc. Petroleum Geologists, Mem. 22, pp. 140-194.
Muir, M.D. (1976) Alcheringa 1, pp. 143-158.
Oehler, D.Z. (1978) Alcheringa 2, pp. 269-309.
Playford, P.E. (1980). West Aust. Geol. Survey, Ann. Rept. 1979, pp. 73-77.
Playford, P.E., and Cockbain, A.E. (1976). In: "Stromatolites, Contribution to Sedimentology" (M.R. Walter, ed.), Vol. 20, Elsevier, Amsterdam-Oxford-New York, pp. 389-411.
Semikhatov, M.A., Gebelein, C.D., Cloud, P., Awramik, S.M., and Benmore, W.C. (1979). Can. Jour. Earth Sci. 16, pp. 992-1015.
Torunski, H. (1979). Senkenbergiana marit. 11, pp. 193-265.
Zhang, Y. (1981). J. Paleont. 55, pp. 485-506.

EARLY STROMATOLITE LITHIFICATION - ORGANIC CHEMICAL ASPECTS

Jaap J. Boon* and J.W. de Leeuw+
*School of Pharmacy, University of California, San Francisco, CA 94143, U.S.A.
+Delft University of Technology, Dept. of Chemistry and Chemical Engineering, Organic Geochemistry Unit, Delft, 2628 RZ, The Netherlands.

ABSTRACT

Curie-point flash pyrolysis methods (Py-MS and Py-GC-MS) were used to analyse the organic matter compositions of two samples from Shark Bay *Entophysalis major* rock samples at different stages of lithification. The organic matter of the original cyanobacteria has changed considerably before calcification. Comparison of the organic matter compositions of the initial soft rock stage and the later hard rock stage indicates that the calcification process inhibits further decomposition of the organic matter.

INTRODUCTION

In the preceding paper Golubic (1982) describes the early stages of fossilization of the *Entophysalis* cyanobacterial mats from Shark Bay (Australia). These mats are frequently subaerially exposed and especially during summer in danger of complete dissication. The microorganisms can survive short periods of dryness because they are surrounded by a thick sheath or glycocalyx, which serves as their hydrosphere. During prolonged exposure and dryness, the dissolved ions in the interstitial water precipitate to a poorly crystalline calcium carbonate precipitate, which totally encapsulates the cyanobacteria. As a result the dried microorganisms and the remains of their glycocalyx are canned in a mineral matrix. According to Golubic (1982) this process of fossilization of these cyanobacteria must have taken place over and over again since the first appearance of the species billions of years ago.

It was our interest to study the organic matter in these carbonates, because the results could serve as a good reference point for comparison with older rocks built up by *Entophysalis*. Several occurrences have been reported, e.g. Gaoyuzhuang formation (North China), Zhang (1981), $1.5x10^9$ years BP; Bitter Springs formation (Australia, Knoll and Golubic (1979), $0.7-1.0x10^9$ years BP. We studied whole rock samples of the soft initial stages of lithification and the later hard and brittle stage, from

P. Westbroek and E. W. de Jong (eds.), Biomineralization and Biological Metal Accumulation, 327–334.

Entophysalis domes of Hamelin Pool (Shark Bay, Australia) using pyrolysis mass spectrometry (Py-MS) and pyrolysis gas liquid chromatography mass spectrometry (Py-GC-MS). The objective was to compare the organic matter composition in the rocks as revealed by pyrolysis data with pyrolysis data on cyanobacteria and to determine possible diagenetic effects before and after initial encapsulation.

EXPERIMENTAL

Two samples of carbonate encrusted *Entophysalis* were recieved for analysis from S. Golubic. One sample was an initial stage of lithication (coded: soft carbonate sample), the other sample was very hard dense lithified carbonate rock (coded: hard carbonate). Both samples had been kept in polyethylene storage bags prior to recieval. The outer coloured skin of the rock samples was removed before further homogenisation. In practice the rock sample was cracked and a few fragments from the inner parts were "hand"picked with forceps. About 0.25 cm^3 were considered suitable for analysis and free of possible contaminants. Both samples were powdered with mortar and pestle. Aliquots of about 20 μg were used for the pyrolysis analyses.

Py-MS was carried out as described elsewhere (Boon *et al.*, 1982). The samples were pyrolysed on ferromagnetic wires with a Curie-point of 510°C. Samples were applied from Sodium phosphate buffer solution (pH = 7.0) to the wires by micropipet. Py-GC-MS was performed on a Varian 3700 MAT 44 instrument equipped with a glass capillary column coated with CP-Sil 5 (50 m, I.D. 0.5 mm, 1.2 μ film thickness). Helium was used as carrier gas. The temperature was programmed from 0°-275°C at a rate of 3°C/min with help of a cryogenic unit. The mass spectrometer was operated under standard conditions at 80 eV. Py-GC was performed on a Packerd Becker instrument as described by van der Meent *et al.* (1980). The gas-chromatograph, equipped with a cryogenic unit, was programmed from 0°C to 275°C at a rate of 3°C/min. The capillary column was the same as described under Py-GC-MS. Helium was used as a carrier gas. In this case the ferromagnetic wires used had a Curie temperature of 610°C and the suspension liquid was methanol.

RESULTS AND DISCUSSION

Py-MS.

Fingerprinting of organic matter by Curie point pyrolysis mass spectrometry has been shown to be a very sensitive method to determine minute differences in composition of mixtures of macromolecules (Irwin, 1979). A wide range of samples has been studied by several investigators. The results have shown that it is possible to draw preliminary conclusions about organic matter composition from the mass peak distributions in the spectra.

Figures 1A and 1B show the pyrolysis mass spectra of the soft and hard carbonate sample. No difference between these mass spectra is observed, which indicates that the organic matter composition of both samples must be very similar.

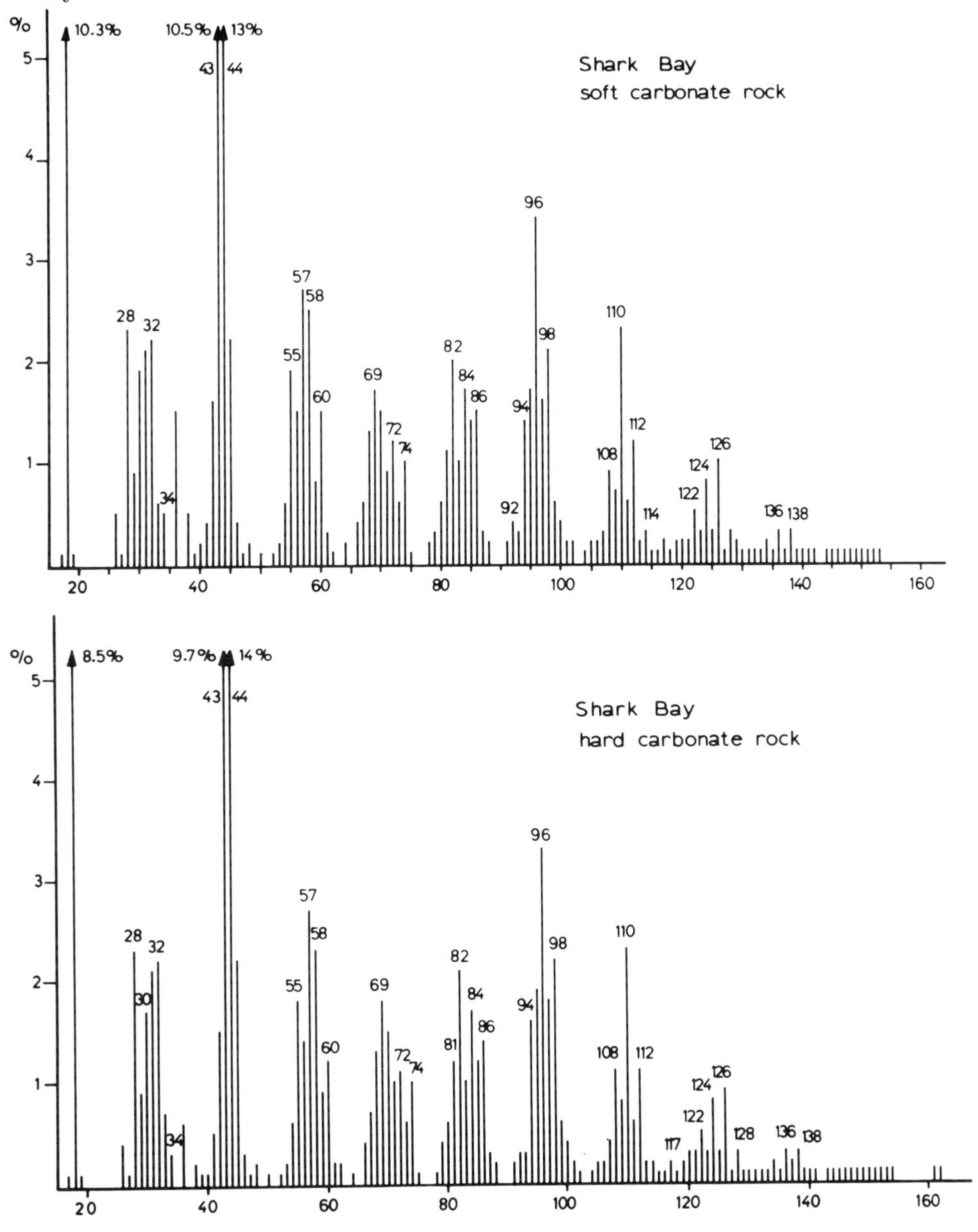

Fig. 1. Pyrolysis mass spectra of the soft and hard carbonate sample (for conditions: see text).

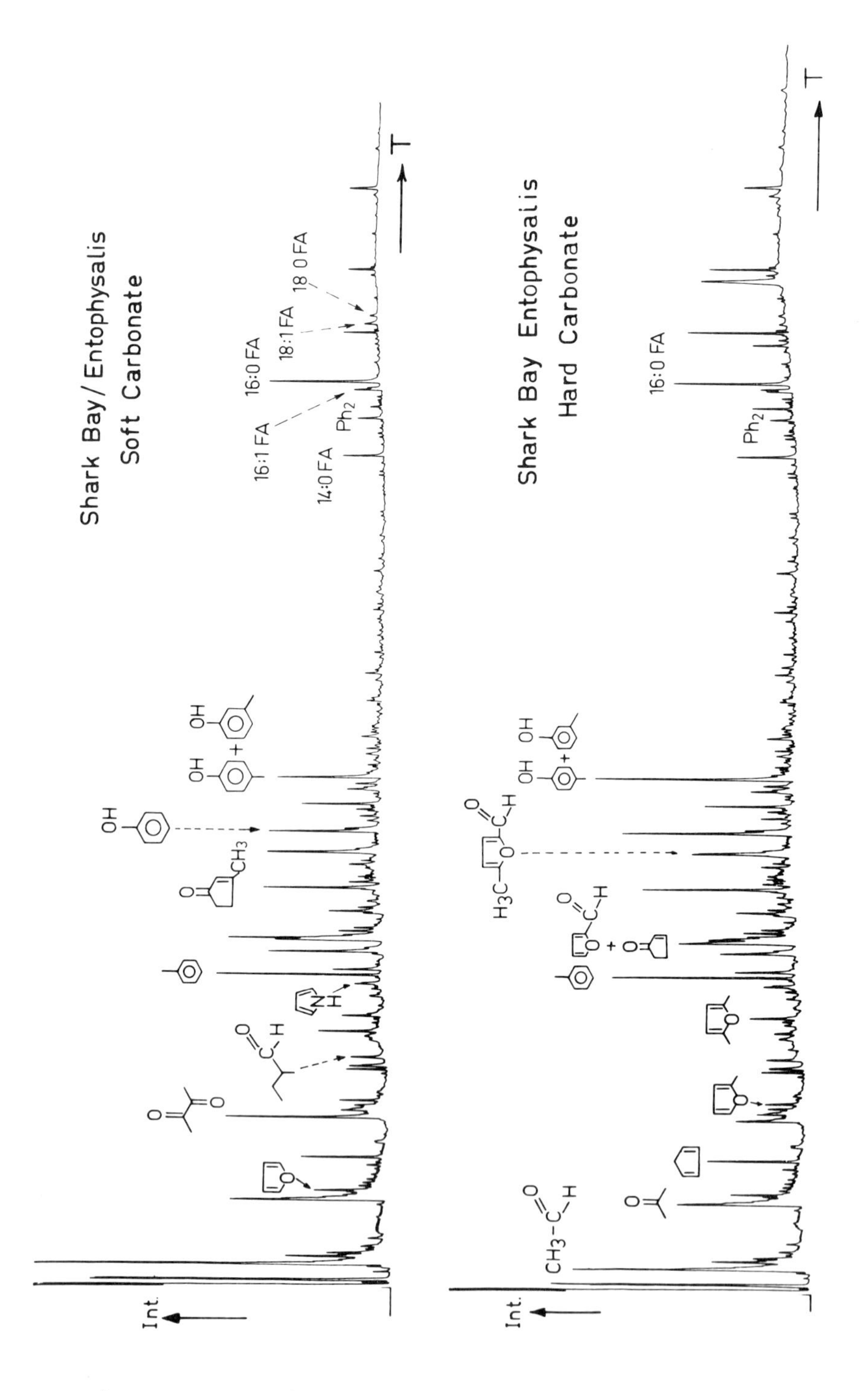

Fig. 2. Pyrolysis gas chromatograms of the soft and hard carbonate sample (for conditions: see text).

Both fingerprints show a number of mass peaks which are often seen in carbohydrate rich samples, for example m/z 55, 58, 60, 69, 72, 74, 82, 84, 96, 98, 110, 112, 114, 126. The peaks at m/z 34, 48, 67, 92, 94, 108, 117, 131 are often present in protein rich samples. The intensities of m/z 34 (H_2S) and 48 (CH_3SH) are very low in these rock samples however.

The high intensity m/z 44 (CO_2) and 28 (CO) are most probably derived from pyrolytic decomposition of the calcium carbonate.

Py-GC

Pyrolysis gas liquid chromatography separates the mixture of volatile compounds evolved from the samples by the pyrolysis process. Contrary to the pyrolysis mass spectrometry, which ideally shows the distribution of all isobaric molecular ions up to mass m/z 160, the gas liquid chromatogram shows the completely separated low boiling fraction and higher boiling fractions. The latter are not detectable by the pyrolysis mass spectrometry method used. Figs. 2A and B are the Py-GC traces of the soft and hard carbonate sample. Again very little difference is seen between both compound profiles, which again points to a very similar organic matter composition. The mixture consists of a low boiling fraction ranging from methane to cresol and significant signals in the higher boiling range. In the latter fraction several fatty acids (nC_{14}:0, nC_{16}:1, nC_{16}:0, nC_{18}:1 and nC_{18}:0) were detected and phytadiene, a pyrolysis product of phytyl moieties in chlorophyll pigments (v/d Meent *et al.*, 1980).

Py-GC-MS

The hard rock sample was subjected to Py-GC-MS. The chromatographic conditions were the same as those used for Py-GC. The reconstructed total ion current trace is shown in Fig. 3.

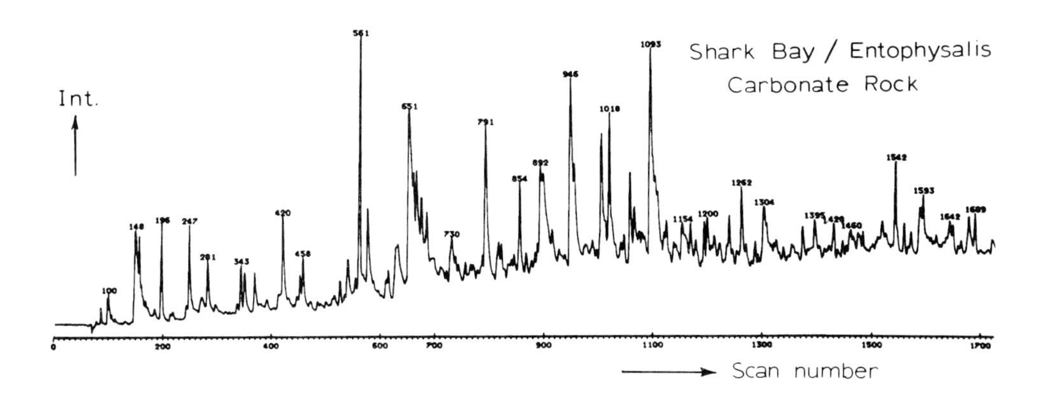

Fig. 3. Py-GC-MS trace of the hard rock sample (for conditions: see text).

scan nr.	identification
89	methylene chloride
100	butene
148	acetone
158	furan
196	cyclopentadiene
247	2,3-butandione
270	unknown (M^+=72)
281	2-methyl-furan
343	methyl-cyclopentadiene
349	methyl-cyclopentadiene
351	3-methyl-butanal
368	2-methyl-butanal
369	benzene
370	hydroxypropanone (tailing peak)
420	2,3-pentadione
453	unknown (M^+= 88)
458	2,5-dimethyl-furan
526	dimethylcyclopentadiene
540	pyrrole
561	toluene
575	cyclopentanone
609	acetamide (tailing peak)
614	unknown (M^+=100)
628	furan-3-aldehyde
651	cyclopentenone
669	furan-2-aldehyde
684	benzylalcohol
730	methylcyclopentenone
744	methyl-cyclopentenone
791	methyl-cyclopentenone
815	trimethylfuran (?)
854	unknowns (M^+=110; M^+=112)
892	methyl-cyclopentenone
898	5-methyl-furan-2-aldehyde
946	phenol
1004	unknown (M^+= 112)
1018	unknown (M^+= 110, basepeak 67)
1057	dimethyl-cyclopentandione
1092	cresol
1304	unknown (basepeak 91)
1395	indole
1460	amino acid dimer (M^+= 195)
1517	methylindole
1542	unknown (basepeak 58)
1593	amino acid dimer (M^+= 209)

Table 1. Pyrolysis products identified in *Entophysalis* carbonate rock from Shark Bay (Australia).

Table 1 lists a number of identified pyrolysis products. The instrument started data acquisition from mass m/z 50 to 400, so molecules with fragments and molecular ions below m/z 50 are not recorded. The total ion current trace is quite comparable to the gas chromatographic trace despite the differences in mass spectral response for the different compounds. Most of the major compounds were readily identified by comparison their mass spectra with mass spectra of standards and with published literature spectra (Stenhagen *et al.*, 1974). The relative retention times of most compounds are also known. No retention times were available for a number of cyclopentenones, which are tentatively identified from the mass spectra. The compounds differ in their methyl substitution pattern. The compounds corresponding with scan numbers 351, 368, 540, 561, 946, 1092, 1304, 1395, 1460, 1517 and 1593 are also observed in Py-GC-MS runs of proteins (unpublished results). The furan derivatives are evolved from carbohydrates under dehydrating conditions. The 2,3-butandione, cyclopentadiene, hydroxypropanone and 2,3-pentadione are evolved from carbohydrates under basic catalytic conditions which promote chain fission (Shafizadeh and Lai, 1972).

Interestingly, we observed no evidence for slightly dehydrated aglycon units such as levoglucosan, dianhydro-hexoses or levoglucosenone, which are major products evolved from hexosans under the pyrolysis conditions used (van der Kade *et al.*, in prep.). Instead, furans and cyclopentenones are major products which may point to rapid dehydration of the carbohydrates in the glycocalyx either before or shortly after the mineralisation process in the carbonate rocks. The aromatic compounds, presumably evolved from aromatic amino acid moieties in a residual protein fraction, - *s.i.* including melanoidins - are relatively intense in both samples. They are more intense in the hard carbonate sample relative to the furan derivates and cyclopentanones.

The striking similarity of the organic matter composition in both the initial soft rock stage and the later hard rock stage, as demonstrated by these pyrolysis data, indicates that the calcification process inhibits further progress in the decomposition of the organic matter. The organic matter composition of these cyanobacterial rock samples differ greatly from intact healthy cyanobacteria (van Eijkelenburg, 1978, Boon and Haverkamp, 1979) and from living cyanobacterial mats (Boon *et al.*, 1982). Therefore we are forced to assume that major chemical transformation processes have taken place before the encapsulation. One could envisage that the maintenance of the hydrosphere around the organisms, threatened by the subaerial dissication of the glycocalyx during summer, has a great priority. Initially the cyanobacteria may be forced to metabolise themselves in an attempt to supply their sheath with metabolic water. This behavior ultimately results in complete starvation. It is clear that the organic matter composition is severely altered before it is preserved in the mineral matrix. A detailed study of the intact *Entophysalis* mat, the starved mats in increasing stages of dehydration and the mats in initial stages of calcification, should shed more light on the changes in organic matter composition.

ACKNOWLEDGEMENTS

This paper is dedicated to our friend Stjepko Golubic who appeared on the right moments in space and time.

This study was partly supported by NASA Grant NGL-05-003-003.

REFERENCES

Boon, J.J., and Haverkamp, J. (1979). Neth. J. Sea Res. 13, pp. 479-486.

Boon, J.J., Hines, H., Burlingame, A.L., Klok, J., Rijpstra, W.I.C., de Leeuw, J.W., Edmunds, K.E., and Eglinton, G. In: "Advances in Organic Geochemistry 1981" (M. Bjørroy, ed.), in press.

van Eijkelenburg, C. (1978). Antonie van Leeuwenhoek 44, pp. 321-327.

Golubic, S. (1982). This Volume, preceding paper.

Irwin, W.J. (1979). J. Anal. Appl. Pyrol. 1, pp. 3-25 and pp. 89-122.

Knoll, A.H., and Golubic, S. (1979). Precambrian Research 10, pp. 115-151.

van de Meent, D., de Leeuw, J.W., and Schenck, P.A. (1980). In: "Advances in Organic Geochemistry 1979" (J.R. Maxwell and A. Douglas, eds.), Pergamon Press, Oxford, pp. 469-474.

van de Meent, D., Brown, S.C., Philp, R.P., and Simoneit, B.R.T. (1980). Geochim. Cosmochim. Acta 44, pp. 999-1013.

Shafizadeh, F., and Lai, Y.Z. (1972). J. Org. Chem. 37, pp. 278-284.

Stenhagen, E., Abrahamsson, S., and McLafferty, F.W. (1974). Registry of Mass Spectal Data, J. Wiley & Sons, New York.

Zhang, Y. (1981). J. Paleont. 55, pp. 485-506.

THE ROLE OF ENAMEL TUBULE AND THE EVOLUTION OF MAMMALIAN ENAMEL

Y. Kozawa and M. Tateishi
Nihon University School of Dentistry at Matsudo
Matsudo, Japan

The fine structure of early amelogenesis is studied in the pig. The developing ameloblast has long apical processes (about 10 μm) which touch with the odontoblastic process within the predentin. Enamel tubules, containing the processes of the ameloblast and the odontoblast, are observed in early enamel formation. Later this tubule is mineralized, so that it cannot be found in the mature enamel. These findings are compared with those of primitive mammalian enamel, and their possible functional significance is discussed.

MATERIALS

Pig : *Sus domestica* - about 8 weeks old fetus
Kitten: *Felis domestica* - 20 days after birth

The enamel tubule is commonly observed in primitive mammalian teeth (Mammery, 1924; Sahni, 1979), such as the multituberculate and the marsupial. Previous studies suggest that the enamel tubule is formed from the odontoblastic process (Carter, 1922). However, Lester (1970) describes the Oppossum enamel tubules as containing cell process which are extentions of ameloblast cytoplasm originating at the mineralizing front of the enamel. Similar structures are observed in early pig amelogenesis (Kozawa and Tateishi, 1981).

In the rat and kitten, the preameloblast (differentiating state) has apical processes which penetrate the basal lamina and enter the predentin to a depth of about 1-3 μm (Figure 1) (Kallenbach, 1976), but no organelles can be observed in the cytoplasm. Calcification of matrix vesicles occurs around these processes. The pig preameloblast has two distinct types of process which also penetrate the basal lamina into the predentin. The short type (similar to the kitten's, i.e. Type 1) is up to 2-3 μm in length (Figure 2), and the long type is over 10 μm (Figures 3 and 4, i.e. Type 2). These two types of process contain a variety of organelles, such as free ribosomes, microfilaments, coated vesicles, pinocytotic vesicles and so on, suggesting

P. Westbroek and E. W. de Jong (eds.), Biomineralization and Biological Metal Accumulation, 335–339.

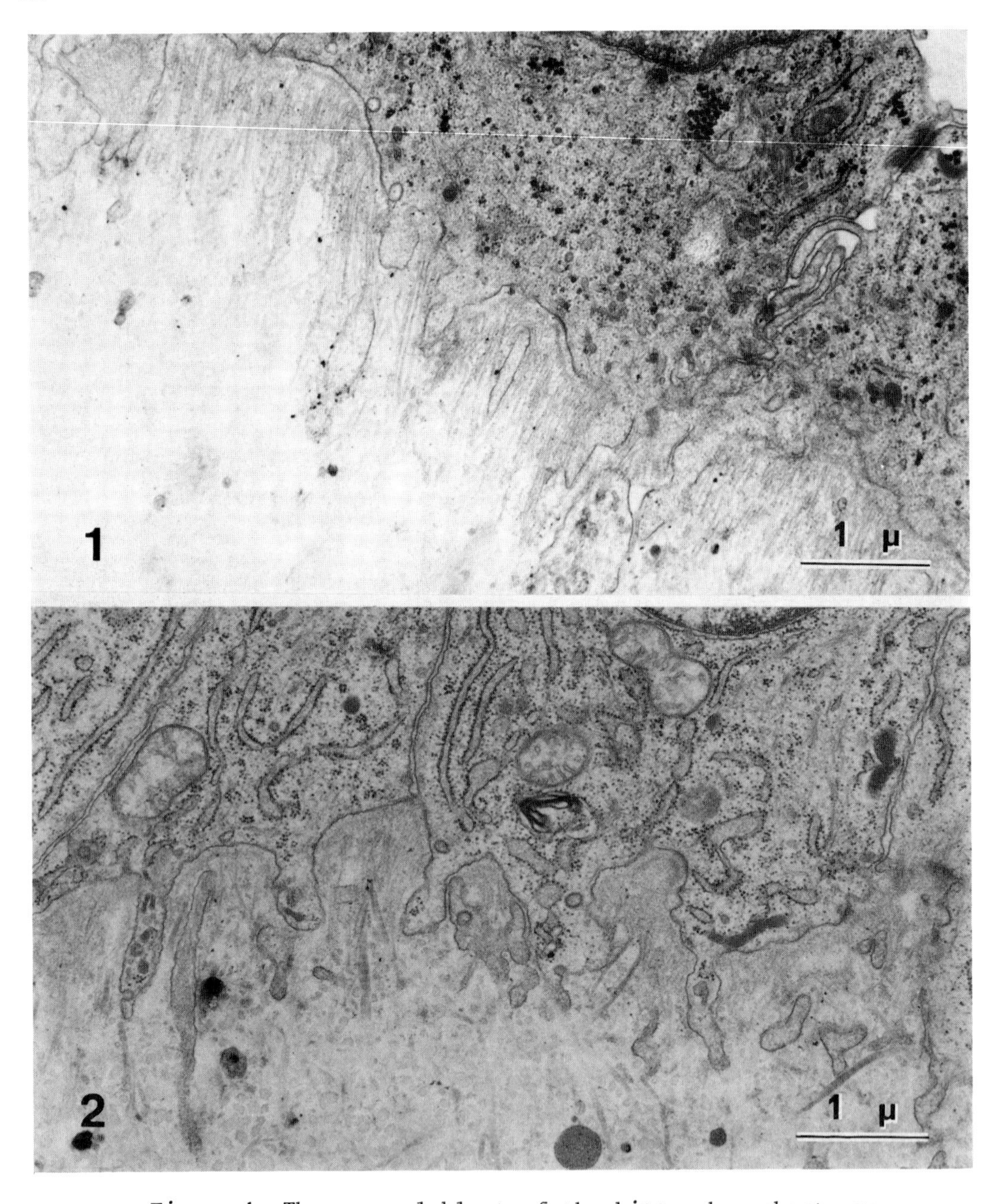

Figure 1. The preameloblast of the kitten has short processes, penetrating into the predentin and making contact with the odontoblastic process (OP).

Figure 2. Short processes (Type 1) of the pig preameloblast might play active roles, since they include some organelles.

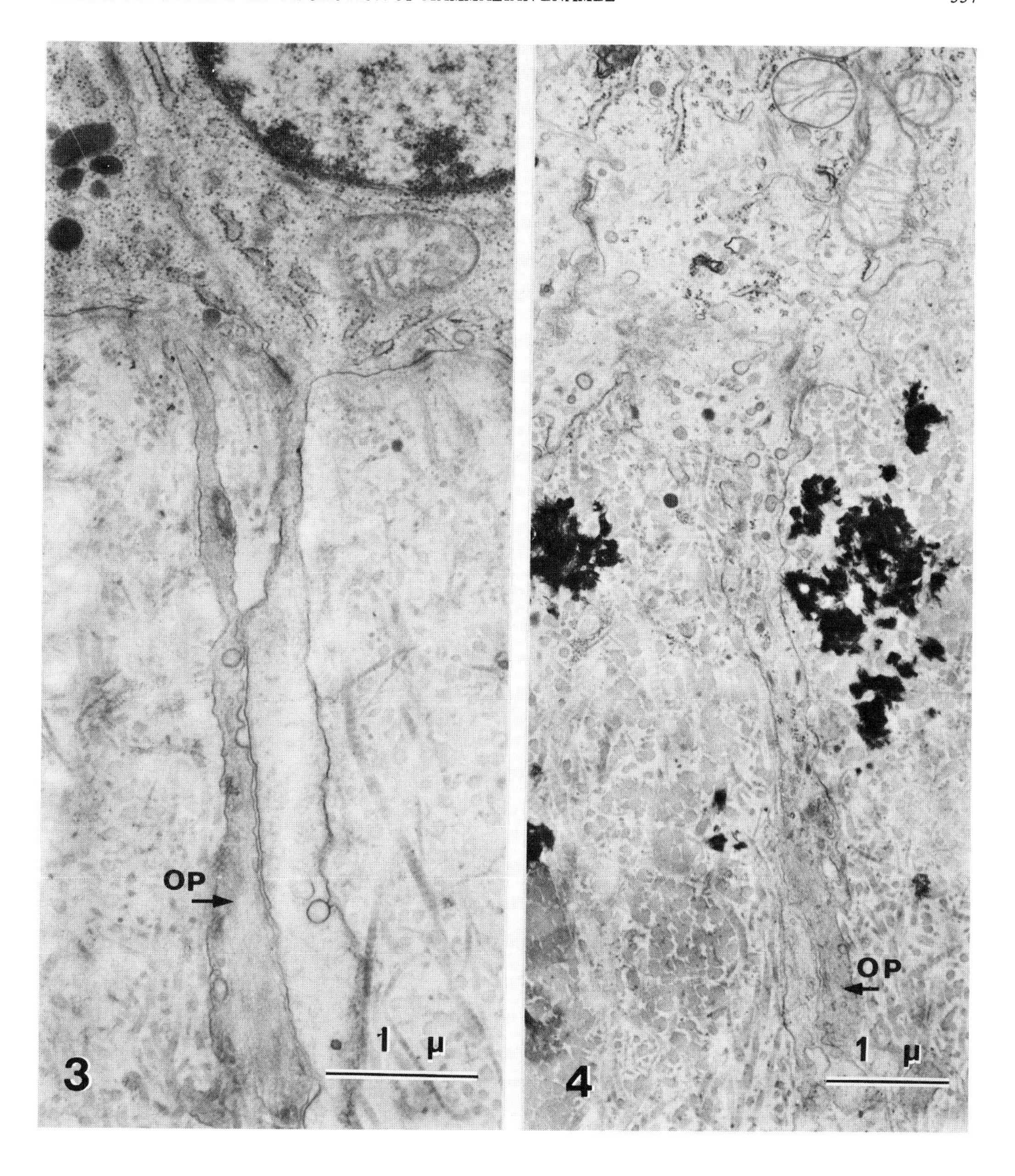

Figure 3. The long processes (Type 2) are attached to the odontoblastic process (OP).

Figure 4. Calcification occurs around the long process which includes many organelles.

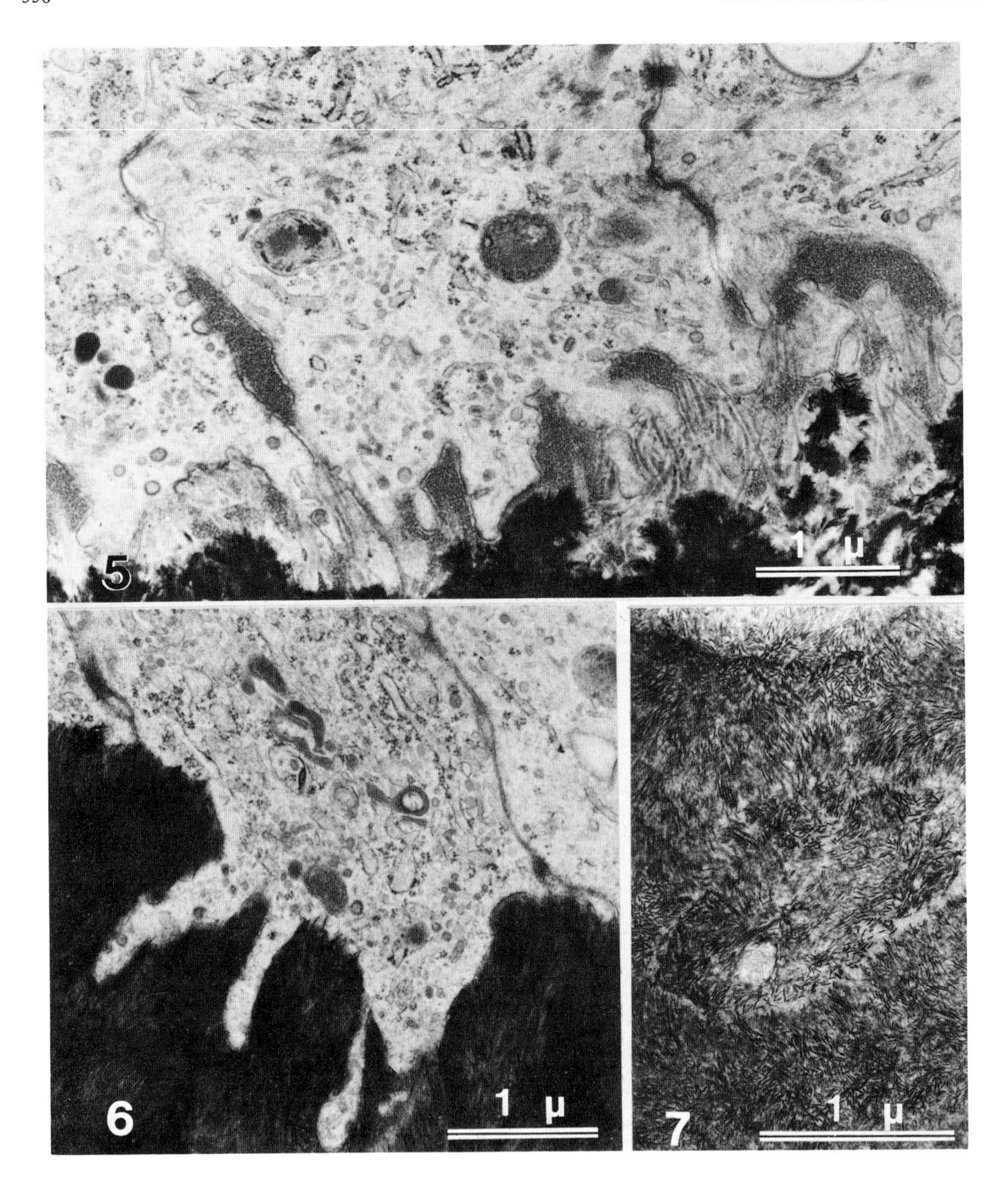

Figure 5. The preameloblast initiates secretion of stippled materials, but the long process cannot be observed.

Figure 6. The ameloblast processes (Type 3) are situated at the tip of the Tomes' process. Some coated vesicles are observed in the cytoplasm.

Figure 7. Enamel crystals are deposited in the enamel tubule, so that it cannot be observed in the mature enamel.

active metabolism within the process. The Type 2 process often makes contact with the odontoblastic process (Figures 3 and 4). It appears likely that interactions are therefore taking place between the ameloblast and the odontoblast. When enamel deposition starts, these processes temporarily disappear. At the early matrix stage, Type 3 processes extending to the enamel matrix appear around the tip of the Tomes' process (Figure 6). These also contain coated vesicles and other organelles in the cytoplasm. Many tubules are observed in the enamel prisms and in the interprismatic enamel. Those in the enamel prisms containing the Type 3 process, occasionally make contact with the odontoblastic process. Those in the interprismatic enamel reach the odontoblastic process. In later developmental stages, enamel crystals are deposited on the enamel tubule surrounding the process, and that tubule finally disappears (Figure 7). The enamel tubules cannot be observed therefore in the mature enamel of the pig. These structures are only transient phenomena that occur during the process of amelogenesis, suggesting that they are similar to the vestiges of tubular enamel found in lower mammals. In many cases, the tubular enamel does not have the Hunter-Schreger bands, which are developed in highly evoluted enamel without the enamel tubule. The contact of the odontoblast and the ameloblast implies the interactions of these cells, and the authors conclude that the odontoblast regulates the ameloblast movement, which forms the Hunter-Schreger bands. The processes containing the coated vesicles may have significant roles to play in crystal formation.

CONCLUSIONS

1. The enamel tubule originated from two processes; the ameloblast and the odontoblast.
2. The ameloblast movement might be regulated by the odontoblastic process.
3. The ameloblastic process contributes to the calcification of the enamel.

REFERENCES

Carter, J.T.: 1922, Proc. Zool. Soc. London, pp. 599-608 (Figures 1-11).
Kallenbach, E.: 1976, Am. J. Anat. 145, pp. 283-318.
Kozawa, Y., and Tateishi, M.: 1981, Nihon Univ. J. Oral. Sci. 7, pp. 223-228.
Lester, K.S.: 1970, J. Ultrastruct. Res. 30, pp. 64-77.
Mammery, J.H.: 1924, The Microscopic and General Anatomy of the Teeth, Human and Comparative (Oxford University Press).
Sahni, A.: 1979, Palaeontographica Abt. A. 166, pp. 37-49 (Figures 1-3).

URINARY STONE FORMATION AS A BIOMINERALIZATION PROCESS

Leo J.M.J. Blomen, Olav L.M. Bijvoet
Clinical Investigation Unit, Bln. 19
University Hospital, 2333 AA LEIDEN
The Netherlands.

Abstract

A short survey is given of urinary stone formation. Being the result of crystallization in a urine-like environment, stone formation is discussed in relation to the physical chemistry of mineralization. Influences of the solid phase (the stone), the liquid phase (urine) and their interaction are described.

Introduction

Urinary stones affect more than one % of the world population (1), being the major example of undesired biomineralization. The affliction may be as old as human memory goes back. An almost 7000 year old stone was discovered in the mummy of an ancient egyptian boy. Ironically, the stone was "pulverized" in London during the 1941 bombings (2). Since ancient times, considerable efforts have been spent on research and treatment of this multifactorial disease. Maybe the first example of the development of a medical specialism, urology, was founded when Hippocrates stated that the operation of bladder stones should be left to specialized lithotomists (3). However, despite the huge research efforts, there still exists no unifying theory about the cause of stone formation, and the number of urinary stones and of stone patients is increasing.

Processes involved during stone formation

The urinary tract is a flow-system where, in order to regulate composition of body fluids, urine is produced by filtration and subsequent absorption processes. The composition of the freshly filtrated "pre-urine" changes during passage through the kidney tubules, and the fluid contents of more than one million tubules per kidney come together in the renal pelvis. The magnitude of the processes is reflected in the fact that two kidneys produce 180 liters of filtrate per day, which is more than four times the total water content of the body, of which only an average amount of 1-2 liters is excreted as urine! Urine composition is therefore dependent upon diet, absorption in the gastro-intestinal

P. Westbroek and E. W. de Jong (eds.), Biomineralization and Biological Metal Accumulation, 341–345.

tract, and kidney function. Many diseases contribute to high urine concentrations of some stone-forming compounds. However, in the majority of the patients, no underlying disease can be found. Due to the flow in the kidneys, urine concentrations will be time- and site-dependent. Renal anatomy, on the other hand, can provide favourable places to form crystals (whether intra-or extra-cellular) and constitutes renal hydrodynamics. In about 8-12 minutes, the urine traverses the whole kidney (4). The figure illustrates this. Possible nucleation sites, hydrodynamics and the concentrations in urine determine the origination of new crystals (nucleation). After that, the crystals can grow in supersaturated urine, they can form aggregates (agglomerates) and they can grow on crystals of another crystal phase, of which "epitaxy" is an example. Due to all these processes, a certain particle size distribution exists, and part of the crystals will be "washed" away from the body by voiding urine. However, when crystals stick anywhere, or when insufficient voiding occurs, there will be a resultant crystal surface area in the urinary tract, which is available for further growth, agglomeration, etc.

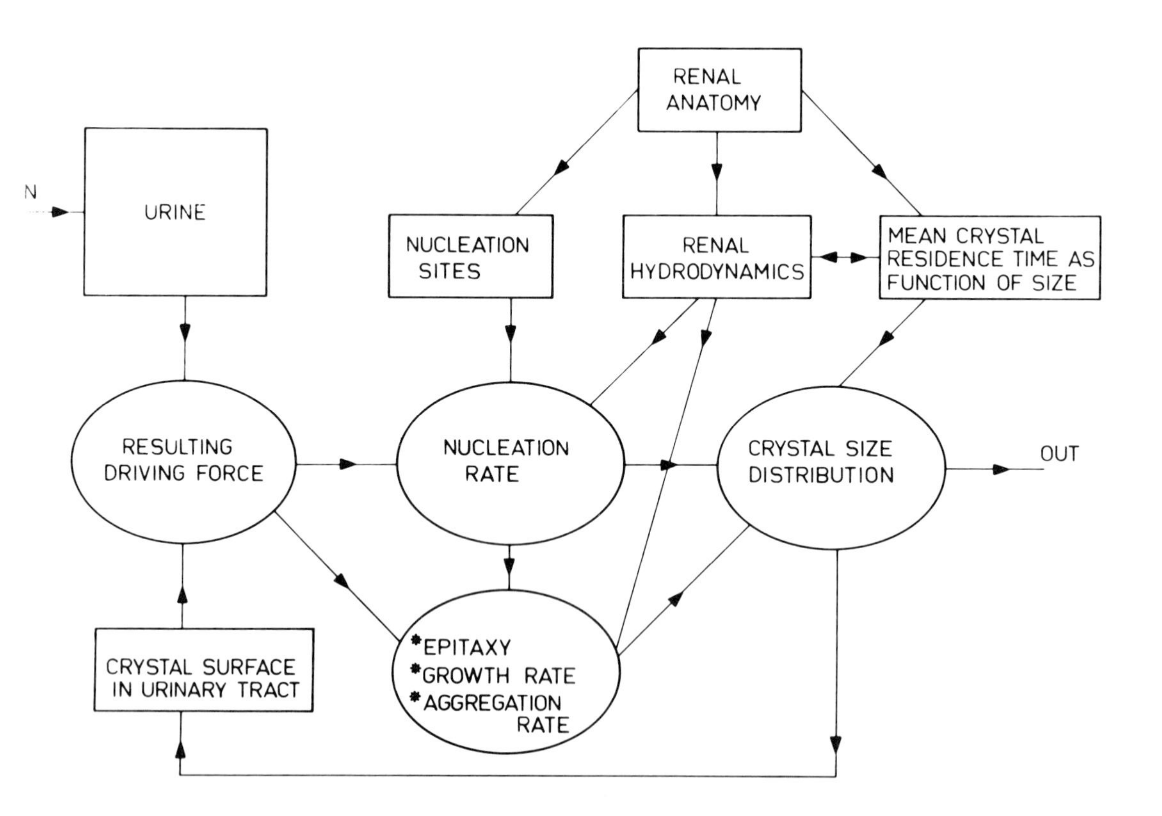

The solid phase: the stone

Urinary stones are composed of one or a mixture of more crystalline substances, together with an organic, colloidal matrix which contributes about 2.5% of the stone mass (5). The inorganic crystal phases consist primarily of calcium oxalate hydrates (mainly whewellite and weddellite), calcium fosfates (hydroxyapatite, $Ca_{10}(PO_4)_6(OH)_2$; carbonateapatite, $Ca_{10}(PO_4,CO_3)_6(OH,CO_3)_2$; brushite, $CaHPO_4.2H_2O$; octacalciumfosfate, $Ca_8H_2(PO_4)_6.5H_2O$; whitlockite, $Ca_3(PO_4)_2$), struvite ($MgNH_4PO_4.6H_2O$) and uric acid and urates. Apart from the occurrence of mixtures of these phases, morphologies may be quite different, and several processes may interact. Solid state transormations between several phases have been reported. Even when stone composition is known, it is impossible to derive conclusions about the genesis of that particular stone. Stone growth is governed by different, changing growth conditions. Like a tree which is exposed to changing climate, a stone builds up in often concentric circles. At the present stage, information about stone genesis is only occasionally employed in order to make therapeutic decisions.

The fluid phase: the urine

The main problem encountered in urine chemistry is the variability of its composition. No two similar urines can be found. Besides, in vivo concentration measurements are difficult and mostly excretions are used to evaluate stone forming propensity. Most important substances that affect stone growth can be measured, though the oxalate ion, which is often decisive for the stone development, is hard to measure (6). Urine contains, apart from the inorganic substances, also important inhibiting substances. Several of these seem to be involved in the process, and some are hard to measure. Important inhibitory effects are measured from sodium, magnesium, citrates, pyrofosfate (7). Other groups, however, (8) attribute a larger effect to higher molecular weight compounds, like glycosaminoglycans (or acid mucopolysaccharides (9)), acid peptides (10) and glycoproteins (11). Besides, some dicussed the importance of urinary amino acids on solubility (12) and of trace elements, especially heavy metal ions (13). Not only inhibitors have been identified, but also crystallization promoting substances, which are also found inside the stones, often in even higher concentrations, and therewith pointing to possibly important roles during stone formation. Among these substances are matrix substance A (14), Tamm-Horsfall mucoprotein (15) and uromucoid (16), albumine and α- and γ-globulines (17). However, the relative importance of these substances has not yet been generally accepted. Our own investigations (18) even indicate that some substances act as inhibitors of growth, but at the same time promote agglomeration of crystals. Most research reports in this field did not use standardized methods. Therefore, many contradictory research results have been reported.

Theories of stone formation

Many theories about the cause of stone formation exist, though no unifying hypothesis has been accepted. Roughly, two groups of theories can be distinguished, the first dealing with stone formation as a pure precipitation-crystallization process of inorganic crystals, the second dealing with the formation of organic matrix at the basis of the formation process. However, whatever the cause of the origination of stones might be, crystals once formed have the opportunity to grow and agglomerate in the normally supersaturated urine. The balance of the processes explained, and indicated in the figure, decides about stone-formation after the unknown, initial origination.

Principle of therapy

A logical concequence of the above mentioned facts is that therapy is mainly based on the best known processes. Indeed, by far the largest number of therapeutic measures is based on a physico-chemical idea: reduction of supersaturation. About 75% of all patients is treated by drinking advices, dietary measures and chemical therapies reducing absorption in the gut, decreasing urinary total concentrations or increasing complex formation of the dangerous compounds in urine. Partly, measures are based on inhibition of crystallization processes. For the remaining 25% of stone patients, therapy consists mainly of removal of the stone, biochemical treatment of stone formation stimulating infections in the urinary tract, and some biochemical treatments. Developments are to be expected from the field of fundamental research. However, at the present time, the last stone, and with it the problem of urolithiasis, has not yet been solved, viewing the increase in formed stones during last years.

Literature

1. W. Vahlensieck, A. Hesse, D. Bach, Fortschr. Urol. Nephrol. 17 (1981) 1.
2. N.B. Cummings, in "Idiopathic urinary bladder stone disease", Fogarty International Center Proceedings No. 37, R. van Reen, ed., U.S. Government Printing Office, Washington, D.C, (1977) 1.
3. E.L. Prien, Sr., J. Am. Med. Assoc. 216 (1971) 503.
4. B. Finlayson, in "Idiopathic urinary bladder stone disease", Fogarty International Center Proceedings No. 37, R. van Reen, ed., U.S. Government Printing Office, Washington, D.C. (1977) 7.
5. W.A. Boyce, Am. J. Med. 45 (1968) 673.
6. W. Dosch, Urol. Res. 7 (1979) 227.
7. D.J. Sutor, J.M. Percival, S. Doonan, Brit. J. Urol. 51 (1979) 253.
8. W.G. Robertson, D.S. Scurr, C.M. Bridge, J. Crystal Growth 52 (1981) 182.
9. W.G. Robertson, F. Knowles, M. Peacock, in "Urolithiasis Research", H. Fleisch, W.G. Robertson, L.H. Smith, W. Vahlensieck, eds., Plenum, New York, (1976) 331.
10. H. Ito, F.L. Coe, Am. J. Physiol. 233 (1977) F455.
11. W.B. Gill, J.W. Karesh, L. Garsin, M.J. Roma, Invest. Urol. 15 (1977) 95.
12. M.G. McGeown, J. Urol. 78 (1957) 318.
13. E.D. Bird, W.C. Thomas, Proc. Soc. Exp. Biol. 112 (1963) 640.

14. W.H. Boyce, J.S. King, Jr., M.L. Fielden, J. Clin. Invest. 41 (1962) 1180.
15. I. Tamm, F. Horsfall, Proc. Soc. Exp. Biol. Med. 74 (1950) 108.
16. P.C. Hallson, G.A. Rose, Lancet, i (1979) 1000.
17. J.E.A. Wickham, in "Scientific Foundations of Urology", vol. I, D.I. Williams, G.D. Chisholm, eds., Heinemann, London (1976) 323.
18. L.J.M.J. Blomen, O.L.M. Bijvoet, W. Blomen-Kuneken, Fortschr. Urol. Nephrol., in press.

THE INFLUENCE OF AN INTENSE FLUORIDE PRETREATMENT ON REMINERALIZATION OF ENAMEL LESIONS

H.E. Boddé, J. Arends
Lab for Materia Technica, State University,
Groningen, The Netherlands

ABSTRACT

Remineralization of enamel is an important factor in the prevention of dental caries. Fluoride may either enhance or reduce remineralization effectiveness, depending on dose, frequency and form of application. This contribution pertains to in-vitro remineralization of demineralized enamel after an intense fluoride pretreatment. This pretreatment inhibits longterm remineralization. To explain this phenomenon it is suggested that the pretreatment induces a loss of growth sites on enamel crystallites.

1. INTRODUCTION

Remineralization is considered an important repair mechanism in relation to dental caries. It is operative both in vivo and in vitro; mineral uptake by demineralized enamel from either natural or synthetic saliva has been observed (Backer Dirks, 1966; Silverstone, 1977). SEM observations indicate growth of enamel crystallites during remineralization (Silverstone and Wefel, 1981; Arends and Jongebloed, 1978); at low supersaturation the stoichiometry of the precipitate generally matches that of calciumhydroxyapatite ($Ca_{10}(PO_4)_6(OH)_2$) (Moreno et al., 1977). Fluoride boosts remineralization when added to the supersaturated liquid at a constant, low level (~ 1 ppm) (Silverstone, 1977; ten Cate and Arends 1977). However, in dental practice fluoride is often supplied by topical application of highly concentrated (~ 10^4 ppm) agents; the cariostatic effect of these agents depends on the partition of fluoride between the dental enamel and the surrounding oral fluid after the treatment. (Fejerskov et al., 1981). Recently attempts have been made to study the effect of topical fluoride treatments on remineralization, both in vivo (Gelhard and Arends, 1981; Lambrou et al., 1981) and in vitro (ten Cate et al., 1981).
Crucial parameters are: the type of enamel defect studied and the intensity and frequency of the fluoride treatment.
The work presented here pertains to *in vitro* remineralization of initial

P. Westbroek and E. W. de Jong (eds.), Biomineralization and Biological Metal Accumulation, 347–352.

enamel lesions with and without an intense pretreatment with potassium-fluoride at pH 7.0.

2. MATERIALS AND METHODS

From 22 central bovine incisors 44 square blocks (~ 25.0 mm^2) were cut. After polishing the labial enamel surfaces with silicon carbide paper all remaining surfaces were covered with sticky wax.
All enamel blocks were *demineralized* for 5 days in a stirred acid buffer (1 $cm^2 . l^{-1}$), containing 50 mM acetic acid, 3.9 mM $CaCl_2 . 2 H_2O$, 3.9 mM KH_2PO_4 and 1 ppm KN_3 at pH 5.0 and 37° C.
28 blocks received a *fluoride pretreatment* at 37° C during 16 hours prior to remineralization with a solution containing 52.6 mM KF and 10 mM $(CH_3)_2AsO_2K$ at pH 7.0; the surface-volume ratio was 0.1 $cm^2.ml^{-1}$.
12 fluoridated blocks were *washed* under running tapwater (240 $l.h^{-1}$) for 48 hours prior to remineralization.
Remineralization took place in a solution containing 1.5 mM $CaCl_2.2 H_2O$, 0.9 mM KH_2PO_4, 130 mM KCl and 20 mM $(CH_3)_2AsO_2K$ at pH 7.0 and 37 °C. In several cases $^{45}CaCl_2$ was added to the remineralizing solution to obtain an activity between 0.1 and 1.0 $\mu Ci.ml^{-1}$. The surface-volume ratio was 0.1 $cm^2.ml^{-1}$; the stirring rate 200 rpm.
Calcium and Phosphate determinations in *unlabelled* solutions were done with atomic absorption and colorimetry (Chen et al., 1956), resp.
Enamel blocks remineralized in *labelled* solutions were dried and subsequently in-depth profiles of labelled Ca* were obtained with a serial sectioning method (Flim, 1976). Depths were determined according to ten Cate (1979).
Microradiography was done according to Groeneveld (1974) on enamel sections cut perpendicular to the anatomical surface. For details on SEM see Jongebloed (1976).
To obtain *fluoride in-depth profiles* serially sectioned (Flim, 1976) enamel samples were weighed and then dissolved in 2 ml 0.15 M perchloric acid; after adding 1 ml 1.2 M trisodiumcitrate, fluoride was measured with a specific F-electrode. For determination of depths see ten Cate (1979).
Calciumfluoride in enamel was determined by extracting whole blocks with 1 N KOH during 24 hours (Caslavska et al., 1975).

Three *unlabelled* experiments were each done in triplicate, using four enamel blocks per run:
control experiment C: dem-rem
fluoride " F: dem-fluor-rem
wash " W: dem-fluor-wash-rem
(rem denotes 14 days remineralization with daily solution changes; for dem, fluor and wash see treatments described above).
Four *labelled* experiments were done with two enamel blocks each:
C_1*: dem-rem*
C_4*: dem-rem-rem*
F_1*: dem-fluor-rem*
F_4*: dem-fluor-rem-rem*
(-denotes solution change; here rem denotes 3 hours *unlabelled* reminera-

lization; rem* denotes one (1st or 4th) hour *labelled* remineralization).

3. RESULTS AND DISCUSSION

3.1. Demineralization

The average microradiogram of the demineralized enamel (Figure 1) shows an enamel defect (hereafter called "lesion" for simplicity) in which the characteristic "surface layer" is beginning to form. The calcium loss is 45 ± 8 μmol.cm^{-2}.

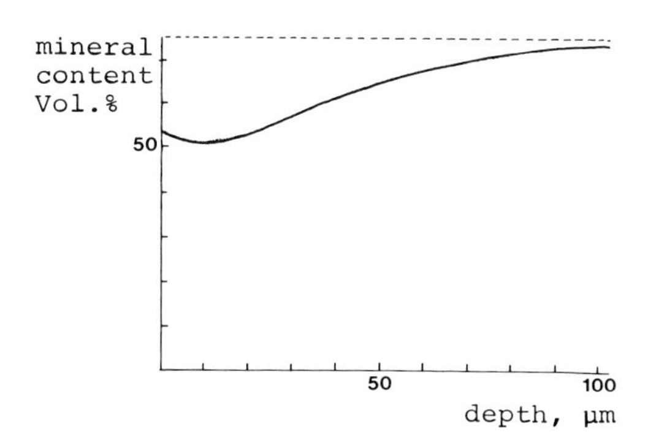

Figure 1. Average microradiogram. Bar = biological S.D.

3.2. Fluoride pretreatment

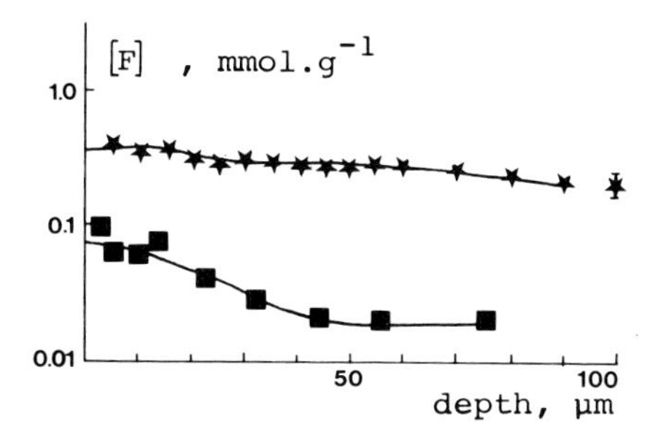

Figure 2. Fluoride in-depth profiles in fluoridated (*) and unfluoridated (■) enamel.

In the course of the fluoride pretreatment the enamel released about 2.5 μmol.cm^{-2} phosphate and took up about 7.3 μmol.cm^{-2} fluoride, of which 4.4. μmol.cm^{-2}was KOH-extractable. The in-depth distributions of fluoride in pretreated and control specimen are compared in Figure 2.
In contrast to crystallites in demineralized enamel (Figure 3 a, b) the ones in subsequently fluoridated specimen (Figure 3 c, d) have "erased" contours and seem to be covered with an amorphous layer. This phenomenon, being observable at various depths, is consistent with the in-depth distribution of fluoride (Figure 2).

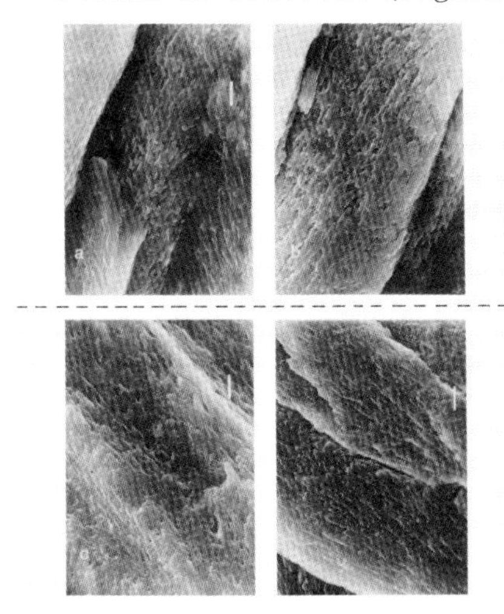

Figure 3. SEM pictures of demineralized (a, b) and fluoridated (c, d) enamel at 30 μm (a, c) and 80 μm (b, d) depth. Bar equals 1 μm.

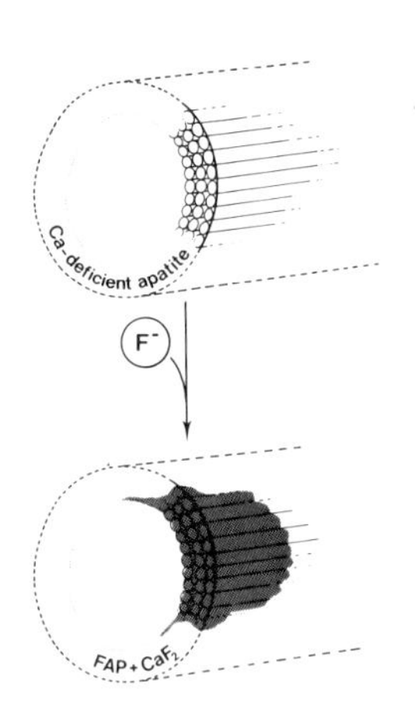

Figure 4. Model for effects caused by fluoride pretreatment.

The foregoing data are compiled in a schematic model (Figure 4), which suggests that during the pretreatment enamel crystallites near interprismatic pores transform into CaF_2 and a fluoridated apatite.

3.3. Remineralization

During 14 days of remineralization both calcium and phosphate were taken up by the lesion in the unlabelled experiments C, F and W (see section 2). Calcium uptake in these experiments is compared in Figure 5.
Comparing experiment F to C, calcium uptake is initially enhanced, but subsequently inhibited. The cross-over point occurs after ~ 2 days. In the washing experiment W inhibition is more pronounced than in F, and a cross-over occurs earlier.
In experiment F the level of fluoride leached from the specimen was initially ~ 3 ppm, and dropped stepwise with each solution change.
In experiment W the fluoride concentration was always below the detection limit. The labelled experiments C_1* - F_4* (Figure 6) confirm the inhibition effect. They also show in-depth penetration of Ca* in all cases.

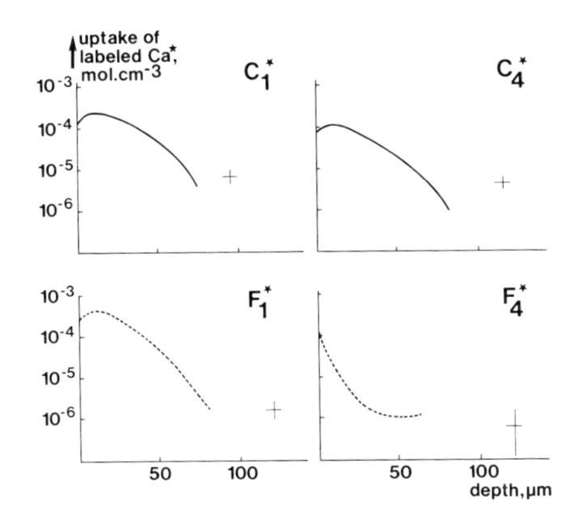

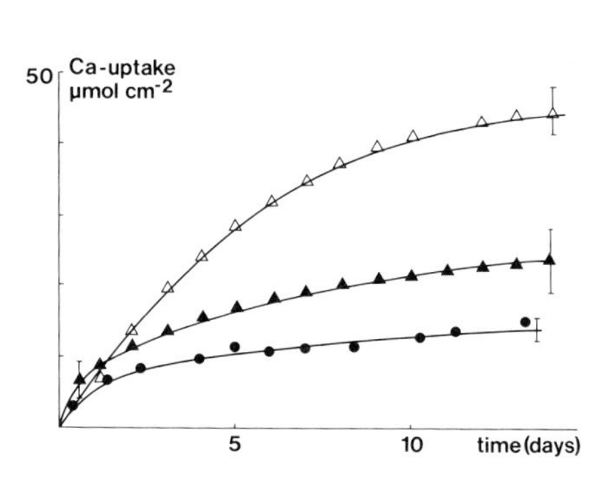

Figure 5. Ca uptake during rem. in exp. C (△), F (▲) and W (●); bars indicate biological variation.

Figure 6. In-depth profiles of Ca*. Each profile is average of two; bars indicate analytical errors.

All data discussed indicate that inhibition of mineral uptake following a fluoride pretreatment occurs when most of the dissolved fluoride is either redeposited or washed away.
The negative effect upon calcium uptake caused by the prewash excludes blocking of enamel pores (ten Cate, 1979) as an explanation for this effect. In conclusion we can say, that the inhibition is caused mainly by an alteration of crystallite surfaces as illustrated in Figures 2 and 4. The fluoride pretreatment most likely destroyes growth sites on enamel crystallites.

REFERENCES

Arends, J. and Jongebloed, W.L.: 1978, J. Biol. Bucc. 6, pp. 161 - 171
Backer Dirks, O.: 1966, J. Dent. Res. 45, pp. 503 - 511
Caslavska, V., Moreno, E.C. and Brudevold, F.: 1975, Arch. Or. Biol. 20, pp. 333 - 339
Cate, J.M. ten: 1979, Ph.D. thesis, Groningen
Cate, J.M. ten and Arends, J.: 1977, Caries Res. 11, pp. 277 - 286
Cate, J.M. ten, Jongebloed, W.L. and Arends, J.: 1981, Caries Res. 15, pp. 60 - 69
Chen, P.S. jr., Toribara, T.Y. and Warner, H.: 1956, Analyt. Chem. 28, pp. 1756 - 1758
Fejerskov, O., Thylstrup, A. and Larsen, M.J.: 1981, Acta Odont. Scand. 39, pp. 241 - 249
Flim, G.J.: 1976, Ph.D. thesis, Groningen
Gelhard, T.B.F.M. and Arends, J.: 1981, Abstract 95, IADR, Groningen

Groeneveld, A.: 1974, Ph.D. thesis, Utrecht
Jongebloed, W.L.: 1976, Ph.D. thesis, Groningen
Lambrou, D., Larsen, M.J., Fejerskov, O. and Tachos, B.: 1981, Caries Res. 15, pp. 341 - 345
Moreno, E.C., Zahradnik, R.T., Glazman, A. and Hwu, R.: 1977, Caries Res. 24, pp. 47 - 57
Silverstone, L.M.: 1977, Caries Res. 11 (suppl. 1), pp. 59 - 84
Silverstone, L.M. and Wefel, J.S.: 1981, J. Cryst. Growth 53, pp. 148 - 159

MICROSTRUCTURAL FEATURES OF TELEOST OTOLITHS

J.M. Dean, C.A. Wilson, P.W. Haake and D.W. Beckman
Belle W. Baruch Institute for Marine Biology and Coastal Research, Marine Science Program and Dept. of Biology, University of South Carolina, Columbia, South Carolina, 29208, United States of America

We have examined the macro- and microstructural features of teleost otoliths using light and scanning electron microscopy. There is a wide range in otolith morphology within the teleosts with little apparent relationship between otolith size and structure and the biology of the species. Once otolith formation began, the microstructural features, which include the crystals, organic matrix, incremental zone and the associated discontinuous zone of sagittal otoliths, were similar. Increments were observed to begin formation at different times in different species. Examples of such features from freshwater, brackish water, and marine species; active and passive species; egg layers and live bearers; and benthic and pelagic species are shown. The otolith appears to be a permanent physiological record of the fish and applications of the information it contains will be discussed.

INTRODUCTION

Fishes are of great economic and ecological significance. Knowledge of the age of individual organisms is critical to understanding many fundamental processes (growth, reproduction, life span) and this is especially true for fish ecology.

Calcified tissues have been used for age estimation of fishes for more than 200 years (Hederstrom, 1759; Bagenal, 1974). Some studies have demonstrated the formation of an annulus in temperate fish, which is the result of slow and fast growth periods. The presence of these annuli in otoliths has been validated through work with fish of known histories and tagged and recaptured fish.

There are three calcified otoliths (lapillus, asteriscus, and sagitta) located in the semicircular canal on each side of the cranial cavity of fishes. These structures are aragonite in teleosts, but vaterite has been found in other groups of fishes (Carlstrom 1963).

A significant breakthrough occurred with Panella's finding (1971) that otoliths contained "daily rings" and that these "rings" could be

P. Westbroek and E. W. de Jong (eds.), Biomineralization and Biological Metal Accumulation, 353–359.

counted. Thus, one could estimate the age of a fish even if it lacked an annulus as is the case in some tropical species and very young fishes. This hypothesis was validated (Brothers et al. 1976; Taubert and Coble 1977) with observation of increments in fishes of known age. Increment formation has since been demonstrated in several species from many different families and habitats (Barkman 1978; Townsend and Graham, 1981; Brothers 1982; Radtke and Dean 1982).

The validity of daily increment formation in the otoliths of fishes has provided fishery scientists with a very powerful tool. The temporal history of the fish is recorded in a permanent physiological calendar, the otolith, making it possible to analyze seasonal and short-term events in the daily increment record. It is now possible to make otolith preparations rapidly and with excellent replication. The use of otoliths for estimating age, particularly in previously unaged groups, (Wilson and Dean, in press) enables population biologists and resource managers to refine existing models and develop new ones.

There are still fundamental questions that remain unanswered as this is a recent area of research. The vocabulary for the structural features under examination is still developing but there is consistency in the terminology used by Tanaka et al. (1981), Watabe et al. (1982) and Radtke and Dean (1982). It is often speculated that short-term events (temperature change, photoperiod, diet, pressure fronts, tidal rhythms, exercise, feeding activity, daily growth rate) are recorded in otoliths, but the critical experiments have not been done. It would be most interesting if the "instantaneous" growth rate of the scales as observed by Ottaway and Simkiss (1977) could be correlated with the width of the daily growth increment. Validation of the structural features for age estimation has too often been ignored.

RESULTS AND DISCUSSION

In this paper we show there is consistency in the microstructure of the sagittae of several families of fishes. Otolith microstructural features have provided us with a technique whereby we can estimate age for larval and juvenile fish, individuals which are less than one year old, groups which reproduce throughout the year, and pelagic fishes which have not been aged with other methodologies. The fishes listed represent some of those we have examined and are examples of fishes from different families, habitats, modes of reproduction, and sizes. All preparations discussed are of the sagittae which were removed from the fish and prepared as in Haake et al. (1982), and Radtke and Dean (1982) (Table 1).

An initial examination of the external morphology and surface features of various families of fishes shows the great range in size, shape, and surface texture of the otolith. There is no apparent correlation between the size of the otolith and the size of the fish. For example, the sagitta of a 1.53 cm bluegill (Fig. 1) is 0.08 cm long while the sagitta of a 280 cm marlin (191 kg) (Fig. 4) is 0.20 cm long. There are obvious differences in the surface features as well: the otolith of the marlin (Fig. 4) has a very deep sulcus (S) with a prominent rostrum (R) and antirostrum (AR) which gives it a saddle or butterfly appearance that is unique to the Istiophoridae and contrasts markedly with the bluegill (Fig. 1).

Table 1. Some Characteristics of Fishes in this Study

Family	Genus Species	Common Name	Repro.	Hab.	Fig. Num.
Cyprinodontidae	*Lucania goodei*	Bluefin killifish	E	FW	9
	Fundulus heteroclitus	Mummichog	E	BW	11
Poeciliidae	*Heterandria formosa*	Least killifish	LB	FW	8
Centrarchidae	*Lepomis macrochirus*	Bluegill	E	FW	1,2,3
Sciaenidae	*Leiostomus xanthurus*	Spot	E	BW	10
Coryphaenidae	*Coryphaena hippurus*	Dolphin	E	SW	7
Xiphiidae	*Xiphias gladius*	Swordfish	E	SW	6
Istiophoridae	*Makaira nigricans*	Blue Marlin	E	SW	4,5

E=egg layer; LB=live bearer; FW=freshwater; BW=brackish water; SW=oceanic; Repro.=mode of reproduction; Fig. Num.= Figure number

Although the external morphology is highly varied (Figs. 1, 4), there is a similarity in the internal microstructure of otoliths from these families. Internal features of the sagittae such as the spherule, the core, and the increment (made up of the incremental zone and discontinuous zone) are well described by Tanaka et al. (1981) and Watabe et al. (1982), and are readily visible in Fig. 7. The crystalline and morphological features of the incremental zone, which begin at different times in different species are very similar among fishes (Fig. 3 and 5-11). The crystals that make up the incremental zone are acicular and branched crystallites that are oriented radially from the core to the perimeter of the otolith. The variation in increment width between species, between individuals of the same species, and between different otolith regions in a single fish is a function of environmental conditions such as temperature and diet (Dean and Haake, unpubl., Brothers, pers. comm.).

The otolith organic matrix also showed incremental features as distinct as the crystals. When intact otoliths were decalcified in a solution of gluteraldehyde and EDTA, the organic matrix was left as an envelope around the structure. The incremental features were readily visible in the remaining matrix (Fig. 12). The matrix was totally decalcified as it was not birefringent.

We do not presume that the consistency of these microstructural features will hold for all fishes. With more than 20,000 different species of teleosts, one would expect exceptions. For example, increments were present in mature blue marlin (Fig. 5), but we have not vali-

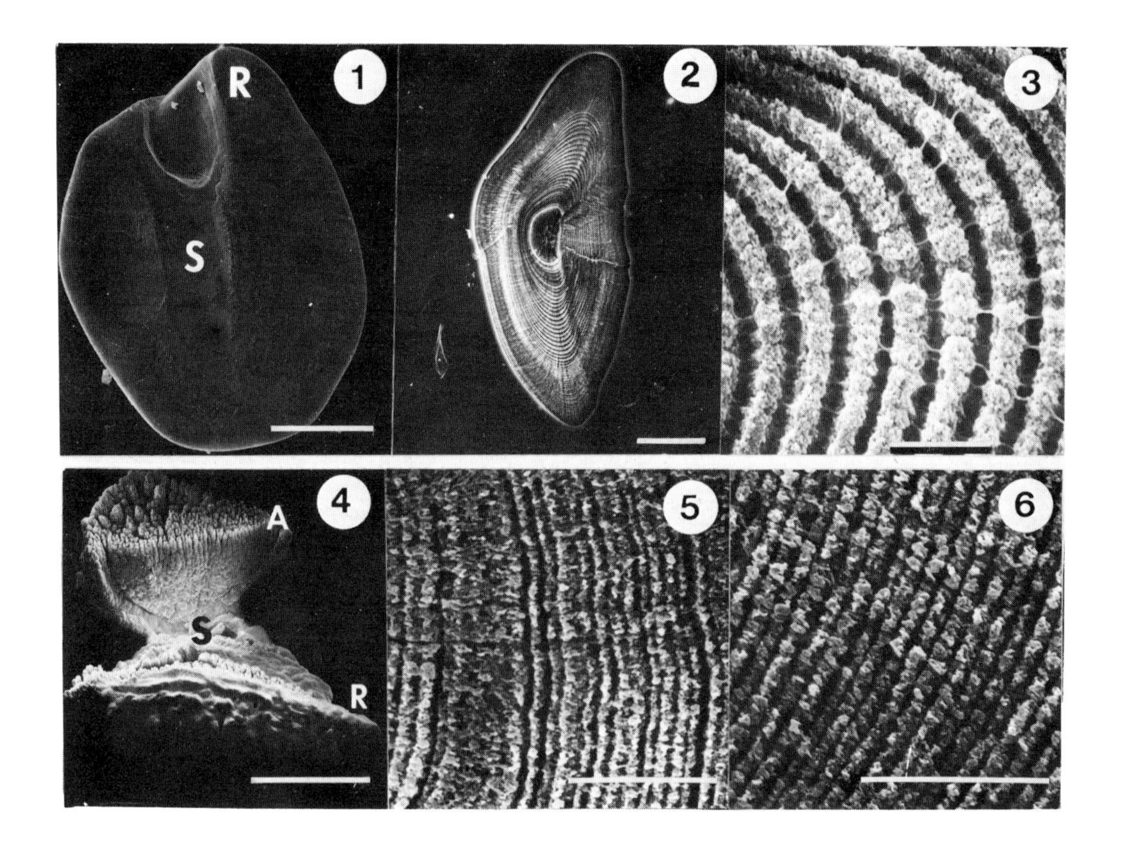

Fig. 1. Medial surface of bluegill sagitta; Bar = 200 μ

Fig. 2. Transverse section of bluegill sagitta; Bar = 100 μ

Fig. 3. Enlargement of Fig. 2; Bar = 10 μ

Fig. 4. Medial surface of Pacific blue marlin sagitta; Bar = 1 mm

Fig. 5. Transverse section of fig. 4; Bar = 10 μ

Fig. 6. Transverse section of swordfish sagitta; Bar = 10 μ

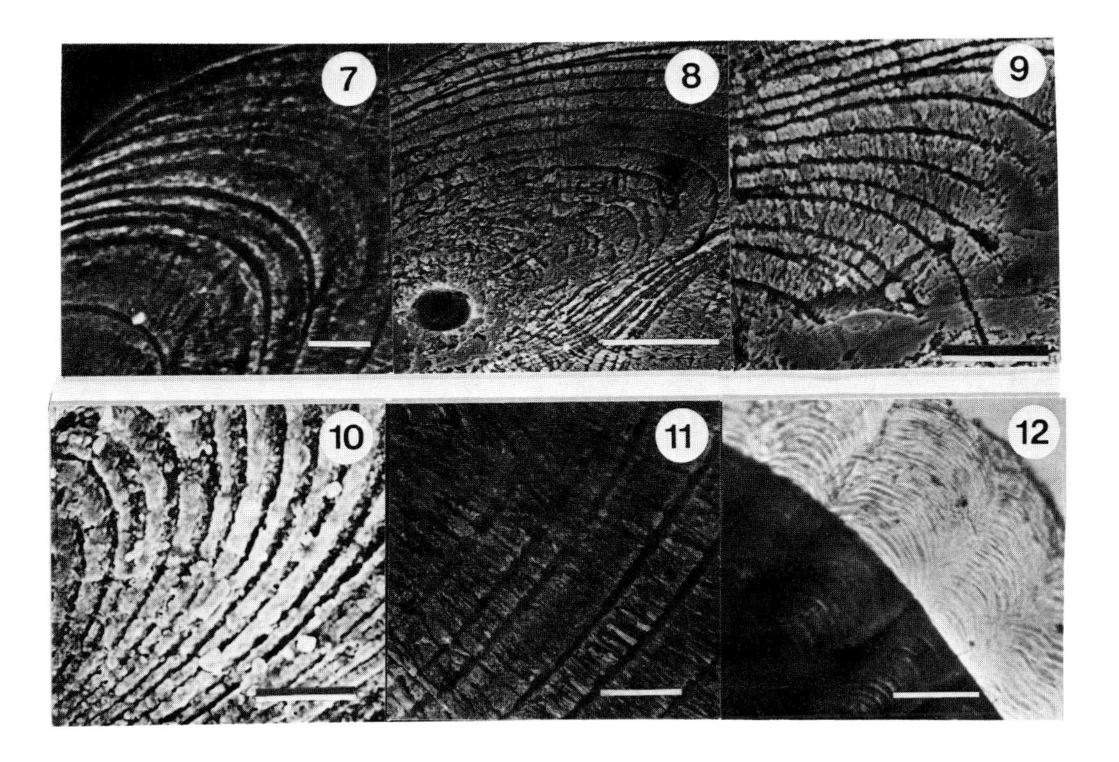

Fig. 7. Transverse section of dolphin sagitta; Bar = 10 μ

Fig. 8. Transverse section of least killifish sagitta; Bar = 10 μ

Fig. 9. Transverse section of bluefin killifish sagitta; Bar = 10 μ

Fig. 10. Transverse section of spot sagitta; Bar = 10 μ

Fig. 11. Transverse section of mummichog sagitta; Bar = 10 μ

Fig. 12. Partially decalcified sagitta of spotted sea trout (*Cynoscion nebulosus*) showing increments in matrix.

dated them as daily increments nor have we found daily increments in freshwater and marine catfish (Ictaluridae and Ariidae). The species observed in our laboratory that do show these microstructural features and daily increments far exceed the exceptions (Brothers, 1982).

CONCLUSION

Studies to date have been primarily descriptive with only a few experiments on otolith formation in teleosts (Mugiya et al. 1981; Tanaka et al. 1981; Radtke and Dean 1982). Future efforts must be directed at elucidating the mechanism of otolith formation (Dunkelburger et al. 1980) and increment formation (Mugiya et al. 1981). For example: What are the control mechanisms of calcification and organic matrix formation? Is increment formation controlled by an environmental cue and hormonal mediation with a zietgeber as suggested by the results of Tanaka et al. (1981) and Mugiya et al. (1981)? Is photoperiod the principal external cue for increment formation as indicated by Tanaka et al. (1982), Mugiya et al. (1981) and Radtke and Dean (1982)? Can temperature also act as a primary control feature (Brothers 1982)? Much remains to be done, but the elements are now exposed, the technology is available, and the questions are before us.

ACKNOWLEDGEMENT

This work was supported by the South Carolina Sea Grant Consortium (NOAA) and is contribution number 452 of the Belle W. Baruch Institute for Marine Biology and Coastal Research, University of South Carolina, Columbia, South Carolina, U.S.A. We would like to thank Dr. N. Watabe and D. Dunkelberger of the Electron Microscopy Laboratory for their discussions and criticisms.

REFERENCES

Bagenal, T.B. 1974. The ageing of fish. Unwin Brothers Limited. Surrey, England. pp. 234.

Barkman, R.C. 1978. The use of otolith growth rings to age young Atlantic silverside, _Menidia menidia_. Trans. Am. Fish. Soc. 107(6): 790-792.

Brothers, E.B., C.P. Mathews, and R. Lasker. 1976. Daily growth increments in otoliths from larval and adult fishes. Fish. Bull. 74: 1-8.

Brothers, E.B. 1982. Aging reef fishes. _In_: The Biological Bases for Reef Fishery Management. NOAA Tech. Memorandum. NMFS-SEFC-80. NMFS Beaufort, N.C., U.S.A. Ed. G.R. Huntsman, W.R. Nicholson and W.W. Fox, Jr. pp. 3-23.

Carlstrom, D. 1963. A crystallographic study of vertebrate otoliths. Biol. Bull. 124: 441-463.

Dunkelberger, D.G., J.M. Dean, and N. Watabe. 1980. The ultrastructure of the otolithic membrane and otolith in the juvenile mummichog, _Fundulus heteroclitus_. J. Morph. 163: 367-377.

Hederström, H. 1759. Rön om Fiskars Alder. Handl. Kungl. Vetenskapsakademin (Stockholm), Vol. XX: 222-229. In: Inst. Freshwater Research. Drottinghölm, 1959, 11-14.

Haake, P.W., C.A. Wilson, and J.M. Dean. 1982. A technique for the examination of otoliths by SEM with application to larval fishes. In: Proc. of the 5th Ann. Larval Fish Conf. Ed. C.F. Bryan, J.V. Conner, and F.M. Truesdale. Louisiana State University Press, Baton Rouge, Louisiana. pp. 12-15.

Mugiya, Y., N. Watabe, J. Yamada, J.M. Dean, D.G. Dunkelberger, and M. Shimizu. 1981. Diurnal rhythm in otolith formation in the goldfish, Carrassius auratus. Comp. Biochem. Physiol. 68A: 659-662.

Ottaway, E.M. and K. Simkiss. 1977. "Instantaneous" growth rates of fish scales and their use in studies of fish populations. J. Zool.(Lond.). 181: 407-419.

Panella, G. 1971. Fish otoliths: daily growth layers and periodical patterns. Sci. 173: 1124-1127.

Radtke, R. and J.M. Dean. In press. Increment formation in the otoliths of embryos, larvae, and juveniles of the mummichog (Fundulus heteroclitus). Fish. Bull.

Tanaka, K., Y. Mugiya, and J. Yamada. 1981. Effects of photoperiod and feeding on daily growth patterns in otoliths of juvenile Tilapia nilotica. Fish. Bull. 79(2): 459-466.

Taubert, B.D. and D.W. Coble. 1977. Daily rings in otoliths of three species of Lepomis and Tilapia mossambica. J. Fish. Res. Bd. Can. 34: 332-340.

Townsend, D.W. and J.J. Graham. 1981. Growth and age structure of larval Atlantic herring, Clupea harengus harengus, in the Sheepscot River Estuary, Maine, as determined by daily growth increments in otoliths. Fish. Bull. 79(1): 123-130.

Watabe, N., K. Tanaka, J. Yamada, and J.M. Dean. 1982. Scanning electron microscope observations of the organic matrix in the otolith of the teleost fish Fundulus heteroclitus (Linnaeus) and Tilapia nilotica (Linneaus). J. Exp. Mar. Biol. Ecol. 58: 127-134.

Wilson, C.A. and J.M. Dean. In press. The potential use of sagittal otoliths for estimating age of Atlantic swordfish (Xiphias gladius). In: Age Determination of Oceanic Pelagic Fishes --- tunas, billfishes, sharks. NOAA-Technical Circular. Ed. E. Prince. NMFS, Miami Florida.

Part III

Biological Accumulation of Metals Other than Calcium

TRACE ELEMENTS AS PROBES OF BIOMINERALIZATION

K. Simkiss
Department of Zoology, University of Reading,
Whiteknights, Reading RG6 2AJ, England

Abstract
The presence of trace elements in molluscan shells has been the subject of extensive studies. These investigations have established that frequently there is a correlation between the environmental levels of these metals and those present in the shells of animals from these sites. Intravascular injections of radioisotopes have therefore been used to test Wilbur's proposal that there are three basic cellular systems of calcification. The results of ^{54}Mn and ^{85}Sr studies indicate that there is a physiological distinction underlying the morphological basis of this classification.

* * * * * * *

Molluscan shells contain a great variety of trace elements. Analyses for the metals Ag, Al, Cd, Co, Cu, Fe, Mg, Mn, Ni, Sr and Zn are available for a whole range of animals (Segar, Collins and Riley, 1971; Carriker, Palmer, Sick and Johnson, 1980) but the reasons for undertaking these studies are almost as varied as the experimental regimes involved. They include investigations into palaeoclimatology, geochemical cycles, diagenesis, taxonomy, evolution and various aspects of pollution monitoring and comparative physiology. The geological interests in these studies have been reviewed by Dodd (1967) and Lowenstam (1954, 1980) but the question which I would like to consider here is whether such studies can throw any light on the physiological mechanisms of biomineralization.

Despite considerable effort the mechanism(s) of molluscan shell formation remain as a matrix of tantalising schemes in which are embedded only a few hard facts. It has generally been possible to measure satisfactorily the rates of shell deposition, the types of crystal formed and to inhibit various components of the system (Wilbur, 1972; 1980). What has generally evaded investigation is an analysis of the driving forces involved, the identification of the cells responsible and an elucidation of their involvement in the nucleation and growth of the crystals. One of the basic problems is undoubtedly the fact that it is difficult to obtain a preparation that continues to calcify under the constraints required for the experimental

P. Westbroek and E. W. de Jong (eds.), Biomineralization and Biological Metal Accumulation, 363–371.

study of this process. The chemical properties of the trace elements that are found in shells provide, however, a series of natural probes for the calcification system and their study should, therefore, be capable of providing fundamental information about the processes involved.

This approach to trace element chemistry is not entirely new. The accumulation of magnesium and strontium in calcareous shells was reviewed by Dodd (1967) who elaborated on the concept of a distribution coefficient (d.c.) defined as:

$$\text{d.c.} = \frac{\left({}^{M}/_{Ca} \right) \text{skeleton}}{\left({}^{M}/_{Ca} \right) \text{water}}$$

where M/Ca is the molar ratio of the metal to calcium in the skeleton of the animal and the water in which it lives. Using this approach Odum (1957) argued that the Sr/Ca ratios of shells taken from waters containing different levels of these metals showed that the calcification site discriminated against Sr. He also made the interesting comment that because of this discrimination Sr would accumulate in the surrounding tissues. If it was not cleared from these sites it would raise the Sr/Ca ratio of the fluids until the cellular discrimination broke down. The strontium level of the newly deposited shell would then increase. Results in keeping with this concept were obtained by Likins, Berry and Posner (1963) who kept the freshwater gastropod *Australorbis glabratus* in various concentrations of the isotope ^{85}Sr. Uptake into the shells of living snails was 50 to 100 times greater than into isolated shells. At equimolar concentrations the Ca:Sr discrimination was approximately 3:1 but the ability to discriminate against Sr decreased as its concentration in the water increased.

The partitioning coefficient (k_{Sr}^{arag}) between a solution and the crystal aragonite is a function of the solubilities of $SrCO_3$ and $CaCO_3$, the activities of Sr and Ca in solution and the activity coefficients of the two ions in the crystal. The value for K_{Sr}^{arag} is 1.12 and although this occurs in many natural aragonites molluscan shells usually have a much lower value. In an experimental study on the freshwater gastropod *Limnaea stagnalis* Buchardt and Fritz (1978) compiled data over a wide range of Sr/Ca ratios of water levels and found k_{Sr}^{arag} values of shell aragonite to be 0.23 (fig. 1) from which they concluded that there must be a biochemical discrimination against strontium during the formation of the molluscan shell.

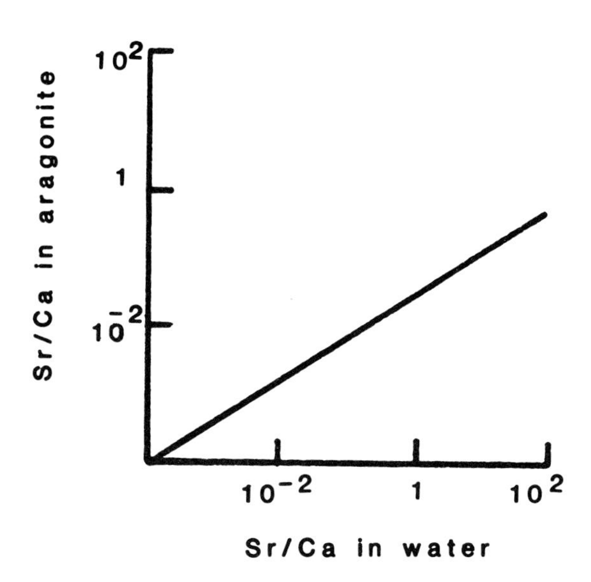

Figure 1. Sr/Ca ratios of aragonite shells in relation to the Sr/Ca ratio of seawater (data from Buckardt and Fritz, 1978)

Somewhat similar results were obtained in studies of Mg incorporation into the calcite layers of *Mytilus edulis* shell (Lorens and Bender, 1977). Normal sea water has a Mg/Ca ratio of about 5.2 but under these conditions the calcite which is formed by the mussel has a magnesium content of only 5% of what would be expected from the partition coefficient. If, however, the Mg/Ca ratio of the seawater was raised above 7 the magnesium exclusion mechanism broke down and there was a rapid increase in magnesium content of the shell.

It appears, therefore, that two phenomena can be identified. For many elements increasing the environmental availability of trace elements increases the content of that metal in the shell (fig. 2). This has been shown for lead (Sturesson, 1976), cadmium (Sturesson, 1978), strontium (Likins et al, 1963), magnesium (Lorens and Bender, 1977) and zinc (Romeril, 1971). If the metal concentration is raised to abnormally high levels the proportional rate of incorporation into the shell changes indicating a break down in the ion exclusion system (fig. 3). This has been shown for strontium (Dodd, 1967) and magnesium (Lorens and Bender, 1977). The problem in interpreting these observations in terms of shell formation arises from the fact that all the trace elements were manipulated at the environmental level. A large number of biological epithelia show ion selectivity and it is not clear therefore whether the ion discrimination effects that were observed were due to the mantle epithelium or to general discrimination by respiratory, cutaneous, digestive or renal tissues. In order to elucidate this one therefore needs to have investigations of the composition of the extrapallial fluid.

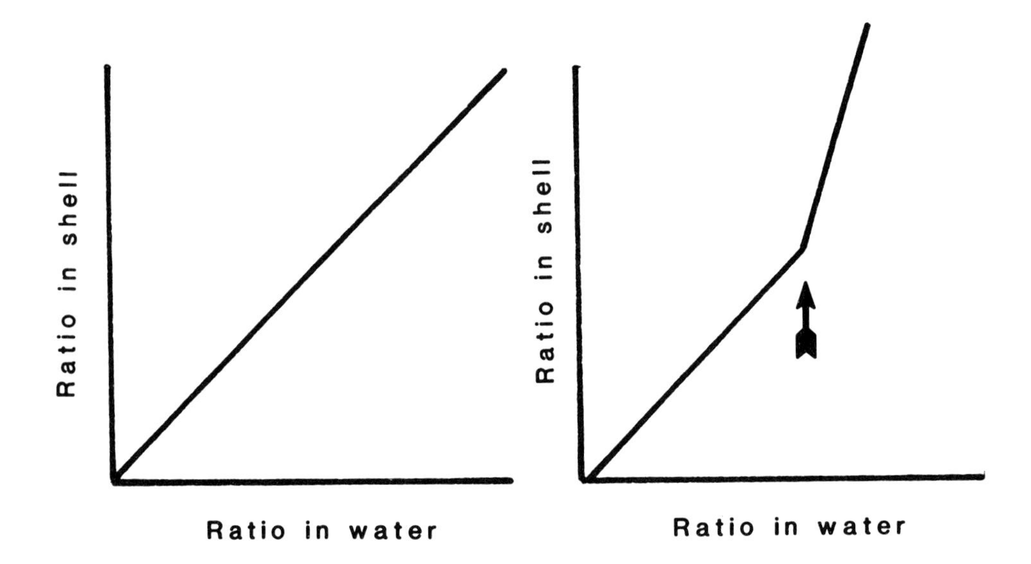

Figure 2. Effects of increasing the ratio of metal/Ca in water on the composition of the shell.

Figure 3. Effects of abnormally high concentrations of metals on the metal/Ca content of the shell suggesting the breakdown (arrow) of ion exclusion systems from sites of calcification.

Analyses of extrapallial fluid are relatively rare, partly because the volumes that can be collected are small and partly because the interpretation of the results in terms of ionic and protein bound fractions is difficult. The analyses of Crenshaw (1972) concentrated upon normal physiological ions but Wada and Fujinuki (1976) included Mg, Sr, Cu, Zn, Fe and Mn. All of these ions occurred at much higher concentrations in the extrapallial fluid of marine (*Pinctada fucata, Pinna attenuata, Crassostrea gigas* and *Chlamys nobilis*) and freshwater molluscs (*Hypiopsis schlegeli, Cristaria plicata*) than in their surrounding water. Most of the heavy metals appeared to be bound to phosphorus and sulphate ions as judged by the effects of filtration or dialysis. The only work that appears to have studied the composition of extrapallial fluid in relation to changes in environmental water is that of Pietrzak, Bates and Scott (1976) who studied 4 species of Unionids in 4 different solutions. They found that the magnesium concentration of the extrapallial fluid varied with the water concentration ($[Mg]_{EPF} \simeq 0.25[Mg]_{water}$); that enriching the water with zinc had no effect on the composition of extrapallial fluid, and that the manganese concentration in this fluid remained constant over a wide range of manganese concentrations in the water. It has previously been argued (Simkiss, 1965) that the extrapallial fluid must have at least a partially regulated ionic composition in order to avoid the effects of various crystal modifiers and crystal poisons but it is also clear that this, and other sites of calcification, are difficult to sample directly.

The microenvironments in which biomineralization occurs are one of the keys to understanding how the process of calcification is regulated and in the past few years Karl Wilbur has proposed that there are basically 3 main types (Wilbur and Simkiss, 1979; Wilbur, 1980). The most common of these are:

1. The extracellular epithelial microenvironment, and,

2. The intracellular vesicle.

The third type, involving the extracellular deposition of mineral by single cells is relatively rare but occurs in, for example, spicule formation by the echinoderm larva.

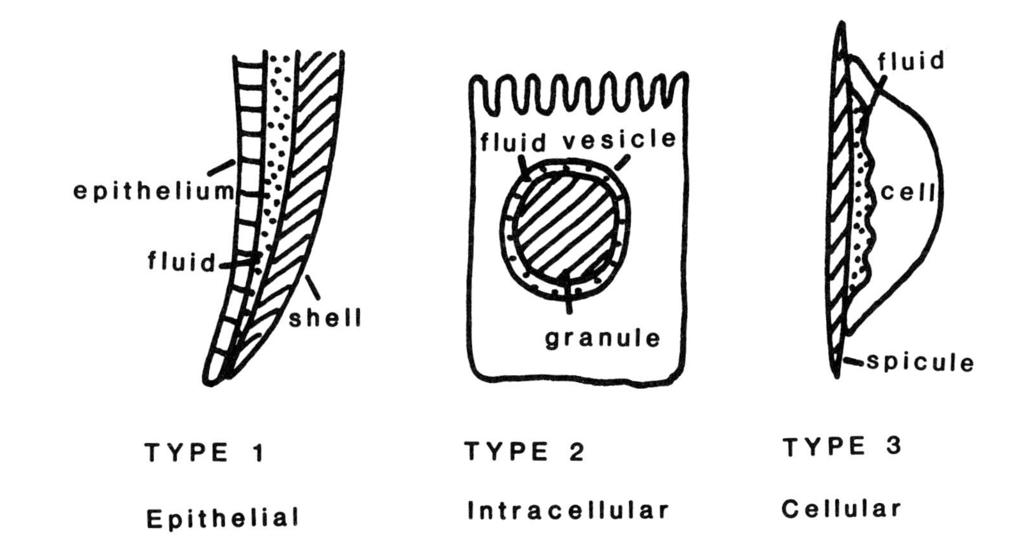

Figure 4. The 3 morphological types of calcification system proposed by Wilbur (1980)

Examples of both the common types of microenvironment involved in calcification occur in molluscs. Shell formation is generally considered to be induced by the mantle epithelium regulating the composition of the extrapallial fluid from which calcium carbonate crystallizes, i.e. it is an example of morphological type 1. Molluscs also contain a number of cell types which form intracellular deposits. These are the phosphate-depositing basophil cells of the hepatopancreas and the carbonate-depositing cells of the connective tissues (Simkiss and Mason, 1982) both of which are morphological type 2. The connective tissue cells are known, however, to be capable of forming pores in their cytoplasm so that the mineral deposits may come into contact with the body fluids after calcification has occurred (Sminia, de With, Bos, Nieuwmegen, Witter and Wondergen, 1977). This produces problems in the interpretation of labelling experiments and dissolution may occur as well as mineralization. For that reason the granules of the basophil cells were chosen for study rather than those of the connective tissue cells. In this paper, therefore, I wish to present a new technique for investigating calcification by studying the incorporation of trace elements into mineralized deposits. In order to test this approach I have applied it to the two main morphological systems of calcification described by Wilbur to see if there is an underlying physiological basis for this classification.

Many of the problems of studying biomineralized structures arise from biological variation in the rates of mineral deposition. This can be overcome by measuring relative uptake rates. There are problems in using ^{45}Ca in this way since large deposits of calcium occur naturally in these animals and it is difficult to distinguish between exchange and adsorption effects. The concentrations of calcium in the blood of snails is very variable (Burton, 1968) and for all these reasons it is difficult to work in specific activities for this ion. For these reasons I have preferred to use Mn^{2+} and Sr^{2+} as probes for the system of calcification. In order to avoid cellular sites of discrimination other than those involved in biomineralization the metals were given intravascularly. Furthermore, since the metals are not only deposited at sites of mineral formation but are also excreted it is necessary to perform acute experiments. This has the advantage that the metals are deposited on the surfaces of the minerals where crystallographic discrimination should be minimal over this time scale. It also means, however, that the quantities involved are very small and it is therefore necessary to use radioisotopes to measure them accurately.

The experimental technique which has been developed on the snail *Helix aspersa* (Simkiss, 1981) involves injecting 200μl of saline containing 10mmol.dm^{-3} of $MnCl_2$ and $SrCl_2$ spiked with $^{54}Mn^{2+}$ and $^{85}Sr^{2+}$ into 7→8g animals. Injections are made through a cannulated optic tentacle which gives direct access to the haemocoel. Animals are killed after 1h or 6h and samples of blood, tissues and shell are dried, weighed and counted in a dual channel γ counter (LKB model 1280) with automatic correction for "spill-over" between channels. It is possible to separate the mineral deposits from the cells which form them and also to incubate spiked blood with shell _in vitro_. By this means it is possible to identify cellular discrimination from mineral incorporation and also to compare _in vivo_ and _in vitro_ adsorption of the metals. Some preliminary results on the application of this technique to shell and basophil granule formation are shown in table 1. Metal ions that are injected into the blood are, of course, partly bound to blood proteins. The extent of this binding can be estimated by determining the ultrafilterable fraction for each of the metals being considered and the relevant values are shown in table 2.

Table 1. Sr/Mn molar ratios of various organs and shell of *H. aspersa* 1h and 6h after an injection of equimolar ions. Values are means ± S.D.

Tissue	Sr/Mn molar ratios	
	1h	6h
Blood	0.84 ± 0.37	0.81 ± 0.42
Kidney	0.87 ± 0.25	0.39 ± 0.09
Digestive gland	0.62 ± 0.31	1.05 ± 0.11
Mantle	0.94 ± 0.29	0.76 ± 0.11
Shell in vivo	4.11 ± 1.70	5.66 ± 2.30
Shell in vitro	-	5.08 ± 1.12

Table 2. Ultrafilterable and non ultrafilterable fractions of blood of *H. aspersa* after the addition of various ions. (% of total metal) (after Howard and Simkiss, 1981)

Metal ion	Ultrafilterable	Non ultrafilterable
Ca^{2+}	42.6 ± 1.9	57.4 ± 1.9
Sr^{2+}	44.3 ± 2.4	55.7 ± 2.4
Mn^{2+}	32.3 ± 2.6	67.7 ± 2.6

It will be apparent from these results that the ionic levels of Sr^{2+} and Mn^{2+} induced in the blood of the snail are roughly similar but that there is considerable partitioning of these ions between the extracellular and the intracellular sites of mineralization. The epithelial system of shell formation shows a preferential deposition of Sr^{2+} compared to Mn^{2+} whereas the opposite situation tends to occur in the intracellular deposits (Simkiss, 1981). In the case of the shell the ratio of Sr:Mn that is deposited in vivo is roughly 5:1 and a similar ratio is found for shell labelled in vitro from snail blood. The mantle epithelium does not show any similar trend. This raises the possibility that the deposition of Sr and Mn onto the shell occurs directly from the blood without any cellular discrimination. This implies an intercellular rather than a transcellular route of these ions and suggests perhaps that the mantle is only involved in removing inhibitor ions rather than in driving the mineralization process by the secretion of cations. Calcification may, therefore, be driven by anion secretion. An entirely different set of results occurs for the intracellular granules of the digestive gland which contain relatively little Sr in relation to Mn. It appears from these results therefore that the morphological classification of calcification systems proposed by Wilbur (1980) embodies in this particular case a comparable physiological distinction and that the further study of trace elements in mineralized deposits may provide some important approaches for the understanding of biomineralization.

References

Buchardt, B., and Fritz, P. 1978. Strontium uptake in shell aragonite from the freshwater gastropod *Limnaea stagnalis*. Science, 199, pp.291-292.

Burton, R.F. 1968. Ionic balance in the blood of Pulmonata. Comp. Biochem. Physiol. 25, pp.509-516.

Carriker, M.R., Palmer, R.E., Sick, L.V. and Johnson, C.C. 1980. Interaction of mineral elements in sea water and shell of oysters (*Crassostrea virginica*) cultured in controlled and natural systems. J. exp. mar. biol. ecol., 46, pp.279-296.

Crenshaw, M.A. 1972. The inorganic composition of molluscan extrapallial fluid. Biol. Bull. mar. biol. Lab. Woods Hole, 143, pp.506-512.

Dodd, J.R. 1967. Magnesium and strontium in calcareous skeletons: a review. J. Paleontol., 41, pp.1313-1329.

Howard, B., and Simkiss, K. 1981. Metal binding by *Helix aspersa* blood. Comp. Biochem. Physiol., 70A, pp.559-561.

Likins, R.C., Berry, E.G., and Posner, A.S. 1963. Comparative fixation of calcium and strontium by snail shell. Ann. N.Y. Acad. Sci., 109, pp.269-277.

Lorens, R.B., and Bender, M.L. 1977. Physiological exclusion of magnesium from *Mytilus edulis* calcite. Nature, Lond., 269, pp.793-794.

Lowenstam, H.A. 1954. Factors affecting the aragonite and calcite ratios in carbonate secreting marine organisms. J. Geol., 62, pp.284-322.

Lowenstam, H.A. 1980. Bioinorganic constituents of hard parts, In: Hare, P.E., Herring, T.C., and King, K. (ed). Biogeochemistry of amino acids. J. Wiley, pp.3-16.

Mason, A.Z., and Simkiss, K. 1980. Sites of mineral deposition in metal accumulating cells. Experimental Cell Res. (in press).

Odum, H.T. 1957. Biogeochemical deposition of strontium. Pub. Inst. mar. Sci. Univ. Texas, 4, pp.38-114.

Pietrzak, J.E., Bates, J.M., and Scott, R.M. 1976. Constituents of Unionid extrapallial fluid II pH and metal ion composition. Hydrobiologia, 50, pp.89-93.

Romeril, M.G. 1971. The uptake and distribution of ^{65}Zn in oysters. Mar. Biol., 9, pp.347-354.

Segar, D.A., Collins, J.D., and Riley, J.P. 1971. The distribution of the major and minor elements in marine animals. II Molluscs. J. mar. biol. Ass. U.K., 51, pp. 131-136.

Simkiss, K. 1965. The organic matrix of the oyster shell. Comp. Biochem. Physiol., 16, pp.427-435.

Simkiss, K. 1981. Cellular discrimination processes in metal-accumulating cells. J. exp. Biol., 94, pp.317-327.

Simkiss, K. and Mason, A.Z. 1982. Metal ions: metabolic and toxic effects, In: Saleuddin, A.S.M., and Wilbur, K.M. (ed), Biology of Mollusca, vol. 6. Academic Press (in press).

Sminia, T., de With, N.D., Bos, J.L., Nieuwmegen, M.E., Witter, M.P., and Wondergern, J. 1977. Structure and function of the calcium cells of the freshwater pulmonate snail *Lymnaea stagnalis*. Neth. J. Zool., 27, pp.195-208.

Sturesson, U. 1976. Lead enrichment in shells of *Mytilus edulis*. Ambio, 5, pp.253-256.

Sturesson, U. 1978. Cadmium enrichment in shells of *Mytilus edulis*. Ambio, 7, pp.122-125.

Wada, K., and Fujinuki, T. 1976. Biomineralization in bivalve molluscs with emphasis on the chemical composition of the extrapallial fluid. In: Watabe, N., and Wilbur, K.M. (ed). The mechanisms of mineralization in the invertebrates and plants. Univ. South Carolina Press, Columbia, pp. 175-190.

Wilbur, K.M. 1972. Shell formation in molluscs. In: Florkin, M., and Scheer, B.T. (ed). Chemical Zoology VII Molluscs. Acad. Press, pp.103-145.

Wilbur, K.M. 1980. Cells, crystals and skeletons. In: Omari, M., and Watable, N. (ed). The mechanisms of biomineralization in animals and plants. Tokai Univ. Press, Tokyo, pp.3-11.

Wilbur, K.M., and Simkiss, K. 1979. Carbonate turnover and deposition by metazoa. In: Trudinger, P.A., and Swaine, D.J. (ed). Biogeochemical cycling of mineral-forming elements. Elsevier Scientific Publ. Amsterdam, pp.69-106.

STRUCTURE OF GRANULES IN *HELIX ASPERSA* BY EXAFS AND OTHER PHYSICAL TECHNIQUES

M. Taylor, †G.N. Greaves, and K. Simkiss
Department of Zoology, University of Reading,
Whiteknights, Reading RG6 2AJ, England
†SERC, Daresbury Laboratory, Daresbury,
Warrington WA4 4AD, England

Abstract

The hepatopancreas of the common garden snail, *Helix aspersa*, contains a large number of intracellular granules which in their natural state consist largely of Mg, Ca and phosphates with some water and a small quantity of organic matter. A wide range of environmental metals can be incorporated into these deposits with Mn being the most readily assimilated. The granules have been characterised using a variety of techniques. Since they are amorphous to X-ray diffraction an EXAFS study of the Ca and Mn edges has been undertaken to assess and compare the local structure of the metals in the granules.

* * * * * * *

Intracellular granules from the garden snail *Helix aspersa* have been characterised using a variety of analytical and physical techniques (Howard *et al*, 1981). The presence of inorganic granules in a variety of cells in many species of invertebrates is well known. The granules have a variable composition with Ca^{2+} and Mg^{+} as the major cations and CO_3^{2-} and PO_4^{3-} as the anions (Burton, 1972; Becker *et al*, 1974). More recently it has been found that many trace metals are incorporated into the granules (Simkiss, 1977).

The hepatopancreas of the snail contains a large number of these granules which act as sites for accumulation of a wide range of environmental metals (Coughtrey and Martin, 1977). Manganese is assimilated most readily by the snail. Under the electron microscope the granules appear as concentrically layered deposits up to 10μm in diameter that occur in membrane bound vesicles often associated with the endoplasmic reticulum or Golgi system. Analyses have shown the presence of an organic component amounting to ca 5% (w/w). The granules are, however, amorphous to X-ray diffraction so the organic component does not have an epitaxial role.

Analyses of the snail granules showed that they contained Mg, Ca

P. Westbroek and E. W. de Jong (eds.), Biomineralization and Biological Metal Accumulation, 373–377.

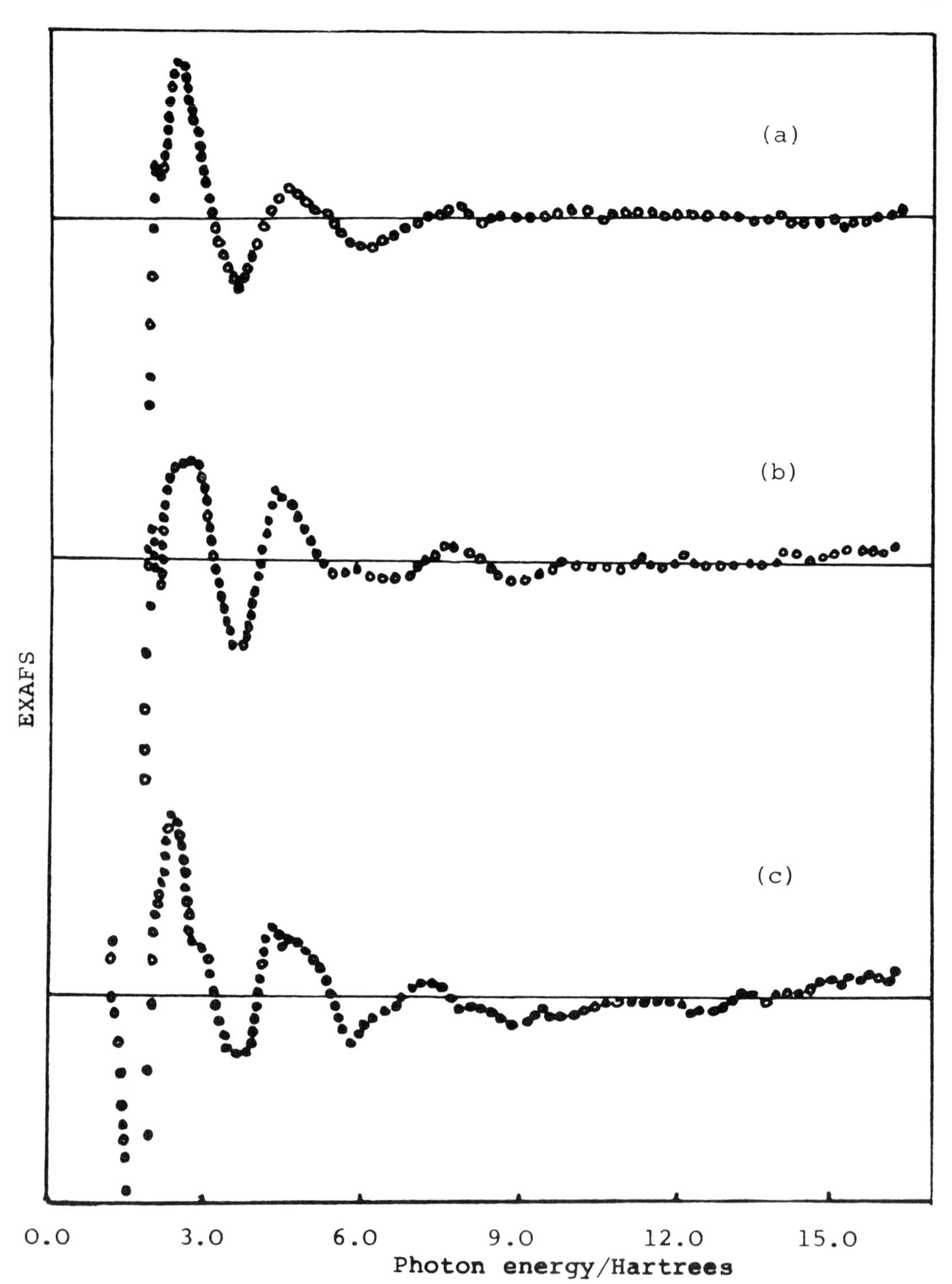

Figure 1. EXAFS spectra of (a) granules, (b) $\beta\ Ca_2P_2O_7$, (c) $CaHPO_4.2H_2O$.

Table 1. Ca-X bond lengths (Å) in the granules and some calcium phosphates

Ligand atom.X	EXAFS Best fit values: Granules 3 shells	EXAFS Best fit values: Granules 6 shells	EXAFS Best fit values: $\beta Ca_2P_2O_7$	Literature values: $\alpha Ca_2P_2O_7$	Literature values: $\beta Ca_2P_2O_7$	Literature values: $Ca_2P_2O_7.2H_2O$	Brushite $CaHPo_4.2H_2O$
O	2.46 (5)	2.43 (5)	2.43 (5)	2.39 (3.5)	2.38 ± 0.04 (5)	2.35 (3.5)	2.16 (1)
O	2.62 (2)	2.59 (2)	2.62 (3)	2.54 (2.5)	2.64 ± 0.10 (2.5)	2.47 (2)	2.46 (4)
O	2.79 (1)	2.73 (2)	2.94 (1)	2.73 (4)	3.07 ± 0.15 (1)	2.64 (1.5)	2.80 (3)
P		3.46 (6)	3.25 (6)		3.43 ± 0.16 (6)		
O		3.73 (2)	3.78 (3)		3.71 ± 0.13 (3)		3.90 (3)
Ca		4.13 (1)	3.97 (4)		3.78 ± 0.08 (4)		

α-$Ca_2P_2O_7$ - low T phase

β-$Ca_2P_2O_7$ - high T phase

(traces of K) and phosphate. The infra red spectrum showed the absence of CO_3^{2-} and the presence of a band at 745cm^{-1} which is assigned (P-O-P) as found in calcium pyrophosphate (Corbridge and Lowe, 1954; Palmer, 1961). The spectrum was unlike that of hydroxyapatite. An enzyme assay specific for pyrophosphate (Drake *et al*, 1979) confirmed that more than half the total phosphate was present as pyrophosphate. Absolute levels of pyrophosphate are difficult to determine since some pyrophosphate may be hydrolysed prior to analysis as the procedure involved dissolving the granules in perchloric acid. Thermal analyses (thermogravimetric and differential thermal analyses) confirmed the absence of carbonate.

To assess the local structure of the metals in the granules an EXAFS (extended X-ray absorption fine structure) study of the Ca and Mn edges has been undertaken. The first results of the Ca edge studies are reported.

The measurements were made at the Synchrotron Radiation Source at the SERC Laboratories, Daresbury. The spectra of the calcium K-edge were recorded in transmission mode from finely powdered samples. Spectra of brushite ($CaHPO_4.2H_2O$) and the β- form of calcium pyrophosphate were recorded in addition to those of the granules. The reduced spectra i.e. with background subtracted are shown in Figure 1. A hydroxyapatite spectrum was also run but comparison with the granule spectrum confirmed our earlier findings that the granules did not have an apatite structure so it was not analysed further.

The spectra were curve fitted using X-ray crystallographic data for brushite and calcium pyrophosphate. A feature of the phosphate structures is the range of calcium-oxygen bond lengths and coordination numbers found (Calvo, 1968; Webb, 1966; Mandel, 1975; MacLennan and Beevers, 1956).

Bond distances and coordination numbers for the granules and β-calcium pyrophosphate from the EXAFS analysis are compared with the literature data from X-ray crystal determinations for α- and β- calcium pyrophosphates, the dihydrate and brushite. Although the oxygen shells made the major contribution to the EXAFS it was found that good fits were only obtained when phosphorous and calcium shells were added. The EXAFS experiments show that calcium is in a unique environment in the granules which most resembles that of pyrophosphate.

References

Becker, G.L., Chen, C.H., Greenawalt, J.W., and Lehninger, A.L. 1974. J. Cell Biol., 61, pp.316-326.
Burton, R.F. 1972. Comp. Biochem. Physiol., 43A, pp.655-663.
Calvo, C. 1968. Inorg. Chem., 7, pp.1345-1351.
Corbridge, D.E.C. and Lower, E.J. 1954. J. Chem. Soc., pp.493-502.
Coughtrey, P.J. and Martin, M.W. 1977. Oecolgia, 27, pp.65-74.
Drake, H.L., Gross, N.H., and Wood, H.G. 1979. Anal. Biochem. 94,

pp.117-120.
Howard, B., Mitchell, P.C.H., Ritchie, A., Simkiss, K., and Taylor, M. 1981. Biochem. J., 194, pp.507-511.
MacLennan, G., and Beevers, C.A. 1956. Acta Cryst., 9, pp.187-190.
Mandel, N.S. 1975. Acta Cryst., 31B, pp.1730-1734.
Simkiss, K. 1977. Calcif. Tissue Res., 24, pp.199-200.
Webb, N.C. 1966. Acta Cryst., 21, pp.942-948.

APPLICATIONS OF MICROINCINERATION IN LOCALISING BIOMINERALIZED INORGANIC DEPOSITS IN SECTIONED TISSUES

A. Z. Mason
Department of Zoology, University of Reading,
Whiteknights, Reading RG6 2AJ, Berkshire,
England

Abstract
The applications of four different modifications of the basic technique of microincineration to localise and analyse biomineralized deposits in sections from dissected tissues are described. In combination, these techniques offer a range of procedures which can show either high structural and analytical fidelity or permit rapid and routine evaluation of large pieces of tissue. Under experimental situations, the nature of the analytical problem will obviously dictate which of the techniques is chosen. The major advantages of viewing microincinerated sections of aqueous fixed and cryofixed materials by transmission and scanning electron microscopy are compared, in outline, to viewing conventionally prepared sections.

* * * * * * * *

One of the major applications of microincineration lies in the general field of 'biomineralization' and it is somewhat surprising therefore that it has been generally neglected by workers involved in monitoring the accumulation of metals and mineralized products in biological tissues.

It has been known since the 1800's that heat can be used to remove the organic component from biological material to leave a recognisable non-combustable inorganic residue (Raspail, 1833). However, it was not until recently that this technique was applied to electron microscopy (Fernández-Morán, 1952). It was further developed in the sixties, particularly by Thomas (1964, 1969, 1974). However, the full potential of the technique was not realised until the advent of commercially available apparatus which could ash material by using an active oxygen plasma generated by radio-frequency at low temperature. In replacing high temperature ashing in muffle furnaces, these units avoided charring, melting, coalescence, fusion effects, convection disturbance and reduced the extent of volatilization of many elements.

In the past decade, as a result of the development of X-ray microanalysis, microincineration has been used increasingly to assess the physical and chemical constitution of mineralized components in soft

P. Westbroek and E. W. de Jong (eds.), Biomineralization and Biological Metal Accumulation, 379–387.

biological tissues. The ashing procedure produces residues rich in Na_2O, CaO, MgO, Fe_2O_3, SiO_2, $Ca_3(PO_4)_2$, $Mg_3(PO_4)_2$ and other oxides and phosphates. Attenuation of the organic component during the incineration process improves the sensitivity of detection by X-ray microanalysis of certain elements by a factor of 2-4 (Barnard and Thomas, 1978). This results from both an increase in the relative concentration of these elements and a reduction in the level of the non-specific background radiation in the spectra which is normally generated by the organic bulk of the specimen. Thus, the ashing procedure increases the peak to background ratio of the elements within the spectrum facilitating the detection of small peaks.

Most of the earlier microanalytical work on incinerated sections has used the transmission electron microscope (TEM). More recently, however, sections have been investigated in the scanning electron microscope (SEM). In the following article, the relative merits of four different applications of low temperature microincineration in TEM and SEM are described for localising and analysing biomineralized deposits in sections of tissue from gastropod molluscs.

APPLICATIONS IN TEM

The technique and applications of incineration for the X-ray microanalysis of thin resin embedded sections in the TEM are described in reviews by Thomas (1969; 1974), Hohman and Schraer (1972) and Hohman (1974). Applied to thin sections, microincineration produces ash patterns (spodographs) which reflect the organisation of the tissues and show considerable ultrastructural detail. One of the critical factors affecting spatial resolution of fine structure within the ash pattern is the distance between the bulk of the specimen and the supporting film. Thus, lateral migration of the ash increases with an increase in the vertical 'collapse distance'. For microanalytical studies it is necessary therefore to choose an optimum section thickness which not only displays the ultrastructural organisation of the section but also produces sufficient quantities of ash for effective analysis. Sections 120-150nm in thickness appear to achieve this compromise. Fine structure can be resolved in the ash and lateral migration of the residues is minimal.

Many of the preparative steps leading up to viewing and analysing the ashed sections in the TEM are identical to those of conventional techniques. There are, however, certain details which are quite different. One of the primary considerations is to select a grid and support film that will remain unoxidised by heat or active oxygen. Aluminium grids and silicon monoxide films conform to all these requirements and, in addition, they do not contain contaminating elements that are detectable by X-ray microanalysis. However, one of the undesirable effects of silicon monoxide is that during incineration it is converted to the dioxide. This oxidative process induces pronounced sagging of the films between the grid bars. As a result, the changes in the take-off angle of the X-rays restricts the quantitative evaluation of elements in the ash.

Conventionally fixed materials

There are a number of advantages associated with ashing conventionally fixed ultrathin sections. Apart from increased analytical resolution, one of the principal advantages resulting from the removal of the organic matrix is that the remaining inorganic residues show high contrast. Consequently, in introducing optical contrast to resin embedded sections, ashing provides an alternative to the use of heavy metal stains. Clearly, the removal of peaks originating from extrinsic elements such as metal stains introduced during preparation is advantageous when undertaking microanalytical studies. In this respect, it is also valuable to note that under certain conditions, microincineration can reduce the levels of osmium from the fixative and chlorine from the embedding resin below the limits of detection by X-ray microanalysis. In general, therefore, spodographs produced by this procedure show high inherent contrast (figure 1) and produce spectra showing little background radiation (figure 2).

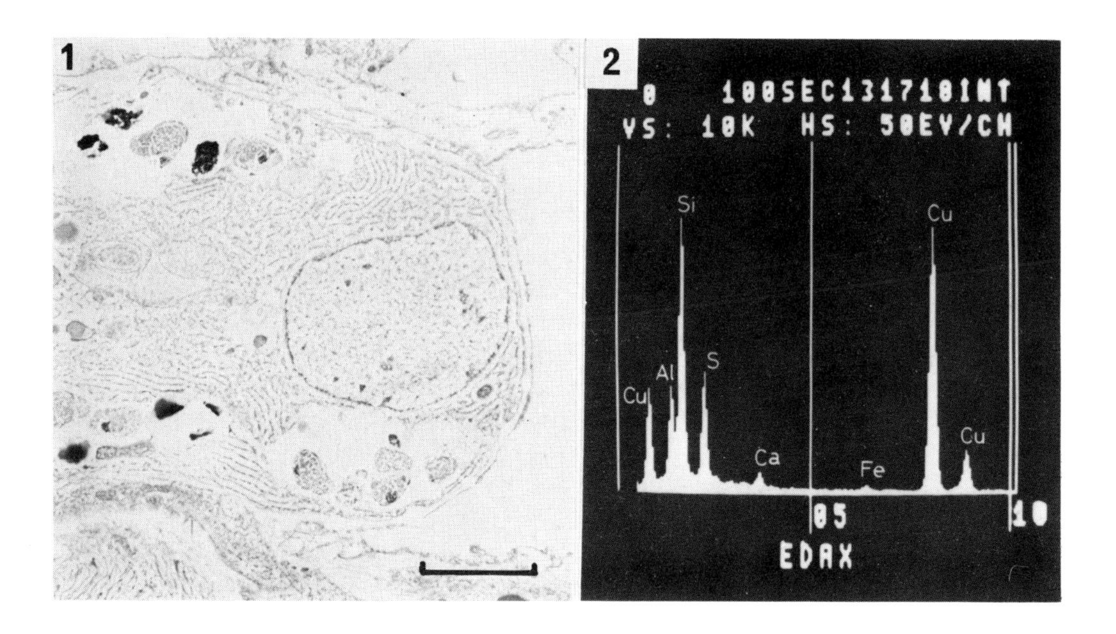

Figure 1. Spodograph of an ultrathin resin embedded section of a pore cell in the connective tissue of *Littorina littorea*. Residues originating from the nucleus, cytoplasmic granules and rough endoplasmic reticulum can be identified. Scale bar represents 2µm.

Figure 2. Analyses of the granules give spectra showing low background radiation and large peaks for copper and sulphur in association with smaller peaks for iron and calcium. Peaks for aluminium and silicon arise from the support film and grid respectively.

Cryofixed materials

Many of the major advantages of combining microincineration with ultracryomicrotomy are similar to those outlined above for conventionally prepared materials. Primarily, ashing produces recognisable ultrastructural detail into thin cryosections which normally contain poor inherent contrast. The value of incineration is evident in figures 3 and 4 where many of the cytological features within the central area of the micrograph are only apparent after the specimen is ashed. Moreover, an improvement in the peak to background ratio as a result of incineration results in the identification of a number of minor constituents in the structures such as Mn, Fe, Co, Ni, Cu and Zn (figure 5). These elements cannot be identified in the unashed section.

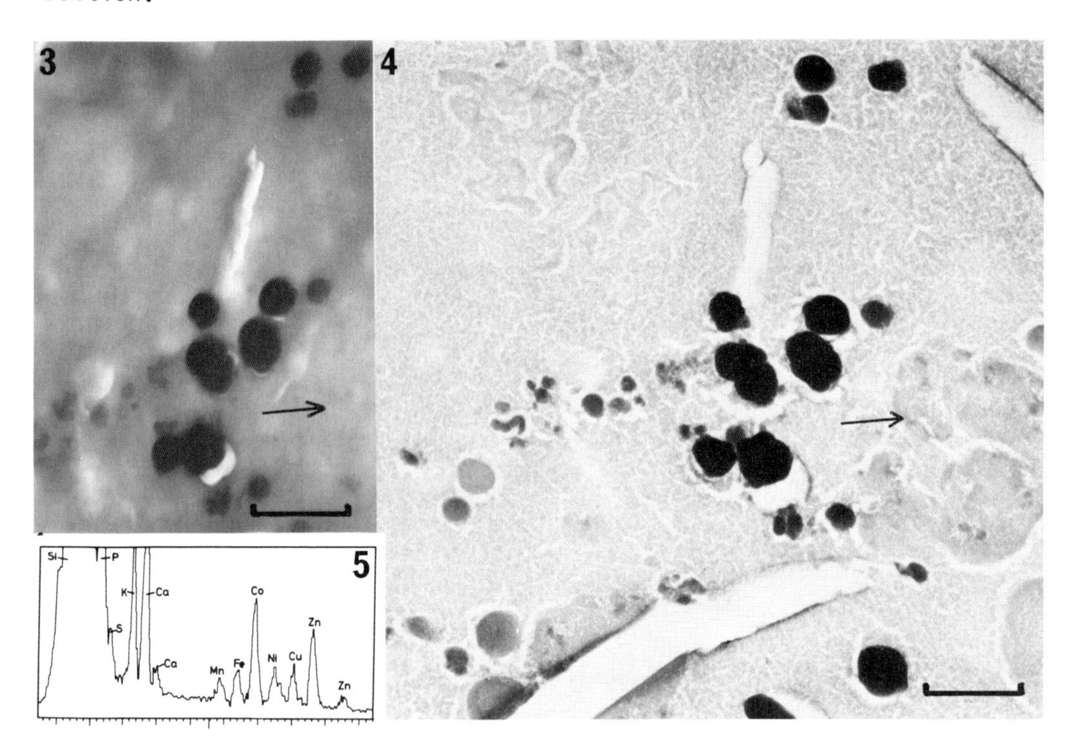

Figures 3, 4 and 5. TEM micrograph of an ultrathin cryosection of the digestive gland of *Littorina littorea* before and after incineration in an oxygen plasma (figures 3 and 4 respectively). The ashed section shows additional contrast. Analyses on cytological components in the ashed cells (arrow) show the presence of Mn, Fe, Co, Ni, Cu and Zn (figure 5). These elements cannot be identified in the unashed section. Scale bar represents 3μm.

APPLICATIONS IN SEM

Most of the work describing the use of microincineration for ashing thin biological sections for electron microscopy has involved the use of TEM. However, for a number of reasons, more recent work has used the SEM. Thus, although the spatial and analytical resolution of SEM studies are generally inferior to those obtained by ashing ultrathin sections for TEM studies, the versatility of the instrument with its wide range of magnifications and a depth of focus which extends to several mm's, allow rapid evaluation of the specimen. In addition, the sections and support base need not transmit electrons and there are no grid bars to mask areas of the section. Aluminium supports are ideal for analytical purposes and unlike silicon monoxide films, they are robust, flat and ensure uniform contact with the section.

Conventionally prepared materials

When 1μm resin embedded sections are etched in an oxygen plasma, differential surface volatilization produces residues which form a recognisable 3 dimensional skeleton of ash and are derived from membranes and other cytoplasmic features (Humphreys and Henk, 1979). The spatial resolution of the etched patterns permits the identification of profiles derived from plasma membranes, mitochondrial cristae and nuclei. The emergence of these structures with respect to each other and the embedding matrix, is related directly to their inherent resistance to etching.

Figures 6 a-e illustrate the gradual emergence of topographical relief within resin embedded sections of stomach tissue after an increasing length of exposure to the oxygen plasma. Unetched sections have uniform contrast and display little or no surface relief. The first structures to become prominent during incineration are usually cytoplasmic inclusions with a high mineral content which have strong primary and secondary electron emission. Thus, after only 15 second incineration, biomineralized deposits emerge in the mid-region of the stomach as white areas upon a dark, almost featureless background (figure 6b). After an additional 15 seconds, the profiles of numerous cytological features emerge (figure 6c) and after 1 minute exposure, the organisation of the etch patterns corresponds with the histological features in light micrographs (cf. figure 6a with figure 6d). However, the higher resolution of the SEM permits the identification and analysis of intracellular components that cannot be resolved in the light microscope. After two minutes exposure to the oxygen plasma, the sections show additional contrast but this is offset by a marked loss of the finer details of structure within the section (figure 6e). Incineration times in excess of three minutes produce little or no change in the residues and it can be proposed that at this stage, oxidation of the sample is complete. The transformation from the etched to the ashed state is demonstrated in the lipid deposits and the mucus secretions in the stomach cells.

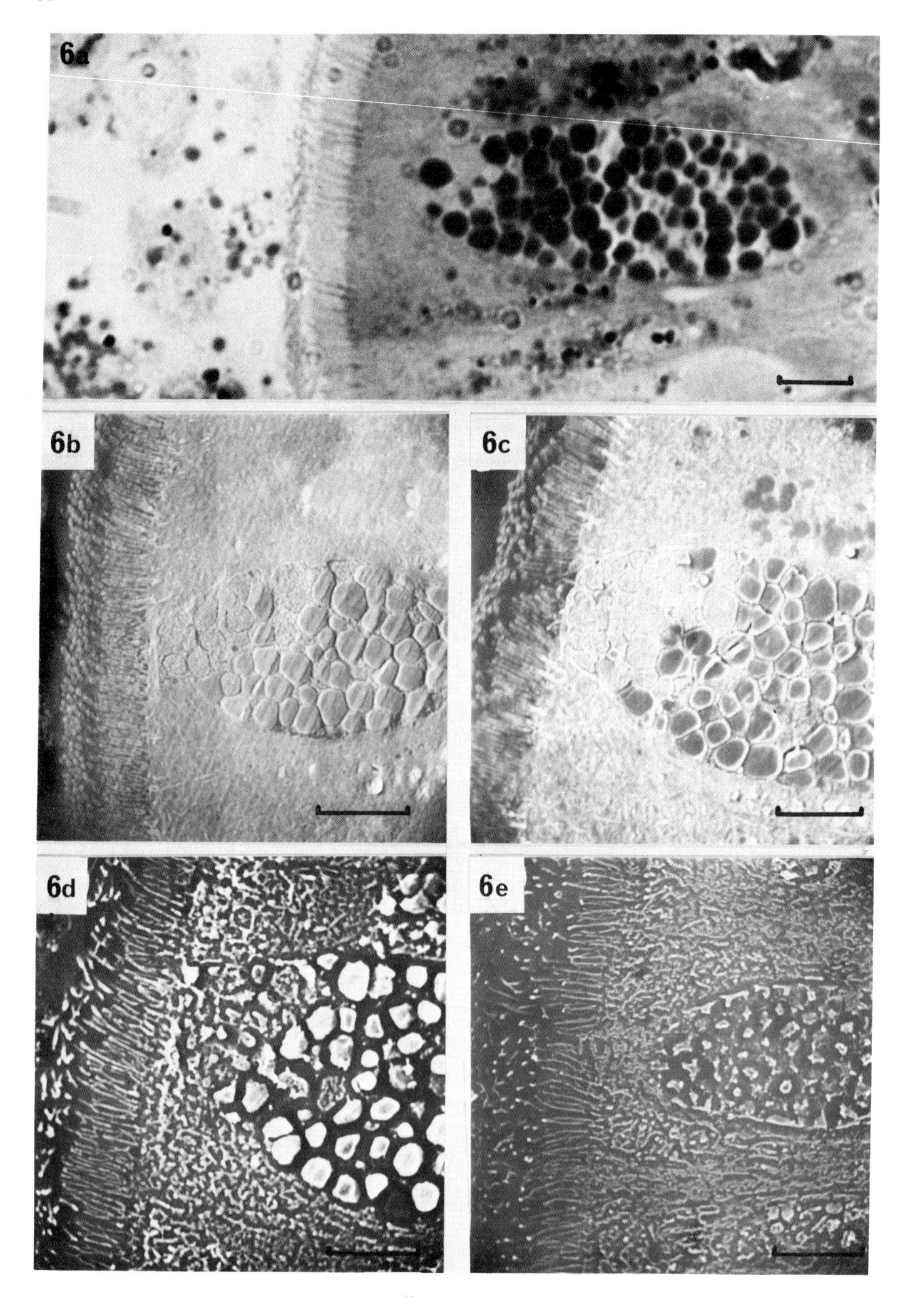
6a
6b
6c
6d
6e

Figures 6a-e. Series of 1μm thick resin embedded sections of the stomach epithelium of *Littorina littorea* at different stages of ashing. Figure 6a shows an unashed section stained with toluidine blue and viewed by bright field illumination. SEM micrographs showing the portion of the cell displayed in figure 6a are shown after etching in an oxygen plasma for 15 seconds (6b), 30 seconds (6c), 1 minute (6d) and 2 minutes (6e). Many of the cytological features recognisable in figure 6a can be localized and identified as lipid droplets, mitochondria and inorganic deposits in the etched sections.
Scale bars represent 4μm.

After 15 seconds exposure, the lipid and mucus droplets appear as flat discs (figure 6b). Additional etching causes the discs to become increasingly prominent with respect to adjacent features (figure 6c). After two minutes exposure to the oxygen plasma, the residues shrink to form a clear zone (figure 6e).

The results clearly emphasize some of the virtues of viewing microincinerated sections by SEM. Thick sections can be cut from large block faces and this allows small quantities of inorganic materials to be localised and analyses within large areas of tissue as well as within cells. Moreover, unlike analytical TEM studies, where a compromise specimen thickness is required, in SEM studies, the section thickness is immaterial since the degree of etching can be controlled to produce residues suitable for either analysis or assessment of fine structure. Consequently, it is theoretically possible to preview etched sections or block faces in the SEM for ultrastructural evaluation and then complete the oxidation process and relocate areas of interest for X-ray microanalysis. Clearly with this mode of preparation the degree of incineration is critical and attempts have been made to standardize the process by utilizing certain materials for internal standards such as polystyrene latex spheres of a given size (Thomas *et al*, 1976). However, direct estimates of this kind only provide empirical information on the combined effects of a number of parameters which all have a direct effect on the process; exposure time, wattage, levels of O_2, the composition and chemical properties of the specimen and its location in the specimen chamber. Many of these factors appear to be subject to considerable variation within a given study, so more precise methods of evaluation are obviously necessary to analyse and resolve the effects each parameter contributes to the etching procedure.

Cryofixed materials

In the past, thin cryosections have provided the only means for the assessment of diffusible materials in biological sections. Their application in this context has, however, been limited since one of the major problems associated with thin cryosections is that they have to be prepared by a relatively complex procedure which requires experience in the use of specialized equipment. Recently, however, microincineration has been applied to thick freeze-dried cryosections cut on steel knives in a conventional cryostat to reveal surface relief and enable the identification of certain cytological features and the analysis of the residual inorganic components by SEM (Mason and Nott, 1980). Naturally, the fidelity of the residues to the original ultrastructure is generally poorer than those obtained by viewing ashed ultrathin cryosections by TEM but it does avoid many of the technical preparative problems associated with the latter mode of viewing. Consequently, despite some limitations in analytical and spacial resolution, the technique offers a rapid, routine method for localising and analysing biomineralized deposits in materials untreated by aqueous solutions of chemicals. These merits are evident in figures 7 and 8 where relatively small intracellular mineral deposits have been localized within large areas of the section. Comparative analyses with conventionally prepared specimens show that diffusible components within the mineral deposits shown in figure 8 contain significant quantities of potassium and that this element appears to be lost during preparation with aqueous chemicals.

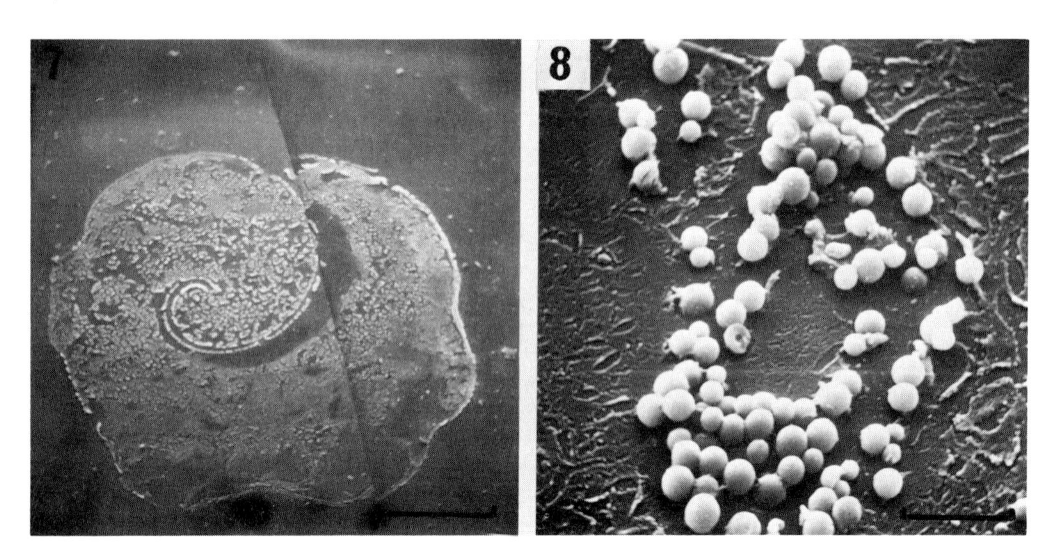

Figures 7 and 8. SEM micrographs illustrating the inorganic residues remaining in a transverse cryosection across the visceral whorls of *Helix aspersa* after incineration at low temperature in an oxygen plasma. The location of the mineral containing cells in figure 7 are clearly marked by clusters of spherical inorganic granules in figure 8.
Scale bars represent 1mm and 10μm in figures 7 and 8 respectively.

References

Barnard, T., and Thomas, R.S. 1978. X-ray microanalysis of Epon sections after oxygen plasma microincineration. J. Microscopy, 113, pp.269-277.

Fernández-Morán. 1952. The submicroscopic organization of vertebrate nerve fibres. An electron microscope study of myelinated and unmyelinated nerve fibres. Exp. Cell Res., 3, pp.282-329.
Hohman, W.R. 1974. Ultramicroincineration of thin sectioned tissue. Principles and techniques of electron microscopy - Biological applications, vol. 4. (1974). Ed. M.A. Hayat. Van Nostrand, Reinhold, Cincinnati, USA. pp.129-158.
Hohman, W., and Schraer, H. 1972. Low temperature microincineration of thin sectioned tissue. J. Cell Biol., 55, pp.328-354.
Humphreys, W.J., and Henk, W.G. 1979. Ultrastructure of cell organelles by scanning electron microscopy of thick sections surface etched by an oxygen plasma. J. Micr., 116, pp.255-264.
Mason, A.Z., and Nott, J.A. 1980. A rapid, routine technique for the X-ray microanalysis of microincinerated cryosections: An SEM study of inorganic deposits in tissues of the marine gastropod *Littorina littorea* (L). J. Histochem. Cytochem., 28, pp.1301-1311.
Raspail, V.F. 1833. Noveau System de Chimie Organique Fondé sur des Méthods Nouvelles d'Observation. Bailliere, Paris, pp.528.
Thomas, R.S. 1964. Ultrastructural localisation of mineral matter in bacterial spores by microincineration. J. Cell Biol., 23, pp.113-133.
Thomas, R.S. 1969. Microincineration techniques for electron microscopic localisation of biological minerals. In: Advances in Optical and Electron microscopy, R. Barer and V.E. Cosslett (eds). 3, Academic Press, New York. pp.99-154.
Thomas, R.S. 1974. Use of a chemically reactive gaseous plasmas in preparation of specimens for microscopy. In: Techniques and Application of Plasma Chemistry. J.R. Hollahan and A.T. Bell (eds). John Wiley and Sons Inc., New York, USA. 1974. pp.255-346.
Thomas, R.S., Millard, M.E., and Scherrer, R. 1976. Electron microscopy and X-ray photoelectron spectroscopy of oxygen plasma-etched bacterial spores and cells. 34th Annual Proc. E.M.S.A. 1976. Miami Beach, Florida. (ed. G.W. Bailey) pp. 134-135.

ASPECTS OF SILICIFICATION IN BIOLOGICAL SYSTEMS

Benjamin E. Volcani
Scripps Institution of Oceanography
University of California, San Diego,
La Jolla, California 92093, U.S.A.

ABSTRACT

A very condensed overview of the biological mineralization of silicon is presented. Siliceous structures in chrysophytes, silicoflagellates, diatoms, testaceous amebae, choanoflagellates, radiolarians, sponges, and vascular plants are described. The systems involved in the origin of the structures and possible mechanisms of silicification are discussed.

INTRODUCTION

Next to oxygen, silicon occurs in greater abundance than any other element, and in a variety of forms ranging from silicic acid ($Si(OH)_4$) to opal, quartz, and glass. It is orthagonal to carbon and phosphorus in the Periodic Table, both of which play central roles in the metabolism of terrestrial organisms, and shares some characteristics of these elements. Silicon might therefore be expected to participate actively in biological processes, but until the last two decades it was believed that silicon is biologically inactive and hence of no biological interest. It is now evident that silicon is essential for a variety of biological processes ranging from DNA synthesis (in diatoms) (1) to bone formation in mammals (2) and it is no longer possible to ask whether there is such a thing as "silicon biochemistry". (For an extensive review see (3)).

It has been known for generations that many primitive organisms (chyrysophytes, diatoms, amebae, silicoflagellates, and radiolarians) as well as multicelled organisms such as sponges and some vascular plants produce siliceous structures such as internal and external skeletons, cell walls, scales, spicules, spines, and phytoliths (in plants). Many of these structures have fascinated microscopists with their intricacy and diversity and almost all serve as the basis for the taxonomy of their groups, but it was held that the structures and their complex patterning are the result of physicochemical reactions. Within

P. Westbroek and E. W. de Jong (eds.), Biomineralization and Biological Metal Accumulation, 389–405.

the last 20 years, however, it has been conclusively demonstrated that the siliceous structures of such interest to taxonomists are in fact produced by a system of biological mineralization involving the deposition of $Si(OH)_4$.

This paper will present a glimpse of the wide variety of ways in which silicification is manifested, describe, all too briefly, the little that is known about the mechanisms involved, and offer some idea of the speculations, hypotheses, and models that have been advanced on the subject. To try to do justice to so large a subject, I have drawn most of the material for this paper from the very comprehensive review of the field, "Silicon and Siliceous Structures in Biological Systems" (4) and will use corresponding references.

SILICEOUS STRUCTURES IN LIVING ORGANISMS

In the structures described below, the silicon occurs as amorphous silica gel or opaline silica with hydration ranging from $SiO_2 \cdot 2H_2O$ to $SiO_2 \cdot nH_2O$. The diversity of structures is astonishing, and the great preponderance of patterns are species-specific.

Among the algae, two groups possess siliceous walls. The first group, the chrysophytes (5), produce a characteristic resting stage, the cyst, the walls of which are silicified though as a rule more or less unpatterned. In a few species, however, the cell is covered by a single layer of siliceous scales (Fig. 1) which have a "pectin-like" organic component that retains the shape of the scale when the silica is dissolved in HF. Another group, sometimes included with the chrysophytes, the silicoflagellates (6), differs notably from the chrysophytes in that the siliceous skeleton is internal, usually composed of a basal tubular "ring" supported by tubular struts and spines (Figs. 2, 3).

The second group of silicified algae consists of the diatoms (7,8, 9) which are characterized by a box-and-cover-like cell wall comprised of valves and one or more girdle bands; there are two major groups, ovoid (pennate, Fig. 4) and circular (centric, Figs. 5,6).

Fig. 1. A whole cell and scattered scales of the chrysophyte *Synnura petersenii*. x2835. Shadowcast, TEM. (Courtesy of C.B. McGrory and B.S.C. Leadbeater).
Figs. 2,3. Skeleta of silicoflagellates. Fig. 2. *Dictyocha*. x810. (Courtesy of F.E. Round). Fig. 3. *Rocella gemma*. x605. (Courtesy of E.C. Bovee).
Figs. 4-6. Diatoms. Fig. 4. Frustule of *Navicula monilifera* (pennate). x810. (Courtesy of R.M. Crawford). Fig. 5. *Thalassiosira eccentrica* (centric). x740. (Courtesy of A.M. Schmid). Fig. 6. View from inside of a species of *Coscinodiscus*; note chambers' septa and sieve plates. x3000. (Courtesy of R.M. Crawford). V = valve; G = girdle bands. SEM.

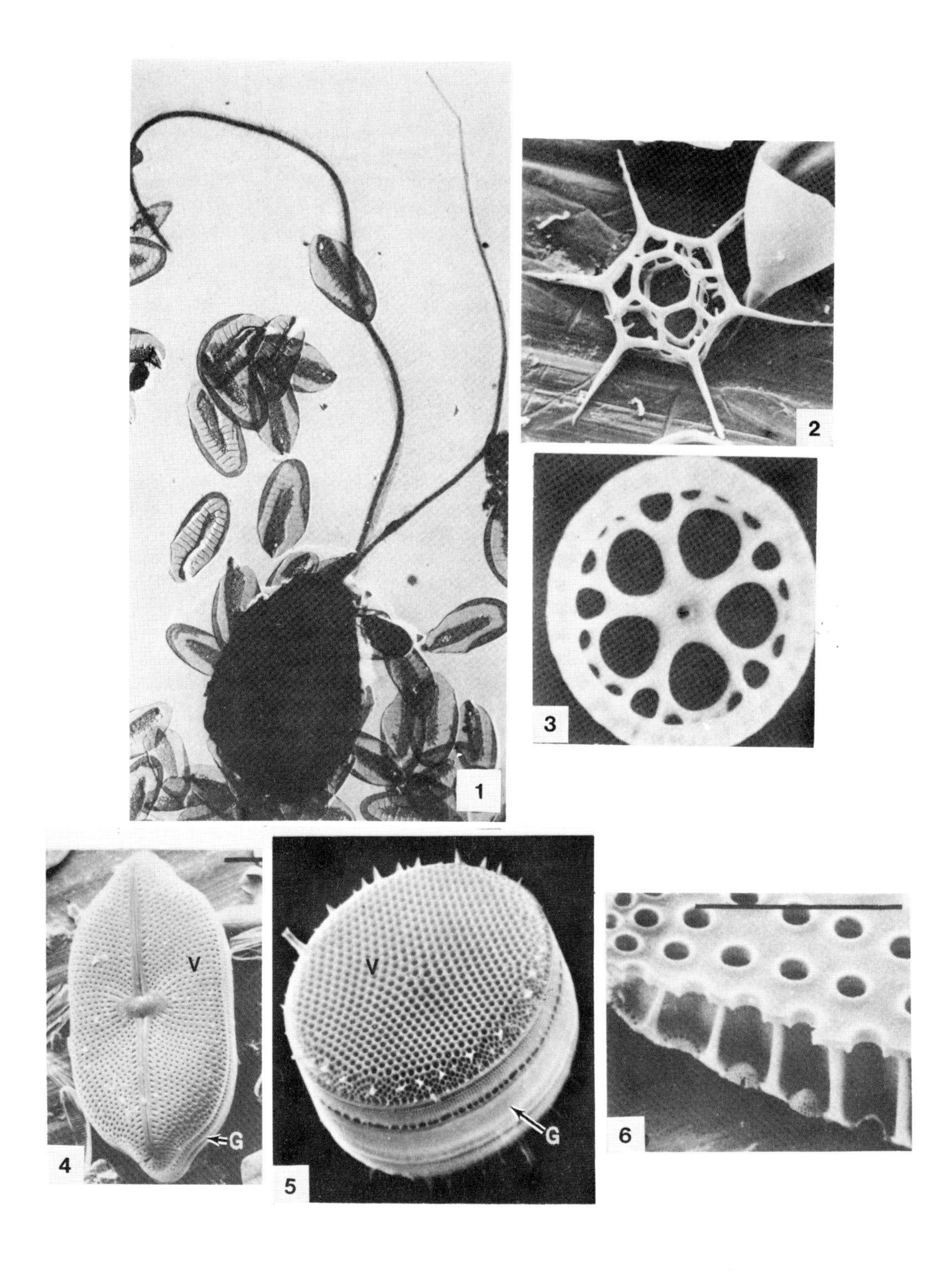
1
2
3
V
G
4
V
G
5
6

The siliceous elements are interlocked with, and enveloped by, a closely adhering organic component or "casing"; the two components are so firmly adherent that they cannot be separated except by dissolving the silica with HF vapor; the patterning of the siliceous component is retained by the organic casing (Figs. 7-9).

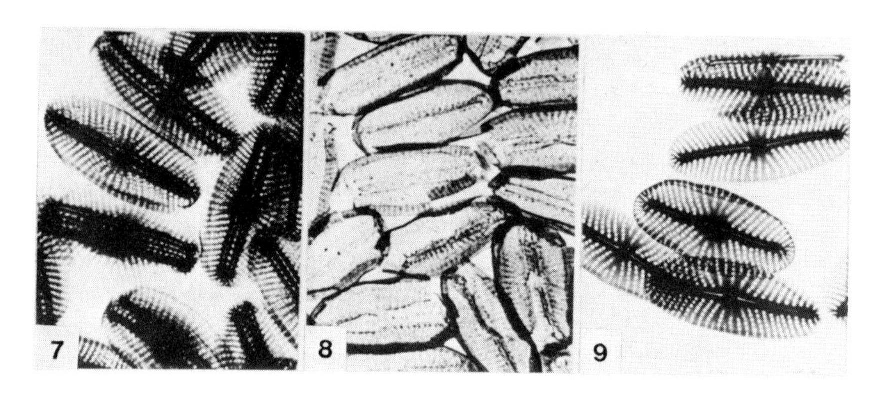

Figs. 7-9. Electron micrographs of *Navicula pelliculosa*, isolated cell walls and their components. Fig. 7. Intact walls (frustules). Fig. 8. Organic casing, obtained by treating walls with hydrofluoric acid. Fig. 9. Silica shells, obtained by treating walls with concentrated sulfuric acid. x 3370

Among the protozoans (6,10), the majority of silicon-using species "secrete" siliceous structures in the form of scales on the body and/or flagella, but some species produce spines, endoskeletons, or tests ("armored" cell walls) (Figs. 10,11). Many others do not produce their own silica but ingest silica particles, such as quartz that they "excrete" and then cement together to form an exoskeleton. For example, some testacid amebae, e.g., *Difflugia*, arrange quartz sand particles atop the test on an inner, organic layer. Spaces between the particles are then filled with "idiosomes". Other *Difflugia* species select smooth round granules of quartz sand and arrange them in rosettes according to size, and then fill in the spaces with smaller granules; the particles are cemented together with protein secretions that resist HF.

Among the choanoflagellates (11), the Acanthoecidae are distinctive in possessing a basket- or cage-like "lorica" composed of siliceous ribs or strips termed "costae" that are cemented together enclosing the protoplast (Figs. 12-14). The morphologies of the lorica and its strips are taxonomic criteria, but the arrangement of the strips, which can vary in number from 7 to more than 300, is probably the most important diagnostic feature. A thin "investment" (membrane) lies between the protoplast and the adjacent parts of the lorica and secures one to the other; observations suggest that the costal strips are held together by organic material.

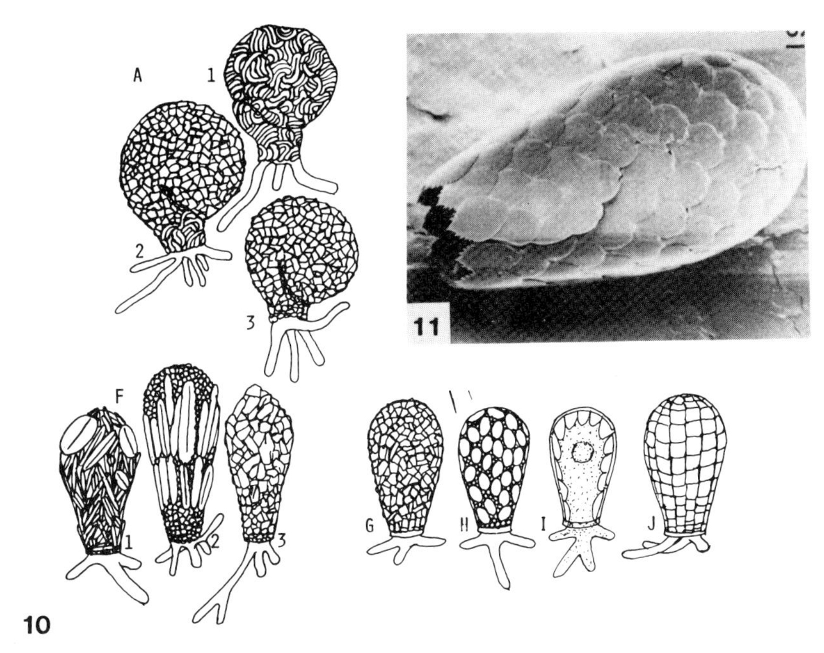

Fig. 10. Tests of some testacid amebae that incorporate siliceous material. A. *Lesquereusia* sp. A1. With intrinsically made particles. A2. With siliceous sand granules, except at the neck. A3. Entirely covered with sand granules. x110 F1. *Difflugia oblonga* with diatomaceous frustules as the attached and embedded siliceous particles. F2. *D. oblonga* with both siliceous sand granules and diatomaceous frustules; F3. with only sand granules. x162. G. *Heleopora picta*, entirely covered with sand granules. x162. H. *Nebela collaris* with small and large scales. x162. I. *Quadrulella symmetrica* with intrinsic rectangular scales. x243. (Courtesy of E.C. Bovee). Fig. 11. *Euglypha* showing theca of siliceous plates. x740. SEM. (Courtesy of F.E. Round).

Competing with the diatoms for the seemingly infinite range of their geometric forms, are the radiolarians (12,13). Their shapes and patterns are of remarkable delicacy and regularity, each species-specific and each bearing its own intricate patterning (Figs. 15-17).

The cytoplasm or "central capsule" in this organism consists of two components: an organic envelope or "capsular wall" surrounding the cytoplasm, which in turn is enveloped in a transparent gelatinous coat or casing. In one group, the *Spumellaria*, the spheroidal endoskeleton is composed of a fine siliceous network supported on slender branching radial spines; in another group, the *Nasselaria*, the skeleton is thimble-shaped; a very complex arrangement of microtubules within the central capsule provides continuity between the intro- and extracapsular cytoplasms. In both groups, the gross patterns of the skeleton and

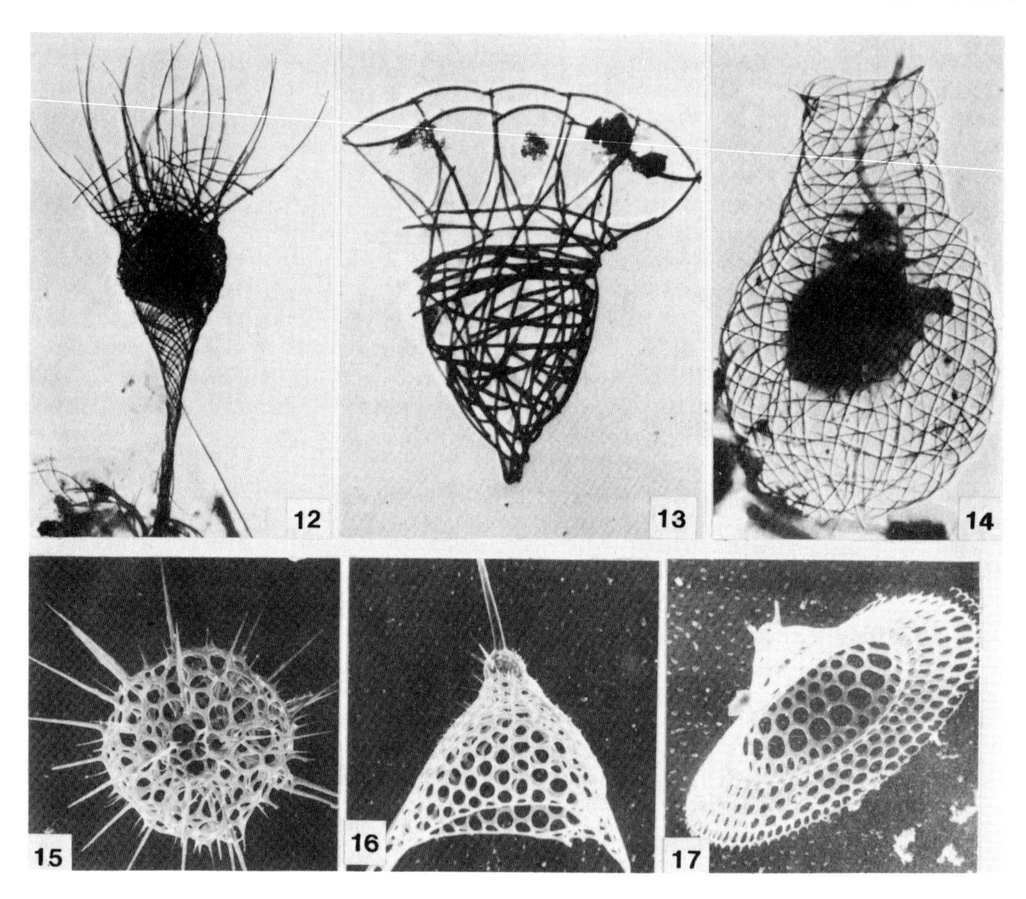

Figs. 12-14. Choanoflagellates lorica. Fig. 12. *Acanthoeca spectabilis*. Fig. 13. *Saepicula pulchra*, empty lorica. Fig. 14. *Savillea parva*. x3770. Shadowcast, TEM. (Courtesy of B.S.C. Leadbeater).
Figs. 15-17. Radiolaria. Fig. 15. Unidentified larcoidian sp. x207. Fig. 16. *Dictyophimus clevei*. x114. Fig. 17. *Eucecryphalus* sp. x101. (Courtesy of E.C. Bovee).

cytoplasm complement each other and the latter appears to provide a "ground plan" for the skeleton. All skeletal elements are enclosed by a thin cytoplasmic sheath, the "cytokalymma" (Fig. 25); recent studies suggest that the cytokalymma is synthesized prior to silica deposition and may thus serve as a template for the skeleton.

In the kingdom of multicellular organisms, the only phylum that produces biologically mineralized silica is the sponges (14-16). A majority of sponges are characterized by skeletons composed of opaline spicules or "sclera" of very diverse forms that are species-specific, although eight or even more kinds of spicules may occur within a given species (Figs. 18-22).

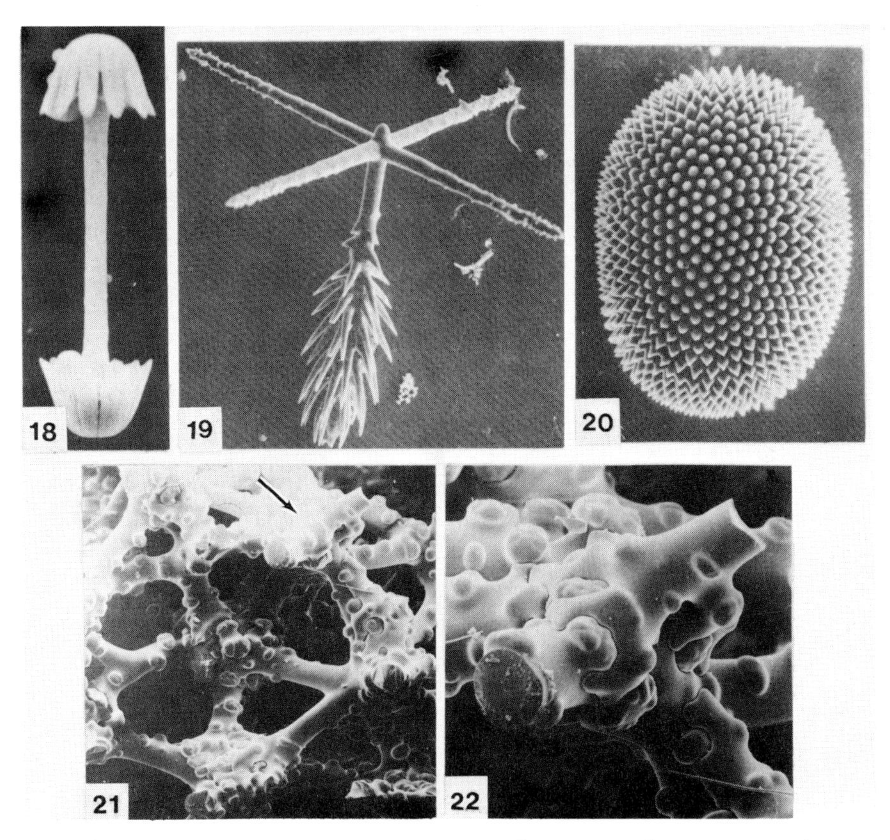

Figs. 18-20. Sponge spicules. Fig. 18. Birotulate microscleres in *Iotrochota birotulata*. x4050. Fig. 19. Dermal pentactinal pinule with heavily spined distal ray in *Sympagella nux*. x550. Fig. 20. Sterrasters, late development, in *Placospongia* sp. x810.
Fig. 21. Skeletal framework of the sponge *Racodiscula* sp. x81. Fig. 22. Enlargement of arrow-marked region in Fig. 21, showing articulations between desmas. x243. SEM. (Courtesy of W.D. Hartman).

In some species secondary deposits of silica on the basic spicule result in a rigid skeletal framework of fused or interlocking spicules, although as a rule the spicules appear to serve as a flexible support for the soft "tissues" of the organism. Spicules are of two types: the large megascleres, held together in a network by collagenous fibrils or "spongin", and the microscleres (one hundredth the size of the megascleres) that are scattered throughout the superficial layers. Their function, however, is not clear; they may provide some additional strength to the tissues, and in some species are aggregated at the surface, providing a sort of protective armor.

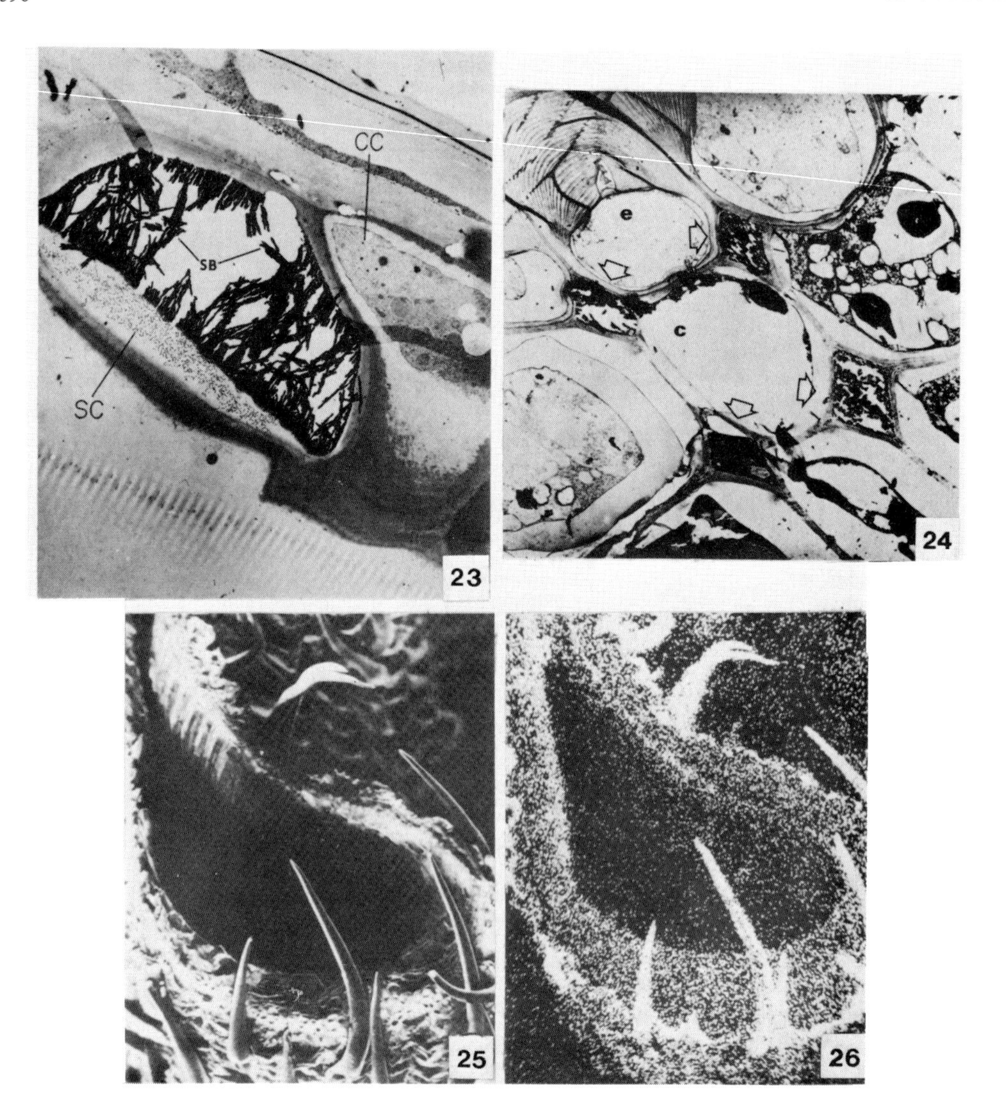

Fig. 23. A radial view of sectioned mature pairs of silica cell (SC) and cork cell (CC) of *Avena* internodal epidermal cells. The lumen of the silica cell has become nearly filled with amorphous silica bodies (SB). x6100. TEM. (Courtesy of P.B. Kaufman).

Fig. 24. Silica deposits in mature roots of the grass *Molinia caerulea* located in intercellular spaces (arrows) of the persistent cortical cells immediately external to the endodermis (e). x3372. (Courtesy of A.G. Sangster and D.W. Parry).

Figs. 25,26. SEM view and Si x-ray map of the cut surface of a mature lemma in "Reimei" rice spikelet. Fig. 26 shows significant silica deposition, most prominent in the outer side of the outer epidermis, especially in the tips of the tubercles and in the trichomes. x122. (Courtesy of P.B. Kaufman).

The vascular plants (17,18) comprise the other very large taxa in which opaline silica is an important structural component; it is found in many groups ranging from equisitum to many families of seed-bearing plants, and while it occurs preponderantly in the aerial organs (Figs. 23,25,26), may also be found in the roots (Fig. 24). Three types of silica deposits have been reported from the grasses: (a) opal phytoliths, i.e., deposits that fill the cell and take on its form; (b) deposits in the cell wall in close association with cellulose; and (c) extracellular deposits in the form of thin sheets or as occlusions of the intercellular spaces. In roots, the silica deposits occur outside the cytoplasm, usually between the cells in association with the cell wall, or in spaces within the cell where they may occur either as silica aggregates or as silica layers in the cell wall. And although the form of the silica is the same as that in the other biological systems, the silicification process in plants differs in several respects from that in single-celled organisms and in the sponges.

SILICA DEPOSITION

In almost all siliceous organisms except plants, silica is deposited intercellularly within a silica deposition vesicle, the SDV, that is bounded by a specialized membrane, the silicalemma. In chrysophytes, choanoflagellates, and testaceous amebae, after formation the siliceous structures of specific morphology are "secreted" to the exterior of the cell and then "glued" together in species-specific patterns. Generally speaking, silica deposition in the chrysophytes takes place in a previously formed SDV which may first appear as a flattened vesicle or cisterna and which, as a rule, is closely associated with the vesicles of the endoplasmic reticulum (ER) either of the chloroplast or of the cytoplasm (Fig. 27). In several cases, densely staining fibrils are found between the outer surface of the chloroplast ER and the proximal surface of the SDV, and microtubules can be seen close to the distal surface of the SDV or of the scale. In all cases, during early stages of deposition the silicalemma is thick and more electron-dense, becoming thinner and less opaque as deposition is completed.

In most diatoms, on the other hand, deposition and cell wall formation are initiated immediately after cytokinesis with the appearance in the cell of a pair of tubular vesicles or SDVs. Additions to the SDV are rapidly and sequentially formed on either side of the original pair by extension, and are simultaneously filled with polymerized silica (Fig. 28).

In the pennate diatoms (having bilateral symmetry), silica deposition may be centrifugal, expanding to the margins, or may be centripetal, and in some genera, both types may occur in different species.

Various types of fibrils have been found associated with wall formation in several species, for example, as electron-dense, feathery

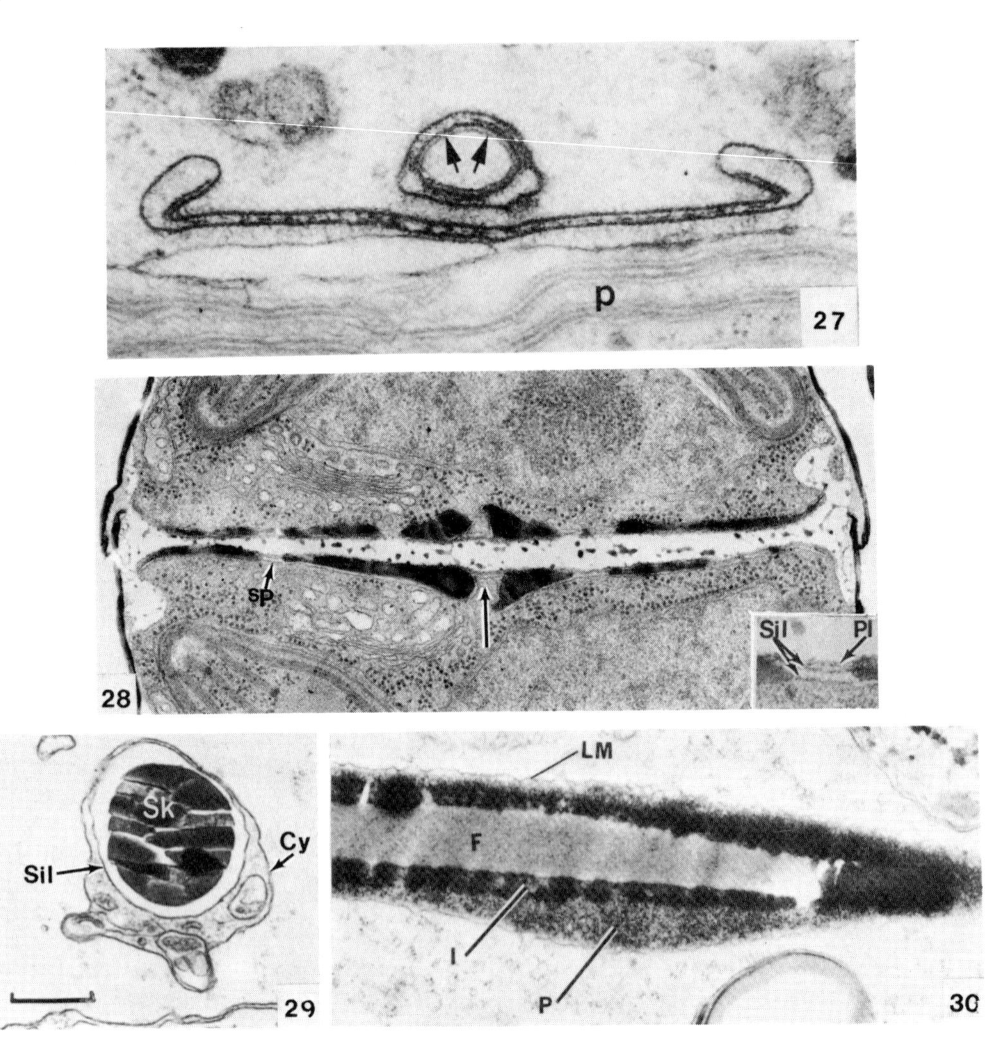

Figs. 27-30. Silicalemma and silica deposition vesicles in various organisms. Fig. 27. Early stage of silica deposition in silica deposition vesicle of the chrysophyte *Synura petersenii*. p = plastid; arrow points to membrane of chloroplast ER cisterna. x49,600. (Courtesy of C.B. McGrory and B.S.C. Leadbeater). Fig. 28. Late stage of valve development in the diatom *Navicula pelliculosa* in which silica deposition is almost complete. Note the continuity of the silicalemma across the raphe fissure (arrow), and formation of the sieve plate (sp). x27,589. Inset: high magnification of the developing sieve plate within the silicalemma (Sil). Pl = plasmalemma. x101,570. (See ref. 9). Fig. 29. A section through a segment of a siliceous skeleton (Sk) of a radiolarian showing the surrounding silicalemma (Sil) and cytokalymma (Cy). Marker = 0.5 µm. (Courtesy of O.R. Anderson). Fig. 30. Longitudinal section of a young spicule of *Spongilla lacustris* showing the two-component system of silicification: axial filament (F) and silicalemma (LM). The peripheral silica (P) is much less dense than the internal silica (I). x18,225. (Courtesy of T.L. Simpson).

structures within the SDV at the start of deposition, as extending from both surfaces of the silicalemma; as well-organized organic structures within the SDV prior to silica deposition; or as thickening fibrils that extend from both surfaces of the silicalemma and merge with the central mass of the previously formed wall. Microtubules have also been reported to occur during wall formation, and in some species appear to play an important role in wall formation if not in silica deposition as such. There have been suggestions that the fibrils constitute a reticular matrix component in which silica is deposited.

In the rhizopod amebae, formation of comma-shaped, membrane-enclosed siliceous platelets that compose the test is apparently related to activities of the many Golgi complexes of the cell, though no direct association with the Golgi has been found (10). Each Golgi complex consists of six to nine smooth cisternae or vesicles, one atop the other, and containing either silica or fibrous material. Dense vesicles containing silica are also present and are used, as well as the platelets, in forming the test. When fully formed, the platelets are "secreted" to the exterior of the cell, en route to the test where they are positioned in a silicon-containing matrix that has resulted from the fusion of adjacent dense vesicles.

The choanoflagellates are particularly interesting in that the costal strips are formed prior to cytokinesis. Deposition takes place in narrow, membrane- (silicalemma) bound vesicles within the peripheral cytoplasm. Each of the vesicles is associated with two parallel microtubules that are adjacent to the interior surface of the vesicles and remain in close association until strip formation is complete, at which time they disappear. Prior to assembly of the lorica, the strips are accumulated outside the protoplast; assembly then takes place with great rapidity.

The cytokalymma that encloses the internal skeletal structures of radiolaria contains vesicles but few other organelles (Fig. 29). There is considerable evidence that silica is deposited within an inner membrane or silicalemma, beneath the cytokalymma, that surrounds the skeleton and that may serve as a template for skeleton morphology. Aggregates of silicon granules can be seen very close to the previously deposited electron-dense segments. In contrast to the diatoms and sponges in which SDV formation takes place within the cell, in radiolaria the cytokalymma is formed outside the capsular wall and is connected to the rhizopodial system, or to cytoplasmic lobes, by rhizopodial strands. It is suggested (13) that these strands extend and produce branches that fuse on contact, creating the cytoplasmic template; the silicalemma could be produced by vacuolization, thus providing space in which silica is deposited.

Deposition of the megascleres of sponges is initiated with the appearance within single cells of an organic axial filament surrounded by a silicalemma (Fig. 30); microtubule structures are sometimes seen between the axial filament and the silicalemma, and areas of the silicalemma are adjacent to the plasmalemma. Silica is deposited upon the filament as it increases in length. Interestingly, a biochemical study has shown that protein synthesis of the axial filament occurs during silica deposition (19). The whole spicule has the same morphology as the axial filament suggesting that the filament determines, in large part at least, the eventual configuration of the spicule.

Silica deposition in higher plants appears to involve aggregations of silica granules. Once silicon has entered the conduction tissues it appears to move passively up the plant shoot in the transpiration stream of the xylem. Silica first occurs in the form of very small silica spheres, about 100 nm in diameter, called "primary units of silica". These coalesce to form either "silica aggregates", specialized conical or spherical deposits which only partly fill the open space of the cell, or phytoliths which completely fill the cell and assume its shape. In some cases individual "silica lenses" are formed in which the primary spheres are not visible; the lenses coalesce to form "aggregates".

The cork-silica cell (CSC) pairs found in the aerial organs of some grasses are most interesting. In oats (18), the pairs originate when young epidermal cells, in the intercalary meristem just above the base of an internode, divide unequally producing alternate long and short daughter cells. The short cells may then become either hairs or CSC pairs. The cork cell is always found close to the base of the internode, the silica cell is always uppermost. The potential silica cell develops an enormous nucleus which later disintegrates along with all of the cytoplasmic organelles; fibrillar elements accumulate in the resulting open spaces of the cell (the lumen) which then fills with "bodies" of silica gel (Fig. 23). The specific function of these pairs has not been delineated, but there is evidence that in sugar cane the silica cells in the epidermal system might serve as "windows" through which more light can be transmitted to photosynthetic tissue than can penetrate through neighbouring epidermal cells.

THE SILICALEMMA, ITS ORIGIN AND FATE

The silicalemma, which bounds the SDV, is apparently a functionally unique cytoplasmic membrane which in diatoms, appears immediately beneath the plasmalemma and is so closely attached at all times to the forming siliceous structure that the two cannot be readily distinguished. It is thought to serve as a "mold" or template for the scales of chrysophytes, and for the sponge spicules, and a number of workers have suggested that it serves similarly in diatoms. On the other hand, some workers have suggested that the silicalemma and/or the SDV provides a matrix for deposition: in testaceous amebae and chrysophytes, the SDVs do indeed possess a fibrillar matrix; fibrils are associated with the silicalemma in diatoms; and in vascular plants deposition is often associated with cellulosic cell wall components. In sponge spicules, however, the pattern of both the silicalemma and the deposited silica appears to be due to a underlying pattern of matrix molecules.

Though the origin of the silicalemma has been much investigated, particularly in diatoms, studies in several organisms have failed to produce unequivocal data. In chrystophytes and perhaps in choanoflagellates there is evidence that the silicalemma is a product of the Golgi membranes, and in testaceous amebae it appears that siliceous scales are formed directly within the Golgi vesicles. There are many theories as to the origin of the silicalemma in diatoms (9), but none has as yet been confirmed. One hypothesis is that it derives from the endoplasmic reticulum; another, that it originates in the vesicular activity of the Golgi apparatus; still another, that it arises from interactions among the endoplasmic reticulum, the Golgi vesicles, the microtubules (often found in association with the silicalemma) and the plasmalemma. Somewhat different theories propose that the silicalemma is derived from the fusion of small dictyosome-derived vesicles or that the SDV itself may be a very large vesicle corresponding to the Golgi complex but not necessarily derived from it or associated with it.

Two hypotheses have been advanced with respect to the radiolaria silicalemma (13). One suggests that the silicalemma is synthesized by extension, branching, and fusion of rhizopodia; this is followed by vacuolization of the resulting rhizopodal network, thus providing a space within which silicification can take place. The other hypothesis proposes that the membrane arises within a layer of cytoplasm surrounding the cell and is then separated by vacuolization. So far, evidence favours the first theory.

The fate of the silicalemma is equally a matter of conjecture and theory, particularly in diatoms. In a newly released daughter cell, the original plasmalemma and the silicalemma cannot be distinguished from each other, but a new plasmalemma is visible beneath the valve, and there is data suggesting that the organic casing of the new valve is composed of the distal portion of the silicalemma and the "old" plasmalemma, plus organic material synthesized toword the end of wall development. On the other hand, it has been reported that both the "old" plasmalemma (surrounding the daughter valves) and the distal silacalemma are sloughed off as the cell reaches maturity; the proximal silicalemma then becomes the plasmalemma of the daughter cell. However, the possibility that the "remnants" of the "old" membranes may be artifacts has not been ruled out, and such a transformation of the proximal silicalemma would entail a radical change in character for so specialized a membrane, requiring it to carry out the quite different and varied functions of a plasmalemma.

SILICIFICATION

The mechanism by which living organisms mineralize $Si(OH)_4$ involves two major sets of processes: transport of the $Si(OH)_4$ into the cell and thence into the deposition sites, and the silicification (i.e., polymerization) of the monomer to silica. Here, only the questions surrounding silicification will be discussed. And like the silicalemma, this is the subject of much theory and speculation but as yet little conclusive data. Nevertheless, since almost all siliceous structures are species-specific, even those that are "self-assembled" from ingested arenaceous silica particles, the mechanisms must be under genetic control, and there must be specialized sites for the initiation of silicification as well as specialized cytoplasmic components that are involved in the process.

In several chrysophytes, the endoplasmic reticulum near the chloroplast, and in others, the Golgi, have been suggested as the sites of scale production, particularly in view of the fact that unmineralized scales are produced within vesicles of the Golgi apparatus; in several amebae and in the choanoflagellates the Golgi apparatus appears to be involved in producing the siliceous structures. In one protozoan species, the site of scale production and storage is near the nucleus, and in certain diatoms there is evidence that the position of the primary silicification site depends on the position of the nucleus. In most siliceous organisms, including plants, the presence of silica spheres or granules is reported. There have been several suggestions that the dense material of siliceous skeletons, scales, spines and cell walls is formed as a result of the organized aggregation of small silica spheres into larger spheres and their subsequent compaction, in a process similar to that by which precious opal is formed. If such were the case, the smallest silica spheres, or "primary silica particles", would result, according to one model, not from a physicochemical reaction at the surfaces, as earlier researchers thought, but from a two-stage process whereby silicic acid, polymerized via condensation, forms a colloidal sol suspension of primary silica particles that then aggregate due to physical attraction.

A very different model of silicification is based on the biochemistry of the silicalemma in diatoms. According to this theory, it is plausible to assume (a) that there is a complementary relationship between the molecular structure of the silicalemma and the deposited silica (analogous to the collagen-calcium apatite relation in calcification), and that the silicalemma serves as the matrix for silicification; (b) that the silicalemma is permeated by $Si(OH)_4$ or a modified form, whether by diffusion, active transport, or some other method, and that there are interactions between the silicalemma and $Si(OH)_4$ at specified sites; and (c) that the chemistry of the silicalemma is unusual, allowing the membrane to participate in silicification reactions.

There are indications that the chemistry of the silicalemma may indeed be unusual, though this cannot be proved because so far it has not been possible to separate silicalemma and organic casing (9). Nevertheless, it was hypothesized that very early stage ("embryonic") walls would consist chiefly of almost unmodified silicalemma, and that their chemical composition might reflect to some degree that of the silicalemma. Indeed, it was found (9) that the chemistry of the embryonic walls and the mature casing do differ notably: e.g., protein content of the embryonic walls is markedly high while carbohydrate content is low; this situation is reversed in the casing; the presence of sugars in the embryonic wall is evidence that polysaccharides are synthesized early in wall development, and the young walls bind polysilicic but not monomeric acid; the polysilicic acid-binding polysaccharide is synthesized during wall formation.

These findings are relevant to a third model for silicification in the diatom (20) (and in sponges, since trends found in comparing the diatom casing and cytoplasm have also been found in sponge spicules and whole tissues). The polymerization of $Si(OH)_4$ to a silica gel could result from changes in $Si(OH)_4$ concentration or in pH; it could result from hydrogen bonding to specific sites or ionic interaction with them, or it could result from condensation by hydroxyl groups as the third model proposes. This is not inconsistent with findings on the protein and polysaccharides composition of embryonic walls and supports the model, which assumes that the proximal surface of the silicalemma consists of one or more mineralizing template proteins high in glycine and in the hydroxyl-containing serine and threonine. Presenting a layer of hydroxyl groups, such protein(s) would provide a surface on which silicic acid molecules could condense in a molecular arrangement favouring further condensation. Although the embryonic walls contain little serine or threonine, these compounds do occur in the mature casing, and the silicalemma itself may have more of them than was found in the embryonic walls. Moreover, spicules of two sponge species are higher in serine and glycine than in any other amino acids; the fact that the casing-cytoplasm and spicule-tissue trends mentioned above are similar suggests that silicification, i.e., polymerization, in the two taxa might be similar. Yet a fourth silicification model (21) is based on the assumption that the silicalemma in diatoms creates fine strands of polysaccharide which serve as a template on which the siliceous matrix is precipitated.

There are no firm data on the mechanism of silicification in plants, but an interesting two-piece model has been proposed (18) for which there is some evidence. It is based on two different systems: the first involves a physicochemical mechanism, in which $Si(OH)_4$ is carried in the xylem sap and is silicifed at specific sites. The second system involves a biological mechanism to account for deposition in areas, such as siliceous hairs or thorns, where there is no sap movement.

The conditions for polymerization (concentration of $Si(OH)_4$ or pH) required for the first system are present, at least in rice shoots. At the upper end of the plant internode, the tissues act like a very fine-pored membrane, filtering out all silica polymers while allowing the monomer to pass. Thus, an "upside-down filter-cake" of silica is built up, first in the epidermal tissue and later in internal tissues. The deposited silica may be interspersed with organic polymers, e.g. cellulose, or may fill the intercellular spaces.

The second, biological model assumes the presence of a membrane of some sort onto which silica particles are induced by ionic attraction to form a silica layer and onto which, in turn, monomeric $Si(OH)_4$ is deposited from a supersaturated solution. It is suggested that since there is no transpiration that could increase the concentration of the $Si(OH)_4$ by evaporation, soluble chelates of silicon are formed and are concentrated by physiological processes; when the pH is changed or the chelating agent is enzymatically destroyed, silicon is freed in specific areas. This, of course, would involve metabolic (and hence genetic) control.

Findings within the last 10 years have demonstrated a relationship between calcium and silicon in bone formation, in connection with aging of skin and aorta, and in the development of atherosclerotic calcium plaques; there appears also to be a relationship between silicon, endocrine balance and mineralization. Silicon is essential for calcification in the early-forming bone and speeds the rate of mineralization, but its major role is in the formation of organic bone matrix by affecting the synthesis of collagen, and of glycosaminoglycans (3).

Various data, including the above, suggest the possibility that there may be an evolutionary connection between silification, matrix formation, and calcification; a remarkable illustration of this is a group of sponges, the *Sclerospongiae*, some species of which produce siliceous spicules partially enclosed in an organic matrix; both are eventually embedded in a calcareous skeleton. In another taxa, the testaceous amebae, one genera uses silica whereas another closely related genera uses calcium, although the tests of each are remarkably similar.

This newly found system, the biological mineralization of silicon, poses a host of questions for researchers; the biology of silicon has now to be reckoned with, and in the study of the siliceous organisms will be found clues to understanding this ubiquitous, important, but enigmatic element and its many roles in life processes.

ACKNOWLEDGEMENTS

I wish to thank Drs. O.R. Anderson, E.C. Bovee, R.M. Crawford, W.D. Hartman, P.B. Kaufman, B.S.C. Leadbeater, C.B. McGrory, D.W. Parry, F.E. Round, A.G. Sangster, and T.L. Simpson for providing me the micrographs.

REFERENCES

1. Sullivan, C.W., and Volcani, B.E. (1981). Silicon in the cellular metabolism of diatoms. In "Silicon and Siliceous Structures in Biological Systems" (T.L. Simpson and B.E. Volcani, eds.), Springer-Verlag, New York, Heidelberg, Berlin, pp. 15-42.

2. Carlisle, E.M. (1981). Silicon in bone formation. Ibid., pp. 69-94.
3. Bendz, G. and Lindqvist, I. (eds.) (1978). Biochemistry of Silicon and Related Problems. Plenum Press, New York, p. 591.
4. Simpson, T.L., and Volcani, B.E. (eds.) (1981). Silicon and Siliceous Structures in Biological Systems. Springer-Verlag, New York, Heidelberg, Berlin, p. 587.
5. McGrory, C.B., and Leadbeater, B.S.C. (1981). Ultrastructure and deposition of silica in the Chrysophyceae. Ibid., pp. 201-230.
6. Bovee, E.C. (1981). Distribution and forms of siliceous structures among protozoa. Ibid., pp. 233-279.
7. Round, F.E. (1981). Morphology and phyletic relationship of the silicified algae and the archetypal diatom-morphology or polyphyly. Ibid., pp. 97-128.
8. Crawford, R.M. (1981). The siliceous components of the diatom cell wall and their morphological variation. Ibid., pp. 129-156.
9. Volcani, B.E. (1981). Cell wall formation in diatoms: morphogenesis and biochemistry. Ibid., pp. 157-200.
10. Harrison, F.E., Dunkelberger, D., Watabe, N., and Stump, A.B. (1981). Ultrastructure and deposition of silica in rhizopod amebae. Ibid., pp. 281-294.
11. Leadbeater, B.S.C. (1981). Ultrastructure and deposition of silica in loricate choanoflagellates. Ibid., pp. 295-315.
12. Riedel, W.R., and Sanfilippo, A. (1981). Evolution and diversity of form in radiolaria. Ibid., pp. 323-346.
13. Anderson, O.R. (1981). Radiolarian fine structure and silica deposition. Ibid., pp. 347-380.
14. Hartmann, W.D. (1981). Form and distribution of silica in sponges. Ibid. 453-494.
15. Garrone, R., Simpson, T.L., and Pottu-Boumendil, J. (1981). Ultrastructure and deposition of silica in sponges. Ibid., pp. 495-526.
16. Simpson, T.L. (1981). Effect of germanium on silica deposition in sponges. Ibid., pp. 527-550.
17. Sangster, A.G., and Parry, D.W. (1981). Ultrastructure of silica deposits in higher plants. Ibid., pp. 383-408.
18. Kaufman, P.B., Dayanandan, P., Takeoka, Y., Bigelow, W.C., Jones, J.D., and Iler, R. (1981). Silica in shoots of higher plants. Ibid., pp. 409-450.
19. Shore, R.E. (1972). Axial filament of siliceous sponge spicules, its organic components and synthesis. Biol. Bull. 143, pp. 689-698.
20. Hecky, R.E., Mopper, K., Kilham, P., and Degans, E.T. (1973). The amino acid and sugar composition of diatom cell-walls. Mar. Biol. 19, pp. 323-331.
21. Pickett-Heaps, J.D., Tippet, D.H., and Andreozzi, J.A. (1979). Cell division in the pennate diatom *Pinnularia*. IV. Valve morphogenesis. Biol. Cellulaire 35, pp. 100-203.

HIGH RESOLUTION ELECTRON MICROSCOPY STUDIES OF THE SILICA LORICA IN THE CHOANOFLAGELLATE *STEPHANOECA DIPLOCOSTATA* ELLIS

S. Mann and R.J.P. Williams
Inorganic Chemistry Laboratory, South Parks Road, Oxford
OX1 3QR U.K.

SUMMARY

The silica lorica of *Stephanoeca diplocostata* Ellis has been studied by ultra-high and high resolution electron microscopy. Ultra-high resolution electron microscopy revealed an extremely amorphous silica structure within costal strips. T-junctions between strips in intact loricae have been examined and in several cases strips of viscid connective material containing in part amorphous silica have been observed running between the two strips. Silica demineralisation initiates from the central axis of the costal strips leaving tubular brittle rods. The surface of costal strips has been shown to be active in binding low concentrations of cations (Co^{2+}, Fe^{3+}) from external solutions. Full details of this work are reported elsewhere (Mann, Williams, 1982).

1. INTRODUCTION

Recent advances in ultra-high resolution electron microscopy combined with energy dispersive analysis have had a remarkable impact on the study of the nature of solids. The great advantage of ultra-high resolution electron microscopy is that it reveals structure directly and is thus an excellent technique for studying local architectural anomalies and structural microheterogeneities. This is in direct contrast to the classical approaches of structural determination (X-ray, electron and neutron diffraction) in which the volume of sample interacting with the incident radiation is microscopically large yielding statistical data averaged out over the area analysed. The real-space imaging of solid structure is then crucial in any investigation of so-called amorphous materials such as biological silica.

Any thorough investigation of biological solids must combine ultrastructural information obtained at low levels of magnification with that observed at higher degrees of spatial resolution. Here we have chosen to study the siliceous nature of costal strips in the intact lorica of the choanoflagellate *Stephanoeca diplocostata* Ellis using both ultra-high and high resolution electron microscopy.

P. Westbroek and E. W. de Jong (eds.), Biomineralization and Biological Metal Accumulation, 407–412.

Stephanoeca diplocostata Ellis is a marine choanoflagellate which has been successfully established in culture (Leadbeater 1979a). The protoplast is lodged in a basket-like casing (lorica) constructed of 150-180 silica costal strips and new costal strips are produced in advance of mitosis within long thin vesicles in the peripheral cytoplasm. Previous investigations of Stephanoeca diplocostata have used light microscopy and low resolution electron microscopy to study lorica ultrastructure and cell division (Leadbeater and Manton 1974, Leadbeater 1979a, 1979b). Our aim has been to obtain information on the chemical and structural nature of costal strips in intact loricae.

2. MATERIALS AND METHODS

Stephanoeca diplocostata Ellis was grown in culture as described elsewhere (Leadbeater 1979a). One drop of a concentrated suspension of loricae was placed on a formvar-coated copper grid, and left to dry on filter paper in the air at room temperature. No staining methods were applied to any of the samples.

The electron microscope used for ultra-high resolution imaging was a JEOL 200CX instrument operating at 200KeV with a high brightness LaB_6 cathode and a point to point resolution of 2.5Å. For normal resolution work a JEOL 100CX Temscan analytical electron microscope fitted with a Kevex Li-drifted Si detector, operating at 100KeV was used.

Silica demineralisation experiments were carried out by drying samples of loricae from cells taken from old cultures (6-8 days after inoculation into subculture) onto electron microscope grids.

Binding studies of costal strips with the aquated cations Co^{2+} and Fe^{3+} were performed by treating the intact loricae for 24 hours at room temperature with 2mM solutions of $CoCl_2$ and $FeCl_3$ respectively.

3. RESULTS

3.1 Silica Microstructure of Costal Strips

Figure 1(A) shows an intact lorica of Stephanoeca diplocostata Ellis with the protoplast removed by osmotic shock. Thin areas of costal strips in intact loricae were examined using ultra-high electron microscopy capable of imaging short range order within the structure greater than 15Å in dimensions. Careful examination of this and other similar images revealed irregular fringes in all areas of the costal strips. No short range order above 15Å could be observed. Hence the costal strips are composed of silica of an extremely amorphous nature. (Figure 2)

3.2 T-joins Between Costal Strips in intact loricae

The structure of T-junctions consisted of smooth rod-shaped costal strips with rounded tips attached at right angles to each other. The

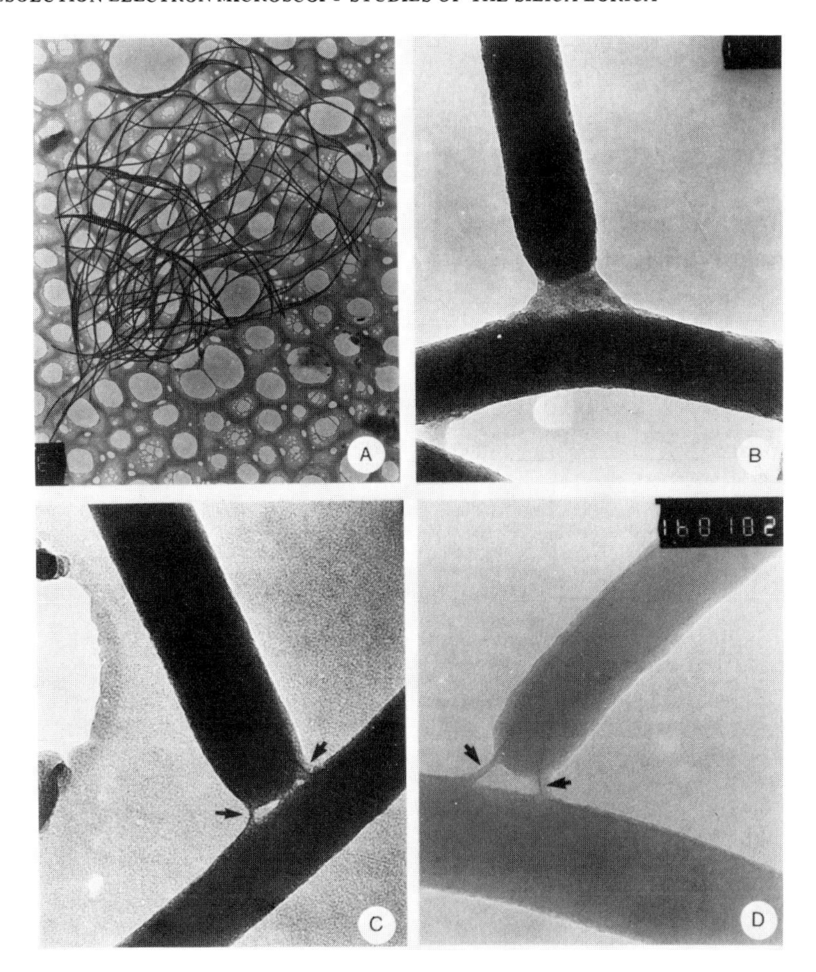

Figure 1: (A) Whole mount of an empty silica lorica of *Stephanoeca diplocostata* Ellis (x3,600).
(B) T-junction between costal strips showing the presence of connective material within the join (x171,000).
(C),(D) T-joins between costal strips showing the presence of fine structure (filaments) within the connective material.
Figure(C),(x175,500). Figure(D),(x180,000).

costal strips adhered to each other by means of connective material of unknown total composition. Where costal strips were slightly pulled apart, presumably due to the drying of samples onto the support film, the connective material was observed to be viscid such that two strands of this material could be seen running between the costal strips.

In T-junctions suitably orientated on the support film, we have occasionally observed fine structure within the connective material between costal strips (figures 1(C) and 1(D). In each electron micrograph two continuous filaments of electron dense material can be observed within the connective material between the costal strips (see arrows on figures). The width of these opaque threads is 1.5nm (figure 1(C)) and 1.65nm (figure 1(D)).

T-joins were also imaged under conditions of ultra-high resolution and the material traversing the join was found to be stable in the electron beam and showed a similar but less dense image pattern as that for the siliceous material of the costal strips. We conclude that the material traversing the join contains in part a continuum of amorphous silica, but this does not exclude the presence of some organic material in the original join which is subsequently destroyed by the electron beam under these experimental conditions.

3.3 Demineralisation of Costal Strips

The silica-containing costal strips are meta-stable and slowly redissolve in the aqueous medium. We have followed the process of demineralisation using electron microscopy. The initial stage of demineralisation takes place at localised centres along the central axis of the strip. These centres become increasingly demineralised until they extend and join together along the central axis forming tubular, brittle roads (figure 3). As demineralisation increases the hollowing becomes more extensive and at later stages the outer edges of the costal strips also show signs of demineralisation becoming rough and pitted.

3.4 Binding Studies of Aqueous Ions to Costal Strips

The detailed nature of the surface of costal strips is not known. Evidence from demineralisation studies (section 3.3) suggests that the surface of costal strips may be stabilised in some way.

X-ray microprobe analysis was performed on many costal strips treated with Co(II) and detected the elements Si, Co and Cu (from the grid). No Co was detected in background analyses. Hence Co(II) ions were bound preferentially to the surface of the costal strips.

Similar results for Fe(III) binding were obtained. Thus, Fe(III) ions were also found to be preferentially bound to costal strips.

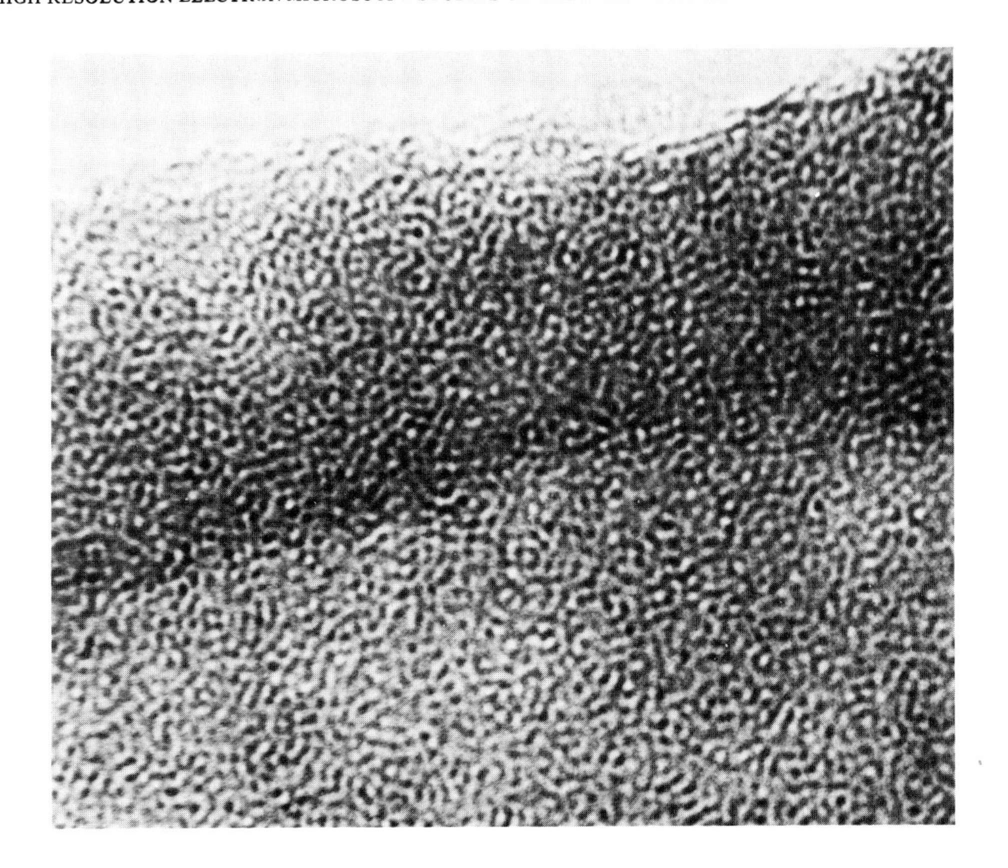

Figure 2: Ultra-high resolution electron micrograph of a costal strip in an intact lorica showing the extremely amorphous nature of the silica (x2,970,000).

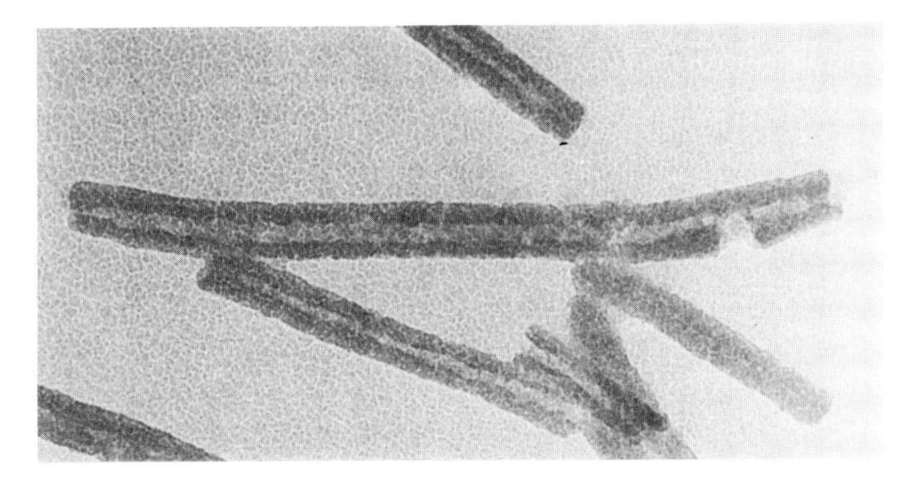

Figure 3: Demineralised costal strips showing tubular, brittle rods (x137,800).

4. DISCUSSION

The real-space images obtained by ultra-high resolution electron microscopy show very irregular atomic fringes in all regions of the costal strips. These observations favour a random network structure over a microcrystalline model for amorphous silica since the degree of short range order is very low. The structure would then comprise a random array of SiO_4 tetrahedra connected through Si-O-Si bonds of variable bond angle. It is this wide variation in bond angle which marks an important distinction between amorphous and crystalline silica. That the silica is deposited in an extremely amorphous form may be due to the high activation energies required (of the order 200kcal mol^{-1} [Degens 1976]) for crystalline silica deposition.

Demineralisation of costal strips shows that there is a central zone within the strip which is destablised at a faster rate than the surface silica. This suggests that the central silica has a different chemical environment to the silica at the surface of the strip. This could be due to the presence of a central core of organic material (organic template?) which degrades faster than the silica phase and/or that the surface silica is stabilised by an external coating of some kind (organic/metal ions). Binding experiments indicate that the surface of costal strips is active towards cations present in low concentrations in the external solution. The presence of a 'buffer' layer of such ions may be important in the stabilisation of the surface silica. Circumstantial evidence from micrographs recorded at normal and ultra-high resolution suggests that T-joins are constructed around two filaments running between the strips.

The composition of the filaments is not known. The greater electron scattering from these structures indicates regions of material of higher density than in the surrounding costal strips. A possible explanation is that the silica deposition has been altered in these regions because of a localised organic phase which binds silica in a network of greater density than in a structure containing only one component of amorphous silica.

These images were not observed in all T-joins which may imply that there are different mechanisms of join formation in different T-junctions. Also, such images may be difficult to resolve when the connective material is very opaque or when the costal strips have undergone partial demineralisation.

REFERENCES

Degans, E.T. 1976, Topics in Curr. Chem. 64, 1-112.
Leadbeater, B.S.C., Manton, I. 1974, J. Mar. Biol. Ass. UK 54, 269-276.
Leadbeater, B.S.C. 1979a, Protoplasma 98, 241-262.
Leadbeater, B.S.C. 1979b, Protoplasma 98, 311-328.
Mann, S., Williams, R.J.P. 1982, Proc. R. Soc. London, Ser. B. in the press.

A BIOINORGANIC VIEW OF THE BIOLOGICAL MINERALIZATION OF IRON

J. Webb
School of Mathematical and Physical Sciences,
Murdoch University, Murdoch, Western Australia,
Australia, 6150.

ABSTRACT

A variety of mineralized iron deposits is now known to occur in a range of living organisms, extending from microorganisms to higher vertebrates. Minerals reported include magnetite, maghemite, goethite, lepidocrocite, ferrihydrite and amorphous ferrihydrates. The biochemical events that lead up to and accompany mineralization of iron are not well understood. In contrast a fairly detailed picture can be drawn of the cellular and molecular events involved in iron metabolism in mammals, particularly man. This review seeks to relate our general understanding of iron absorption, transport, storage and utilization to account for biogenic mineralization of iron. Particular systems discussed include a lamprey species, magnetotactic bacteria and chitons.

INTRODUCTION

Biological mineralization of iron is now known to occur fairly widely (1,2), yet studies of the complete process and its relationship to the overall biology of particular organisms are few in number. To achieve such a complete description requires a multi-disciplinary approach, ranging across the many sub-specialities of contemporary science. In this perspective, mineralization is seen as a process that involves a variety of structural and biological components, more commonly encountered in biology and inorganic chemistry, but which in mineralization can be intimately inter-related, particularly with respect to the control of mineralization itself. A general aim of such detailed bioinorganic studies is a better understanding of how mineralization can be related to the functioning of the mineralizing organism, be it animal, plant or microbe.

The promise of such a multi-disciplinary approach is accompanied by the frustration of working with a wide range of concepts, terminology and literature (3). Thus, as described below, the mineralization of iron impinges on such diverse fields as geology, inorganic

P. Westbroek and E. W. de Jong (eds.), Biomineralization and Biological Metal Accumulation, 413–422.

chemistry, soil science, haematology, biochemistry and physiology. In turn, these disciplines offer a wide range of experimental techniques to probe the structures and reactivities of compounds involved in the mineralization process.

In this paper, the biological mineralization of iron is considered in the light of these general comments. A comprehensive review of such a broad topic cannot be easily or briefly achieved. Consequently, the bioinorganic perspective is used to highlight significant aspects of this complex and fascinating process that are, as yet, incompletely understood.

In a recent review, Lowenstam has gathered together much of the diverse literature on biogenic minerals (1). This survey showed that, while carbonates are the most widely used bioinorganic constituents, iron oxides have now been reported in each of the five kingdoms of living organisms. These data are summarized in the Table, supporting Lowenstam's observation that ferrihydrite and related ferric oxide minerals are the third most extensively formed biogenic mineral, and that magnetite may well rank fourth. Biomineralization of iron is then a rather widespread phenomenon, even if it has been recognized as such only rather recently. Perhaps the most striking observations from the Table are the widespread occurrence of magnetite and of ferrihydrite. Each of these will be considered in more detail later in this paper.

TABLE*

DISTRIBUTION OF BIOGENIC IRON-CONTAINING MINERALS

	Mineral	Kingdom/Phylum
Oxides	- magnetite (Fe_3O_4)	bacteria mollusca arthropoda chordata
	maghemite (γFe_2O_3)	possibly in bacteria
	goethite ($\alpha FeOOH$)	mollusca
	lepidocrocite ($\gamma FeOOH$)	porifera mollusca
	ferrihydrite ($5Fe_2O_3 . 9H_2O$)	bacteria fungi annelida mollusca chordata plantae
	amorphous ferrihydrates	foraminifera annelida chordata
Sulfides	- pyrite	bacteria
	hydrotroilite	bacteria

* after Lowenstam, ref 1.

From the perspective of the organism, biomineralization of iron forms just one part, albeit a significant one, of the various pathways involved in iron metabolism. A quite detailed picture can be sketched of many of these aspects of iron metabolism, particularly in the case of vertebrates and mammals (4-6). Iron is known to be essential for all cells both aerobic and anaerobic, with the possible exception of some lactate bacteria (7). Furthermore, the clinical management of iron in man is a major medical concern. A range of human disorders is known, from iron deficiency anemia to iron overload. This latter pathology can include iron mineralization in tissues, as is seen in idiopathic hemachromatosis and transfusional siderosis such as accompanies the clinical treatment of thalassemia (8). However, before considering metabolic aspects, the chemistry of iron, especially iron(III), will be reviewed.

SOME BIOINORGANIC CHEMISTRY

The bioinorganic chemistry of iron is dominated by only two oxidation states, Fe(II) and Fe(III) (3,9,10). In aqueous solution, the maximum solubility of $Fe(H_2O)_6{}^{2+}$ is 10^{-1} M, much higher than that of $Fe(H_2O)_6{}^{3+}$, which is only 10^{-17} M. Polymerization of iron(III) begins at quite low pH, the dimeric (i.e., binuclear) complex $[(H_2O)_4Fe(OH)_2Fe(H_2O)_4]^{4+}$ being predominant at pH less than 2. Continued hydrolysis of iron(III) solutions yields eventually brown $Fe(OH)_3$ whose low solubility ($K_{sp} = 10^{-38.7}$) has been alluded to above. Although often an amorphous precipitate, a number of these hydrolysates have been reported with sufficient crystallinity to give limited X-ray diffraction patterns (3).

An unexpected hydrolysis product of $Fe(NO_3)_3$ solutions is the remarkably homogeneous soluble iron(III) polymer, spherical in shape and approx. 70Å in diameter, that incorporates about 1200 iron(III) ions in a mixed oxide-hydroxide lattice (9). Similar polymers have been isolated in the presence of potential chelating agents such as citrate and fructose, although their role in the polymer structure is unknown. Inclusion of strong chelating agents e.g., EDTA, inhibits hydrolytic polymerization, although dimeric complexes are formed.

When considering the chemistry of iron(II) and iron(III) in the presence of potential ligand atoms a useful generalization is provided by the classification of metal ions and ligand atoms as either "hard" or "soft" (11). "Hard" summarizes, at least qualitatively, properties such as small size, high ionic charge and low polarizability, with "soft" being the inverse of these properties. This approach leads to the classification of iron(III) as a 'hard' metal ion and iron(II) as 'borderline' between 'hard' and 'soft'. Iron(III) is most commonly bound by 'hard' ligand atoms, such as oxide and hydroxide, which can also be present as part of an organic ligand e.g., the anion of catechol (1,2-dihydroxybenzene) i.e., $C_6H_4O_2{}^{2-}$ which is known to form strongly-bound complexes with iron(III) (12). Similarly, it is not surprising that microbes produce chelates able to extract iron(III) from environmental iron that occurs as various insoluble iron(III) oxyhydroxides, using such 'hard' ligand atoms as occur in catechol and the hydroxamic acids. The synthesis, excretion and reabsorption of these siderophores by microorganisms have been studied in considerable detail (7). In addition, the iron(III) oxide and hydroxide systems contain a variety of crystalline phases, many of which are listed as biomineralization products in the Table. The structures and in particular their interconversions have been studied in detail (13-23).

Some of these transformations are topotactic in character (τοπος place; ταξις - arrangement), i.e., the transformation can occur by simple rearrangement of atoms such that the two phases have a definite structural relationship to each other. Such transformations when they occur involve no major crystallochemical changes (13-16). For the biomineralized iron oxides and hydroxides, their topotactic relationships are shown in Figure 1.

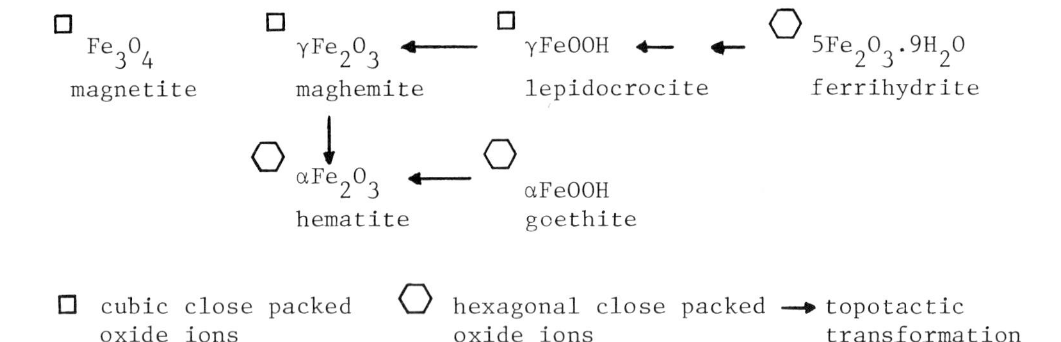

Figure 1. Structural transformations among some of the iron oxide/hydroxides (after ref.15). All shown except hematite have been reported as biogenic minerals.

Magnetite, Fe_3O_4 is a mixed Fe(II).Fe(III) oxide with a cubic inverse-spinel structure. As such it differs considerably from the iron oxyhydroxides, yet in the case of the radula of chitons, a group of marine molluscs, it appears to be formed from them as precursors, as discussed below. Ferrihydrite, the most widely distributed of the biomineralized iron oxides (see Table) frequently occurs also in precipitates from iron-bearing waters and in some soils (17). The formulation, $5Fe_2O_3.9H_2O$ is that first given for a synthetic preparation (18) but others have been proposed: $Fe_5(O_4H_3)_3$ (19) and $Fe_2O_3.2FeOOH.\ 2.6\ H_2O$ (20). Natural ferrihydrites often contain silica and/or organic compounds, which appear to retard their transformation to the thermodynamically more stable goethite (21-2).

The occurrence of ferrihydrite in biological organisms is in association with the iron-binding protein ferritin. This protein is made up (24) of a series of polypeptide chains that are arranged in a shell structure, inside of which can be accommodated a ferrihydrite core of about 70Å diameter, similar in size to the synthetic polymer referred to earlier (9). The crystallinity of these cores which show limited X-ray diffraction effects has been confirmed by electron microscopic studies (24). The structural model proposed by Towe and Bradley (18) has been further supported by electronic absorption (26) and extended X-ray absorption fine structure (27) spectra. The ferrihydrite structure is a disordered or defect structure related to that of hematite, containing hexagonally close-packed oxide ions, with iron(III) present in octahedral sites. An earlier structure for the ferritin core, with tetrahedrally coordinated iron(III), cannot be sustained (24). The topotactic transformation of ferrihydrite to lepidocrocite (Figure 1) occurs through a precipitation sequence that includes the green rusts I and II, whose nature has been discussed elsewhere (13-16,23).

When bound to ligand atoms and so protected from hydrolysis, iron can be involved in a variety of compounds of considerable biochemical importance (28). When proteins are the ligands, iron can be involved in reversible binding of oxygen (e.g., myoglobin, hemoglobin and the rather less-well known hemerythrin (29,30)) and in enzymatic activity concerned with redox reactions. These latter exploit the availability of the two oxidation states of iron whose relative stability can be modified considerably by interaction with appropriate ligands. For example, in the case of two quite similar proteins from the microorganism *Chromatium* the redox potentials are -0.49v and +0.35v. To achieve this difference by varying the concentrations of oxidized or reduced forms would require concentration changes of 10^{14} (see 28). Functions of iron-containing enzymes include direct electron-transfer e.g., the iron-sulfur proteins, cytochrome *c*; reactions with molecular oxygen e.g., tyrosine hydroxylase; and other rather poorly characterized processes e.g., as occur in some acid phosphatase enzymes. In addition to the above iron-proteins, there are well-known proteins for transport and storage of this essential mineral nutrient.

IRON METABOLISM

The bioinorganic chemistry of iron metabolism can be discussed in the framework of Figure 2, which outlines a general scheme for the management of iron by any organism. The food sources vary with the organism, but all organisms have some molecular mechanism for obtaining iron (31). In general, low molecular weight complexes are involved in membrane transport of iron, although colloidal iron hydroxide can be absorbed by endocytosis in some organisms e.g., the common mussel *Mytilus edulis* (32). There can be competition among various chelating agents for the dietary iron, such as in the acidic environment of the gut. A feature of iron absorption that is well documented in many vertebrates including man is the feedback interaction between iron absorption and total body load of iron. For man, there is no significant means of excreting iron (apart from bleeding) and control of iron levels occurs at the absorption step.

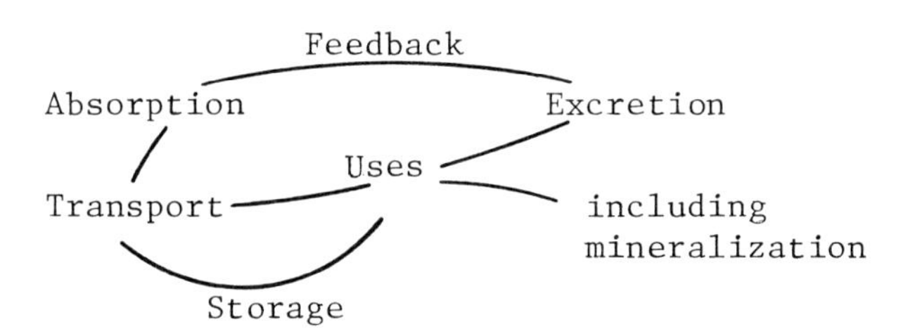

Figure 2. General scheme of iron metabolism.

Once absorbed, iron is transported in the bloodstream of almost all vertebrates by the protein, transferrin. One of the most intensely studied of all proteins, the molecular structure and biological functions of transferrin are being rapidly revealed (33-4). While specific cell surface receptors for binding transferrin (the first key step in the delivery of iron by transferrin to the cell) are now known, the chemistry of this interaction is still unclear, as is the mechanism of binding and release of iron by the protein. Controversy persists over the relative physiological importance of binding by competitive chelates or anions and of reduction of iron(III) to a more weakly bound iron(II). Other mechanisms for blood plasma iron-transport appear to be present in some species of the *Urchordata* (35), in a larval form of a species of *Agnatha* (36) and in the *Polyplacophora* (37).

The structure and function of ferritin, a protein commonly encountered in tissue storage of iron are being similarly revealed. In addition to studies of the ferrihydrite core, the detailed molecular features of the polypeptide shell are under investigation (24,34). The protein is also considered to serve (as is transferrin) to bind iron(III) ions which, unbound by organic ligands, are toxic to cells. A detailed model of iron uptake and hence growth of the ferrihydrite has been suggested, involving oxidation of incoming Fe(II) on the crystallite surface. Delivery of iron as iron(III) chelates has also been suggested. Release of iron from ferrihydrite is thought to occur by reduction via an oxidoreductase and diffusion of a small molecule reductant through channels in the protein shell. Ferritin occupies a central position in iron biomineralization because of its wide occurrence (Table) often as a precursor of other biomineralization products.

Another form of tissue iron that can be considered as a product of biomineralization is hemosiderin (38-9). Detailed molecular studies are limited, for hemosiderin has a variable composition, especially of organic constitutents. It appears under the light microscope as granules that react strongly with Perl's prussion blue stain, $[Fe(CN)_6]^{4+}$. The presence of electron dense particles of iron(III) oxyhydroxide is revealed by X-ray diffraction, electron microscopy, and Mössbauer and infrared spectroscopies. The particles are of a slightly smaller size but have a similar atomic structure to the ferrihydrite core of ferritin. Other biochemical data add further support to the general view that hemosiderin is, in general, a degradation product of ferritin catabolism. However, the terminology distinguishing between ferritin and hemosiderin is not used consistently particularly in discussions of iron-overload pathology where hemosiderin is commonly observed.

MINERALIZING ORGANISMS

(i) A lamprey species

An interesting animal model that offers the opportunity of distinguishing the different types of iron-containing granules is the Southern Hemisphere lamprey *Geotria australis* Gray (36,40). In the larval stage, the animal accumulates iron to a remarkable degree. Plasma iron levels of nearly 20,000 μg Fe/100 ml are vastly in excess of the normal vertebrate range of 50-250 μg Fe/100 ml. Tissue iron deposits of hemosiderin and ferritin, with its ferrihydrite core, are extensive. Other larval lampreys studied do not show these deposits. It appears that for *G. australis* regulation of iron uptake within the intestinal epithelium does not occur. Nor does the animal have the mechanism for excreting excess iron that has been reported for the vampire bat, *Desmodus rotundus* (41). An obligatory sanguinophore, this bat has a dietary intake of iron estimated to be 800 times that of man. Iron is accumulated as granules (presumably hemosiderin in nature, but possibly formed directly from dietary iron) present in vacuoles of the intestinal epithelial cells that are subsequently extruded into the faeces.

(ii) Magnetotactic Bacteria

As mentioned earlier, magnetite and ferrihydrite are the most widely distributed biogenic iron minerals. Magnetite has now been reported in a wide range of organisms (42-46), two of which are of particular interest for studies of the process of mineralization: the magnetotactic bacteria and the chitons. Thus, in the magnetotactic bacteria, magnetite is present as a sequence of small particles whose dimensions and attendant magnetic behaviour are consistent with a role in sensing direction, in particular, of the earth's magnetic field and hence of the sediment, their preferred environment (46-50). Falling in the size range 400-1200Å, this biogenic magnetite exists as thermally stable single-domain particles. In contrast, the ferrihydrite core of ferritin and synthetic ferrihydrite exhibit more complex magnetic behaviour, acting as superparamagnetic particles that are antiferromagnetically ordered (2). In the bacteria the magnetite particles are enclosed by membrane structures in an organelle termed the magnetosome. How these particles are formed and grow to their specific size is not known, but may well be controlled by the magnetosome membrane structure. Recently magnetite particles of similar dimensions have been formed in laboratory syntheses (51-2). Experimental conditions were severe (e.g., temperatures of 400°C) when compared to those in which the bacteria synthesize magnetite.

The possible role of ferritin in bacterial synthesis of magnetite is of particular interest, since ferritin has been reported in several bacteria e.g., *E. coli* (53). In addition, as mentioned above, the mechanisms of bacterial iron uptake are fairly well delineated. From these established data, studies of the growth, metabolism and genetics of magnetotactic bacteria (one species of which can be grown in pure culture (47,54)) could well serve to elucidate the relationship of the magnetite particles to the overall biology of these bacteria.

(iii) Chitons

A rather different sequence of magnetite synthesis is found in the radula teeth of the *Polyplacophora*, or chitons. A series of elegant studies by Lowenstam and colleagues has greatly clarified our understanding of this system (42,55-59). Nevertheless, important questions remain unanswered. Ferrihydrite present as ferritin, and perhaps hemosiderin, in the superior epithelial cells of the radula sac is transformed to magnetite in the teeth. This change is not topotactic. It involves reduction of one third of the iron(III) atoms to iron(II), removal of oxygen and water, and appreciable changes in crystal packing. In addition, iron is transferred from the ferritin core surrounded by protein to form an extensive array of magnetite crystals. The biochemical control of this process is not known. The delivery of iron to the mineralizing tissue is thought to occur via high levels of plasma ferritin, as described elsewhere in this volume (60). New teeth, i.e., new deposits of iron minerals, are continually being formed. Hence the chiton teeth represent a sequence of different stages of mineralization. Magnetite, with a Moh hardness of $5\frac{1}{2}$-$6\frac{1}{2}$ (cf. apatite as 5) is useful as a hard-wearing tooth capping. But how chitons manage their high levels of iron in relation to biomineralization and their total metabolism of iron remains a fascinating but unanswered question.

And, as the English poet Rudyard Kipling expressed it:

Gold is for the Mistress
Silver for the Maid
Copper for the Craftsman
Cunning at his trade
"Good", said the Baron
Sitting in his hall
"But, iron, cold iron
is master of them all".

REFERENCES

1. Lowenstam, H.A.: 1981, Science 211, 1126.
2. Lundgren,D.G. and Dean, W.: 1979, in Biogeochemical Cycling of Mineral-Forming Elements. Eds. P.A. Trudinger and D.J. Swaine, Elsevier, Amsterdam.
3. Webb, J.: 1975, in Techniques and Topics in Bioinorganic Chemistry Ed. C. McAuliffe, Macmillan, London p.270.
4. Jacobs, A. and Worwood, M., Editors: 1980, Iron in Biochemistry and Medicine, II. Academic Press, London.
5. Jacobs, A. and Worwood, M.: 1981, Disorders Mineral Metabolism 1,1.
6. Bezkorovainy, A.: 1980, Biochemistry of Nonheme Iron. Plenum Press, New York.
7. Neilands, J.B.: 1980 in ref.3, p.529.
8. Jacobs, A.: 1982, Recent Adv. Haematology, 3,1.
9. Spiro, T.G. and Saltman, P.: 1969, Struct. Bond. 6, 116.
10. May, P.M. and Williams, D.R.: 1980, in ref.3, p.1.
11. Pearson, R.G.: 1972, Hard and Soft Acids and Bases. Dowden, Hutchinson and Ross, Pennsylvania.
12. Anderson, B.F., Buckingham, D.A., Robertson, G.B., Webb, J., Murray, K.S. and Clark, P.E.: 1976, Nature 262, 722.
13. Bernal, J.D., Dasgupta, D.R. and Mackay, A.L.: 1957, Nature 180, 645.
14. Francombe, M.H. and Rooksby, H.P.: 1959, Clay Min. Bull. 4,1.
15. Bernal, J.D., Dasgupta, D.R. and Mackay, A.L.: 1959, Clay Min. Bull. 4, 15.
16. Dasgupta, D.R.: 1961, Indian J. Phys. 35, 401.
17. Carlson, L. and Schwertmann, U.: 1981, Geochim. Cosmochim. Acta 45, 421.
18. Towe, K.M. and Bradley, W.F.: 1967, J. Colloid Interfac.Sci. 24, 384.
19. Chukhrov, F.V., Zvyagin, B.B., Ermilova, L.P. and Gorshkov, A.I.: 1972, Proc. Int. Clay Conf. (Madrid), p.333.
20. Russell,J.D.: 1979, Clay Min. 14, 109.
21. Schwertmann, U. and Taylor, R.M.: 1972, Clays Clay Min. 20, 159.
22. Schwertmann, U. and Fischer, W.R.: 1973, Geoderma 10, 237.
23. Schwertmann, U. and Thalman, H.: 1976, Clay Min. 11, 189.
24. Harrison, P.M., Clegg, G.A. and May, K.: 1980, in ref. 3, p.29.
25. Massover, W.H. and Cowley, J.M.: 1973, Proc. Nat. Acad. Sci. U.S.A. 70, 3847.
26. Webb, J. and Gray, H.B.: 1974, Biochim. Biophys. Acta 351, 224.
27. Heald, S.M., Stern, E.A., Bunker, B., Holt, E.M. and Holt, S.L.: 1979, J. Amer. Chem. Soc. 101, 67.
28. Wrigglesworth, J.M. and Baum, H.: 1980, in ref. 3, p.29.
29. Loehr, J.S. and Loehr, T.M.: 1979, Adv. Inorg. Biochem. 1, 235.
30. Clark, P.E. and Webb, J.: 1981, Biochemistry 20, 4628.
31. Morgan, E.H.: 1980, in ref.3, p.641.
32. George, S.G. and Coombs, T.L.: 1977, J. Exp. Mar. Biol. Ecol. 28, 133.
33. Aisen, P.: 1980, in ref.3, p.87.
34. Aisen, P. and Listowsky, I.: 1980, Ann. Rev. Biochem. 49, 357.

35. Webb, J. and Chrystal, P.; 1981, Mar. Biol. 63, 107.
36. Macey, D.J., Webb, J. and Potter, I.C.: 1982, Comp. Biochem. Physiol., in press.
37. Webb, J. and Saltman, P., unpublished results.
38. Wixom, R.L., Prutkin, L. and Munro, H.N.: 1980, Int. Rev. exp. Pathol. 22, 193.
39. Jacobs, A.: 1980, in ref.3, p.428.
40. Macey, D.J., Webb, J. and Potter, I.C.: 1982, Acta Zool., in press.
41. Morton, D. and Wimsatt, W.A.: 1980, Anat. Rec. 198, 183.
42. Lowenstam, H.A.: 1962, Geol. Soc. Am. Bull. 73, 435.
43. Gould, J.L., Kirschvink, J.L. and Deffeyes, K.S.: 1978, Science, 201, 1026.
44. Walcott, C., Gould, J.L. and Kirschvink, J.L.: 1979, Science 205, 1027.
45. Zoeger, J., Dunn, J.R. and Fuller, M.: 1981, Science 213, 892.
46. Blakemore, R.: 1975, Science 190, 377.
47. Frankel, R.B., Blakemore, R.P. and Wolfe, R.S. 1979, Science 203, 1355.
48. Balkwill, D.L., Maratea, D. and Blakemore, R.P.: 1980, J. Bacteriology 141, 1399.
49. Blakemore, R.P. and Frankel, R.B.: 1981, Sc. Amer. 245, 42.
50. Towe, K.M. and Moench, T.T.: 1981, Earth Planet, Sci. Lett. 52, 213.
51. Regazzoni, A.E., Urrutia, G.A., Blesa, M.A. and Maroto, A.J.G.: 1981, J. inorg. nucl. Chem. 43, 1489.
52. Sugimoto, T. and Matijevic, E.: 1980, J. Coll. Interface Sci. 74, 227.
53. Bauminger, E.R., Cohen, S.G., Dickson, D.P.E., Levy, A., Ofer,S. and Yariv, J.: 1980, Biochim.Biophys. Acta. 623, 237.
54. Blakemore, R.P., Maratea, D. and Wolfe, R.S.: 1979, J. Bacteriol. 140, 720.
55. Towe, K.M., Lowenstam, H.A. and Nesson, M.M.: 1963, Science, 142, 63.
56. Lowenstam, H.A.: 1967, Science 156, 1373.
57. Towe, K.M. and Lowenstam, H.A.: 1967, J. Ultrastruct. Res. 17,1.
58. Kirschvink, J.L. and Lowenstam, H.A.: 1979, Earth Planet. Sci. Lett. 44, 193.
59. Nesson, M.H.: 1968, Ph.D. Thesis, California Institute of Technology, Pasadena, California.
60. Webb, J. and Macey, D.J.: 1983, this volume, p. 423.

PLASMA FERRITIN IN *POLYPLACOPHORA* AND ITS POSSIBLE ROLE IN THE BIOMINERALIZATION OF IRON.

J. Webb, D.J. Macey,[†]
School of Mathematical and Physical Sciences, and [†]School of Environmental and Life Sciences, Murdoch University, Murdoch, Western Australia, Australia, 6150.

ABSTRACT

A complex sequence of mineralization steps leads to the formation of magnetite (Fe_3O_4) crystals in the tooth denticles of chitons (*Polyplacophora*), a group of marine molluscs. The iron-containing protein ferritin has been isolated and partially characterized from the blood plasma of three species: two tropical, *Acanthozostera gemmata* and *Acanthopleura spinosa*; and one sub-tropical, *Clavarizona hirtosa*. In all three cases high levels of plasma iron (range 3600-18,000 μg/100 ml) and plasma ferritin (comprising 75% of total iron) were observed. We suggest that ferritin serves to deliver iron to rapidly mineralizing radula tissue.

INTRODUCTION

The radula teeth of the *Polyplacophora*, or chitons, are composed of a complex mixture of mineralized iron oxides including in particular, magnetite (Fe_3O_4) (1-5). This latter mineral has also been identified in magnetotactic bacteria (6-8), honey bees (9), homing pigeons (10) and at least one species of mammal, the dolphin (11), and may well comprise the fourth most extensively formed biogenic mineral (12).

In chitons, the dynamics and biochemical control of iron mineralization are poorly understood, as is its relationship to the overall metabolism and physiology of iron in these animals (13). A series of studies by Lowenstam and colleagues has implicated ferritin, the well-known iron-binding protein normally involved with iron storage, in the initial mineralization stages in the chiton radula (14,15). Following earlier observations of Nesson (16) we have characterized the plasma of three species of chitons with respect to iron levels and the major iron-binding protein. In all cases very high levels of both iron and ferritin were found. The species studied were collected from the inter-tidal region of the Indian Ocean coastline

P. Westbroek and E. W. de Jong (eds.), Biomineralization and Biological Metal Accumulation, 423–427.

of Western Australia: *Acanthopleura spinosa* and *Acanthozostera gemmata* from Pt. Hedland (20°S, 119°E) and *Clavarizona* (*Liolophura*) *hirtosa* at Fremantle (32°S 116°E).

RESULTS

In all three species the blood, which was obtained from shallow incisions along the foot, had a negligible cellular content and was similar in appearance, ranging in colour from brown to blue-grey. This latter colour is due to the presence of large molecular weight aggregates ($> 10^6$) of the copper-containing respiratory protein hemocyanin while the brown coloration is indicative of the presence of the iron-binding protein ferritin. The yield of blood obtained varied with the species: specimens of *C. hirtosa* (on average: mass 8 g; length 4 cm) yielded 400 μl blood; *A. gemmata* (av.: 20 g; 8 cm) and *A. spinosa* (av.: 30 g; 10 cm) both yielded 600 μl blood.

Analysis of the blood and plasma (obtained by centrifugation at 12,000 g for 10 min.) showed no differences in the concentration of iron, as would be expected from the small cell population. The plasma iron content for the three species is shown in Table 1. The data indicate an exceptionally high level of iron in all cases, and suggest that 10,000 μg Fe/100 ml, i.e., 10 mg Fe/100 ml, or 0.01% Fe by weight can be taken as a normal level for chiton plasma iron.

Up to 15 distinct protein components were obtained from chiton plasma on polyacrylamide gel electrophoresis. Heat treatment of the plasma at 70°C for five minutes denatured many of the proteins which were then removed by centrifugation. For example, hemocyanin was precipitated by heat treatment, as indicated by the loss of the visible absorption band at 345 nm. Heat-treated plasma exhibited a broad featureless absorption across the visible spectrum, similar to that of horse spleen ferritin (17). In this regard it should be noted that mammalian ferritin is well-known to withstand such heat treatment, remaining in solution with only slight modification to its three-dimensional structure (18).

Confirmation of these changes was obtained by analysis for iron, using atomic absorption spectrophotometry. Three-quarters of the total plasma iron was found in the heat-treated plasma. Plasma which had been labelled with iron-59 radioisotope and then subjected to heat treatment still contained 70% of the label in the supernatant.

The composition of heat-treated plasma was further elucidated using polyacrylamide gel electrophoresis and gel filtration chromatography on a Sepharose 6B column (740 mm x 22 mm) which had been calibrated with thyroglobulin (molecular weight 670,000), horse spleen ferritin (443,000) and catalase (232,000). For *C. hirtosa*, only a single peak which absorbed in the U.V. spectrum was obtained on passage through the column. Plasma labelled with ^{59}Fe and treated similarly was found to give an identical radioisotope peak. The fact that the chiton plasma ferritin eluted before horse spleen ferritin indicated that it behaves as a globular protein somewhat larger than horse spleen ferritin. Analysis of these data as described elsewhere (19) indicates a molecular weight for *C. hirtosa* plasma ferritin of approx. 550,000.

Electrophoresis of heat-treated plasma in polyacrylamide gels showed unambiguously the presence of ferritin. The components were stained for protein using Coomassie Brilliant Blue G-250 and for iron with Perl's prussian blue (W. Massover, pers. comm.). In the case of *A. spinosa*, two distinct ferritin bands were observed, while for *C. hirtosa* and *A. gemmata* only single bands were detected.

Further electrophoresis in the presence of carrier Ampholines to create a pH gradient was used to determine the isoelectric point (pI) of plasma ferritin. Using modified cytochrome *c* preparations to calibrate the gel over the pH range of 4-11, only one band was detected. For *C. hirtosa*, pI is 4.7 ± .1 and for *A. gemmata*, pI is 4.5 ± 1.

While the mature teeth in all three species contain magnetite, the colour of the teeth towards the immature end of the radula (white as opposed to red and black) indicate that they are not mineralized. This transition from the younger non-mineralized to the older mineralized teeth is remarkably sharp in all three species, occurring at times in the space of a single row of teeth (Table 2).

DISCUSSION

These results indicate that chiton plasma possesses high levels of both iron and ferritin. For comparison, plasma iron concentrations in a wide range of vertebrates fall within the range 50-250 μg/100 ml (20), much less than are reported here for chitons. The only animal with comparable plasma levels is the larval form of the Southern Hemisphere lamprey *Geotria australis* Gray which has a mean iron level of 19,760 μg/100 ml. However, while some of this iron has been shown to be bound to ferritin, it appears that other iron-binding components are present in the lamprey plasma (17). These have not yet been characterized. In contrast, the heat denaturation experiments show that most of the chiton plasma iron is bound to ferritin. While the chiton ferritin has a slightly different molecular weight, the pI is very similar to that of ferritin isolated from mammalian systems. Mammalian ferritin has the capacity to bind up to 4500 atoms of iron per molecule of protein (18) very many more than the putative iron-transport protein transferrin (21), and thus could serve far better as a source of high levels of iron for mineralizing tissue. The replacement of teeth means that mineralization of iron is a continuous process for the chiton, requiring a continuing supply of iron from the plasma. We suggest that this supply is maintained by the exceptionally high levels of plasma ferritin. In this regard, reports of ferritin binding to reticulocytes and erythroblast cells are of particular interest (22,23).

This unusually high level of iron in the plasma resembles the pathological condition of iron overload in vertebrates, including man (24), and as such both lampreys and chitons can then serve as intriguing comparative animal models for studies of iron overload. In addition, the mechanism and control of the release of iron from ferritin and its incorporation into various iron mineral phases, including magnetite remains to be determined. The chiton radula offers a sequence - in space and in time - of the process of iron mineralization. As such it is a promising system for both the study of iron deposition and of magnetite formation, a process of considerable importance in a wide range of living organisms.

TABLE 1

The mean, range and sample size (N) for plasma iron levels in 3 species of chitons.

	Metal	Mean µg/100 ml	Range	N
Acanthozostera gemmata	Fe	8,800	6400-13,300	12
Acanthopleura spinosa	Fe	10,200	5800-18,000	11
Clavarizona hirtosa	Fe	11,600	3600-16,200	15

TABLE 2

Radula characteristics in three species of chitons.

		No. of rows of teeth			
	Length (mm)	white	transition	black	Total
Acanthozostera gemmata	25-30	12	2	70	84
Acanthopleura spinosa	27-35	14	4	100	118
Clavarizona hirtosa	25-30	10	1-2	60	71-2

REFERENCES

1. Tomlinson, J.T. (1959) Veliger 2, 36.
2. Lowenstam, H.A. (1962) Geol. Soc. Am. Bull. 73, 435.
3. Lowenstam, H.A. (1967) Science, 156, 1373.
4. Lowenstam, H.A. (1972) Chem. Geol. 9, 153.
5. Kirschvink, J.L. and Lowenstam, H.A. (1979) Earth Planet. Sci. Lett. 44, 193.
6. Blakemore, R. (1975) Science 190, 377.
7. Frankel, R.B., Blakemore, R.P. and Wolfe, R.S. (1979) Science 203, 1355.
8. Towe, K.M. and Moench, T.T. (1981) Earth Planet. Sci. Lett. 52, 213.
9. Gould, J.L., Kirschvink, J.L. and Deffeyes, K.S. (1978) Science 201, 1026.
10. Walcott, C., Gould, J.L. and Kirschvink, J.L. (1979) Science 205, 1027.
11. Zoeger, J., Dunn, J.R. and Fuller, M. (1981) Science 213, 892.
12. Lowenstam, H.A. (1981) Science 211, 1126.
13. Boyle, P.R. (1977) Oceanogr. Mar. Biol. Ann. Rev. 15, 461.
14. Towe, K.M., Lowenstam, H.A. and Nesson, M.H. (1963) Science 142, 63.
15. Towe, K.M. and Lowenstam, H.A. (1967) J. Ultrastruct. Res. 17, 1.
16. Nesson, M.H. (1968) Ph.D. Thesis, California Institute of Technology, Pasadena, California.
17. Webb, J. and Gray, H.B. (1974) Biochim. Biophys. Acta. 351, 224.
18. Harrison, P.M., Clegg, G.A. and May, K. (1980) in Iron in Biochemistry and Medicine, II (Eds: Jacobs, A. and Worwood, M.) Academic Press, London p.131.
19. Macey, D.J., Webb, J. and Potter, I.C. (1982) Comp. Biochem. Physiol., in press.
20. Morgan, E.H. (1980) in Iron in Biochemistry and Medicine, II (Eds: Jacobs, A. and Worwood, M.) Academic Press, London, p.661.
21. Aisen, P. and Listowsky, I. (1980) Ann. Rev. Biochem. 49, 357.
22. Tanaka, Y. (1970) Blood 35, 793.
23. Pollack, S. and Campana, T. (1981) Biochem. Biophys. Res. Commun. 100, 1667.
24. Jacobs, A. (1982) Recent Adv. Haematology 3, 1.

MAMMALIAN METALLOTHIONEIN: EVIDENCE FOR METAL THIOLATE CLUSTERS

Milan Vašák and Jeremias H.R. Kägi
Biochemisches Institut der Universität Zürich
Zürichbergstrasse 4
CH-8032 Zürich, Switzerland

Metallothioneins are metal-binding proteins playing a major role in the metabolism and the detoxification of the essential trace metals Zn and Cu and of nonessential d^{10} metals such as Cd, Hg, Ag, Au and Bi. These functions are related to the capacity of these proteins to bind 7 bivalent metal ions per polypeptide chain of 61 amino acid residues. This capacity derives from the presence of 20 cysteine residues, an amino acid known to be a good ligand for polarizable metal ions. The spatial structure of metallothionein is still unknown. However, spectroscopic studies have now established that each metal ion is bound to 4 thiolate ligands, that the symmetry of each complex is tetrahedral, and that the complexes are grouped to form two oligonuclear metal thiolate clusters with 3 and 4 metal ions, respectively. The clusters are proposed to have an adamantane-related structure.

Metallothioneins are widely occurring nonenzymic metal-containing proteins which have recently become the object of intensive interest in many branches of the life sciences (1). Initially identified as tissue components responsible for the intracellular binding of cadmium (2,3), they are now thought to play a major role in the metabolism and the detoxification of essential and nonessential trace metals. In mammals, their principal physiological purpose appears to lie in the regulation of the flow of zinc and copper through the cells and in supplying these metals for the synthesis of proteins and cellular structures requiring zinc and copper. However, the same proteins also bind a number of other metals such as cadmium, mercury, nickel, silver, gold, bismuth, and lead and may thus function in their recognition and sequestration (4). Since many transition and post-transition metals when present in sufficiently high concentration also accelerate the biosynthesis of the metallothioneins, it appears that these proteins are part of a fast-responding feedback mechanism controlling the intracellular concentration of chemically reactive free metal ions (5).

P. Westbroek and E. W. de Jong (eds.), Biomineralization and Biological Metal Accumulation, 429–437.

The covalent structure of the metallothioneins is highly adapted to form metal complexes. Their metal-binding properties are linked to the presence of 20 cysteine residues in the polypeptide chain (6). Each molecule binds 7 bivalent metal ions yielding a stoichiometric ratio of nearly 3 cysteine residues per metal (7, 8). All cysteine side chains participate in metal binding through the formation of metal thiolate complexes. In the present paper a short review is given on some recent studies which established the structure and coordination geometry of these metal thiolate complexes and which led to the discovery of metal thiolate clusters in metallothionein.

COMPOSITIONAL AND STRUCTURAL ASPECTS

The chemistry of the mammalian metallothioneins is in many respects unconventional (9, 10). All forms are single polypeptide chain proteins with 61 amino acid residues. With the metal included, their molecular weight ranges from 6500 to 7000 depending on the metal ion species present. Another special feature is the abundance of sulfur. Mammalian metallothioneins have the largest sulfur content of any known protein (11%) which is accounted for fully by the sulfur-containing amino acid residues (20 Cys and 1 Met). Conversely, they are totally lacking aromatic amino acids and histidine.

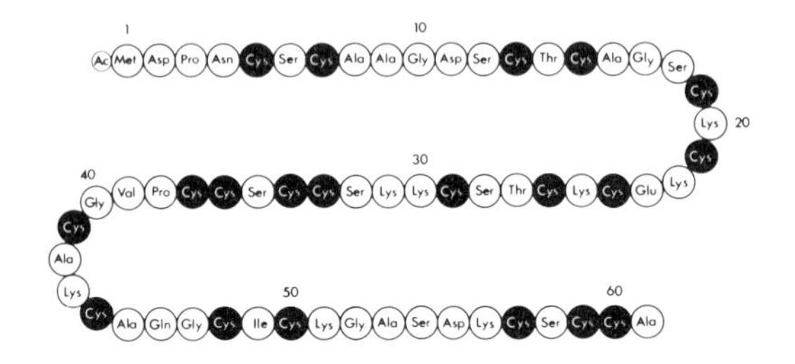

Figure 1. Amino acid sequence of human hepatic metallothionein-2. (Adapted from 11).

Complete amino acid sequences are presently known for seven mammalian forms (9). A representative example is shown in Fig. 1. In every species there are at least two major variants of the protein (isometallothioneins) which differ in a number of amino acid residues. They are thought to have arisen from two or more gene duplications during vertebrate evolution (12). Comparison of all known sequences reveals identical positions for 75% of the amino acid residues. Completely conserved are, among others, the positions of the 20 cysteine residues.

Cysteine occurs 14 times as part of -Cys-X-Cys- sequences (where X stands for an amino acid residue other than Cys), structures which have been suggested to serve as primary chelation sites in metal binding (6).

The metallothioneins range among the proteins with the highest metal content. However, they differ from other metalloproteins by their unusually broad metal ion specificity (13). Thus, preparations judged to be pure by a number of analytical criteria are often highly heterogeneous in metal composition. Beside the most abundant components zinc, cadmium and copper, they may contain small quantities of mercury and, more rarely, of nickel, silver, gold, bismuth, and lead, all in nonintegral proportions (9). In part, the metal composition and its variation are governed by the extent of exposure of the organisms to the particular metals. However, large compositional differences are also noted between preparations isolated from different tissues of the same organism. Thus, the metallothioneins from human and equine liver tend to contain more zinc and less cadmium than the corresponding preparations obtained from kidney cortex (Table 1). Analogous differences exist with respect to the relative zinc and copper contents of the hepatic and renal metallothioneins of rodents (14). Since the same isometallothioneins are formed in liver and kidney, it is improbable that these compositional differences are caused by structural differences in the proteins (13). They are more likely the result of differing local physiological circumstances controlling the availability of the metals within different compartments of the organism. It must be mainly for such reasons, too, that the metal composition varies with age as exemplified by the fact that copper is a major metallic component in human fetal liver while it is nearly absent in the protein isolated from adult liver (15,16).

Table 1. Typical Metal Composition of Human and Equine Metallothioneins Isolated from Liver and Kidney (3,7,16)

Metal	Human		Equine	
	liver	kidney[a]	liver	kidney
Zinc	6.40	3.05	5.82	3.02
Cadmium	0.14	3.52	0.34	3.09
Copper	0.13	0.34	0.18	0.14
Mercury	b	0.23	b	b
Total	6.67	7.14	6.34	6.25

Moles of metal per 6100 g of apometallothionein

[a]Recalculated on the basis of 20 Cys per molecule (3)
[b]Not determined

Spectrophotometric titration and x-ray photoelectron spectroscopy (ESCA) indicate that all cysteine residues participate in metal binding and that the stability of the thiolate complexes of the metal ions studied increases in the order of Zn(II) < Cd(II) < Cu(I), Ag(I), Hg(II) (7,17). Accordingly, the less firmly bound zinc is readily displaced by an excess of cadmium ions and both of them by copper, silver and mercury ions. Zinc and cadmium are also readily removed on exposure to low pH yielding the metal-free apometallothionein. By adding appropriate metal salts to the latter, homogeneous forms of metallothionein containing seven moles of Zn(II), Cd(II), Hg(II), Co(II) or Ni(II) can be prepared (18,19).

METAL COORDINATION

Metallothionein has not as yet been crystallized. Hence, information on the geometry of metal coordination is derived exclusively from spectroscopic studies. The absorption spectra of the metal-free apometallothionein (thionein) and of metallothioneins containing either Zn(II), Cd(II), Cu(I) or Hg(II) are shown in Fig. 2. Apometallothionein exhibits a plain absorption spectrum with a massive single absorption band at 190 nm which is mainly due to transitions of the primary and secondary amide chromophores of the polypeptide chain (20). Complex formation with group-2B metal ions and Cu(I) generates additional absorption in the far-UV region which terminates at the low energy side by a characteristic shoulder whose position is specific for the particular metal complex.

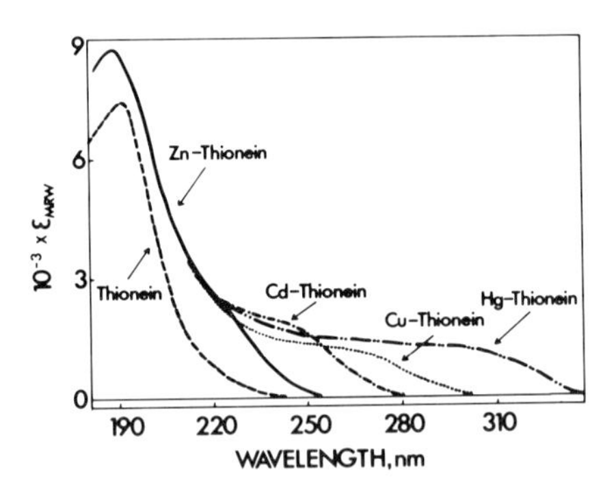

Figure 2. UV-absorption spectra of apometallothionein (thionein) and of metallothionein containing cadmium (Cd-thionein), zinc (Zn-thionein), copper (Cu-thionein), and mercury (Hg-thionein). The spectra were obtained at pH 2 for thionein and at pH 7 for the metallothioneins. ε_{MRW} is based on a mean residue weight of 99.

This metal-dependent shift is known to be diagnostic of an electron transfer origin of the bands underlying the shoulder (7). It is seen in tetrahedral halide and thiolate model complexes of the same metals and, hence, is a strong indication that the geometry of the metal thiolate complexes in metallothionein is also of tetrahedral symmetry (18).

The simplicity of the absorption spectra of the various metal derivatives suggests, moreover, that the various metal-binding sites of metallothionein are both chemically and structurally similar. The same conclusions come from studies on metallothionein reconstituted with paramagnetic Co(II), a derivative whose physical features are much more sensitive to the mode of metal coordination than those of metallothioneins containing d^{10} metal ions. The absorption spectrum of the intensely green-colored Co(II)-metallothionein shows, in addition to the Co(II)-thiolate electron transfer bands in the near-UV region, characteristic d-d bands in the visible and near-IR region with maxima at 600, 690 and 743 nm, and at 1150 and 1275 nm, respectively (Fig. 3). These features are virtually indistinguishable from those of tetrahedral Co(II)-thiolate complexes and of Co(II)-derivatives of crystallographically defined metalloproteins where the metal is known to be bound to four cysteine residues, e.g., in rubredoxin and in the structural metal sites of horse liver alcohol dehydrogenase (19). One can conclude, therefore, that in Co(II)-metallothionein each cobalt ion is bound to groups of four thiolate ligands, forming high-spin tetrahedral complexes. The latter inference is also documented by magnetic circular dichroism (MCD) and electron spin resonance (ESR) measurements (19).

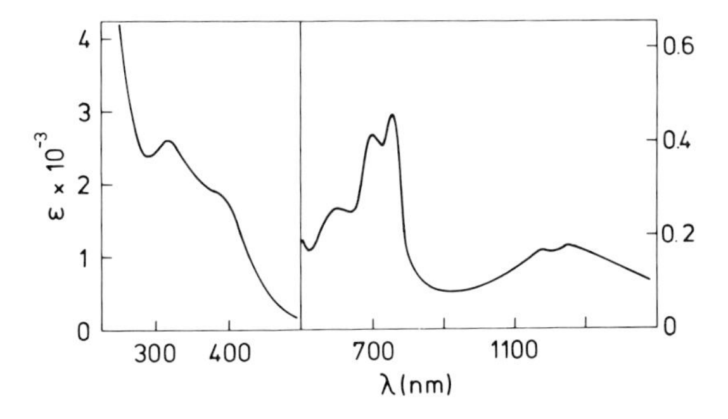

Figure 3. Absorption spectrum of rabbit liver $Co(II)_7$-metallothionein recorded in 0.05 M Tris-HCl, pH 7.0. The molar absorbancy refers to the metal (19).

METAL THIOLATE CLUSTERS

The binding of each of the seven metal ions in metallothionein to four thiolate ligands requires that 8 of the 20 cysteine residues are coordinated to two metal ions, thereby forming sulfur-bridged metal clusters. That such oligonuclear aggregates exist is now also supported by spectroscopic evidence. Thus, in an elegant study by Otvos and Armitage (21), it was shown that in the ^{113}Cd NMR spectrum of ^{113}Cd-enriched metallothionein the ^{113}Cd thiolate sites display well-resolved resonances which are split into multiplets as a consequence of spin-spin interactions of each ^{113}Cd nucleus with neighbouring ^{113}Cd nuclei via bridging thiolate ligands. By selective crossirradiation the multiplets can be decoupled, thereby allowing unambiguous identification of the coupled resonances. By such homonuclear decoupling studies the signals are shown to arise from two separate thiolate clusters containing 4 and 3 metal ions each (22).

Independent evidence for the building-up of cluster structures in metallothionein was obtained also by comparing the ESR properties of samples of Co(II)-metallothionein reconstituted from apometallothionein to contain the metal in different Co(II)-to-protein ratios (23). While the ESR spectra of the various stoichiometric forms all display the typical profile of tetrahedrally bound Co(II) (Fig. 4, left), they manifest a peculiar loss of signal intensity when the protein is getting saturated with the metal (Fig. 4, right).

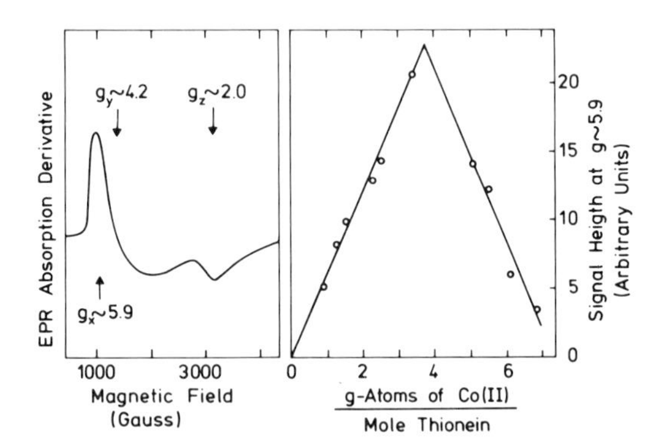

Figure 4. Left: Electron spin resonance (ESR) spectrum of rabbit liver Co(II)$_7$-metallothionein at 4 K. Right: Dependency of ESR signal size on Co(II)-to-protein ratio. As a representative measure of the magnitude of the high-spin Co(II) spectrum the amplitude at $g_x \sim 5.9$ is plotted (23).

Up to binding of about four equivalents of Co(II), the intensity of the signal increases in proportion to the metal present but it is drastically diminished on further addition. This loss of paramagnetism which was also confirmed by magnetic susceptibility measurements is attributed to antiferromagnetic coupling of neighbouring paramagnetic metal ions. The spin cancelling is thought to be mediated via the thiolate bridging ligands generated upon filling-up of the cluster.

Adamantane Cluster Model

The actual organisation of the inferred cluster structures in metallothionein is not as yet known. However, the fixed sulfur-to-metal ratio, the rules of stereochemistry and the spectroscopic information at hand limit the choice of models considerably. A very attractive possibility is the $Me(II)_4S_6$ adamantane cluster structure proposed by Dance (24) or variations of it (Fig. 5a). This regular cage-like polyhedral structure is built up entirely from metal thiolate units of tetrahedral symmetry to result in an arrangement of the metal and sulfur atoms that is known to occur in non-molecular form in zinc blende (ZnS). In the molecular complex under discussion, such a structure is characterized by a partitioning of the set of liganding cysteine residues into a group in which the sulfur is coordinated to two metal ions and is forming part of the cage (bridging thiolates), and into a group in which the sulfur is coordinated to only one metal ion (terminal thiolates). A somewhat related arrangement occurs in the cubane clusters of 4 Fe ferredoxins (Fig. 5b). In this case it is, however, inorganic sulfide rather than cysteine thiolate that contributes the ligands bridging adjacent metal ions. The cysteine residues function solely as terminal ligands anchoring the iron-sulfur core to the polypeptide chain.

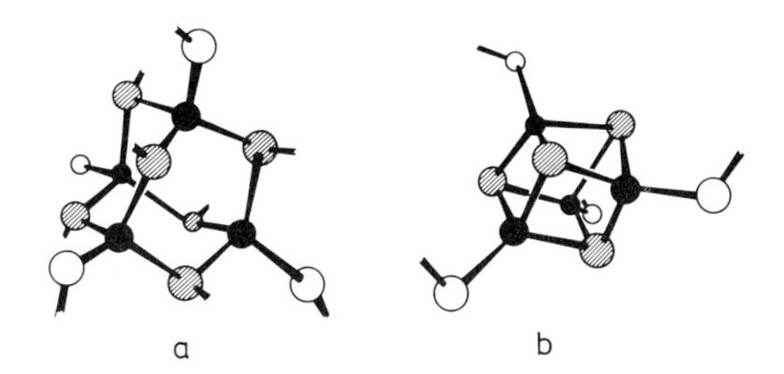

Figure 5. Adamantane-type (a) metal-thiolate cluster proposed for metallothionein and cubane-type (b) cluster of clostridial 4 Fe ferredoxins. The filled circles represent the metal, the empty circles the terminal sulfur ligands and the hatched circles the bridging sulfur ligands.

The results of the ^{113}Cd NMR homonuclear decoupling measurements and of the ESR and magnetic studies cited above are compatible with the existence of two incomplete adamantane-type cluster structures in metallothionein, one of them containing 4 bivalent metal ions and 11 cysteine residues, the other 3 bivalent metal ions and 9 cysteine residues (22). Most recently, this model of two separate clusters has also received independent support from the observation that the 4-metal cluster can be obtained in pure form following proteolytic cleavage of metallothionein in the middle of the polypeptide chain (25). The 4-metal cluster was found to be associated with the COOH-terminal fragment of the molecule, thus suggesting that metallothionein has a segmental structure with separate domains for each cluster. The spontaneous formation of structurally well defined oligonuclear thiolate complexes with d^{10} metal ions in metallothionein is clearly dependent on the very special distribution of its cysteine residues that is typical for this protein. The stringency of the stereochemical requirements of bioinorganic structures such as the proposed adamantane cluster may explain, in fact, the remarkable preservation of the positions of the cysteine residues in all metallothioneins throughout vertebrate evolution. In what way these unique metal thiolate clusters condition the still somewhat elusive function of the metallothioneins remains to be established. Clearly, further studies on the organization of these clusters and on the dynamics of their formation and degradation are necessary for a proper assessment of the role played by these proteins in cellular processes.

ACKNOWLEDGMENT

This work was supported by Swiss National Science Foundation Grant No. 3.495-0.79.

REFERENCES

1. "Metallothionein" (Kägi, J.H.R. and Nordberg, M., eds.), 1979, Birkhäuser Verlag, Basel.
2. Margoshes, M. and Vallee, B.L.: 1957, J. Am. Chem. Soc. 79, pp. 4813-4814.
3. Pulido, P., Kägi, J.H.R., and Vallee, B.L.: 1966, Biochemistry 5, pp. 1768-1777.
4. Cherian, M.G. and Goyer, R.A.: 1978, Life Sci. 23, pp. 1-10.
5. Durnam, D.M. and Palmiter, R.D.: 1981, J. Biol. Chem. 256, pp. 5712-5716.
6. Kojima, Y., Berger, C., Vallee, B.L., and Kägi, J.H.R.: 1976, Proc. Natl. Acad. Sci. USA 73, pp. 3413-3417.
7. Kägi, J.H.R. and Vallee, B.L.: 1961, J. Biol. Chem. 236, pp. 2435-2442.

8. Kägi, J.H.R., Himmelhoch, S.R., Whanger, P.D., Bethune, J.L., and Vallee, B.L.: 1974, J. Biol. Chem. 249, pp. 3537-3542.
9. Nordberg, M. and Kojima, Y., in "Metallothionein" (Kägi, J.H.R. and Nordberg, M., eds.), 1979, Birkhäuser Verlag, Basel, pp. 41-116.
10. Webb, M., in "The Chemistry, Biochemistry and Biology of Cadmium" (Webb, M., ed.), 1979, Elsevier/North-Holland Biomedical Press, Amsterdam, pp. 195-266.
11. Kissling, M.M. and Kägi, J.H.R.: 1977, FEBS Lett. 82, pp. 247-250.
12. Kissling, M.M.: 1979, Ph.D. Thesis, University of Zürich.
13. Kojima, Y., Berger, C., and Kägi, J.H.R., in "Metallothionein" (Kägi, J.H.R. and Nordberg, M., eds.), 1979, Birkhäuser Verlag, Basel, pp. 153-161.
14. Suzuki, K.T.: 1979, Arch. Environm. Contam. Toxicol. 8, pp. 255-268.
15. Riordan, J.R. and Richards, V.: 1980, J. Biol. Chem. 255, pp. 5380-5383.
16. Bühler, R.H.O. and Kägi, J.H.R.: 1974, FEBS Lett. 39, pp. 229-234.
17. Weser, U., Rupp, H., Donay, F., Linnemann, F., Voelter, W., Voetsch, W., and Jung, G.: 1973, Eur. J. Biochem. 39, pp. 127-140.
18. Vašák, M., Kägi, J.H.R., and Hill, H.A.O.: 1981, Biochemistry 20, pp. 2852-2856.
19. Vašák, M., Kägi, J.H.R., Holmquist, B., and Vallee, B.L.: 1981, Biochemistry 20, pp. 6659-6664.
20. Bühler, R.H.O. and Kägi, J.H.R., in "Metallothionein" (Kägi, J.H.R. and Nordberg, M., eds.), 1979, Birkhäuser Verlag, Basel, pp. 211-220.
21. Otvos, J.D. and Armitage, I.M.: 1979, J. Am. Chem. Soc. 101, pp. 7734-7736.
22. Otvos, J.D. and Armitage, I.M.: 1980, Proc. Natl. Acad. Sci. USA 77, pp. 7094-7098.
23. Vašák, M. and Kägi, J.H.R.: 1981, Proc. Natl. Acad. Sci. USA 78, pp. 6709-6713.
24. Dance, I.G.: 1980, J. Am. Chem. Soc. 102, pp. 3445-3451.
25. Winge, D.R. and Miklossy, K.-A.: 1982, J. Biol. Chem. 257, pp. 3471-3476.

BACTERIAL INTERACTIONS WITH MINERAL CATIONS AND ANIONS: GOOD IONS AND BAD

Simon Silver
Biology Department, Washington University,
St. Louis, MO 63130 USA

INTRODUCTION

Bacterial cells interact with the cations and anions formed by the elements of the Periodic Table. For essentially each and every cation and anion found, there are specific bacterial systems--and this applies, in general, for trace elements and toxic mineral ions as well as for the more abundant types. Basically the Periodic Table forms ions that fall into three classes:

A. Essential nutrient ions are required by all living cells. These include Mg^{2+}, K^+, PO_4^{3-}, and SO_4^{2-} (although organic phosphorus and sulfur may replace inorganic nutrients), which can be considered "macro-nutrient" minerals. "Trace" or micronutrient minerals include Mn^{2+}, Fe^{2+}, Zn^{2+} plus ions of perhaps 20 other elements. For each and every essential nutrient ion, one or more highly specific membrane transport system exists, to assure the cell of an adequate supply in the face or variable or limited amounts in the external environment.

B. Unessential, but common or abundant cations and anions are frequently used for cellular functions. Examples are Ca^{2+}, Na^+ and Cl^-, for which there is no measurable requirement for bacteria such as *Escherichia coli*. Other bacteria have requirements for examples of this second class; and they play essential roles in limited situations. For example, Na^+ gradients are generated by primary Na^+ pumps in bacterial membranes or by Na^+/H^+ exchange mechanisms. These Na^+ gradients (outside high; inside low) can be "coupled" to drive amino acid transport systems which function as Na^+ amino acid couples (1,2). Calcium is not needed for growth of many bacteria, including *E. coli* and *Bacillus subtilis*. Ca^{2+} is, however, accumulated in massive amounts during bacterial sporulation. Ca^{2+} is deposited within the developing spore and plays a role in the transition from the heat-resistant metabolically-inactive spore into a metabolically active, heat sensitive vegetative cell. As with essential nutrients, intracellular levels of common unessential ions are regulated by specific transport systems in the cellular membrane. The transport systems for essential and for unessential but abundant ions are determined by genes on the bacterial chromosome.

P. Westbroek and E. W. de Jong (eds.), Biomineralization and Biological Metal Accumulation, 439–457.

More detailed reviews of bacterial cation and anion transport systems (3) and briefer summaries (4,5) have appeared. Interested readers are referred to these.

C. Toxic ions of no known use constitute the third class of minerals for which bacterial cells have highly specific systems. I will not dwell here on the mechanisms of toxicity (which have been widely studied and which generally share mechanisms with cells of higher organisms; ref. 6). Rather, we will be concerned with mechanisms of resistance. For many cations and anions that are highly toxic, bacterial cells have elaborated specific resistance mechanisms. Such ions include Hg^{2+} (and organomercurials), AsO_4^{3-}, AsO_2^- (and organoarsenicals), Cd^{2+}, Ag^+, Pb^{2+}, SbO^+, BiO^+, and others. For each toxic ion for which resistance occurs, a highly specific mechanism is determined by genes on small extrachromosomal bacterial plasmids. This non-chromosomal basis allows movement of the resist-ance determinant from resistant to sensitive microorganisms and then the spread throughout a population--at time of exposure. A detailed review of such plasmid systems was published by Summers and Silver (7) and briefer more recent summaries have appeared (8-10).

NUTRIENT CATIONS AND ANIONS

Figure 1 summarizes many of the essential nutrient systems found in bacterial cells. Nitrogen is certainly essential for all cells. In addition to utilizing organic nitrogen (such as amino acids or urea), bacterial cells utilize inorganic nitrogen. There are N_2 fixing bacteria, and indeed all atmospheric nitrogen that is incorporated into higher plants is obtained initially by means of bacterial reduction from N_2 to NH_4^+. Extracellular NH_4^+ can be

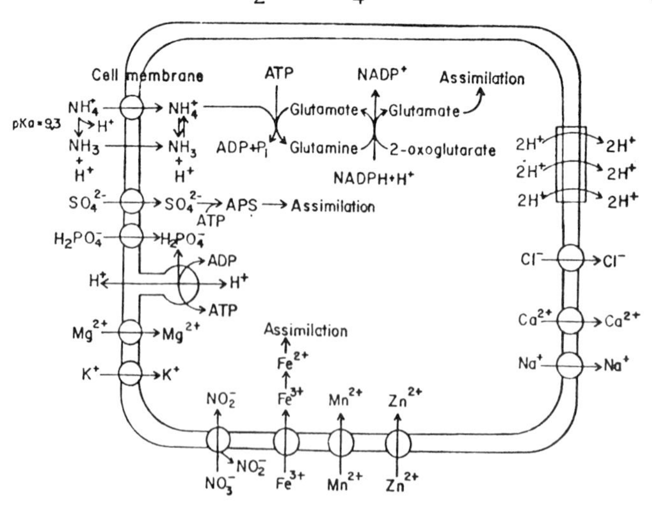

Figure 1. Some known membrane transport systems for cations and anions (from ref. 4).

assimilated itself. There remains some question as to whether bacterial cells acquire ammonia as NH_3, which is lipid soluble and can passively diffuse across biological membranes, or whether most bacterial cells have highly specific NH_4^+ transport systems. The pK_a for the NH_3 to NH_4^+ transition of 9.3 favors lipid insoluble NH_4^+. The frequent environmental limitation for nitrogen for growth certainly favors the selection of highly specific NH_4^+ transport systems over slower passive diffusion. Evidence of the nature of these systems with the radioactive analog $^{14}CH_3NH_3^+$ was obtained first with eucaryotic and later with procaryotic microbes. More recently, direct studies with the short half-life (9.96 min) radioactive $^{13}NH_4^+$ also indicate the existence of ammonium transport systems. Such evidence was summarized by Silver and Perry (4).

Sulfate transport systems have been found in a variety of bacteria including *E. coli* and *Salmonella typhimurium*. The *S. typhimurium* system is carefully regulated. Unfortunately sulfate transport systems have not been thoroughly studied in recent years.

E. coli has two well-studied phosphate transport systems (called Pit, for Pi, inorganic phosphate, transport; and Pst, for phosphate specific transport). The kinetic parameters of these systems are listed in Figure 2. The Pit system has about an equal K_m for PO_4^{3-} of 25 μM and a K_i for AsO_4^{3-} as a competitive inhibitor (and as a substrate for transport). The Pit system is synthesized at all times (i.e. constitutively) and has a V_{max} of about 60 μmol $PO_4^{3-}$$min^{-1}$ g^{-1} cells. Only a single chromosomal gene is known to be required for Pit system. Mutants of that gene (lacking the Pit system) are relatively resistant to arsenate. The Pit system is "chemiosomotic" (1), in that it is dependent upon the cell membrane potential (Δψ) and pH gradient (internal alkaline) that are generated by the respiratory chain and by

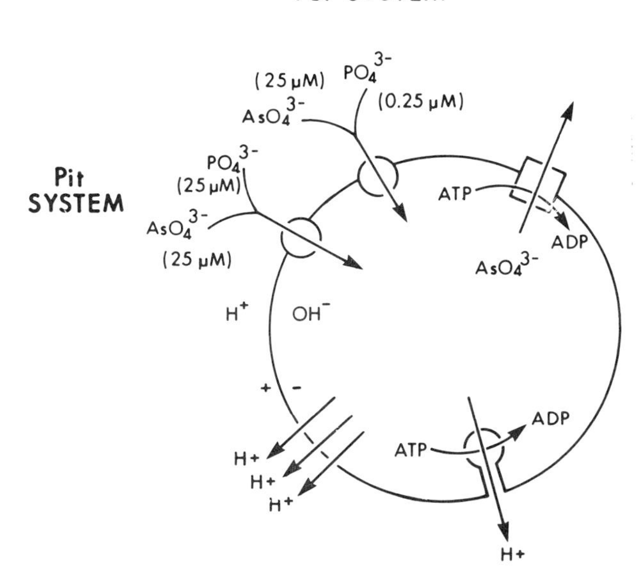

Figure 2. The phosphate (arsenate) transport systems and the arsenate efflux system of *E. coli*. 0.25 μM or 25 μM, where indicated, is the K_m (for PO_4^{3-}) or the K_i (for AsO_4^{3-} as a competitive inhibitor of PO_4^{3-} transport) for the Pit and Pst phosphate transport systems.

the reversible ATP synthetase ATPase (both shown as H^+ transporters in Figure 2).

Whereas the Pit system with its relatively low affinity and relatively relaxed substrate specificity can be considered a transport system for "good times", the other phosphate transport system, Pst, has the properties characteristic of a starvation or "hard times" system. The Pst system is more highly specific, with a K_m for PO_4^{3-} 100-times lower than that for the Pit system. The Pst system is synthesized in a carefully regulated manner and increases in amount during either phosphate starvation or upon the addition of toxic levels of arsenate. The Pst system is not chemiosmotic but has the properties of a membrane ATPase system, including the existence of a phosphate-binding protein in the space between the cell membrane and the outer membrane cell-wall. In addition to the _phoS_ gene, which determines the PO_4^{3-} binding protein, at least three additional genes, _pstA_, _pstB_ and _phoT_ have been tentatively identified (11). Some of these genes are expected to code for transport proteins of the ATPase in the bacterial inner membrane. A clearer understanding of the functioning of the Pst system appears imminent. Figure 2 also diagrams the arsenate efflux pump that will be described below under the heading of arsenate resistance. The eucaryotic yeast _Saccharomyces cerevisiae_ also appears to have two phosphate transport systems, one of which functions as a Na^+ -dependent co-transport system (12). With _Saccharomyces_ also, PO_4^{3-} transport function increases with phosphate starvation, so we may hope that one can generalize from _E. coli_ conclusions with only a reasonable level of caution.

The first direct evidence for NO_3^- transport systems in bacteria (13) appeared this year. I hypothesized the existence of such systems four years ago (3). Bacteria contain two types of nitrate reductase enzymes: (i) a low level assimilatory reductase synthesized during nitrate-dependent growth; and (ii) a dissimilatory nitrate reductase that functions as an electron sink in anaerobic respiration. In the first case, either the NO_3^- must be transported across the membrane prior to reduction or, less likely, the product (NO_2^-) must be transported subsequently to continue the enzymatic reduction to NH_4^+ and incorporation into amino acids. There appear to be two such systems (13). Since NO_2^- is toxic, it is likely (14,15) that the dissimilatory nitrate reductase functions at the membrane surface (as diagrammed in Figure 1) so that the NO_2^- never enters the cell. Eucaryotic microbes also have NO_3^- and NO_2^- transport systems (16) whose function is regulated by the availability of NH_4^+ (17).

E. coli has two K^+ transport systems, one of which functions constitutively and appears to be chemiosmotic in energy coupling and the other of which is induced by K^+ starvation and consists of a membrane ATPase. There are a total of nine known genes that govern K^+ transport and accumulation. The high K_m (about 1 mM K^+) TrkA system is only moderately specific and accepts Rb^+ and Tl^+ in addition to K^+ (18). Mutations in three genes, _trkA_, _trkE_, and trkG affect TrkA activity. trkB and trkC mutants have normal uptake of K^+ but are unable to retain it; they appear as "slippy K^+ clutch" mutants. A fifth gene trkD has mutants that reduce residual activity of trkA mutants. The second K^+ transport system (Kdp, for K^+ dependent growth) is the only bacterial transport ATPase that has been well studied. The Kdp system is made only under conditions of K^+ starvation and functions with a high affinity (K_m about 2 μM or 1000-times higher affinity than the TrkA system). The Kdp system consists of four genes, three of which kdpA, kdpB and kdpC determine proteins in the inner bacterial membrane (19,20) and the fourth gene kdpD, is a regulatory gene which responds to changes in osmotic turgor pressure (21). The kdpA protein appears to be the outer surface recognition site for K^+; and the kdpB protein has ATPase activity and appears to be phosphorylated. (The Kdp system can accumulate K^+ against a concentration gradient of up to 10^6 to 1 (0.2 M K^+ inside, less than 0.1 μM K^+ outside) and such an extreme gradient could not be obtained by a chemiosmotic mechanism, which would be limited to gradients of 10^3 for monovalent cations.

There appear to be two transport systems for Mg^{2+} in E. coli as well, one of which is less specific and synthesized constitutively and the other of which is more substrate specific (discriminates better against Mn^{2+}, Co^{2+} and Ni^{2+}) and is synthesized inducibly (22). Unfortunately, this system has not been studied in recent years.

There are more numerous iron transport systems in E. coli than there are for any other cation or anion (3,23,29). The total of five systems are named for the small extracellular carrier molecules (called siderophores) that are involved. The best known system functions with a catecholic molecule, 2,3 dihydroxybenzoylserine cyclic trimer (variously called enterochelin or enterobactin). There are 13 known genes in the cycle of synthesis, uptake and utilization of Fe^{3+}-enterochelin (3). Clearly space considerations preclude an adequate description. The second system functions with hydroxomate siderophores synthesized by eucaryotic microbes and is referred to as the ferrichrome system after one such compound. The third system uses citrate as an iron carrier and is synthesized only during iron starvation in the presence of citrate. The iron enters the cells without citrate movement across the membrane. (E. coli lacks a citrate transport system and cannot grow on citrate.) The fourth system is called the "low affinity iron transport system" because it is represented by residual iron uptake activity in the absence of the first three systems (because of mutation and/or lack of siderophore). Finally, in recent years, a fifth system has been found (25-27) that utilizes a different hydroxymate siderophore called aerobactin (because it was first isolated from Aerobacter aerogenes). The genes for aerobactin synthesis and transport reside on small extrachromosomal ColV plasmids that also have genes for synthesis of the antibacterial protein colicin V and genes affecting pathogenicity.

Much less is known about the highly specific Mn^{2+} transport systems that occur in all bacterial cells thus far tested (3,28). These high affinity trace nutrient systems discriminate more than 10^5:1 in favor of Mn^{2+} and against the macronutrient Mg^{2+}. Mn^{2+} transport is highly regulated in _B. subtilis_: a 50-fold increase in rate occurs after Mn^{2+} starvation and the addition of exogenous Mn^{2+} causes an inactivation of Mn^{2+} transport (28). For another trace nutrient, Zn^{2+}, which is needed by all microbial cells, the amounts found as contaminants in most media suffice and there is no direct evidence for a specific transport system in bacteria (3,29). With eucaryotic microbes, there appears to be a greater requirement for Zn^{2+} and transport systems have been demonstrated. For other trace nutrients that are known to be essential for various enzymatic activities, in general no effort has been made to find specific transport systems (3).

UNESSENTIAL COMMON CATIONS AND ANIONS

Three examples in this class are diagrammed in Figure 1. For Na^+, Ca^{2+} and Cl^-, we think there are outwardly oriented transport systems that maintain the intracellular concentrations below the external level. The properties of the Ca^{2+} efflux system were earlier summarized in refs. 3 and 30 and most recently in ref. 31. All bacteria have Ca^{2+} efflux systems that function either chemiosmotically as $Ca^{2+}/2H^+$ exchange systems or as ATPases (32). A single bacterium may have more than one Ca^{2+} efflux system. A similar picture is emerging for Na^+ efflux systems (2,3,33-35).

Chloride, the major inorganic anion found in growth environments but not generally used for metabolism, is probably governed by a separate, outwardly oriented transport system, whose function is to regulate osmotic pressure. Although there is evidence for a Cl^- transport system in eucaryotic microbes (36,37), no such experiments have been done with bacteria. Perhaps the first break in this direction consists of studies of a direct light-dependent chloride pump in the salt-brine living photosynthetic bacterium, _Halobacterium halobium_ (J. K. Lanyi, personal communication).

TOXIC CATIONS AND ANIONS

Highly specific bacterial resistance systems exist for a wide range of toxic cations and anions. We have reviewed understanding of these systems (7-10). For this report, I will briefly describe the three best known examples: The mechanism of mercuric (and organomercury) resistance is the enzymatic detoxification of these compounds into volatile Hg°; the mechanisms of AsO_4^{3-} and Cd^{2+} resistances involve separate efflux transport systems that pump the toxic ions out from the cells as rapidly as the ions enter. However, the energy coupling mode for AsO_4^{3-} and for Cd^{2+} efflux appear to be different.

A. Mercury and mercurial resistances.

The mercury cycle in the environment is the best known case of microbial metabolism affecting the distribution and chemical form of a toxic heavy metal.

Microbial activity is associated with mercury transformations including methylation and demethylation of Hg^{2+} and oxidation and reduction of inorganic mercury. Since this discussion is limited to resistance mechanisms, we will deal only with the transformations from highly toxic methylmercury (as found in fish) to less toxic ionic Hg^{2+} (the predominant form in sea water) to least toxic Hg°. Both of these transformations are carried out by enzymes governed by bacterial plasmids and transposons--not by normally chromosomal genes.

The earliest studies of enzymatic detoxification of Hg^{2+} were with a multiply drug-resistant *E. coli* (38) and with a soil pseudomonad (39,40) for which a plasmid-determinant has still not been demonstrated. Plasmid-determined resistance to Hg^{2+} and to organomercurials occurs in both *Staphylococcus aureus* (41) and *E. coli* (42). The frequency of Hg^{2+} resistance determinants among clinical isolates can be over 50% (43,44) and among a collection of over 800 plasmids introduced into *E. coli* K-12; about 25% conferred Hg^{2+} resistance (45). More recently, mercuric and organomercurial-resistant strains with very similar properties have been found in a wide variety of bacterial species from soil, water and marine environments (46-50).

A small number of predictable resistance patterns for organomercurials was found among strains with plasmids (51): (a) In *E. coli* about 96% of the plasmids conferring mercuric resistance also confer resistance to the organomercurials merbromin and fluorescein mercuric acetate but to no other tested organomercurial. The other 4% of known plasmids conferred additional resistances to phenylmercuric acetate and thimerosal (Figure 3). (b) The plasmids in *Pseudomonas*

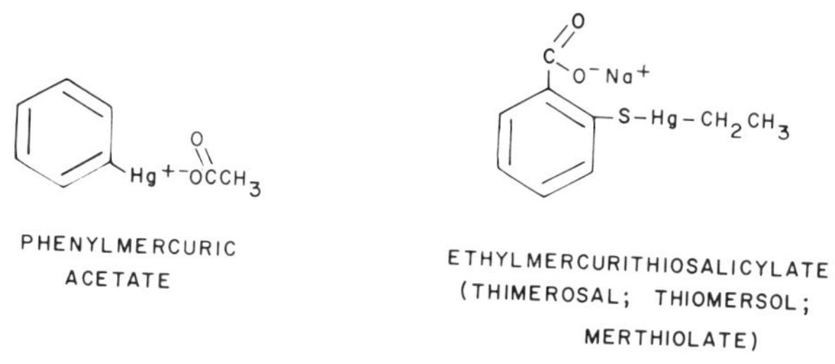

PHENYLMERCURIC ACETATE

ETHYLMERCURITHIOSALICYLATE (THIMEROSAL; THIOMERSOL; MERTHIOLATE)

Figure 3. Structures of phenylmercuric acetate and thimerosal: organomercurials that are degraded by plasmid-mediated enzymes.

aeruginosa also form two classes with regard to resistance to organomercurials (52); however, about 50% fell into each class. Furthermore, all *Pseudomonas* plasmids also confer resistance to *p*-hydroxymercuribenzoate and some *Pseudomonas* plasmids show additional resistance to methylmercuric and ethylmercuric compounds (51,52). (c) Only a single pattern was initially reported with S. auerus plasmids (51,53). This pattern is different from those with the Gram-negative bacteria because all the S. aureus plasmids confer resistances to phenylmercuric acetate, p-hydroxymercuribenzoate and fluorescein mercuric acetate, but not to thimerosal or to merbromin. In the last year, the first thimerosal-resistant S. aureus were found (54). These new clinical isolates can volatilize mercury more rapidly from thimerosal (Figure 4), than do the previous S. aureus strains. Of course, the plasmid-less sensitive strain can not volatilize mercury from thimerosal at all (Figure 4). The new strains also

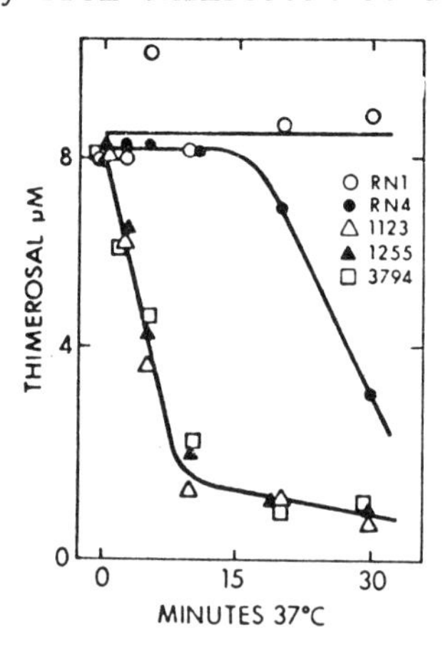

Figure 4. Detoxification of (loss of mercury from) thimerosal by a plasmid-less sensitive strain (RN1), by a previously studied (ref. 53) thimerosal-sensitive strain (RN4) and by new thimerosal-resistant isolates (1123, 1255, and 3794) (from ref. 54).

showed a greater activity for thimerosal as an "inducer" of synthesis of this mercuric detoxification system, which is only synthesized upon exposure to low levels of Hg^{2+} or organomercurials (53,54). The conclusion is that there are a very limited number of patterns of resistance, and that these can be understood in terms of the biochemistry of the enzymes involved.

1. Enzymatic mechanism of mercurial detoxification.

Hg^{2+} resistance in bacteria results from enzymatic detoxification of mercury compounds leading to the volatilization of mercury. This was discovered independently in Japan (38,40,55,56) and in St. Louis (57). The volatile mercury is metallic Hg° in each case, and the enzyme responsible is called mercuric reductase (Figure 5A).

Several organomercurials are also enzymatically detoxified to volatile compounds. These organomercurials include methylmercury, ethylmercury, phenylmercury, p-hydroxymercuribenzoate and thimerosal (Figure 5B); benzene is produced from phenylmercury, methane from methylmercury, and ethane from ethylmercury. The enzymes responsible for cleaving the Hg-C bond are organomercurial lyases. Tezuka and Tonomura (58,59) separated two small soluble lyase enzymes. Both have molecular weights of about 19,000 and require thiol reagents such as thioglycolate. One enzyme cleaved phenylmercuric acetate, p-hydroxymercuribenzoate and methylmercury, while

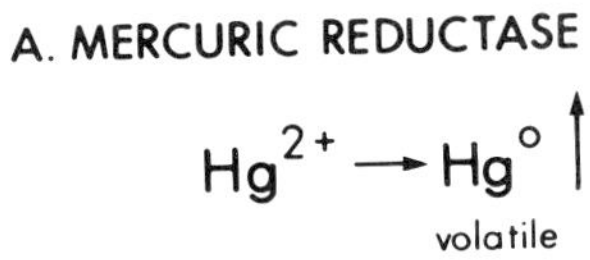

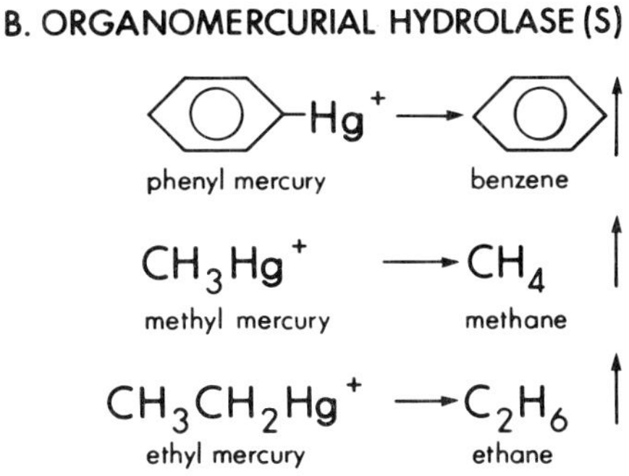

A. MERCURIC REDUCTASE

$Hg^{2+} \rightarrow Hg^{0}$ ↑

volatile

B. ORGANOMERCURIAL HYDROLASE (S)

phenyl mercury $\rightarrow$ benzene ↑

$CH_3Hg^{+} \rightarrow CH_4$ ↑

methyl mercury, methane

$CH_3CH_2Hg^{+} \rightarrow C_2H_6$ ↑

ethyl mercury, ethane

Figure 5. Enzymatic detoxifiication of (A) Hg(II) and (B) organomercurials.

the other enzyme cleaved only phenylmercuric acetate and p-hydroxymercuribenzoate (59). With a plasmid-containing E. coli, there was no evidence for cleavage of p-hydroxymercuribenzoate (51), and Schottel (60) was unable to separate the two lyases. Nevertheless, kinetic analysis indicated that there were two enzymes active toward phenylmercuric acetate and only one active toward methyl- and ethyl-mercury. The general properties of the enzymes from the soil pseudo-monad and E. coli were similar, except the E. coli organomercurial lyases had a somewhat greater molecular weight (60).

Mercuric reductase has been studied in greater detail both with plasmid-bearing E. coli (56,60,61) and with the soil pseudomonad (55). We once thought the enzyme was strictly NADPH-dependent, but it now appears that some mercuric reductases can utilize either NADPH or NADH (e.g. the Streptomyces strain used in Figure 6C).

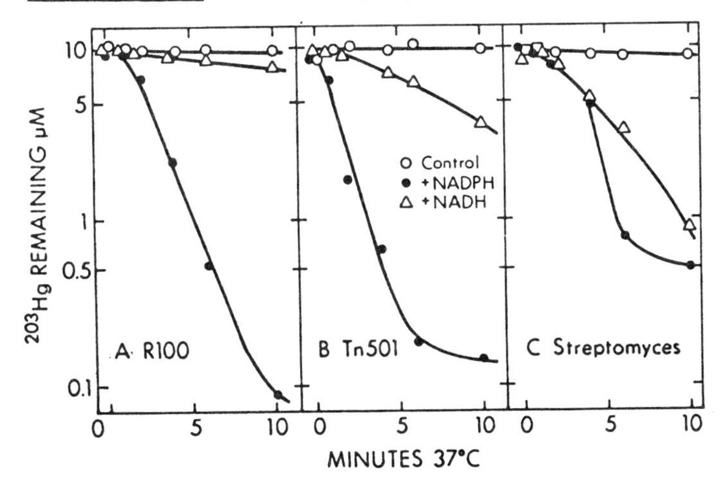

Figure 6. Volatilization of mercury from Hg^{2+} by enzyme preparations from cells bearing plasmid R100 (A), transposon Tn501 (B), or a new Streptomyces strain(C). Conditions were as given in ref. 60, except 25 µM NAD(P)H was added with 10 µM $^{203}Hg^{2+}$ (unpublished experiment).

The molecular weight and subunit structure of the mercuric reductase enzyme from various sources is currently under reevaluation. Furukawa and Tonomura (62) reported that the enzyme from the soil pseudomonad had a molecular weight of about 65,000. Schottel (60), however, reported that the enzyme from plasmid R831 that originated in Serratia was a trimer of 170,000 daltons containing identical monomer subunits, each approximately the size of the Pseudomonas enzyme. Each subunit contained a single bound FAD, for a total of 3 FAD's per 170,000 molecular weight. Recently, the situation became more complex, as Fox and Walsh (61) reported that the molecular weight of still another mercuric reductase (from transposon Tn501, which originated in a clinical P. aeruginosa isolate) was about 125,000, and the enzyme appeared to be dimeric. We have confirmed the dimeric structure of the mercuric reductase of Tn501 using both gel filtration and migration through polyacrylamide gels of varying porosity as measures of molecular size (63).

Antibodies have been prepared against purified mercuric reductases coded by two plasmids in E. coli (63). All reductases obtained from different Gram-negative sources reacted with these antibodies, as shown by inhibition of enzyme activity and by formation of precipitin bands on double-diffusion gels. The enzymes fell into two major subclasses, based on only partial immunological identity. The prototype of the first enzyme class is coded by transposon Tn501. This enzyme class also includes mercuric reductases governed by a variety of plasmids found in clinical isolates of enteric bacteria and P. aeruginosa, in marine pseudomonads, and in Pseudomonas putida (the MER plasmid, ref. 46). One strong conclusion from studies of plasmid-determined mercuric resistance is that the same system appears widely in clinical isolates and in bacteria from other environments. However, newer Hg^{2+}-(and in one case phenylmercuric-) resistance transposons from soil microbes show different patterns of digestion by DNA restriction endonucleases (50), indicating that although the systems are related immunologically, they do not have identical DNA sequences.

The second immunological subgroup of Gram negative mercuric reductases has as its prototype the enzyme coded by plasmid R100, one of the earliest and most thoroughly studied of the antibiotic resistance plasmids. It is with R100 that the genetic structure of the mercuric resistance operon was studied in detail (64,65). The enzyme from R100 appears to be trimeric (63), as was the enzyme from plasmid R831 (60). This subgroup includes enzymes from plasmids of a wide variety of incompatability groups and also the enzyme determined by a second Pseudomonas mercury transposon Tn502 (63; V. Stanisich, personal communication). Although all of the mercuric reductases from Gram-negative bacteria were immunologically related, the antibodies prepared against the two classes of Gram-negative enzymes did not cross react with mercuric reductases from S. aureus strains and marine and soil bacilli. These enzymes from Gram-positive sources showed similar functional requirements to those from the Gram-negative bacteria (53,63), yet they are immunologically distinct.

To summarize briefly the current understanding of plasmid-determined mercuric and organomercurial resistances: (a) They occur widely in both Gram positive and Gram negative species and are the best understood of all plasmid-coded heavy metal resistances. (b) Resistance is due to enzymatic detoxification of the mercurials to volatile compounds of lesser toxicity. (c) The enzymes responsible (mercuric reductases and organomercurial lyases) have been purified and studied in vitro.

B. Arsenic and antimony resistances.

Arsenic and antimony resistances are governed by plasmids that also code for antibiotic and other heavy metal resistances (41,66,67). Arsenate, arsenite and antimony(III) resistances are coded for by an inducible operon-like system in both *S. aureus* and *E. coli* (68). Each of the three ions induces all three resistances. In *E. coli*, Bi(III) is a gratuitous inducer of arsenate resistance, since it causes reduced $^{74}AsO_4{}^{3-}$ accumulation even though the plasmid system does not confer Bi(III) resistance. *S. aureus* has a genetically separate plasmid-mediated Bi(III) resistance determinant (41,69) of unknown mechanism.

The mechanism of arsenate resistance is a reduced accumulation of arsenate by induced resistant cells. Arsenate is normally accumulated via the cellular phosphate transport systems, of which many bacteria appear to have two (Figure 2). The distinction between arsenate and arsenite resistances was shown by finding that phosphate did not protect against arsenite inhibition of growth (68). Genetic studies with *S. aureus* plasmids have also demonstrated that the arsenate resistance gene is different from but closely linked to the arsenite resistance gene; additionally, arsenite and antimony resistances may be determined by separate genes (69). The presence of the resistance plasmid does not alter the kinetic parameters of the cellular phosphate transport systems; even the K_i for arsenate as a competitive inhibitor of phosphate transport (Figure 2) is unchanged. This finding, along with direct evidence for plasmid-governed energy-dependent efflux of arsenate indicated that the block on uptake of arsenate resulted from rapid efflux. The energy dependence of the efflux process was shown by its sensitivity to "uncouplers" such as CCCP and tetrachlorosalicylanilide and ionophore antibiotics such as nigericin and monensin (70). Hg^{2+} also inhibits the efflux system, and the Hg^{2+} inhibition is readily reversed by mercaptoethanol (70). Recently, Mobley and Rosen (71) transferred arsenate-resistance plasmid R773 (66,68) into an *E. coli* strain that could not synthesize ATP from respiratory substrates and in that strain demonstrated that glucose but not succinate could energize arsenate efflux. Although all of these data are indirect, our tentative conclusion is that $AsO_4{}^{3-}$ efflux is directly coupled to a pH-sensitive ATP-linked transport system, similar to the Na^+ efflux system found in *Streptococcus faecalis* (34,35).

An interesting question about the arsenate efflux system concerns its specificity. Arsenate generally functions as a phosphate analogue and is accumulated by bacteria via phosphate transport systems (3). The arsenate-resistance efflux system should not excrete phosphate, since the cells would then become phosphate starved, a situation no more advantageous than being arsenate inhibited! A basic conclusion from our work on arsenate (and also on Cd^{2+} resistance; see below) is that toxic heavy metals often get into cells by means of transport systems for normally-required nutrients (3,7). Energy-dependent efflux systems functioning as resistance mechanisms should be highly specific for the toxic anion or cation to prevent loss of the required nutrient.

The mechanism(s) of arsenite and of antimony resistances are not known. Arsenicals and antimonials are toxic by virtue of inhibiting thiol-containing enzymes (72). Some dithiol reagents such as BAL (British anti-Lewisite) protect against arsenical and antimonial toxicity. Resistant cells did not excrete soluble thiol compounds into the medium, since pre-growth of resistant cells in medium containing these toxic ions does not allow subsequent growth of sensitive or of uninduced resistant cells (68). Arsenite is not oxidized to the less toxic arsenate by plasmid-bearing _E. coli_ or _S. aureus_ (68). The absence of "detoxification" measured by experiments inocculating sensitive cells into medium "preconditioned" by growth of resistant cells eliminates all other mechanisms involving changes in extracellular chemical states.

C. Cadmium and zinc resistance.

Plasmid-determined cadmium resistance has been found only in _S. aureus_ (41). In some collections, Cd(II) resistance is the most common _S. aureus_ plasmid resistance, exceeding in frequency both mercury and penicillin resistances (43). Gram-negative cells without plasmids are just as resistant to Cd(II) as are staphylococci with plasmids (44), probably because of relatively reduced Cd(II) uptake (Silver, unpublished). However, there are occasional Gram-negative bacteria that are sensitive or even "hypersensitive" to Cd(II) (47; T. Barkay, unpublished data). The basis of Cd(II) sensitivity and resistance in other bacterial species is not known.

In staphylococci, Cd(II) is accumulated by a membrane transport system utilizing the cross-membrane electrical potential (73-75; Figure 7). This uptake system is highly specific for Cd(II) and Mn(II) with respective K_m's of 10 μM and 16 μM in whole cells (75) and 0.2 μM and 0.95 μM in membrane vesicles (73).

Two separate plasmid genes are responsible for the Cd(II) resistance of _S. aureus_ strains (41,69; Figure 8). The _cadA_ and _cadB_ genes confer respectively, a large and a small increase in Cd(II) resistance (Figure 8). When both genes are present, the effect of _cadA_ masks the _cadB_ gene effect; _cadA_$^+$ _cadB_$^+$ strains are no more resistant than are cad$^+$ cadB$^-$ strains (77). Both genes confer increased Zn(II) resistance (Figure 8). This and the genetic linkage of Cd(II) and Zn(II) resistance (41) indicate that the cadA and cadB genes are also responsible for Zn(II) resistance.

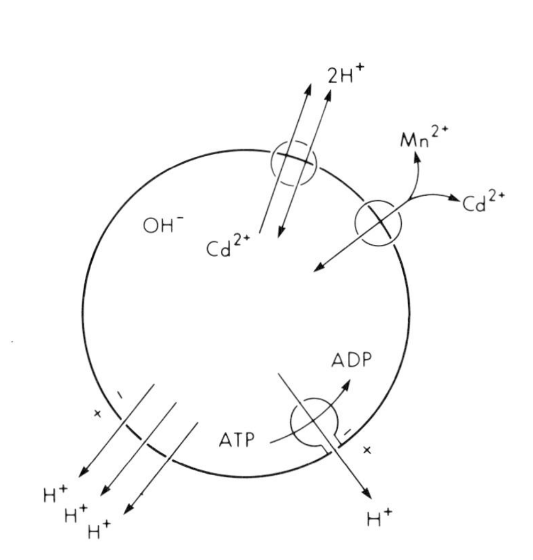

Figure 7. Model for Cd^{2+} uptake and efflux systems.

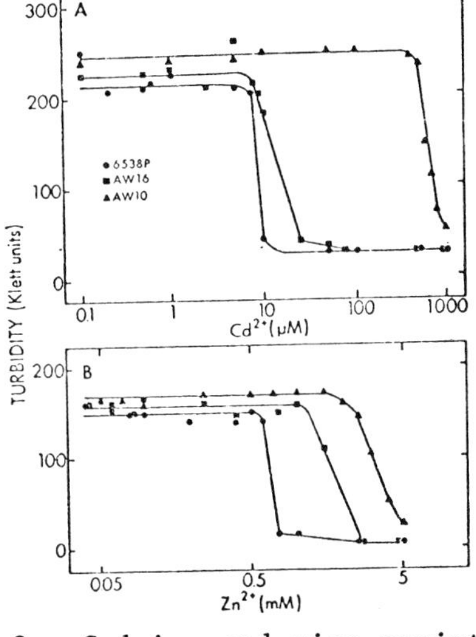

Figure 8. Cadmium and zinc resistance of strains with and without plasmids. Overnight cultures were diluted 1:200 into broth containing varying concentrations of (A) $CdCl_2$ or (B) $ZnSO_4$. Culture turbidities were measured after 7 h of growth at 37°C. Strains 6538P (sensitive), AW10 (cadA$^+$ resistant) and AW16 (cadB$^+$ resistant (from ref. 73).

Cd(II) resistance is due to a constitutive reduction in Cd(II) accumulation by resistant cells (74,75). The cadA gene product causes this reduced Cd(II) accumulation (73, 75; Figure 9). Reduced Cd(II) accumulation is due to a plasmid-encoded efflux system which rapidly excretes Cd(II) (77; Figure 10). CadA$^+$ resistant strains, but neither sensitive nor cadB$^+$ resistant strains possess this efflux system (10). Although not directly demonstrated, it seems plausible that the cadA gene product might also cause Zn(II) efflux. However, the presence of the cadA$^+$ gene does not reduce Mn(II) uptake nor cause rapid efflux of accumulated Mn(II) (73).

The cadA-encoded efflux system is energy dependent. Cd(II) efflux was abolished by dinitrophenol and at low temperature (Figure 10). Although efflux directly energized by ATP has not been definitely eliminated, inhibitor studies indicate that the efflux system is a $Cd^{2+}/2H^+$ antiport (Figure 7).

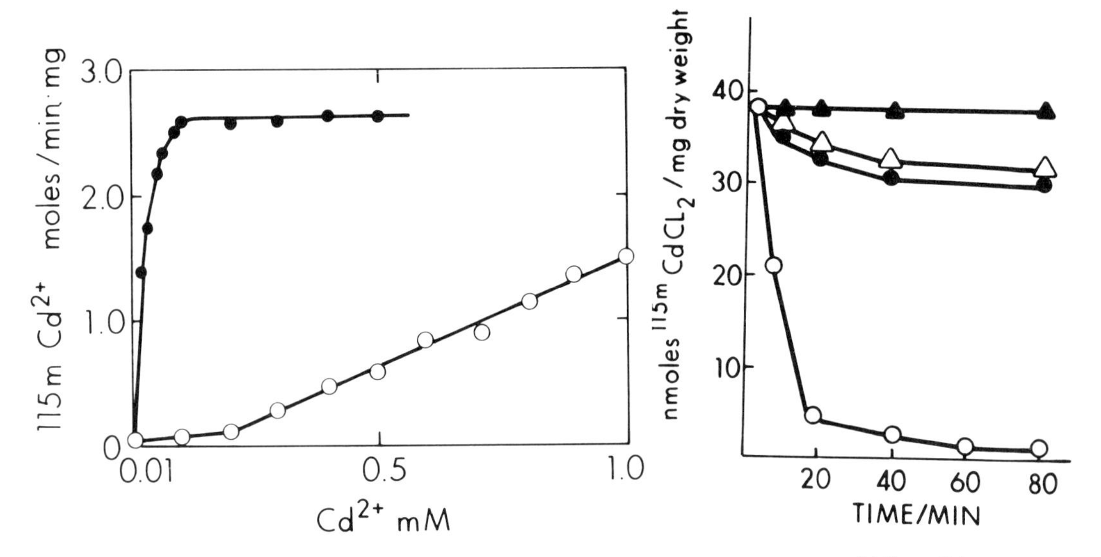

Figure 9. Cadmium uptake by sensitive and by resistant S. aureus. Initial rate of accumulation of $^{115m}Cd^{2+}$ by a resistant plasmid-containing strain (O) and its sensitive plasmid-less variant (●) (ref. 74).

Figure 10. Retention of $^{115m}Cd^{2+}$ by cadmium-resistant S. aureus. Cell suspensions were preincubated with 1 mM $^{115m}Cd^{2+}$ for 10 min at 37°C. Cells were washed free of Cd^{2+}, and then $^{115}Cd^{2+}$ efflux was assayed. O, control cells at 37°C; ▲, cells at 4°C; △, cells with 10 mM dinitrophenol; and ●, cells with 100 μM DCCD (ref. 77).

We do not know the mechanism of cadB gene function. Our clues at the moment include the significant resistance to Zn^{2+} conferred by cadB (Figure 8) and an inducible Cd^{2+} binding activity governed by this gene (10). This binding activity is not energy-dependent and we hypothesize that it might be due to a membrane component analogous to metallothionein, a small Cd^{2+} binding protein made by both mammals (78) and microbes (79,80). This is no more than a working hypothesis at this time. When seeking resistant-mutants of CHO mammalian cells that grow on higher Cd^{2+}, Corrigan and Huang (81) found what appear to be transport mutants that accumulate less Cd^{2+} and more Zn^{2+} than wild type cells. Whether the metallothionein studies or mammalian transport studies will provide a basis for understanding cadB resistance is an open question.

D. Silver resistance.

Microbial silver toxicity is found in situations of industrial pollution, especially associated with the use of photographic film. In hospitals, silver salts are preferred antimicrobial agents for burns (82). It is thus not surprising that silver-resistant bacteria have been found in urban and industrial polluted sites (83) or that silver resistant bacteria (84-86) and silver resistance plasmids have recently been described (87). Silver resistance is constitutive in *E. coli*. The plasmid-determined resistance is very great, and the ratio of minimum-inhibitory concentrations (resistant: sensitive) can be greater than 100:1 (8,10). The level of resistance is strongly dependent upon available halide ions; without Cl^-, there was relatively little difference between cells with or without resistance plasmids (10). Br^- and I^-, at concentrations far below those used for Cl^-, conferred resistances on both sensitive and resistant cells. Both sensitive and resistant cells bind Ag^+ tightly and are killed by effects on cell respiration and other cell surface functions. Once bound extracellularly, Ag^+ enters the cell and is found in high speed centrifugal supernatant fluids (unpublished data). Our current hypothesis is that the sensitive cells bind Ag^+ so tightly that they extract it from AgCl and other bound forms, whereas cells with resistance plasmids do not compete successfully with Ag^+-halide precipitates for Ag^+.

E. Other heavy metal resistances.

There are other plasmid heavy metal resistances (7,83), but nothing today is known about the mechanisms of resistance to bismuth, boron, cobalt, lead, nickel or tellurium ions. Chromate resistance in a pseudomonad isolated from river sediment seems to be due to reduction of toxic Cr(VI) to less toxic Cr(III) (Bopp and Ehrlich, Abstract Q111, 1980 American Society for Microbiology Meetings), and this resistance appears to be plasmid determined (A.M. Chakrabarty, personal communication). Caution on this point is needed. Bacteria capable of oxidizing toxic As(III) to less toxic As(V) are also known; however, this has not turned out to be the mechanism of plasmid-governed resistance.

Over the last twelve years, our laboratory has been studying the mechanisms and genetics of plasmid-mediated heavy metal resistances. In the frequent absence of any obvious source of direct selection, one may ask why these resistances occur with such high frequencies. Selective agents in hospitals and "normal human" environments are only beginning to be examined. It has been suggested that heavy metal resistances may have been selected in earlier times, and that they are merely carried along today "for a free ride" with selection for antibiotic resistances. I doubt that there is such a thing as "a free ride" as far as these determinants are concerned. Where they exist, there is likely to be selection by either non-human or human sources of heavy metals. Radford et al. (50) found Hg^{2+}-resistant microbes in agricultural soil, with no known human mercurial input. In such settings, the prevalence of resistant microbes may be very low, but it may come into much greater quantitative prominence after industrial or agricultural pollution (39,40,47,49). This situation may be closely analogous to that with antibiotic-resistance plasmids, which are found in low frequencies in antibiotic-virgin populations (88,89), but which become dominant with extensive human use of antibiotics.

ACKNOWLEDGEMENTS

Recent work in our laboratory on these topics has been supported by grants from the National Science Foundation PCM79-03986 and the National Institutes of Health AI15672.

REFERENCES

1. Rosen, B.P., and E.R. Kashket. 1978. In "Bacterial Transport" (ed. B.P. Rosen), Marcel Dekker, New York, pp. 559-620.
2. Lanyi, J.K. 1979. Biochim. Biophys. Acta 559: 377-397.
3. Silver, S. 1978. In "Bacterial Transport" (ed. B.P. Rosen), Marcel Dekker, New York, pp. 221-324.
4. Silver, S., and R.D. Perry. 1981. In "Short-lived Radionuclides in Chemistry and Biology" (eds. J.W. Root and K.A. Krohn), American Chemical Society, Washington, D.C., pp. 453-468.
5. Silver, S., and R.D. Perry. 1982. In "Membranes and Transport (ed. A.N. Martonosi), Plenum Press, New York, volume 2, pp. 115-121.
6. Albert, A. 1973. In "Selective toxicity. 5th Ed." Chapman and Hall, London, pp. 392-397.
7. Summers, A.O., and S. Silver. 1978. Annu. Rev. Microbiol. 32: 637-672.
8. Silver, S. 1981. In "Molecular Biology, Pathogenicity and Ecology of Bacterial Plasmids" (eds. S.B. Levy, R.C. Clowes and E.L. Koenig), Plenum Press, New York, pp. 179-189.
9. Silver, S. 1981. In "Environmental Speciation and Monitoring Needs for Trace Metal-Containing Substances from Energy-Related Processes (eds. F.E. Brinckman and R.H. Fish), National Bureau of Standards, Special Publ. 618, Washington, D.C., pp. 301-314.
10. Silver, S., R.D. Perry, Z. Tynecka and T.G Kinscherf. 1982. In "Proceedings of the Third Tokyo Symposium on Drug Resistance in Bacteria" (eds. S.Mitsuhashi and H. Hashimoto), Japanese Scientific Societies Press, Tokyo, in press.
11. Cox, G.B., H. Rosenberg, J.A. Downie and S. Silver. 1981. J. Bacteriol. 148: 1-9.
12. Roomand, G.M., F. Blasco and G.W.F.H. Borst-Pauwels. 1977. Biochim. Biophys. Acta 467: 65-71.
13. Thayer, J.R., and R.C. Huffaker. 1982. J. Bacteriol. 149: 198-202.
14. Garland, P.B., J.A. Downie and B.A. Haddock. 1975. Biochem. J. 152: 547-559.
15. MacGregor, C.H., and A.R. Christopher. 1978. Arch. Biochem. Biophys. 185: 204-213.
16. Schloemer, R.H., and R.H. Garrett. 1974. J. Bacteriol. 118: 259-269, and 270-274.
17. Goldsmith, J., J.P. Livoni, C.L. Norberg and I.H. Segel. 1973. Plant Physiol. 52: 362-367.
18. Epstein, W., and L. Laimins. 1980. Trends Biochem. Sci. 5: 21-23.

19. Laimins, L.A., D.B. Rhoads and W. Epstein. 1981. Proc. Natl. Acad. Sci. USA 78: 464-468.
20. Epstein, W., V. Whitelaw and J. Hesse. 1978. J. Biol. Chem. 253: 6666-6668.
21. Laimins, L.A., D.B. Rhoads, K. Altendorf and W. Epstein. 1978. Proc. Natl. Acad. Sci. USA 75: 3216-3219.
22. Park, M.H., B.B. Wong and J.E. Lusk. 1976. J. Bacteriol. 126: 1096-1103.
23. Rosenberg, H., and I.G. Young. 1974. In "Microbial Iron Metabolism" (ed. J.B. Neilands), Academic Press, New York, pp. 67-82.
24. Neilands, J.B. 1981. Annu. Rev. Biochem. 50: 715-731.
25. Williams, P. 1979. Infect. Immun. 26: 925-932.
26. Stuart, S.J., K.T. Greenwood and R.K.J. Luke. 1980. J. Bacteriol. 143: 35-42.
27. Braun, V. 1981. FEMS Lett. 11: 225-228.
28. Silver, S., and P. Jasper. 1977. In "Microorganisms and Minerals" (ed. E.D. Weinberg), Marcel Dekker Publ., New York, pp. 105-149.
29. Failla, M.L. 1977. In "Microorganisms and Minerals" (ed. E.D. Weinberg), Marcel Dekker Publ., New York, pp. 151-214.
30. Silver, S. 1977. In "Microorganisms and Minerals" (ed. E.D. Weinberg), Marcel Dekker Publ., New York, pp. 49-103.
31. Rosen, B.P. 1982. In "Membrane Transport of Calcium" (ed. E. Carafoli), Academic Press, London, in press.
32. Kobayashi, H., J. Van Brunt and F.M. Harold. 1978. J. Biol. Chem. 253: 2085-2092.
33. Brey, R.N., J.C. Beck and B.P. Rosen. 1978. Biochem. Biophys. Res. Commun. 83: 1588-1594.
34. Heefner, D.L., and F.M. Harold. 1980. J. Biol. Chem. 255: 11396-11402.
35. Heefner, D.L., H. Kobayashi and F.M. Harold. 1980. J. Biol. Chem. 255: 11403-11407.
36. Miller, A.G., and K. Budd. 1975. Canad. J. Microbiol. 21: 1211-1216.
37. Miller, A.G., and K. Budd. 1975. J. Bacteriol. 121: 91-98.
38. Komura, I., and K. Izaki. 1971. J. Biochem. 70: 885-893.
39. Tonomura, K., T. Nakagami, F. Futai and K. Maeda. 1968. J. Ferment. Technol. 46: 506-512.
40. Furukawa, K., T. Suzuki and K. Tonomura. 1969. Agric. Biol. Chem. 33: 128-130.
41. Novick, R.P., and C. Roth. 1968. J. Bacteriol. 95: 1335-1342.
42. Smith, D.H. 1967. Science 156: 1114-1116.
43. Nakahara, H., T. Ishikawa, Y. Sarai and I. Kondo. 1977. Zentralbl. Bakteriol. Parasitenkd. Infektionskr. Hyg. Abt. 1 Orig. A 237: 470-476.
44. Nakahara, H., T. Ishikawa, Y. Sarai, I. Kondo, H. Kozukue and S. Silver. 1977. Appl. Environ. Microbiol. 33: 975-976.
45. Schottel, J., A. Mandal, D. Clark, S Silver and R.W. Hedges. 1974. Nature 251: 335-337.

46. Friello, D.A., and A.M. Chakrabarty. 1980. In "Plasmids and Transposons: Environmental Effects and Maintenance Mechanisms," (eds. C. Suttard and K.R. Rozee), Academic Press, New York, pp. 249-260.
47. Olson, B.H., T. Barkay and R.R. Colwell. 1979. Appl. Environ. Microbiol. 38: 478-485.
48. Olson, G.J., W.P. Iverson and F.E. Brinckman. 1981. Current Microbiol. 5: 115-118.
49. Timoney, J.F., J. Port, J. Giles and J. Spanier. 1978. Appl. Environ. Microbiol. 36: 465-472.
50. Radford, A.J., J. Oliver, W.J. Kelly and D.C. Reanney. 1981. J. Bacteriol. 147: 1110-1112.
51. Weiss, A.A., J.L. Schottel, D.L. Clark, R.G. Beller and S. Silver. 1978. In "Microbiology-1978" (ed. D. Schlessinger), American Society for Microbiology, Washington, D.C. pp. 121-124.
52. Clark, D.L., A.A. Weiss and S. Silver. 1977. J. Bacteriol. 132: 186-196.
53. Weiss, A.A., S.D. Murphy and S. Silver. 1977. J. Bacteriol. 132: 197-208.
54. Porter, F.D., S. Silver, C. Ong and H. Nakahara. 1982. Antimicrob. Agents Chemother., in press.
55. Furukawa, K., and K. Tonomura. 1972. Agric. Biol. Chem. 36: 217-226.
56. Izaki, K., Y. Tashiro and T. Funaba. 1974. J. Biochem. 75: 591-599.
57. Summers, A.O., and S. Silver. 1972. J. Bacteriol. 112: 1228-1236.
58. Tezuka, T., and K. Tonomura. 1976. J. Biochem. 80: 79-87.
59. Tezuka, T., and K. Tonomura. 1978. J. Bacteriol. 135: 138-143.
60. Schottel, J.L. 1978. J. Biol. Chem. 253: 4341-4349.
61. Fox, B., and C.T. Walsh. 1982. J. Biol. Chem. 257: 2498-2503.
62. Furukawa, K., and K. Tonomura. 1971. Agric. Biol. Chem. 35: 604-610.
63. Kinscherf, T.G., and S. Silver. 1982. Manuscript in preparation.
64. Nakahara, H., S. Silver, T. Miki and R.H. Rownd. 1979. J. Bacteriol. 140: 161-166.
65. Foster, T.J., H. Nakahara, A.A. Weiss and S. Silver. 1979. J. Bacteriol. 140: 167-181.
66. Hedges, R.W., and S. Baumberg. 1973. J. Bacteriol. 115: 459-460.
67. Smith, H.W. 1978. J. Gen. Microbiol. 109: 49-56.
68. Silver, S., K. Budd, K.M. Leahy, W.V. Shaw, D. Hammond, R.P. Novick, G.R. Willsky, M.H. Malamy and H. Rosenberg. 1981. J. Bacteriol. 146: 983-996.
69. Novick, R.P., E. Murphy, T.J. Gryczan, E. Baron and I. Edelman. 1979. Plasmid 2: 109-129.
70. Silver, S., and D. Keach. 1982. Proc. Natl. Acad. Sci. USA, in press.
71. Mobley, H.L.T., and B.P. Rosen. 1982. Proc. Natl. Acad. Sci. USA, submitted.
72. Albert, A. 1973. In "Selective Toxicity, Fifth Edition", Chapman and Hall, London, pp. 392-397.

73. Perry, R.D., and S. Silver. 1982. J. Bacteriol. 150: 973-976.
74. Tynecka, Z., Z. Gos and J. Zajac. 1981. J. Bacteriol. 147: 305-312.
75. Weiss, A.A., S. Silver and T.G. Kinscherf. 1978. Antimicrob. Agents Chemother. 14: 856-865.
76. Smith, K., and R.P. Novick. 1972. J. Bacteriol. 112: 761-772.
77. Tynecka, Z., Z. Gos and J. Zajac. 1981. J. Bacteriol. 147: 313-319.
78. Durnam, D.M. and R.D. Palmiter. 1981. J. Biol. Chem. 256: 5712-5716.
79. Lerch, K. 1980. Nature 284: 368-370.
80. Olafson, R.W., K. Abel and R.G. Sim. 1979. Biochem. Biophys. Res. Commun. 89: 36-43.
81. Corrigan, A.J., and P.C. Huang. 1981. Biol. Trace Element Res. 3: 197-216.
82. Fox, C.L., Jr. 1965. Arch. Surg. 96: 184-188.
83. Summers, A.O., G.A. Jacoby, M.N. Swartz, G. McHugh and L. Sutton. 1978. In "Microbiology - 1978" (ed. D. Schlessinger), American Society for Microbiology, Washington, D.C., pp. 128-131.
84. Annear, D.I., B.J. Mee and M. Bailey. 1976. J. Clin. Path. 29: 441-443.
85. Bridges, K., A. Kidson, E.J.L. Lowbury and M.D. Wilkins. 1979. Brit. Med. J. 1: 446-449.
86. Hendry, A.T., and I.O. Stewart. 1979. Canad. J. Microbiol. 25: 915-921.
87. McHugh, G.L., R.C. Moellering, C.C. Hopkins and M.N. Swartz. 1975. Lancet 1: 235-240.
88. Maré, I.J. 1968. Nature 220: 1046-1047.
89. Gardner, P., D.H. Smith, H. Beer and R.C. Moellering, Jr. 1969. Lancet 2: 774-776.

MICROBIAL OXIDATION AND REDUCTION OF MANGANESE AND IRON

Kenneth H. Nealson
Marine Biology Research Division, A-002
Scripps Institution of Oceanography
La Jolla, California 92093 U.S.A.

ABSTRACT

The environmental distribution of iron and manganese, which are required trace metals for virtually all life, is strongly influenced by redox chemistry. The oxidized forms (Mn^{4+}, Fe^{3+}) are usually found as insoluble oxides, while the reduced forms (Mn^{2+}, Fe^{2+}) can be soluble. Biotic mediation of iron and manganese chemistry, which often involves changes in the redox states of these metals. Such activities, which can be indirect (pH/Eh alteration of the environment) or direct (via enzymes or binding components). are common to many bacterial groups. Biotic mediation may thus be important in metal deposition, mobilization and cycling, especially in stratified environments with well-developed oxic-anoxic interfaces. Conversely, the presence of large amounts of iron and manganese, often common in sediments, may be very important to both chemistry and biology of the environment.

INTRODUCTION

Because manganese and iron (usually found in low concentrations in aerobic aquatic environments) are required trace elements of virtually all organisms, it is not surprising that many biological mechanisms have evolved to bind and transport these metals, thus ensuring their supply. For example, because iron is in its insoluble ferric state under most natural conditions, concentrations are so low that a variety of extracellular organic iron chelators (siderophores) are used to scavenge iron as a soluble organic complex (1,2,3). In addition, many organisms, either by direct catalysis or by alteration of environmental conditions, catalyze the oxidation or reduction of these metals. Several books and reviews that describe these biotic interactions with iron and manganese have been published recently, and the reader is referred to these for details not discussed here (4-9). A consequence of these other interactions with iron and manganese is that there can be significant biotic influence on the chemistry and distribution of these metals as discussed here.

P. Westbroek and E. W. de Jong (eds.), Biomineralization and Biological Metal Accumulation, 459–479.

ABUNDANCE AND DISTRIBUTION

Iron and manganese are extremely abundant in the earth's crust, ranking 4th and 12th in total abundance respectively, and 2nd and 5th in abundance as metals in the lithosphere (10). Both metals are similarly distributed, being largely confined to oxidized phases, and found as major or minor components of a large variety of rocks and minerals where they may accumulate to high enough concentrations to become valuable as ore deposits (11). Iron deposits include the massive pre-Cambrian banded iron formations, red beds, bog ores, hydrothermal deposits and manganese nodules, while manganese deposits include manganese ores, manganese nodules, pipeline encrustations, and desert varnish.

Both metals lack atmospheric (volatile) or stable soluble phases in oxic waters, and are thus largely uncoupled from both hydrospheric and atmospheric cycles, so that major global cycling is largely via geological processes such as burial, metamorphosis, uplifting, hydrothermal activity and weathering. Accordingly, on the global scale these elements may be viewed as cycling very slowly, with rates determined by the above geological processes (11).

This is not to imply, however, that iron and manganese are non-reactive on all time scales. In anaerobic environments, reduction and thus solubilization of both iron and manganese occurs, and once reduced, re-oxidation can occur rapidly, giving these metals the potential for representing a major sink for electrons in anaerobic sediments, and the potential to flux rapidly. This potential, coupled with the high concentrations often encountered in sedimentary environments, suggests that these metals play important roles in the chemistry and biology of stratified environments (18). Furthermore, being required elements, both metals are concentrated in biological tissues, and thus participate in a biological cycle, being bound, metabolized, solubilized, and transferred between the various trophic levels of the biota.

CHEMISTRY

Manganese and iron, transition metals located adjacent to each other in the periodic table, share several chemical properties. Like the other transition elements, they can have a variable number of outer shell electrons, and thus several different redox states or valences (12). However, under most conditions at the surface of the earth, only two valences are common for each metal. For manganese, the usual reduced and oxidized states are manganous (Mn^{2+}) and manganic (Mn^{4+}) respectively, while for iron they are ferrous (Fe^{2+}) and ferric (Fe^{3+}). The reduced states of both metals are usually soluble, while the oxidized states readily form a variety of oxides and oxyhydroxides, most of which are quite insoluble (Table 1; 10,13,14). Thus, for both metals, their distribution will be greatly affected by factors that govern their valence states and the redox state of the environment.

Table 1

Some Iron and Manganese Containing Minerals[a]

Iron		Manganese	
Formula	Mineral	Formula	Mineral
$\alpha FeOOH$	Goethite	$Mn(OH)_2$	Pyrochroite
$\gamma FeOOH$	Lepidocrocite		
$\beta FeOOH$	Akaganeite	$MnOOH$	Manganite
αFe_2O_3	Hematite	Mn_3O_4	Hausmannite
γFe_2O_3	Maghemite	MnO_2[b]	Birnessite
			Todorokite
Fe_3O_4	Magnetite		Hollandite
			Pyrolusite
			Rhamsdellite
FeS	Mackinawite	MnS	Alabandite
FeS	Troilite	$MnCO_3$	Rhodochrocite
FeS_2	Pyrite		
FeS_2	Marcasite		
Fe_3S_4	Greigite		
$(Fe, Mg, Mn)CO_3$	Siderite		

[a]Data from 4, 10, 13-15.
[b]The various names assigned to the formula MnO_2 are due to different crystal forms as a result of complexation with a number of alternative cations.

Relative stabilities of some of the phases of iron and manganese compounds are shown in Fig. 1. While such stability diagrams, which are constructed on the basis of thermodynamic data obtained with pure chemicals and simple systems, do not begin to approximate the complexities encountered in natural environments, they are nevertheless of value in predicting in a rough way the forms of these metals one might expect in various environments. For instance, in oxic waters of neutral pH or above, both iron and manganese should be eventually removed to the insoluble oxides, and thus deposited in the sediments. Conversely, in reducing environments solubilization--first of manganese, then of iron--should occur. At the interface between oxidizing and reducing environments, both types of reactions could be coupled and cycling should occur.

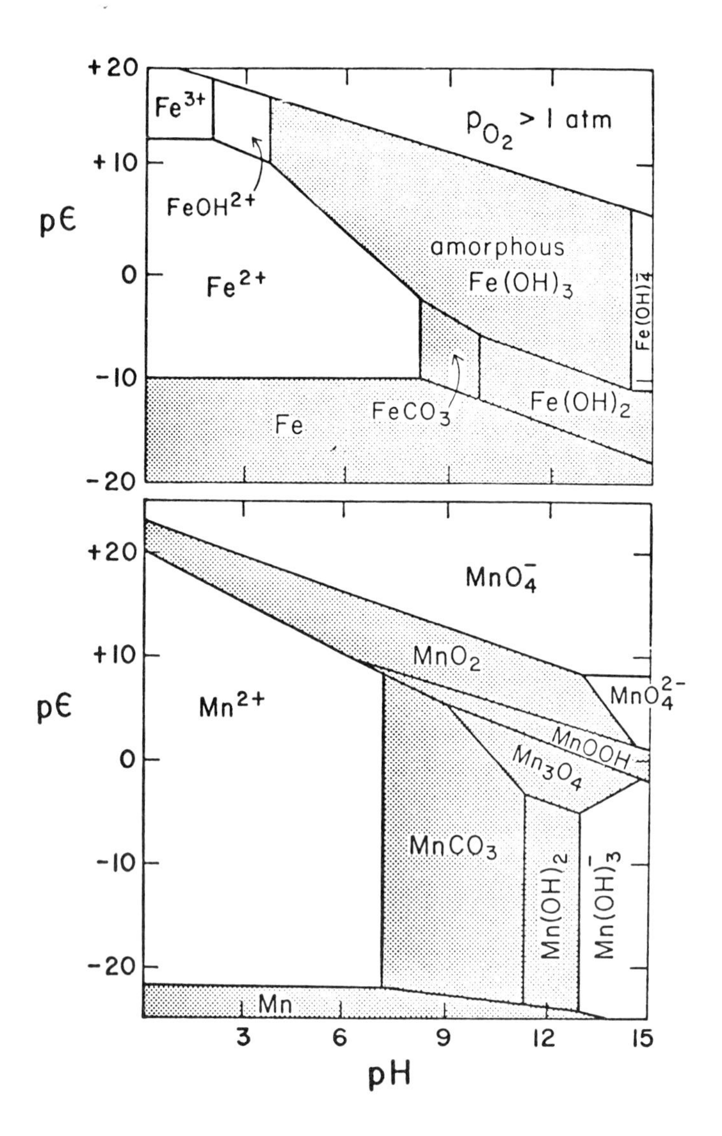

Figure 1. Stability of iron (upper) and manganese (lower) phases at different pε (Eh) and pH values. Such diagrams represent the expected stable phases (based on thermodynamic equilibrium calculations) for these metals under different conditions. Equilibria and equations for construction of these diagrams are in reference 12, pp. 447. Solid phases are represented by shaded areas.

Since the oxidized states are thermodynamically favored for both metals under natural conditions, energy is released upon oxidation. Although calculations can be made and laboratory experiments done, it is difficult to specify exactly the amount of energy available from either of these oxidations in nature. For instance, estimates of the energy available from oxidation of one mole of Mn^{2+} range from 9 to 35 kcal depending on the reaction written and the concentration of other reactants (4,6,8,12,15). Several factors contribute to the confusion. First, the various phases of manganese (and iron) oxides have different free energies themselves, so without knowing the mineral or complexed form of both the products and reactants, accurate predictions cannot be made. Second, the stoichiometry of the reactions have in general not been determined so that the equations cannot be written with any degree of certainty. Finally, for iron, the kinetics of removal are so rapid at neutral pH, that study of its rate of oxidation under these conditions and the energy yield associated with it are not really relevant to biological systems.

The fact that equilibrium considerations are in agreement with overall distribution patterns of both iron and manganese suggests that equilibrium controls are very important on the global scale (15). However, thermodynamic considerations give no rate information, and lead to little or no insight in understanding the mobility of either of these metals in a given environment; kinetic analyses are required for this.

Kinetically, manganese and iron are quite distinct (Fig. 2). For iron, the oxidation kinetics are described by the following equation (12):

$$\frac{-d(Fe^{2+})}{dt} = k(Fe^{2+})\,(OH^-)^2\,p_{O_2}$$

with $k = 8(\pm 2.5) \times 10^{13}$ min^{-1} atm^{-1} mol^{-2} at 20^oC

The reaction is quite dependent on pH; a change of one pH unit causes a 100-fold change in reaction rate. At pH values greater than 6.0, iron is removed rapidly from natural waters. Even in seawater, where the reaction is slowed due to interaction of iron with anions such as SO_4^{2-} and Cl^- (16), half removal time of iron is on the order of minutes.

The kinetics of manganese oxidation are more complicated, and much slower than those of iron. The reaction is autocatalytic, and follows the rate law:

$$\frac{-d(Mn^{2+})}{dt} = k_o(Mn^{2+}) + K_1\,p_{O_2}\,(OH^-)^2\,(Mn^{2+})\,(MnO_2)$$

where MnO_2 is the concentration of solid manganate, which acts as a catalyst by binding free Mn^{2+}. At a pH of 9.5 and p_{O_2} of 1 atm, the following values were calculated (W. Sung, Ph.D. thesis, California Institute Technology, 1981):

$$K_o = 7 \times 10^{-5}\ sec^{-1}$$

$$K'_1 = 4.4\ M^{-1}\ sec^{-1} \qquad (K'_1 = K_1 p_{O_2}[OH^-]^2)$$

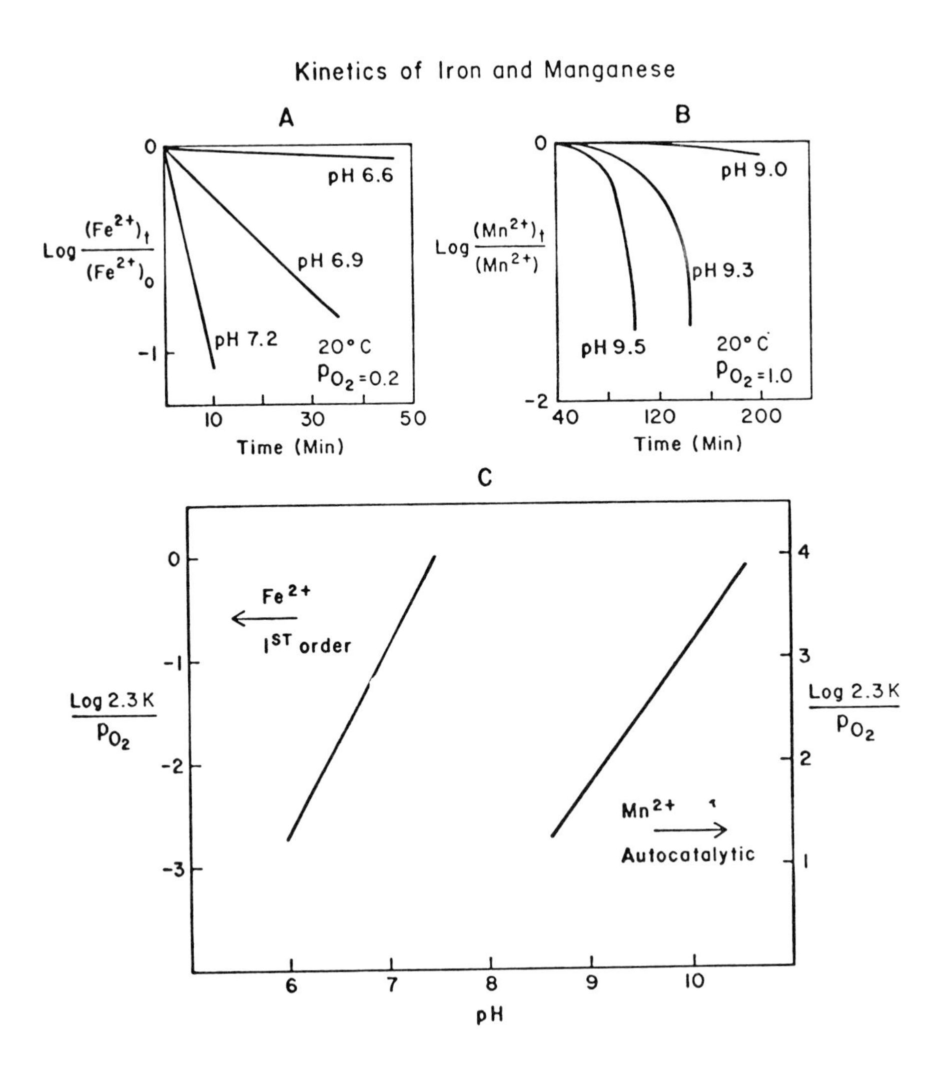

Figure 2. Kinetics of iron and manganese oxidation. (A) First order kinetics of iron oxidation at various pH values. Note that even at low oxygen, by pH 7.2, oxidation is very rapid. (B) Autocatalytic kinetics of Mn^{2+} oxidation at various pH values. Note stability at high oxygen and pH 9.0. (C) Effect of pH on apparent rate constants for iron and manganese oxidation. The figures are drawn after data shown in reference 12, pp. 465-466.

If similar pH and oxygen conditions are applied to iron oxidation, the apparent rate constant for iron oxidation ($K'_{Fe} = K[OH^-]^2 p_{O_2}$) is equal to 1.33 x 10^3 sec^{-1}.

Thus, while manganese oxidation is also second order with respect to pH, the rates are much slower than those for iron. Even at the pH of seawater, roughly 10^3 years would be needed to oxidize 90% of the manganese currently dissolved in seawater by this mechanism (15). Thus, while the oxidation of Mn is thermodynamically favored, at the pH's of natural waters, the removal of Mn is quite slow and biological catalysis could have a significant effect on rates of manganese oxidation.

Reduction of iron and manganese is not favored thermodynamically under oxic conditions, but under low Eh/pH conditions, reduction readily occurs. In sedimentary or stratified aquatic systems, microbial removal of oxygen and production of acids or reductants can result in biologically mediated reduction iron and manganese oxides. The redox potentials of the various compounds of iron and manganese vary somewhat, but overall, oxidized manganese phases have a potential near that of nitrate, while those of iron are located between nitrate and sulfate (Table 2). Thus, in sedimentary environments with high concentrations of iron and manganese, these metals could play major roles in the oxidation of organic matter in the sediments, and should occur predictably after oxygen removal and prior to sulfate reduction. Such sequences commonly occur, as exemplified in marine sediment pore water profiles (17-19; Fig. 3). presumably due to the metabolic activities of the organisms responsible for the various reactions. In contrast in freshwater sediments, low in both NO_3^- and SO_4^{2-}, iron and manganese may thus represent the major electron sinks available after oxygen depletion and before CO_2 reduction (methanogenesis).

Table 2
Electron Potentials of Some Redox Couples[a].

Redox couple	$p\varepsilon^o$ (W)	Process
O_2/H_2O	+13.75	Aerobic respiration
NO_3/N_2	+12.65	Denitrification
$MnO_2/MnCO_3$	+8.9	Mn reduction
NO_3/NH_4	+6.15	Nitrate reduction
$FeOOH/FeCO_3$	.8	Iron reduction
SO_4^{2-}/SH^-	-3.75	Sulfate reduction
CO_2/CH_4	-4.13	Methanogenesis

[a]$p\varepsilon^o$ (W) values are for electron activity for unit activities of oxidant and reductant at pH = 7.0 and 25°C. Data and complete equations can be found on pp. 449-459 (15).

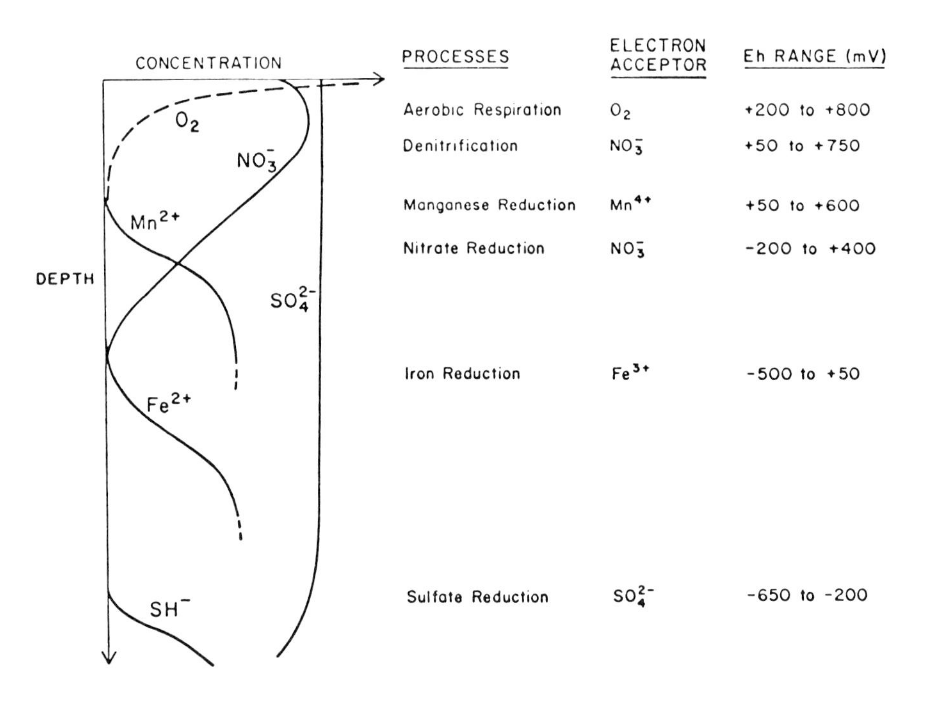

Figure 3. Schematic representation of pore water profiles observed in several marine sediments (17,18,19). Although the spacings vary somewhat, depending on organic input, the sequence of these processes with depth is always as shown.

The kinetics of iron and manganese reduction are less well documented than is the oxidation. Manganese is susceptible to rapid reduction, especially in anoxic marine environments, where SH^-, which will spontaneously reduce MnO_2, often accumulates to high levels. A variety of organic and inorganic reductants will reduce MnO_2, but the kinetics of these reactions are not well characterized. Iron oxides are more stable to reduction than manganates, and iron reduction is often spatially separated from manganese reduction in natural environments (17-19), but like manganese reduction, the kinetics are not well documented.

A final important aspect of the chemistry of both metals is their tendency to form complex with a variety of inorganic and organic compounds and to under reactions such as adsorption or ion exchange at solid surfaces (12,15). Consideration of iron or manganese chemistry without a knowledge of the other components of the system (e.g., carbonates, sulfates, sulfides, organics, solid surfaces) will be incomplete and incorrect. For example, MnO_2 itself catalyzes the further binding and oxidation of manganese. Clearly, if the amount and form of the particulate manganese is not known for a given environment, predicting the rates of manganese oxidation may be futile.

In summary, chemical considerations lead to the conclusion that on the global scale thermodynamic controls of iron and manganese explain the distribution of these metals. In local anaerobic environments, the biota may exert significant effects by reduction of these metals. Here, kinetic controls operate. Iron oxidation is so rapid in natural waters that it may be difficult for the biota to compete, whereas manganese oxidation is sufficiently slow that biological catalysis may be important. The presence of reduced forms of both metals suggests that the biota uses them as electron sinks in anaerobic sediments, although neither the mechanisms nor the kinetics of these processes are well documented.

MICROBIAL MEDIATION OF Mn AND Fe OXIDATION AND REDUCTION

Manganese Oxidizers

A wide variety of different bacteria catalyze manganese oxidation, and many genera of manganese oxidizing bacteria from nearly all environments have been studied (4-6,8,20-24). These microbes bind and oxidize Mn^{2+} producing a variety of manganates, both amorphous and crystalline. Under conditions of slow deposition, it forms a well-crystallized 10Å expandable manganate; under other conditions it forms a highly disordered manganate (Prof. G. Arrhenius, personal communication). Very little work has been done in determining the various mineralogies of biologically formed manganates.

Studies of cell-free manganese oxidation suggest that several different mechanisms exist. A partially purified cell-free bacterial extract that can catalyze manganese oxidation has been reported. (25). Both the bacteria and the cell-free extract were active only in the presence of solid MnO_2. A pronase sensitive metabolic product that catalyzes Mn^{2+} oxidation is excreted by _Sphaerotilus discophorus_ (26), and similar activity for an intracellular protein of a _Pseudomonas_ species was reported (27). In both cases, the oxidation of manganese began only after rapid growth had ceased and it was not dependent on the presence of manganese in the growth medium. Cell-free manganese oxidizing activity in extracts of a soil bacterium isolated from a manganese concretion was reported to exhibit the properties of an enzyme (28,29). The K_m values were in the micromolar range, and maximum rates of manganese oxidation were from 1-2 μmoles/ml h^{-1}. Extracellular acidic polysaccharides may bind Mn^{2+} and hasten its oxidation by _Pedomicrobium_ spp. (30,31). A similar role has been suggested for spore coat protein(s) from a marine _Bacillus_ (32). Thus, there appear to be many biological mechanisms for Mn^{2+} oxidation, and further studies of the mechanisms and relative abundance of the organisms possessing them should lead to insight into the role of bacteria in manganese oxidation in nature.

Only a few studies of the ultrastructure of Mn oxidizers have been reported. SEM analyses of laboratory-formed biological precipitates

(33) revealed that even morphologically simple (rods and cocci) bacterial forms assumed complex morphologies as manganese accretion occurred. As manganese accumulated, SEM was of little use in revealing the bacteria under the precipitates, but TEM analysis of thin sections of the extracellular precipitates left little doubt that bacteria were involved in their formation. Ultrastructural analysis (TEM) of precipitates from the oxic-anoxic interface (where manganese oxidation was rapid) in Saanich Inlet, B.C. revealed manganese-coated bacteria similar to those seen in the laboratory studies (34). Bacterial remains were common inside the manganese precipitates on the ferromanganese concretions from the Batic Sea as shown by SEM and TEM analyses, coupled with specific treatments to remove oxidized manganese (35). In this study, Ghiorse pointed out the usefulness of transmission electron microscopy (TEM) of thin sections for such work. SEM analyses revealed little of the bacteria, which were buried in precipitates of their own making.

The apparent diversity in biochemical mechanisms of manganese precipitation, as well as the taxonomic diversity of the manganese oxidizers precludes easy generalizations about the physiological control of manganese oxidation. Some manganese oxidizers are microaerophiles that prefer low oxygen levels for the oxidation of manganese. The addition of heterotrophic nutrients tends to favor growth and inhibit manganese oxidation in several bacterial strains.

Many manganese oxidizers, but by no means all, exhibit enhanced oxidation with the addition of solid surfaces to the growth medium. Some marine bacteria oxidize manganese only when solid MnO_2 or other components of marine are added to the growth medium (25,36,37). For a marine Bacillus spp., a variety of surfaces, including calcite crystals, sand grains and glass beads, stimulated bacterial oxidation of manganese (38). Further studies with this organism have shown that when a variety of inorganic clays and silica were tested with the same organism, stimulation occurred as a function of the total available surface area added, independent of the nature of the surface (39).

Certainly some bacteria oxidize manganese with no energetic advantage to the organism, yet the possibility that manganese might serve as a energy source for bacteria has intrigued workers for many years (4,8,23,40). Although it has been suggested that manganese might serve as a sole or mixotrophic energy source, at present, no published that unequivocally demonstrate this. In at least one case, manganese enhances growth (41) but the mechanisms by which this enhancement occurs are not specified, and it may not be due to its utilization as an energy source.

Work in progress in our laboratory (Kepkay and Nealson, submitted for publication) involves the growth of marine manganese oxidizing bacteria in chemostats either autotrophically (with Mn^{2+} as the only energy source) or mixotrophically with Mn^{2+} and succinate (Fig. 4). Under these conditions, the cells fix CO_2 in proportion to growth and

Mn^{2+} oxidation (Table 2), further strengthening the argument that facultative Mn^{2+} chemolithotrophs exist.

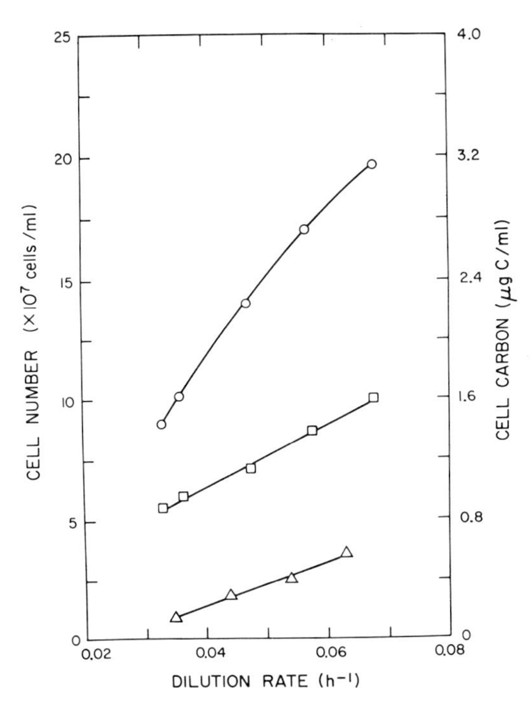

Figure 4. Autotrophic and mixotrophic growth of strain S-36. Steady state cell number and cell carbon of Pseudomonas S-36 in: ( ) autotrophic, Mn-limited environment; ( ) heterotrophic, succinate-limited, and ( ) mixotrophic, succinate plus Mn-limited media at various dilution rates (10 mM succinate, 100 μM Mn).

Manganese Reducers

Although it is known that many different groups of microbes have the ability to reduce manganese (4,20,42,43), only a few studies have been done with manganese reduction in the laboratory (44-46). In one case it was shown that MnO_2 reduction proceeded enzymatically, and involved both an inducible enzyme (MnO_2 reductase) and a specific cytochrome system for electron transport, and it was hypothesized that MnO_2 was used as an alternate electron acceptor in the absence of oxygen (46). An isolate of Arthrobacter that can either oxidize or reduce manganese depending on the growth conditions has also been reported (47). Very little has been reported concerning indirect (Eh, pH, chemical) reduction of MnO_2 by bacteria, although it undoubtedly is a common occurrence.

Iron Oxidizers and Accumulators

Some iron oxidizing bacteria, such as *Thiobacillus ferrooxidans*, are acid tolerant, living at pH's of 3.0 or less, where ferrous iron is stable and soluble. These organisms are chemolithotrophs, using the energy of iron oxidation for growth (5), but are not relevant to this symposium because, as far as in known, they do not participate at all in mineral formation.

The neutral pH iron oxidizers are an enigmatic group, possibly important in mineral formation, and should perhaps be referred to as iron accumulators. These bacteria, almost universal, are commonly encrusted with iron oxides and their association with iron deposits has led to the notion that they are important in minteral formation. The organisms almost certainly accumulate the aready oxidized iron rather than participate in its oxidation. Both laboratory experiments and field studies have suggested that these bacteria are active in the accumulation of iron deposits in nature (20,30,43,48-52).

Of particular relevance to this conference are the recently discovered magnetotactic bacteria (53), so called because they respond to magnetic lines of force. These bacteria contain intracellular "organelles" (54) composed of crystalline magnetite (55). Although the mechanism is not understood, it is known that the magnetite is formed *in vivo* by the bacteria grown on ferrous iron supplied in the growth medium (56). Such bacteria are abundant in a variety of sedimentary environments (56,57), and presumably all are magnetotactic due to the presence of intracellular magnetite. As opposed to most of the biota, where iron rarely exceeds 0.02% of the total dry weight of the magnetotactic bacteria, iron can represent as much as 3-4% of the dry weight.

Iron Reducers

Bacteria that can reduce iron are also taxonomically diverse and widely distributed (42,58-61); iron-reducing *Bacillus* species are particularly widespread and abundant in nature. While some bacteria apparently reduce iron using indirect mechanisms of alteration of the pH and Eh (62), others have the capacity to reduce iron under environmental conditions in which it would remain oxidized (58,59-61,63,79). While the mechanisms of iron reduction have not been elucidated yet, several workers believe that the reduction of iron involves some component of the electron chain to nitrate reductase (59,60,63). For iron-reducing strains that possess nitrate reductase, the addition of nitrate to the medium strongly inhibits iron reduction.

Only limited work has been done with reduction of specific minerals. A marine *Bacillus* was shown to reduce limonite and goethite rapidly, but have very little effect on nematite (80). In all cases, iron reduction was more effective under aerobic than anaerobic conditions.

FIELD DATA

Three kinds of data have helped elucidate the biotic role in the environmental chemistry of iron and manganese: (1) studies of association of microbes with zones of high fluxes or deposition of the manganese and iron; (2) rates of deposition or solubilization that appear to be too rapid to be explained on the basis of chemical kinetics; and (3) activity measurements comparing samples containing live organisms with those using poisoned controls. Taken together these constitute a strong case for microbial mediation in manganese and iron deposition and solubilization.

Manganese oxidizing and reducing microbes are almost always associated with active zones of manganese chemistry, and with manganese precipitates (6). These include fresh water and lacustrine manganese oxidation and reduction zones (20,42,43,61,64); manganese oxides in pipelines (65-67); freshwater manganese layers, rock coatings and ferromanganese nodules (64,68,69); marine manganese nodules (21,22,25,70); manganese cycles in stratified marine environments (35,43); and desert varnish (71).

Associations of bacteria with iron-rich deposits and active areas are less well documented, but include freshwater zones of iron accumulation (20,43,72,73), freshwater iron reduction (20), soils (60,74), and and marine (estuarine) iron reduction (75).

In no case do bacterial numbers alone prove microbial mediation, but it is difficult to escape the fact that in so many instances of metal accumulation or solubilization there are associations of abundant microbes that can catalyze the observed process.

Geochemical measurements of iron and manganese chemistry have been quite useful in defining the distributional patterns and, in some cases, suggesting the participation of the biota. The basic approach involves careful measurement of chemical parameters, including the oxidized and reduced chemical species, under study. The data can then be modelled to separate the physical processes (diffusion, advection, etc.) from _in situ_ chemical reactions. When the rates of these reactions determined by this procedure are inconsistent with rates predicted for abiotic reactions, biological factors are suspected.

A major difficulty in these studies is the surface and chelation chemistry of manganese and iron. Both metals complex with a variety of organic and inorganic compounds and undergo adsorption or ion exchange reactions, and may be either soluble or insoluble in various complexes. Thus, the traditional way of quantitating oxidized and reduced fractions, by separating the particulate and soluble phases respectively, may not necessarily provide what is assumed. The oxidation state of iron and manganese in particulates must be determined in order to better understand the dynamics of environments where these metals are precipitating and dissolving.

One example of such problems can be seen with the siderophores, a varied group of iron-binding organic compounds that bind ferric iron, solubilizing it, and making it biologically available (1-3). These compounds, which are universally distributed in aerobic organisms (1), along with other iron-binding organic compounds, probably account for the fact that "soluble" iron in aquatic environments is almost always somewhat higher than theoretically expected (10,11).

Another example of difficulties encountered in geochemical modelling is provided by studies of manganese iron and reduction, the rates of which are inferred from porewater profiles. Again, a major problem lies in the chemistry of reduced manganese and its tendency to complex with organic and and bind to solid surfaces, thus appearing as particulate or oxidized manganese. Nevertheless, some attempts to determine rates of manganese reduction by modelling reduced sedimentary environments have been made (76,77).

Porewater analyses from many of the deep sea sediments have shown the patterns of manganese and iron participation in the redox cycles that are predicted on chemical grounds (Fig. 3). Manganese is reduced concommitantly with, or shortly after nitrate, and iron is reduced deeper in the sediments, but well before sulfate (17-19). Similarly, studies in stratified marine environments such as Saanich Inlet, B. C., indicate that just above the oxic-anoxic interface there is a large zone of removal (oxidation). Below this zone, manganese reduction occurs (34,35).

From modelling physical and chemical properties of such systems, it was inferred that manganese oxidation was occurring just above the oxic-anoxic interface at rates much faster than could be accounted for on the basis of chemical kinetics alone (18,34). Indeed, in stratified environments such as sediments, meromictic lakes and fjords, manganese oxidation may be primarily due to microbial activities. A compilation of rate constants determined geochemically for a variety of environments in presented in Table 3.

Table 3

Rate Constants for Manganese Oxidation and Reduction

Oxidation

Location	k_{ox}(yr-1)
Lower limit for marine sediments based on laboratory data (81)	3.2
E. Equatorial Atlantic sediments (77)	10.7, 12.0
Long Island Sound sediments (R. C. Aller, Ph.D. thesis, Yale Univ., 1977)	50
Particulate layer of Saanich Inlet (34)	35, 180
Narragansett Bay sediments (82)	1000

Reduction

Location	k_{red}(yr-1)
E. Equatorial Atlantic sediments (77)	0.0015, 0.0021
Chesapeake Bay sediments (83)	0.0173
Long Island Sound sediments (84)	8.2, 25

Activity Measurements

So far, the direct measurement of microbial manganese oxidizing activity has been done in only a few environments: in a freshwater lake (65), a marine fjord (34) and an estuary (79). Microbial activities were assessed by comparison of removal rates of manganese in poisoned versus unpoisoned samples. These experiments utilized the radioactive isotope of manganese, ^{54}Mn, which can be added at levels below those found in the environment, thus allowing study of the system with perturbation. Although not yet quantitated, the rapid oxidation and removal of manganese is clearly due mainly to bacterial activity. In all environments where kinetic controls are thought to operate, poisoned control experiments should help quantitate the bacterial component of the oxidation. A possible defect in such experiments is that the poisons may affect the chemistry of the process being studied. Our laboratory has examined this in detail for manganese (Rosson, Tebo and Nealson, submitted for publication). After testing for (1) the binding of Mn^{2+} to inhibitors, (2) ability of inhibitors to desorb Mn^{2+} from surfaces, and (3) ability of inhibitors to reduce MnO_2, it was concluded that only a few poisons (azide, carefully buffered formalin, penicillin and tetracycline) are acceptable for use in field studies of Mn^{2+} oxidation.

Microbial reduction of iron oxides to soluble iron was demonstrated using poisoned or killed controls in both a freshwater lake G. (Jones, personal communication) and an estuarine sediment (76). No comparable poison control field experiments have been reported for manganese reduction or iron oxidation.

SUMMARY

The study, both in nature and in the laboratory, of microbial effects on iron and manganese chemistry is in its infancy. Recent geochemical and microbiological experiments support the circumstantial evidence that microbial mediation is important. The major sites of rapid production and removal of metal oxides appear to be stratified environments where reducing conditions prevail in lower levels, supplying reduced metals to aerobic areas above them. The rapid cycling that occurs, coupled with the sometimes large amounts of these metals in sedimentary environments, confer a possible significance to these metals that is not obvious at first glance. They may play a major role in the electron balance and chemical dynamics of such environments.

Among the mysteries to be unravelled are the quantitation of the microbial contribution to these redox reactions, the relation of the cycling of the metals to the input and cycling of other biologically modulated elements, and the mechanisms of metal chemistry catalysis.

REFERENCES

1. Neilands, J. B.: 1973. In "Inorganic Biochemistry" (G. L. Eichorn, ed.), pp. 176-202. Elsevier Press, Amsterdam.

2. Neilands, J. B.: 1974. In "Microbial Iron Metabolism" (J. B. Neilands, ed.), pp. 1-31. Academic Press, New York.

3. Neilands, J. B.: 1977. In "Bioorganic Chemistry" (K. N. Raymond, ed.), pp. 3-22. American Chemical Society, Washington, D. C.

4. Marshall, K. C.: 1979. In "Biogeochemical Cycling of Mineral-Forming Elements" (P. A. Trudinger and D. J. Swaine, eds.), pp. 253-292. Elsevier, Amsterdam.

5. Lundgren, D. G. and W. Dean: 1979. Ibid., pp. 211-251.

6. Ehrlich, H. L.: 1981 "Geomicrobiology," pp. 165-249. Marcel

Dekker, New York.

7. Nealson, K. H.: 1982. In "Mineral Deposits and the Evolution of the Biosphere" (N. D. Holland and M. Schidlowski, eds.), pp. 51-66. Springer-Verlag, Berlin.

8. Nealson, K. H.: 1982. Microbiological oxidation and reduction of iron. In "Microbial Geochemistry" (W. E. Krumbein, ed). Blackwell, London (in press).

9. Nealson, K. H.: 1982. The microbial iron cycle. Ibid (in press).

10. Wedepohl, K. H.: 1971. "Geochemistry." Holt, Rinehart and Winston, New York

11. Lepp, H.: 1975. "Geochemistry of Iron." John Wiley and Sons, Inc.

12. Stumm, W. and J. J. Morgan: 1981. "Aquatic Chemistry." Wiley Interscience, New York

13. Burns, R. G. and V. M. Burns: 1979. In "Marine Minerals" (R. G. Burns, ed), pp. 1-40. Mineraogical Society of America, Washington, D. C.

14. Murray, J. W.: 1979. Ibid, pp. 47-98.

15. Crerar, D. A., R. K. Cormic and H. L. Barnes: 1980. In "Geology and Geochemistry of Manganese" (I. M. Varentsov, ed.), pp. 293-334. Hungarian Academy of Sciences, Budapest.

16. Sung and J. J. Morgan: Year. Environ. Sci. Technol. 14, pp. 561-567.

17. Froelich, P. N., G. P. Klinkhammer, M. L. Bender, N. A. Luedtke, G. R. Heath, D. Cullen and P. Dauphin: 1979. Geochim. Cosmochim. Acta 43, pp. 1075-1090.

18. Emerson, S., R. E. Cranston and P. S. Liss: 1979. Deep-Sea Res. 26, pp. 859-878.

19. Klinkhammer, G. P.: 1980. Earth Planet. Sci. Lett. 49, pp. 81-101.

20. Perfil'ev, B. V. and D. R. Gabe: 1965. In "Applied Capillary Microscopy," pp. 9-52.

21. Schutt, C. and J. C. G. Ottow: 1977. Zeitschr. Allgemeine Mikrobiol. 17, pp. 611-616.

22. Schutt, C. and J. C. G. Ottow: 1978. Env. Biogeochem. Geomicrobiol. 3, pp. 869-878.

23. Van Veen, W. L., E. G. Mulder and M. H. Deinema: 1978. Microbiol. Rev. 42, pp. 329-356.

24. Krumbein, W. E.: 1971. Naturwiss. 58, pp. 56-57.

25. Ehrlich, H. L., W. C. Ghiorse and G. L. Johnson: 1972. Dev. Ind. Microbiol. 13, pp. 57-65.

26. Mulder, E. G.: 1972. Rev. Econ. Biol. Sol. 3, pp. 321-348.

27. Jung, W. K. and R. Schweisfurth: 1979. Zeitschr. Allgemeine Mikrobiol. 19, pp. 107-115.

28. Douka, C.: 1977. Soil Biol. Biochem. 9, pp. 89-97.

29. Douka, C.: 1980. Appl. Env. Microbiol. 39, pp. 74-80.

30. Ghiorse, W. C. and P. Hirsch: 1978. Env. Biogeochem. Geomicrobiol. 3, pp. 897-909.

31. Ghiorse, W. C. and P. Hirsch: 1979. Arch. Microbiol. 123, pp. 213-226.

32. Rosson, R. and K. Nealson: 1982. Manganese binding and oxidation by spores of a marine Bacillus. J. Bacteriol. (in press).

33. Nealson, K. H. and B. M. Tebo: 1980. Origin of Life 10, pp. 117-126.

34. Emerson, S., L. Jacobs, S. Kalhorn, R. Rosson, B. Tebo and K. Nealson: 1982. Environmental oxidation rate of Mn-bacterial catalysis. Geochim. Cosmochim. Acta (in press).

35. Ghiorse, W. C.: 1980. In "Biogeochemistry of Ancient and Modern Environments" (P. Trudinger, M. Walter and B. Ralph, eds.), pp. 345-354. Australian Academy of Sciences.

36. Ehrlich, H. L.: 1966. Dev. Ind. Microbiol. 7, pp. 279-286.

37. Ehrlich, H. L.: 1978. Env. Biogeochem. Geomicrobiol. 3, pp. 839-845.

38. Nealson, K. H. and J. Ford: 1980. Geomicrobiol. J. 2, pp. 21-37.

39. Kepkay, P. and K. Nealson: 1982. Surface enhancement of sporulation ad manganese oxidation by a marine Bacillus. J. Bacteriol. (in press).

40. Ehrlich, H. L.: 1978. Geomicrobiol. J. 1, pp. 65-83.

41. Ali, S. H. and J. L. Stokes: 1972. Antonie van Leeuwenhoek 37, pp.

519-528.

42. Troshanov, E. P.: 1968. Mikrobiologya 37, pp. 934-940.

43. Kuznetsov, S. I.: 1970. "The Microflora of Lakes and its Geochemical Activity," University of Texas Press, Austin, Texas.

44. Ghiorse, W. C. and H. L. Ehrlich: 1974. Appl. Microbiol. 28, pp. 785-792.

45. Trimble, R. B. and H. L. Ehrlich: 1968. Appl. Microbiol. 16, pp. 695-702.

46. Trimble,R. B. and H. L. Ehrlich: 1970. Appl. Microbiol. 19, pp. 966-972.

47. Bromfield, S. M. and D. J. David: 1976. Soil Biol. Biochem. 8, pp. 37-43.

48. Harder, E. C.: 1919. U. S. Geol. Survey Professional Paper No. 113. U. S. Government Printing Office, Washington, D. C.

49. Silverman, M. P. and H. L. Ehrlich: 1964. Appl. Microbiol. 6, pp. 153-206.

50. Drabkova, V. G. and E. A. Stravinskaya: 1969. Mikrobiolya 38, pp. 300-309.

51. Pringshein, E. G.: 1949. Biol. Rev. 24, pp. 200-250.

52. Caldwell, D. E. and S. J. Caldwell: 1980. Geomicrobiol. J. 2, pp. 39-53.

53. Blakemore, R. P.: 1975. Science 190, pp. 377-379.

54. Balkwill, D. L., D. Maratea and R. P. Plakemore: 1980. J. Bacteriol. 141, pp. 720-729.

55. Frankel, R. B., R. P. Blakemore and R. S. Wolfe: 1979. Science 203, pp. 1355-1356.

56. Blakemore, R. P., D. Maratea and R. S. Wolfe: 1979. J. Bacteriol. 140, pp. 720-729.

57. Moench, T. T. and W. A. Konetzka: 1978. Arch. Microbiol. 119, pp. 203-212

58. Bromfield, S. M.: 1954. J. Gen. Microbiol. 11, pp. 1-16.

59. Ottow, J. C. G.: 1968. Zeitsch. Allgemeine Mikrobiol. 8, pp. 441-443.

60. Ottow, J. C. G.: 1969. Z. Bakteriol. Parasitenk. Infektions. Hyg. Abt. II 123, pp. 600-615.

61. Ottow, J. C. G.: 1970. Zeitsch. Allgemeine Mikrobiol. 10, pp. 55-62.

62. Troshanov, E. P.: 1969. Mikrobiologya 38, pp. 634-643.

63. Starkey, R. L. and H. O. Halvorson: 1927. Soil Sci. 24, pp. 381-402.

64. Lascelles, J. and K. A. Burke: 1978. J. Bacteriol. 134, pp. 585-589.

65. Chapnick, S., W. Moore and K. Nealson: 1982. Manganese oxidation in a fresh water lake: Geochemical and microbiological studies. Limnol. Oceanogr. (in press).

66. Tyler, P. A. and K. Marshall: 1967. Arch. Microbiol. 56, pp. 344-353.

67. Tyler, P. A. and K. Marshall: 1967. J. Am. Water Works Assoc. 59, pp. 1043-1048.

68. Tyler, P. A. and K. Marshall: 1967. J. Bacteriol. 93, pp. 1132-1136.

69. Shterenberg, L. E., G. A. Dubinina and K. A. Stepanova: 1975. In "Problems of Lithology and Geochemistry in the Precipitation of Rocks and Ores" (A. V. Peive, ed), pp. 166-181. Nauka, Moscow.

70. Mustoe, G. E.: 1981. Bull. Geol. Soc. Amer. 92, 147-153.

71. Ehrlich, H. L.: 1974. Soil Sci. 119, 36-41.

72. Krumbein, W. E. and K. Jens: 1981. Oecologia 50, 25-38.

73. Wolker, H., R. Schweisfurth and P. Hirsch: 1977. J. Bacteriol. 131, pp. 306-313.

74. Jones, J. G.: 1981. J. Gen. Microbiol. 125, pp. 85-93.

75. Bromfield, S. M.: 1954. J. Gen. Microbiol. 11, pp. 1-6.

76. Sorensen, J.: 1982. Appl. Env. Microbiol. 43, pp. 319-324.

77. Berner, R. A.: 1981. "Early Diagenesis, A Theoretical Approach." Princeton University Press, Princeton, N.J.

78. Brudige, D. J. and J. M. Gieskes: 1982. A pore water/solid phase model for manganese diagenesis in marine sediments. Amer. J.

Sci. (in press).

79. Duinker, J. C., R. Wollast, and G. Billen: 1979. Estuar. Coastal Mar. Sci. 9, pp. 727-738.

80. Ottow, J. C. G.: 1970. Zeitsch. Allgemeine Mikrobiol. 10, pp. 55-62.

81. DeCastro, A. F. and H. L. Ehrlich: 1970. Antonie van Leeuwenhoek 36, pp. 317-327.

82. Boudreau, B. P., and M. R. Scott: 1978. Amer. J. Sci. 278, pp. 903-929.

83. Elderfield, N., N. Leudtke, R. J. McCaffrey, and M. Bender: 1981. Amer. J. Sci. 281, pp. 768-787.

84. Holdren, G. R., O. P. Bricker, and G. Matisoff: 1975. In "Marine Chemistry in the Coastal Environment," ACS Symposium Series No. 18, pp. 364-381. Washington, D. C.

85. Aller, R. C.: 1980. Adv. Geophys. 22, pp. 351-415.

AN IN-SITU METHOD FOR DETERMINING MICROBIAL MANGANESE OXIDATION RATES IN SEDIMENTS

David J. Burdige, Paul E. Kepkay, and Kenneth H. Nealson
Marine Biology Research Division
Scripps Institution of Oceanography
University of California, San Diego
La Jolla, California USA

Abstract. An *in-situ* dialysis technique has been developed to quantitatively examine chemical and bacterial rates of manganese oxidation in sediments under the constraint of diffusion-limited conditions. In a laboratory study of the technique using sediments which have been spiked with manganese oxidizing bacteria, the parameters for both abiotic and microbial processes determined directly with this method agree well with independent measurements of these quantities.

INTRODUCTION

The quantitative examination of manganese oxidation in sedimentary environments is useful for a number of reasons. First, accurate estimates of oxidation rates are necessary to describe manganese cycling and the dynamics of manganese deposition (e.g., the growth of ferromanganese nodules) in relation to the flux of Mn^{2+} from reduced to oxidizing sediments. Second, since both manganese oxidation and reduction may be microbially mediated[1-5] understanding these processes may lend insight into the relationship between metal such as manganese and carbon cycling.

Two approaches have been taken in determining rates of diagenetic reactions such as manganese oxidation. The first is the use of steady-state modelling[6], which has led to the development of a series of models for the quantification of sedimentary manganese oxidation rates[7,8]. Some drawbacks of this approach are that the kinetics of manganese oxidation *in-situ* are poorly understood and rate expressions used in diagenetic equations must, by necessity, include certain assumptions. Unless tested, these could lead to serious errors in the reaction rates obtained by fitting pore water and solid phase data. In addition, modelling alone provides no information about the mechanism of manganese oxidation, that is, the relative importance of microbially-mediated as compared to strictly inorganic processes.

The second approach to examining manganese oxidation in nature involves direct activity measurements of microbial manganese oxidation using

P. Westbroek and E. W. de Jong (eds.), Biomineralization and Biological Metal Accumulation, 481–487.

$^{54}Mn^{2+}$ as a tracer. This procedure is not based on assumptions regarding in-situ kinetics and has been used successfully to differentiate biological from non-biological uptake of manganese in the water column of Saanich Inlet, British Columbia[3]. An analagous approach to studying manganese oxidation in sediments has not (to our knowledge) been successful, and we feel there are a number of reasons why. Individual sediment sub-samples taken from a core and incubated separately are isolated from natural gradients within the sediment. As a result, the diffusive flux of reactants and products through the pore waters are effectively eliminated. This is a major consideration when studying manganese oxidation in sediments because the process is confined to a narrow zone where an upward flux of Mn^{2+} interacts with a downward flux of oxygen[8,10].

To overcome these limitations, we have developed a new technique for the study of manganese oxidation in sediments. The method involves the use of a modified dialysis probe or "peeper"[11] that can be emplaced in a sediment, exposing the sediment to diffusional gradients of dissolved manganese. Not only does the technique allow the in-situ examination of manganese oxidation in sediments under the constraint of flux control, but it can also be used to separate bacterially-mediated manganese binding and oxidation from non-biological reactions such as manganese adsorption, ion exchange, or autocatalytic oxidation.

The results presented in this paper are a feasibility study of the technique using laboratory sediments which have been "spiked" with bacteria whose kinetics of manganese binding and oxidation have been examined in detail[5]. The adsorptive, ion exchange and auto-oxidative capacities of the sediments have also been defined independently[12] so that abiotic processes can be directly compared to those that are bacterially mediated.

GENERAL DESCRIPTION AND MATHEMATICAL THEORY OF THE METHOD

The peeper has been designed so that Mn^{2+} can diffuse uniformly in one dimension out of the dialysis cell and into sediments enclosed by an associated core box (Fig.1). Two situations must be considered, where (i)soluble manganese and a poison (such as sodium azide) are allowed to diffuse into the core box, and (ii)only Mn^{2+} is allowed to diifuse out of the dialysis cell. When both poisoned and unpoisoned peepers contain the same sediment plus bacteria, two distinct manganese profiles are developed in the pore waters sampled away from the cell. The direct comparison of these profiles allows adsorption, ion exchange, and autocatalytic oxidation to be separated from microbially mediated manganese binding and oxidation. However this is only valid if azide diffuses faster than Mn^{2+} and therefore creates the necessary abiotic environment. Recent work[9] has shown that azide is a good poison to use in these types of studies as it does not interfere with the adsorption, oxidation or reduction of manganese.

Once sediment is in the core box, a shutter is opened so that manganese can diffuse out of the dialysis cell. During the experiment, the cell is fed from above by a large reservoir. The top of the core box (when closed) notches into the shutter so that the shutter is kept open and allows diffusion into a predetermined section of sediment for a fixed time interval. After this time period, the peeper is dismantled and the sediments plus pore waters sampled as a function of distance away from the dialysis cell.

The window of the dialysis cell is relatively small and its size is governed by the thickness of the manganese oxidizing zone in the sediments of interest. The vertical continuity of the cell means that Mn^{2+} does not diffuse radially or hemispherically into the zone of oxidation from a series of point sources. The matching of the inner walls of the core box with the sides of the dialysis cell also eliminates the possibility of radial diffusion from the edge of the dialysis cell. Such design constraints greatly simplify the mathematical treatment of the system.

Under these conditions, diffusion in the core box will follow a modified form of Fick's second law[6,13],

$$\frac{\partial c}{\partial t} = \frac{D_s}{(1+K)} \frac{\partial^2 c}{\partial x^2} \qquad (1)$$

where c is the concentration of manganese in the pore waters, D_s is the bulk sediment diffusion coefficient, t is time and x is the distance away from the dialysis membrane. Adsorption or ion exchange has been accounted for in (1) by assuming that these processes are rapid, rever- and can be accounted for in (1) by a linear isotherm[6,13,14]. The form of

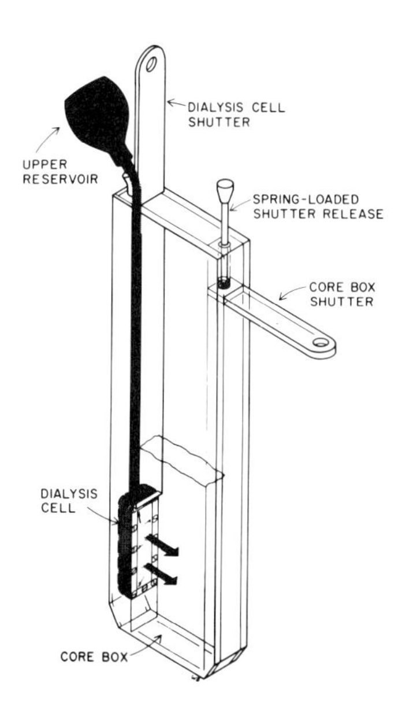

Figure 1. Modified dialysis cell or "peeper" used to study manganese oxidation in sediments. The outer outer dimensions of the core box are 50cm x 16cm x 5cm. The inner dimensions of the box are 50cm x 10cm x 2cm. The dialysis cell is 10cm in height and is situated 5cm from the bottom of the core box.

this isotherm is $\bar{c} = K^* c$, where $\bar{c}$ is the amount of Mn^{2+} sorbed per gram of solid. K^* is related to K (in 1) by the expression $K= \{(1-\phi)/\phi\}\rho K^*$ where ρ is the dry sediment density and ϕ the sediemt porosity. D_S is related to D (the free ion diffusion coefficient) by $D_S=D/\phi F$ where F, the formation factor, is the ratio of wet sediment resistivity to the resistivity of the pore solution alone[15]. This corrects the free ion diffusion coefficient for tortuosity because diffusion in pore waters is hindered by the surrounding particle matrix.

Equation (1) is applicable in systems where there is no microbial activity and can be solved using the following boundary and initial conditions: (i) $x=0, t>0, c=c_o$; (ii) $x>0, t=0, c=c_i$; (iii) $x=\infty, t>0, c=c_i$. The first condition is based on the fact that the dialysis cell serves as a constant source of manganese. In the other conditions, c_i is the background concentration of manganese in the pore waters. With these, the solution to (1) is[13]

$$c(x,t) = c_i + (c_o-c_i)\,\mathrm{erfc}\{x/2\sqrt{D_a t}\} \qquad (2)$$

where $D_a=D_s/(1+K)$ and $\mathrm{erfc}(z)= \{2/\sqrt{\pi}\}\int_z^\infty e^{-y^2}\,dy$.

In peepers where microbial activity is not inhibited by poisons, the microbial uptake of manganese should follow the Michaelis-Menten equation and (1) is modified to become,

$$\frac{\partial c}{\partial t} = \frac{D_S}{(1+K)}\frac{\partial^2 c}{\partial x^2} - \frac{1}{1+K}\left(\frac{V_{max}\,c}{K_m + c}\right) \qquad (3)$$

where V_{max} is the maximum reaction rate and K_m the half saturation constant (or uptake velocity at $\frac{1}{2}V_{max}$). This equation can be solved numerically using finite difference methods[13] and the solution is described in detail in Burdige and Kepkay[12].

MATERIALS AND METHODS

The peepers (Fig.1) used in these experiments were constructed almost entirely out of plastic. The few metal parts (i.e., the screws holding the box together or the pipe fittings to connect the upper reservoir to the dialysis cell) were made of stainless steel to minimize corrosion and contamination. Dialysis membrane (Spectraphor 2 with a molecular weight cutoff of 12,000 to 14,000) was stretched across the cell window and held in place by a thin countersunk stainless steel frame that was fixed to the cell with small screws. A thin film of silicone sealant both sealed the membrane to the cell and minimized leakage around the edge of the overlying frame.

Dialysis cells of unpoisoned peepers were filled with seawater enriched with 300 µM Mn^{2+} that was buffered to pH 7.8 with 10 mM Hepes buffer (N-2-hydroxyethylpiperazine-N'-2-ethanesulfonic acid) and filtered through a 0.2 µm membrane filter (to prevent the chemical precipitation of manganese in the cell). In equivalent poisoned peepers the cells were filled with the same solution supplemented with 20 mM sodium azide.

The sediment used in these experiments consisted of 90% kaolinite (Univ. of Missouri source clay Kga-2) and 10% $CaCO_3$ by dry weight with a final porosity of ϕ=0.7. The solution mixed with these solids was sterile seawater buffered with 10 mM Hepes buffer containing a known number of manganese oxidizing bacteria. The organism used was strain S-36, a marine _Pseudomonas_ sp., whose physiology[16,17] and kinetics of manganese binding[5] have recently been examined. Complete details of the experimental procedures are presented in Burdige and Kepkay[12].

RESULTS

When diffusion was allowed to proceed for 5 days, distinctly different manganese profiles were observed in poisoned and unpoisoned peepers (Figs.2 and 3). The curves shown have been calculated by assuming that the concentration of manganese (or azide) in the dialysis cell is constant over 5 days and equal to the final measured value. While the cell concentration actually decreased between 4 and 16% in these experiments (data not shown), it can be shown[12] that replacing c_o in boundary condition (i) with a range of time dependent functions causes negligible differences in the calculated profiles.

The formation factor of these sediments was 1.65 and with ϕ=0.7 the bulk sediment diffusion coefficient (D_s) for Mn^{2+} was 5.07×10^{-6} cm^2/sec, assuming a free ion diffusion coefficient of $6.88\times10^{-6} cm^2/sec$[18] and $D_s=D/\phi F$. The best fit to the manganese data from the poisoned peeper

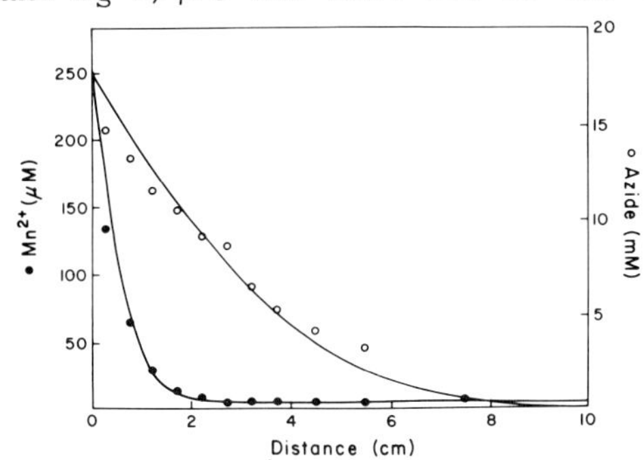

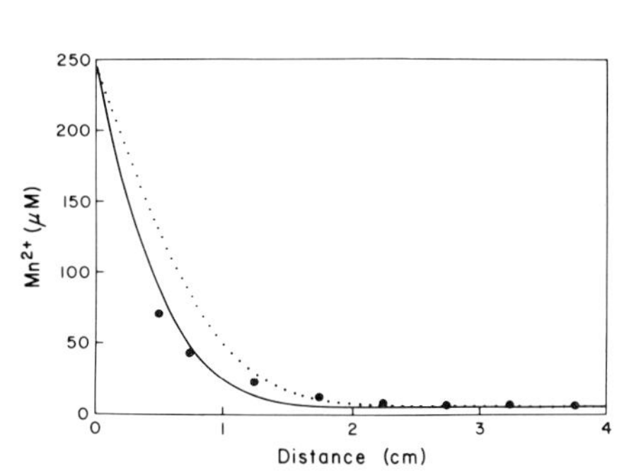

Figure 2. (left) Mn^{2+} and azide profiles in a poisoned peeper after 5 days. The best fit to the manganese data has been calculated using (2) with c_o=249.3 μM, c_i=4.4 μM, $D_s=5.07\times10^{-6} cm^2/sec$ and K=6.9. The best fit azide curve has been calculated with c_o=17.7 mM, c_i=0, K=0, and D_s= $14.1\times10^{-6} cm^2/sec$.

Figure 3. (right) Mn^{2+} profile in an unpoisoned peeper after 5 days. The dotted line has been calculated using the numerical solution to (3)[12] with c_o=244.6 μM, c_i=6.2 μM, $D_s=5.07\times10^{-6} cm^2/sec$, K=6.9 and $K_m=V_{max}=0$. The best fit curve (solid line) has been calculated with K_m=268 μM and V_{max}=24.7 μM/h (all other parameters the same).

(Fig.2) was obtained using (2) with a K of $6.9 \pm 2.2 (1\sigma)$. The azide profile in the same poisoned peeper is also shown in Fig.2. Since kaolinite has a negative surface charge at pH 7.1 (the final pH of the pore waters in these experiments)[19] we have assumed that azide, an anion, does not sorb onto the peeper sediments. With K=0, the best fit azide curve has been obtained using (2) with a bulk sediment diffusion coefficient of $14.1 \pm 2.3 \times 10^{-6} cm^2/sec$. When corrected for porosity and tortuosity the free ion diffusion coefficient was $19.1 \pm 3.1 \times 10^{-6} cm^2/sec$.

The manganese profile (after 5 days) in an unpoisoned peeper is plotted in Fig.3. As this shows, when viable manganese oxidizing bacteria are present, inorganic manganese removal (accounted for by K from the poisoned peeper results) cannot soley explain the observed data. If the additional removal of Mn^{2+} is related to microbial activity then profiles can be calculated using the numerical solution to (3)[12]. Using the K from the poisoned peeper, K_m and V_{max} are sucessively varied to obtain the best fit to the data. The best fit curve shown in Fig.3 has been obtained with a K_m of 268±124 μM and a V_{max} of 24.7±15.4 μM/h. When normalized to a per cell basis (using the initial number of cells present, $9.5 \times 10^8 ml^{-1}$) V_{max} was $2.52 \pm 1.62 \times 10^{-8}$ nmoles $Mn^{2+} cell^{-1} hr^{-1}$.

DISCUSSION

Burdige and Kepkay[12] have recently shown that an equilibrium linear isotherm of the form $\bar{c}=K^* c$ appears to be adequate for describing the abiotic processes which bind manganese to solids. For the sediments used in these experiments, they found that K^* at pH 7.1 ranged from 1.9 to 4.3 ml/gr. When re-expressed as K $(=K^*\{(1-\phi)/\phi\}\rho)$ these values lie between 2.1 and 4.6. The slightly larger K determined from the peeper experiments could be related to the presence of killed bacteria in the sediments since it seems possible that their surfaces might still posess an enhanced capacity for binding manganese. Removal of the is type has been observed by Kepkay et al.[15] in poison controls from their binding experiments.

The K_m for manganese binding by Pseudomonas S-36 in these experiments was 43 to 83 times larger than the K_s of the same organism for growth at the expense of manganese oxidation[17]. If K_m is assumed equal to K_s, Kepkay et al.[5] have shown that V_{max} for manganese binding can be calculated from their data by rearranging the Michaelis-Menten equation. While V_{max} from the peeper results was 3 to 44 times larger than these calculated V_{max} values (when both are normalized to a per cell basis), Kepkay et al.[5] show that the first order kinetics of microbial manganese binding which they observe in their experiments requires that K_m and V_{max} for manganese binding be at least ten times greater than K_s and any calculated V_{max}. The microbial manganese binding parameters obtained from the peeper experiments were consistent with this observation.

When K_m and V_{max} from the peeper experiments were used to generate a Michaelis-Menten curve, first order microbial manganese binding was predicted for Mn^{2+} concentrations less than approximately 15 μM. This implies that in most sedimentary environments microbial manganese binding and oxidation in sediments should be a first order process. The rate constant predicted from the peeper data was 806 yr^{-1}, which is within the range of values reported from a variety of sedimentary and aquatic environments[5].

The results of these experiments indicated that when used together, poisoned and unpoisoned peepers could separate and quantify the enhanced removal of $Mn^{2+}$from pore solutions due to the activity of manganese oxidizing bacteria. Poison peeper results showed that azide diffuses faster than manganese so that the requisite abiotic environment is created. The parameters for abiotic and microbial manganese binding determined with this technique agree well with independent measurements of these quantities and therefore validate the future use of this method for studying manganese binding and oxidation in natural sediments.

ACKNOWLEDGMENTS

The work was supported by NSF(IDOE) grant number OCE 80-27838, a grant from the Scripps Industrial Associates and was carried out while Kepkay was the recepient of an NSERC of Canada post-doctoral fellowship.

REFERENCES

1.Wollast,R.,Billen,G. and Duinker,J.C.,1979,Est. Coastal Mar. Sci. 9, pp.161-169.
2.Froelich,P.N.,et al.,1979, Geochim. Cosmochim. Acta 43,pp.1075-1090.
3.Emerson,S.A., et al.,1982, Geochim. Cosmochim. Acta in press.
4.Nealson,K.H.,1982, in Krumbein,W.E.,ed.,Biogeochemistry, in press.
5.Kepkay,P.E.,Burdige,D.J. and Nealson,K.H.,1982, submitted to Geomicrobiol. J.
6.Berner,R.A.,1980,Early Diagenesis,A Theoretical Approach, Princeton: Princeton Univ. Press,241p.
7.Boudreau,B.P. and Scott,M.R.,1979, Am. Jour. Sci. 278, pp.903-929.
8.Burdige,D.J. and Gieskes,J.M.,1982, Am. Jour. Sci. in press.
9.Rosson,R.A.,Tebo,B.M. and Nealson, K.H.,1982, submitted to Appl.and Environ. Microbiol.
10.Klinkhammer,G.P.,1981, Earth Planet. Sci. Letters 46, pp.361-384.
11.Hesslein,R.H.,1976, Limnol. Oceanogr. 21, pp.912-915.
12.Burdige,D.J. and Kepkay,P.E.,1982, submitted to Geochim. Cosmochim. Acta.
13.Crank,J.,1975,The Mathematics of Diffusion,Oxford:Clarendon Press,414p.
14.Duursma,E.K. and Hoede,C.,1967, Net. J. Sea Res. 3, pp.423-457.
15.McDuff, R.E. and Ellis,R.A.,1979, Am. Jour. Sci. 279, pp.666-675.
16.Kepkay,P.E. and Nealson,K.H.,1982a, submitted to J. Bacteriol.
17.Kepkay,P.E. and Nealson,K.H.,1982b, submitted to Arc. Microbiol.
18.Li,Y.-H. and Gregory,S.,1974, Geochim. Cosmochim Acta 38, pp.703-714.
19.Stumm,W. and Morgan,J.J.,1981,Aquatic Chemistry,New York:Wiley-Interscience,774p.

FE AND MN DEPOSITING BACTERIA IN MARINE SUSPENDED MACRO-PARTICULATES

James P. Cowen
Center for Coastal Marine Sciences
University of California
Santa Cruz, California USA

Fe and Mn precipitating bacteria are known to occur in many diverse environments. Their activity has been studied in soils, streams, lake and marine sediments and in the water column of enclosed inlets and lakes (Gebers and Hirsh, 1978; Nealson et al., 1979). However, the literature contains few examples of corresponding studies from the pelagic zones of the oceans.

This paper describes the discovery of Fe and Mn precipitating bacteria in suspended macro-particulates collected at five stations in the Pacific Ocean. Fe and Mn were also associated with amorphous polymer-like material similar in morphological appearance to the metal-oxide encrusted polymers characteristic of many Fe and Mn precipitating bacteria (Ghiorse and Hirsh, 1979; Nealson and Tebo, 1980). This study was made possible by the availability of specialized methods for the collection of fragile macro-particulates suspended in the ocean and by the Scanning Transmission Electron Microscopy (STEM) - Energy Dispersive X-ray Spectrometry (EDS) analysis technique. The use of thin sectioned material and the STEM-EDS allowed the simultaneous recording of the internal ultrastructure and the identification and microanalysis of submicrometer particles including polymer enshrouded bacteria; SEM-microanalysis of suspended particulates (Honjo, 1980; Lambert et al., 1981), while useful in indicating the presence of particular elemental matrices, is limited in the extent to which it allows the recognition of particles without distinctive surface features. A brief discussion of the potential biogeochemical implications of the relationship between these "metal" bacteria and the suspended macro-particulates will follow a description of the sampling and analytical techniques and resulting data.

Methods

Descriptive data for each sample are summarized in Table 1. Samples were collected by one of three methods: i) Deep Tow (DT), an unmanned submersible equipped with a plankton net and pulled by ship through the nephaloid layer, 10-100 meters above the sea floor, (Wishner, 1980); ii) the *Alvin*, a manned submersible equipped with discrete

P. Westbroek and E. W. de Jong (eds.), Biomineralization and Biological Metal Accumulation, 489–493.

Table 1. The occurrence of Fe and Mn precipitating halo bacteria and Fe, Mn rich polymer-like material in suspended macro-particulates from five Pacific Oceanic Stations.

SAMPLE ID†	DEPTH (m)	HALO BACTERIA Obs.**	HALO BACTERIA Deposits[i] Fe	HALO BACTERIA Deposits[i] Mn	POLYMER MATERIAL Obs.**	POLYMER MATERIAL Deposits[i] Fe	POLYMER MATERIAL Deposits[i] Mn
Floc. Mat. (S-PIT)ǂ 21°51'N, 109°50'W	100	++	+	-	+	+	-
Floc. Mat. (PIT-C)# 35°35'N, 123°50'W	200	++	+	-	no		
Floc. Mat. (PIT-F)# 35°35'N, 123°50'W	600	no			+	+	+
Disc. Floc. (Alvin-D) 32°31'N, 118°03'W	1650	++	(na)		++	(na)	
Disc. Floc. (Alvin-B) 21°46'N, 104°50'W	2500	+	+	-	no		
Floc. Mat. (DT-2) 0°33'S, 85°33'W	neph.*	++	+	+	++	+	+

†Samples are described as 1) floc. mat.(amorphous flocculent material collected in cod end of Deep Tow net (sample DT-2) or in particle interceptor traps (PIT samples) or 2) as disc. floc.(samples collected by Alvin as discreet flocculent aggregate, i.e. marine snow). ǂLarge volume Soutar-type PIT, Bruland et al., 1981. #Knauer-type multi-PIT, Knauer et al., 1979. *Neph.: nephaloid layer 10-100 m above the bottom in depths of 2700-2900 m. **Observed as present (+); common (++); or not observed (no). i peaks significantly above background (+), or not (-); na (not analyzed).

particle samplers (Silver and Alldredge, 1981); or, iii) particle interceptor traps (PITs) which are suspended at depth in the water column and are designed to collect vertically fluxing particles (Knauer et al., 1979; Bruland et al., 1981). Processing for STEM followed the methods of Silver and Alldredge, 1981 or Gowing and Silver, in prep. Thin sections (60-90nm) were placed on nylon grids, carbon coated and analyzed under a Hitachi H-500 STEM equipped with an Ortec 6230 Si(Li) detector for EDS. In some cases replicate sections for photography were collected onto copper grids and stained with lead-citrate and uranyl acetate in order to improve contrast in biological material.

Results and Discussion

Representative EDS spectra and TEM micrographs, on which the areas of analysis are indicated, are presented in Figure 1. Morphologically the bacterial polymer-like haloes closely resemble reported metal encrusted structures associated with Mn and Fe precipitating bacteria (Ghiorse and Hirsh, 1978; Nealson and Tebo, 1980; Tebo, personnel communication). The spectra for the haloes show significant Fe (Fig. 1c)

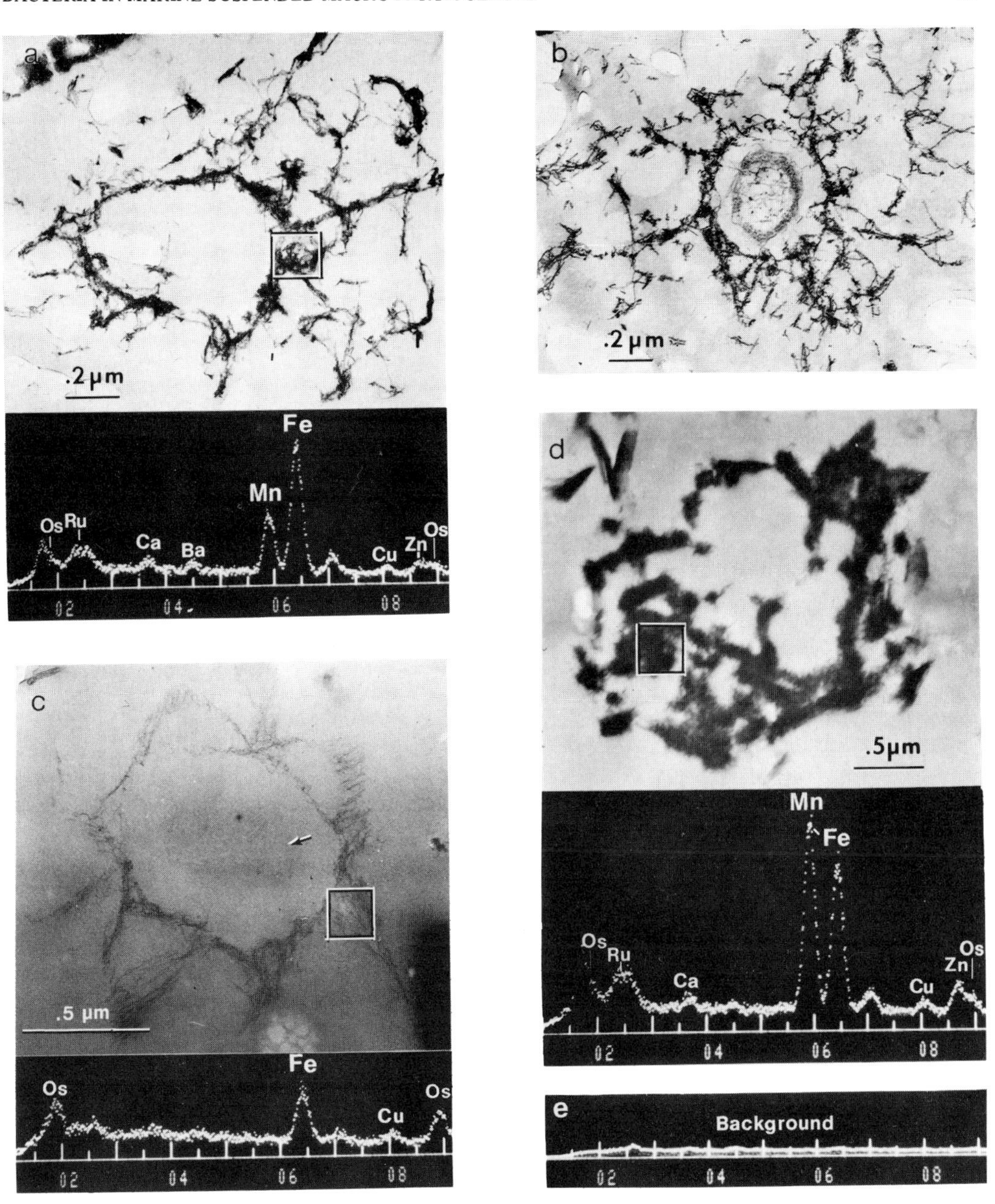

Figure 1. Representative TEM micrographes and EDS spectra:
a) Halo of Fe and Mn rich polymers. Compare to (b). b) Pb citrate and uranyl acetate stained halo bacterium. c) Iron rich halo bacterium; arrow indicates poorly stained cell (no Ru stain used in this prep.) d) Mn and Fe rich polymer-like material; note the close similarity between spectra of (d) and (a). e) Typical background spectra of EDS analysis from an area of the section near (d) but devoid of sample material.

or Mn and Fe (Fig. 1a) peaks relative to background analysis from nearby areas of the section which were devoid of apparent specimen material. The presence of osmium and ruthenium peaks show that these stains were taken up by the halo material. Ru is thought to bind to polymers which are highly substituted with acid residues, such as acid mucopolysaccharides (Luft, 1966), while Os (as OsO_4) tends to bind to lipids (phospholipid membranes, secretions) and proteins (Weakler, 1977). Small variable amounts of Ca, Ba, Cu, and Zn were also present.

Fig. 1d illustrates Fe and Mn rich polymer-like material apparently unassociated with bacteria. Its EDS spectrum is similar to Fig. 1a. Spectra from other polymer-like material approximate those shown in Fig. 1c. A large number of analyses were also made on a variety of both identifiable and unrecognizable amorphous materials within the particulate samples. Significant Fe and Mn peaks were rare, mostly restricted to the halo bacteria and polymer-like material already mentioned, and to probable inorganic clays and Fe, P rich precipitates. Neither Os nor Ru stains were associated with materials from the latter two catagories.

Table 1 summarizes the occurrence of Fe and Mn encrusted halo bacteria and polymer-like materials found in the present particulate samples. However, systematic relative abundance data for Fe and Mn rich material in these sample particulates are not yet available. Cursory observations indicate great variability. Ratios of "halo" bacteria to total bacteria range from 0% to 3-5%. None have yet been found in near surface suspended particulate samples while the 200m PIT-S sample had an abundance of Fe depositing, haloed bacteria. However, bacteria able to induce Mn^{++} oxidation in solid culture media (leukoberbelin blue technique, Krumbein and Altmann, 1973) have been isolated from numerous oceanic water samples, including the surface water (Cowen, unpublished observation). Bacteria with Mn rich haloes have so far been found, *in situ*, only in the sample from 10-100m above the sea floor (DT-2), perhaps reflecting a resuspension of sediment and associated microorganisms up into the nephaloid layer (see Baker and Feely, 1978).

The EDS spectra provide a qualitative assessment of relative element abundances. Substantial heterogeniety of the sample matrix preclude quantitative comparisons between individual analyses on different samples or even within a single sample. However, qualitative evaluation of relative element abundances (eg. peak ratios) for Mn and Fe in conjunction with histological information obtainable from heavy metal stains (Os, Ru) and morphological data from micrographs can be used to describe the metal content and chemical and biological nature of these organic and inorganic substances.

Recent studies have shown that particles greater than 50um are responsible for greater than 90% of the vertical flux in the oceans (Spencer et al., 1978). Important sources of large particles include zooplankton which effectively package unassimilated ingested particles into rapidly settling fecal pellets and marine snow which consist of fragile polymorphous aggregations of living organisms and biotic and abiotic detritus (Silver et al., 1978; Honjo, 1980; Lambert et al., 1981; Bruland and Silver, 1981). Suspended macro-particulate trace element loads include that fraction present in the matrices of its constituent subparticles (abiotic clays, biogenic minerals and debris, living organisms) and a scavenged fraction resulting from adsorption, ion-exchange, and authigenic (de)mineralization processes. Bacterially mediated metal precipitation would most effect the latter "scavenging" fraction. The presence of small but significant peaks for Zn and Cu in the spectra described above (Fig. 1) indicates that concurrent scavenging of these other trace elements may also be important.

Data presented here demonstrate that Fe and Mn precipitating bacteria are present in suspended pelagic particulates suggesting active biomineralization at these sites. Contributions of Fe and Mn to the particulate fraction from this biological source may have important implications with regards to elemental vertical transport, speciation, and recycling in the water column.

Acknowledgements: I gratefully thank M. Silver, R. Franks, P. Davoll, G. Cutter, W. Landing, B. L. Quon, J. Nowell, R. La Board and P. Bolt for their help and comments. A. Alldredge, M. Gowing, M. Silver and K. Wishner generously provided access to samples. This work was supported by the VERTEX program, NSF (80-003200); Alvin program, OCE 80-19525; and the NASA Planetary Biology Intern Program.

References

Baker, E. T. and Feely, R., 1978: Science 200, pp. 533-535.
Bruland, K. W.; Franks, R. P.; Landing W. M. and Soutar, A., 1981: Earth and Planetary Science Letters 53, pp. 400-408.
___ and Silver, M. W., 1981: Marine Biol. 63, pp. 295-300.
Gebers, R. and Hirsch, P., 1978: Environmental Biogeochemistry and Geochemistry Vol.3. Ed. W. E. Krumbein; Ann Arbor Science Publ. Inc.
Ghiorse, W. C. and Hirsh, P., 1979: Arch. Microbiol. 123, pp. 213-226.
Honjo, S., 1980: J. Marine Res. 38, pp. 53-97.
Knauer, G. H.; Martin, J. H. and Bruland, K. W., 1979: Deep Sea Res. 26, pp. 97-108.
Krubien, J. and Altmann, H. J., 1973: Helgol. Wiss Meeresunters 25, pp. 347-356.
Lambert, C. E.; Jehanno, C.; Silverberg, W; Brun-Cottan, J. C. and Chesselet, R., 1981: J. Marine Res. 39, pp. 77-98.
Luft, J. H., 1966: Sixth Intern. Congress for Electron Microscopy KYOTO
Nealson, K. H.; Moore, W. and Chapnick, S., 1979: Proc. Fall Meeting Amer. Geophys. Union.
___ and Tebo, B., 1980: Origins of Life 10, D. Reidel, Holland, pp-117-126
Spencer, D. W.; Honjo, S. and Brewer, P. G., 1978: Oceanus 21, pp. 20-25.
Silver, M. W. and Alldredge, A., 1981: J. Marine Res. 39, pp. 501-530.
___; Shanks, A. L. and Trent, J. D., 1978: Science 201, pp. 371-373.
Weakley, B. S., 1977, in Biological TEM, ch. 4, pp. 69-81.
Wishner, K. F., 1980: Deep Sea Res. 27A, pp. 203-216.

MICROBIAL ROLE IN WITWATERSRAND GOLD DEPOSITION

Betsy Dexter-Dyer Grosovsky
Department of Biology, Boston University, Boston, MA 02215 USA

The 2.4 billion-year-old Witwatersrand system in South Africa includes carbonaceous material closely associated with gold-bearing sand and conglomerates (Reimer, 1975). The environment of deposition is believed to have been a shallow-water alluvial fan, most likely covered with prokaryotic microbial mats (Zumberge, 1978). Microfossils of prokaryotes may be associated with the gold in these sediments. The opacity of the rock and the interference of ionizing radiation have made both thin-section and chemical studies of these microfossils difficult (Nagy, 1976).

The Witwatersrand gold may in some way have been trapped by the microbes, by precipitation from either a dissolved or colloidal state (Hallbauer, 1975); however, the pathways by which the original volcanic gold deposit was solubilized, and the dissolved gold subsequently deposited in the mats, have not been satisfactorily explained. One difficulty is that the mechanisms must have been compatible with an anaerobic atmosphere and with oxygen present only in microenvironments associated with cyanobacteria.

Biogenically produced cyanide may have dissolved ancient volcanic gold deposits, and the resultant gold cyanide could have been trapped by microbial mats (Praetorious, 1974). This hypothesis is compatible with the observation that some prokaryotes are cyanogenic (cyanide producing) and that in some cases this capability may be enhanced by anaerobiosis, or that at least cyanogenesis can occur facultatively under anaerobic conditions (Knowles, 1976; Castric, 1981). It is also compatible with numerous observations that modern cyanogenic eukaryotes (plants and fungi) dissolve gold and in some cases take up the resultant ions (Lakin et al., 1974; Shacklette, et al., 1970; Girling, 1980). The major difficulty with this hypothesis is that oxidation is required to dissolve gold in the presence of cyanide and subsequent precipitation requires reducing conditions (King, 1949). It is difficult to imagine an appropriate combination of anaerobic and aerobic conditions which would have made this mechanism viable.

P. Westbroek and E. W. de Jong (eds.), Biomineralization and Biological Metal Accumulation, 495–498.

Original volcanic gold deposits may have been weathered into colloidal form and then somehow entrapped by microbial mats (Nagy, 1981). Organisms can influence both the formation of colloidal gold and subsequent flocculation (aggregation of a colloid). Fungi have been shown to cause the flocculation of colloidal gold, but the mechanism is unknown (as reviewed by Boyle, 1979). Freise (1931) first noted that under anaerobic conditions, humic acid (degradation products or "protokerogen") causes the dissolution of gold and that in the presence of oxygen, carbonate, and pyrite, this gold becomes deposited.

It has been suggested that acidic breakdown products of microorganisms could have stabilized colloidal gold weathered from rocks (Ling Ong, 1974), and it is believed that gold weathered in the presence of humic acid would be in colloidal form (B. Nagy, personal communication). Indeed, several modern bacteria isolated from gold fields (Parès, 1965a, b) are able to dissolve gold, and this ability seems to be related to their production of organic acids (as reviewed by Rapson, 1982).

Based on these considerations and preliminary data to be presented below concerning the flocculation of colloidal gold and the precipitation of dissolved gold by modern microbial mats, the following model for the deposition of Witwatersrand gold is suggested:

Humic acid, as a byproduct of microbial mat decomposition in the presence of an anaerobic atmosphere influenced the weathering of gold from ancient volcanic deposits. Weathered gold then traveled as a colloid, stabilized by humic acid and the anaerobic conditions, to the sites of microbial mats. These mats, producing oxygen by photosynthesis and actively causing the precipitation of carbonates, provided ideal sites for the gold to be flocculated. The model thus postulates the active influence of microbial communities on the Witwatersrand deposits.

Preliminary experiments have been performed to approach the question: Can modern microbial mats cause the flocculation of colloidal gold, or the precipitation from dissolved gold chloride of metallic gold? These experiments were designed as follows:

A hand sample of microbial mat from Laguna Figueroa, Baja California, Mexico, was placed in a flask of artificial seawater, sealed with parafilm, and left on a north-exposed windowsill for a year. This community, consisting of cyano-, purple and other bacteria was broken into small pieces and placed in the bottoms of 160ml milk dilution bottles, which were filled with artificial seawater and capped. These bottles were incubated under grow lights for two weeks. At this point the bottles were divided into two sets; one set was autoclaved to kill all organisms, while the organisms in the other set were left undisturbed. A solution of gold chloride was then added to half the bottles in each set to give a gold content of 0.08% (w/v) (800ppm, or about 10^{-4}M); the final volume in each bottle was 120ml. Within the first couple of days the gold chloride formed a deep purple opaque colloid, for reasons not well understood. After a week it was obvious to the naked eye that the colloid in the "living" mat bottles had flocculated to such an extent that the fluid above these mat samples was almost clear. In the control bottles in which gold solution had been added to autoclaved mat the supernatant remained colloidally opaque for the 22-day duration of the experiment.

An additional experiment was done by the same design with the exceptions that the organisms used were cyanobacteria and other prokaryotes collected from the edge of the Charles River in Boston, Massachusetts, and the water used in filling the bottles was Charles River water, which was slightly brackish. Initially, all the gold chloride remained in solution (rather than forming a colloidal suspension as in the preceding experiment). After three days the golden-colored solution was still manifest in the autoclaved bottles; however, the water had been cleared above the organisms in the non-autoclaved bottles. A stannous chloride spot test for gold chloride (Wise, 1964) confirmed that the gold chloride in the autoclaved bottles remained in solution above the dead microbes for the 13-day duration of the experiment. No gold chloride was detected by this assay above the living mat samples. This result is consistent with experiments (Lakin, et al., 1974) which demonstrated that gold chloride has a strong tendency to co-precipitate with carbonates and sulphides, as both of these types of compounds are probably being actively produced in the Charles River cyanobacterial community.

To extend these findings, it will be necessary to do further experiments with much smaller quantities of gold (i.e., in the parts per million range) and to use more sensitive methods to assy the metal. The results of this work, involving microbial communities from sources as disparate as Baja California, Mexico, and Boston, Massachusetts, provide preliminary evidence that many cyanobacterial communities possess mechanisms for the active precipitation or flocculation of metallic gold. This supports the proposed model for live microbial involvement in the deposition of Witwatersrand gold.

REFERENCES

Boyle, R.W.: 1979, "The geochemistry of gold and its deposits (together with a chapter on geochemical prospecting for the element)," Geo. Sur. Bull. Canada Geol. Sur., no. 280.

Castric, P.: 1981, "The metabolism of hydrogen cyanide by bacteria," in Cyanide in Biology, ed. B. Vennesland, E.E. Conn, C.J. Knowles, J. Westleg, and F. Wissing, Academic Press, London.

Freise, F.: 1931, "The transportation of gold by organic underground solutions," Econ. Geol. 26, pp. 421-431.

Girling, C.A., and Peterson, P.J.: 1980, "Gold in Plants," Gold Bull. 13(4), pp. 151-158.

Hallbauer, D.K.: 1975, "The Plant Origin of the Witwatersrand Carbon," Min Sci. Eng. 7, pp. 111-131.

King, A.: 1949, Gold Metallurgy in the Witwatersrand a textbook of Rand Mettallurgical Practice, Chas. Griffin and Co. Ltd., London.
Knowles, C.J.: 1976, "Microorganisms and cyanide," Bact. Rev. 40, pp. 652-680.
Lakin, H.W., Curtin, G.C., Hubert, A.E., Shacklette, H.T., and Doxtader, K.G.: 1974, Geochemistry of Gold in the Weathering Cycle, Geo. Surv. Bull. 1330, U.S. Dept. Int. U.S. Govt. Printing Office, Washington, D.C.
Ling Ong, H., and Swanson, V.E.: 1974, "Natural Organic Acids in the transportation, deposition, and concentration of gold," Colo. Sch. Mines Q. 69, pp. 395-425.
Nagy, B.: 1981, The geochemistry of organic matter in ore deposits. Carnegie Inst. of Washington Geophysical Lab extended abstracts and bibliographies of participants 97-99.
Nagy, B., and Nagy, L.A.: 1976, "Interdisciplinary search for early life forms and for the beginning of life on earth," Interdisc. Science Rev. 1(4), pp. 291-310.
Parès, Y., Cuper, J., Geraud, J., Loko, H., Gagnaire, L., Mauvieux, L., and Preira, M.: 1965a, "Utilisation de Milieux de culture de Faible prix de revient pour la mise en solution de l'or par voie bactérienne," Ann. de l'Inst. Pasteur 108, pp. 674-681.
Parès, Y., Hans-Moevi, Loko, H., and Gagnaire, L.: 1965b, "Influence de la concentration du milieu de culture sur la mise en solution de l'or par voie bactérienne," Ann. de l'Inst. Pasteur 108, pp. 815-819.
Praetorious, D.: 1974, "Gold in the Proterozoic sediments of South Africa: Stystems paradigms and models," U. Witwatersrand Econ. Geol. Res. Unit Inf. Circ., 87, pp. 1-22.
Rapson, W.S.: 1982, "Effects of biological systems on metallic gold," Gold Bull. Int. Gold Corp. 15(1), pp. 19-20.
Reimer, T.O.: 1975, "The age of the Witwatersrand system and other gold-uranium placers: Implications on the origin of mineralization," N. Jahrbuch für Mineralogie Monatshafte, pp. 79-98.
Shacklette, H.T., Lakin, H.W., Hubert, A.E., and Curtin, G.C.: 1970, Absorption of Gold by Plants, Geo. Sur. Bull. 1314-B U.S. Govt. Printing Office, Washington, D.C.
Wise, E.: 1964, Gold Recovery Properties and Applications, D. van Nostrand Co. Inc., Torotno, 367 pp.
Zumberge, J.E., Sigler, A.C., and Nagy, B.: 1978, "Molecular and elemental analysis of the carbonaceous matter in the gold and uranium bearing Vaal Reef carbon seams, Witwatersrand sequence," Min. Sci. Eng. 10, pp. 223-246.

BIOLOGICAL ACCUMULATION OF SOME HEAVY METALS-BIOTECHNOLOGICAL APPLICATIONS

J.A. Brierley
Department of Biology, New Mexico Institute of Mining and Technology, Socorro, New Mexico USA and
C.L. Brierley
Advanced Mineral Technologies, Inc., Socorro, New Mexico USA

ABSTRACT

Microorganisms, including algae, bacteria and fungi, accumulate metals by adsorption or absorption. Adsorption occurs when positively charged inorganic ions are attracted to negatively charged ligands on the surfaces of microorganisms. Metal accumulation by microorganisms is viewed as a viable, commercial alternative to costly and often ineffective physical-chemical technologies for the treatment of vast quantities of wastewater containing low concentrations of soluble and particulate metals. Several mining operations have installed algal and bacterial treatment systems for removal of metals from wastewaters. These treatment systems depend on the uptake of ions by algae and cyanobacteria and the precipitation of metals by hydrogen sulfide generated by sulfate-reducing bacteria. A greater sophistication in engineering of these microbial treatment systems may enhance their effectiveness in wastewater renovation and resource reclamation.

INTRODUCTION

Biotechnological processes have been applied to recovery of heavy metals. These processes include the use of microorganisms for extraction of metals from minerals and concentration and recovery of metals from dilute solutions. Such microbial applications are of potential use for either economic recovery of metals or treatment of waste streams for environmental protection.

Microbial Leaching

The most well defined system for application of microorganisms to the metal's area is the hydrometallurgical process of leaching of metals from minerals (Brierley, 1978). Several types of microbes are active in metal's extraction. These include the mesophilic *Thiobacillus ferrooxidans* and *Thiobacillus thiooxidans*, unnamed facultative thermophilic iron-oxidizing bacteria (Norris et al., 1980), and the extremely

P. Westbroek and E. W. de Jong (eds.), Biomineralization and Biological Metal Accumulation, 499–509.

thermophilic *Sulfolobus* species (Brierley, 1980). The leaching bacteria participate in dissolution of metals by processes termed "indirect" and "direct". The bacteria oxidize ferrous iron to ferric iron (1), the latter being an effective oxidant of metal sulfides

$$4FeSO_4 + O_2 + 2H_2SO_4 \rightarrow 2Fe_2(SO_4)_3 + 2H_2O \quad (1)$$

such as chalcocite (2) and uraninite (3). The "direct" process of metal's

$$Cu_2S + 2Fe_2(SO_4)_3 \rightarrow 2CuSO_4 + 4FeSO_4 + S \quad (2)$$

$$UO_2 + Fe_2(SO_4)_3 \rightarrow UO_2SO_4 + 2FeSO_4 \quad (3)$$

solubilization involves contact between microbes and minerals with oxidation of minerals such as pyrite (4) and sphalerite (5). Solubilized

$$4FeS_2 + 15O_2 + 2H_2O \rightarrow 2Fe_2(SO_4)_3 + 2H_2SO_4 \quad (4)$$

$$ZnS + 2O_2 \quad ZnSO_4 \quad (5)$$

metal sulfate is recovered by precipitation or solvent extraction. These processes are primarily applied for recovery of copper and uranium. Copper is reclaimed from mine waste material by the process of dump leaching, and uranium is extracted from mined-out areas and heaps or piles of low grade ore materials. Although the microbes are active participants in metals leaching, no processes have been specifically designed to enhance biological activity.

Biological Accumulation

Microbial processes have potential for use in recovery of heavy metals from dilute solutions. The organisms function by either accumulation of dissolved and particulate metals or by production of by-products which render the metals insoluble. Although algae, bacteria and fungi concentrate a variety of metals from solution, yeasts were found to be more effective than bacteria in accumulating two metals, nickel and copper (Norris and Kelly, 1979). However, bacteria effectively concentrate potassium, manganese, iron and calcium (Eisenstadt, et al., 1973) and produce complexing agents which selectively extract metals from dilute solution (Pollack, et al., 1970).

Two mechanisms have been identified which enable microorganisms to concentrate metals: binding of metals to cell surface components such as cellular ligands and biopolymers (biosorption) and metabolism-dependent intracellular accumulation (Gadd and Griffiths, 1978). Hatch and Menawat (1979) reported that *Sphaerotilus natans*, a bacterium found in waste sludges and polluted waters, accumulated iron, magnesium, copper, cobalt, and cadmium in an external mucilage layer. The accumulation was a mass transfer limited process. Other physiological activities of cells can also lead to insolubilization of metals. Accumulated uranium (VI) was reduced by cell extracts of *Micrococcus lactilyticus* with hydrogen serving as an electron donor. The cell extract also reduced molybdate to molybdenum blue and selenium to selenate (Woolfork and Whiteley, 1962).

Water treatment systems have been developed to exploit the metal-concentrating ability of algae. These systems include the treatment of lead mine effluents (containing lead, zinc, copper and manganese) (Gale and Wixson, 1979), zinc mine effluents (zinc, cadmium and copper) (Jackson, 1978), and municipal wastewaters (copper, chromium and cadmium) (Filip, et al., 1979).

The development of an effective microbial process for metal removal from dilute solutions is dependent not only upon the removal of metals by the organisms but also upon the fate of the metals in the treatment system. The fate of accumulated metals, following the death and lysis of microbes, was assessed by Guthrie and Cherry (1979). Metal concentrations increased in sediments as the accumulator organisms died, but some metals, which became complexed to organic matter, returned to the aqueous system. Both living and dead algal material accumulated heavy metals (Ferguson and Bubela, 1974; Gutknecht, 1965; Jennett, et al., 1979); however, the metals often were not irreversibly bound to the algal material. The retention of heavy metals in treatment systems may be influenced by interactions between the algal-bound metals and sediment. The presence of algal material in sediment reportedly enhanced retention of heavy metals (Jackson, 1978); however, other studies have suggested that algal material may also mobilize some metals from sediment (Laube, et al., 1979) by formation of water soluble organo-metal complexes.

This paper reviews the applications of microorganisms in reclamation of metal-contaminated wastewater and discloses field and laboratory data obtained from an active biological wastewater treatment system.

BIOACCUMULATION OF URANIUM

Microbial technology has been applied to the treatment of uranium-contaminated effluents of the nuclear energy industry. Such effluents often discharge large amounts of water containing small concentrations of metals, but these discharges must frequently comply with environmental regulations. Biotechnology research has focused on the use of both living, active microbial cells, as well as dead cells. Non-active cells have been considered due to the toxic nature of some effluents.

Shumate et al. (1978) and Strandberg et al. (1981) tested *Saccharomyces cerevisiae* and *Pseudomonas aeruginosa* as biosorbants for uranium. Each type of microorganism accumulated from 10 to 15% of its dry weight of uranium. The mechanism of accumulation differed for the two microbes. The cells of S. cerevisiae accumulated the uranium on the cell surface in a layer 0.2 μm thick. Since only 32% of the total cells accumulated uranium, the uranium concentration factor approached 50% of cell dry weight. The selectivity of certain cells to accumulate uranium has not been ascertained. Ps. aeruginosa accumulated uranium internally, and metabolism was not required for uranium to cross the cell membrane. The mechanism of intracellular accumulation has not been defined. The rate at which the bacterial and yeast cells reached an equilibrium concentration of uranium varied. S. cerevisiae were "slow", reaching equilibrium after one hour; Ps. aeruginosa were "fast" with equilibrium achieved within 10 minutes after contact with uranium. A system for removal of uranium from wastewater was explored using a mixed culture of unidentified denitrifying bacteria (Shumate, et al., 1980), derived from treatment of waste-streams to remove nitrate. Use of cellular material produced from other biotechnological processes rather than growing a culture solely for biosorption would be economically advantageous. Uranium biosorption using denitrifying organisms reached equilibrium distribution in 15 minutes with a cellular concentration of 140 mg uranium g^{-1} cell dry weight. A bioreactor was designed using a biofilm on anthracite coal particles (30 to 60 mesh). A coal column (5 cm x 1 m) containing adsorbed microbes decreased the solution concentration of uranium from 25 mg ℓ^{-1} to 0.5 mg ℓ^{-1} with an 8 minute residence time of the uranium solution.

Rhizopus arrhizus biomass, produced as a by-product of industrial fermentations, has potential for use as a biosorbant of uranium and other metal cations (Tsezos and Volesky, 1981, 1982). These fungal cells have a uranium and thorium uptake capacity of over 180 mg g^{-1} dry weight. This exceeds the capacity, by 2.5 times of a common anionic exchange resin (IRA-400) used by uranium production companies for selective separation of uranium from other ions in solution. R. arrhizus was also found to be a better biosorbant than other types of biomass examined. Three mechanisms are apparently involved in biosorption of uranium by R. arrhizus cells. First, uranium coordinates with the amine nitrogen of the chitin component of the cell wall (the cell wall is the location of uranium accumulation by this microbe). This mechanism reportedly accounts for an accumulation of 6 mg U g^{-1} cells. Secondly, the complexed uranium apparently acts as a nucleation site for accumulation of additional uranium; this accumulation fits a Freundlich adsorption isotherm model. These two processes account for rapid accumulation (loading equilibrium plateau within 60 seconds) of 66% of the total capacity. The third mechanism is a slower process, reaching equilibrium after 30 minutes and involves the hydrolysis and subsequent precipitation of uranyl hydroxide ($UO_2(OH)_2$) on the cell wall. The biosorption processes achieve 66% equilibrium within 60 seconds with a loading in excess of 120 mg g^{-1} cell dry weight. The biosorption of uranium by R. arrhizus is a physico-chemical process with potential for technical applications. The authors (Tsezos and Volesky, 1981, 1982) suggest that a fluidized bed reactor would best facilitate a high-rate removal process.

MINE WASTE WATER TREATMENT

Sterritt and Lester (1979) published a short review of microbiological control processes for treatment of waste water produced by the mining industry. Although algae are effective in accumulation of heavy metals from dilute solutions, consortia of microorganisms may be more effective in reclamation of contaminated water. A treatment process, which uses a microbial consortium for removal of uranium, selenium, radium and molybdenum from uranium mine waters, is in effect in the Ambrosia Lake District near Grants, New Mexico USA. Part of the treatment process involves passage of the water through a series of three consecutive "algae-ponds". The role of algae and bacteria in removal of the inorganic pollutants has been evaluated (Brierley and Brierley, 1980; Brierley et al., 1980).

Uranium and molybdenum were each present in the algae pond water at concentrations of 0.8 ppm; the sediments contained 1190 ppm uranium and 310 ppm molybdenum. The pH of the water averaged 7.7 - 7.8 (Brierley and Brierley, 1980). The pond sediments represented a reducing environment (Eh, -350 mV) which contained a large population of sulfate-reducing bacteria of the genera *Desulfovibrio* and *Desulfotomaculum* (Brierley and Brierley, 1980). In the summer months the pond perimeters supported populations of *Typha*, *Juncus*, and sedges, which died back in winter months. The predominant algae in the ponds were the macrophytic green *Chara*, the filamentous green algae, *Spirogyra*, and the filamentous cyanobacteria, *Oscillatoria*. *Rhizoclonium*, another filamentous green alga, was found during the summer months at the inflow of the first algae pond of the series. Unidentified diatoms and unicellular green algae were associated with the floating mats of filamentous algae and with the benthic *Chara*. These associated algae were most abundant in the summer months.

Laboratory studies (Brierley and Brierley, 1981) indicated that the pond algae, in the form of particulate, decaying material, could be instrumental in removing metals from solution. However, the patterns of retention and release of uranium and molybdenum, as the algae decayed in the presence of sediment, indicated that maintenance of reducing conditions in the sediment or in the algal cultures was critical to the sequestering of the metals. In the model systems, the period of algal decomposition was associated with increased bacterial activity and enhanced reducing conditions in the sediment-containing cultures. When sulfate concentration was elevated to that of the pond water, anaerobic decomposition of algae was accompanied by sulfate reduction. Decomposition was associated in the model systems with removal of molybdenum and uranium from solution, but the period following decomposition was characterized by the partial to complete release of the metals to solution. The release of molybdenum was unexpected, since sulfides, which are formed during sulfate reduction, can form highly insoluble molybdenum disulfide. It was possible that molybdenum was precipitated in the reduced Mo (IV) state, which would not form the sulfide mineral. Solubilization may have occurred as molybdenum was oxidized.

In most cases, the addition of algae to sediment increased the sediment's ability to retain uranium. Reduction of uranium to the insoluble U (IV) ion or adsorption of the ion onto organic material (Leventhal, 1980) may have contributed to its disappearance from solution. Taylor (1979) has observed that production of hydrogen sulfide by sulfate-reducing bacteria in sediments may increase the adsorption of uranyl ion onto organic material present in the sediment, since iron and other sulfide-forming metals could be desorbed from the organic material to increase available adsorption sites.

The implication from the laboratory studies is, that while the algae are instrumental in removing metals from solution, the process is reversible unless the system contains substantial organic material. While organic material accelerates the rate of metal removal and sometimes the extent of removal from water, retention of the metals in sediments is also reversible. These conclusions point to major problems in improving uranium and molybdenum removal in the existing pond system.

In the algae ponds of this study, there was no significant reduction of uranium or molybdenum concentrations in solution in any of the algae ponds or between the influent to the algae ponds and the final effluent. Although the concentrations of both metals in the sediments were considerably greater than the concentrations in the water, the amounts of metals removed were not substantial enough to be reflected in the water concentrations. It is important, when considering metal removed by algae, to assess the water volume which passes throught the system yearly. At an inflow of 15 x 10^6 liters per day, the pond system received 5.5 x 10^9 liters every year. Based on the uranium and molybdenum values (0.8 ppm for each metal), this volume amounted to 4.9 x 10^9 mg (4.95 metric tons) of each metal entering the algae ponds yearly. The filamentous algae present in the ponds accumulated an average of 2000 ppm uranium and 130 ppm molybdenum on a dry weight basis. Given this level of removal, it would require (4.95 x 10^9 mg U x (10^3 g algae/2000 mg U)) 2.5 x 10^9 g dry weight algae produced per year to remove the uranium entering the algae ponds. If the ponds were fertilized to increase productivity, could this amount of algae be grown in the existing pond system? Wetzel (1966) found that a hypereutrophic lake produced 570 g C m^{-2} yr^{-1}. Using 40% (dry weight) for carbon content in algae (Ferguson and Bubela, 1974), 570 g C corresponds to an algal dry weight of 1425 g m^{-2} yr^{-1}. To grow 2.5 x 10^9 g algae yr^{-1}, the amount needed to accumulate the influent uranium, the pond system would have to be expanded to (2.5 x 10^9 g x (m^2/1425 g)) 1.75 x 10^6 m^2, or 17.5 hectares. The ponds are presently 0.15 hectares in area.

The existing algal treatment ponds are relatively unproductive, which is typical of the early to middle stages of succession in *Chara* ponds where phosphate levels are low (Crawford, 1979). Later stages of *Chara* succession are characterized by overgrowth of planktonic algae and higher plants, which increase pond productivity. Additions of phosphate and nitrogen-containing fertilizers would accelerate the replacement of *Chara* with more productive, annual green planktonic algae and cyanobacteria. Promoting eutrophication of the pond system would increase the amount of organic substrate available for sulfate-reducing bacteria and could promote reducing conditions in the pond sediments. Laboratory experimentation has indicated that enhancement of growth of Desulfovibrio and Desulfotomaculum, and thereby of sulfate reduction by these bacteria in sediments, resulted in removal of uranium, molybdenum and selenium from solution (Brierley and Brierley, 1980). While the algae alone cannot function to remove excess trace contaminants, the effect of increased algal productivity on water and sediment chemistry may be significant.

Large populations of bacteria were found in the waters pumped from the uranium mines and water passing through the pond systems. The suspended microflora had an abundance of species of the genus Flavobacterium (1.0×10^4 ml^{-1}), which are widely distributed in soil and freshwater. Sulfate-reducing bacteria, tentatively identified as Desulfovibrio and/or Desulfotomaculum, were present in water from the mines at levels of 2.7 to 3.8×10^2 ml^{-1}. Very high numbers of these microbes were found in the pond sediments (2.5×10^6 g^{-1} dry weight).

A laboratory model, effluent-treatment plant, consisting of a glass distillation column (42 cm x 2.3 cm) filled with glass beads (4 mm diameter) on which a mixed bacterial population ($>10^6$ bacteria g^{-1} beads) including sulfate-reducing bacteria resided, was tested for its ability to remove uranium and selenium from solution. Solutions, pumped at a rate of 0.08 to 7.0 ml min^{-1} in an up-flow pattern, contained a carbon source (0.1% glucose or 0.35% lactate), salts, and uranyl sulfate or sodium selenate. Aerobic conditions were attained by sparging air through a capillary tube into the column. Under anaerobic conditions and at the lowest flow rate tested (0.45 ml min^{-1}), 81% of the uranium and 93% of the selenium was removed from the effluent (Table 1). Under aerobic conditions approximately 90% of the selenium could be removed from solution at a flow rate of 0.9 ml min^{-1} (Table 2). Assuming that a column scaled-up to treat 3000 gallons of water per minute would be as effective as the laboratory-scale model, the selenium content of the effluent waters from the uranium mine waters studied could be reduced to 1.25 ppb. An effective field unit could resemble a countercurrent sand bed, similar to that used in drinking water treatment systems. However, studies to date have only assessed technological feasibility and not cost effectiveness. Algae have also been effectively used to treat effluents from lead mining and milling in the "New Lead Belt" of Missouri, USA (Gale and Wixson, 1979). Lead, zinc, copper, manganese and other heavy metals were removed from the contaminated effluents. The algae identified as participants in the treatment process included Cladophora, Rhizoclonium, Hydrodictyon, Spirogyra, and occasionally Oscillatoria and the macrophyte Potamogeton. The mechanism of removal involved physical trapping of particulates and subsequent chelation of metal cations by cell walls. Mixed algae cultures had a cation exchange capacity greater than 640 μ-quivalents g^{-1} dry weight. An artificial stream meander system facilitated algae development and metal removal.

TABLE 1. Biological removal of uranium and selenium from solution using an anaerobic column.

Flow Rate (ml/min)	U (ppm) Influent	U (ppm) Effluent	% U removed
0.45	0	0.5	none
0.45	21	4.0	81
0.90	17	10.5	38
1.80	20.5	25.5	none
	Se (ppb) Influent	Se (ppb) Effluent	% U removed
0.90	27.5	2.0	93
1.80	28.0	3.0	89

TABLE 2. Aerobic selenium removal by biological reactor columns.

Flow Rate (ml/min)	Sample	Se (ppb) Aerobic	% U removed
0.9	Influent	33.5	90
	Effluent	3.5	
1.8	Influent	26.0	63
	Effluent	9.5	

Biological removal of metals from mining effluents can be a complex system in which algae and sulfate-reducing bacteria participate (Jackson, 1978). Heavy metal cations (e.g., Cd, Cu, Zn) pollute surface waters as a result of mining and smelting in the vicinity of Flin Flon, Manitoba, Canada. Metal concentrations in streams were lowered by microbial processes with subsequent accumulation of the metals in the stream and lake sediments. Blooms of green algae, promoted by nutrients in sewage effluent, were found to serve as a site of biosorption of the metals. Subsequent death, settling, and decomposition of the algae enhanced microbial generation of hydrogen sulfide. The greatest metal accumulation occurred in sediments with the highest organic matter and hydrogen sulfide concentrations. Sulfide was more effective than organic matter in suppressing remobilization of metals. However, some unidentified organic-metal-complexes may be stable in the presence of free sulfide and result in release of metals from the sulfide-rich muds. The partitioning of a metal as a sulfide or organic complex was found to be related to the stability of the metal sulfide, with stability as follows: Hg > Cd > Cu > Fe > Zn. Jackson's (1978) study indicated that contaminated water could be treated by a series of settling ponds in which algae and sulfate-reducing bacterial activities were promoted.

Recovery of metals from mine drainage water may also be accomplished by using *Zoogloea ramigera* flocs as a biosorption process (Norberg, 1981). These flocs rapidly establish equilibrium with heavy metals in solution by a process of adsorption. It may be possible to use the flocs in a fluidized bed for continuous metal recovery. The metals may be recovered by elution with acid from saturated flocs. However, no system, based on this process, has yet been developed.

SILVER RECOVERY

Biological processes have been proposed for recovering silver from industrial effluents such as photographic waste or sulfidic minerals containing silver. A stable consortia of bacteria, consisting of *Pseudomonas maltophila*, *Staphylococcus aureus*, and an unidentified coryneform, reportedly accumulate silver (Charley and Bull, 1979). This community accumulated up to 30% of its dry weight in silver ion at a rate of 21 mg Ag^+ h^{-1} (g biomass)$^{-1}$ in a solution concentration of 12.15 mM Ag^+.

Bacterial oxidation of sulfidic minerals can result in bioaccumulation of silver with subsequent recovery of the silver by a chemical process (Pooley, 1982). When a mixed culture of acidophilic *T. ferrooxidans* and *T. thiooxidans* leached mixed metal sulfide minerals containing silver, the leaching organisms accumulated acanthite (Ag_2S) on their surfaces. Up to 25% of the cell dry weight was found to be accumulated silver, which could subsequently be released by cyanidation. Although *T. ferrooxidans*, when growing on ferrous iron, can accumulate 0.5% of its weight in ionic silver, this concentration was found to be toxic (Norris and Kelly, 1978). However, the accumulation of the silver sulfide does not pose a toxicity problem.

SUMMARY

Biotechnological processes for the sequestration of metals is an alternative to physico-chemical methods for waste water treatment and metals concentration; however, the potential for use of biological systems is yet to be fully realized. Microbial treatment processes operate by two distinct mechanisms: sorption of metals on non-growing or dead biomass and accumulation of metals by living cells. The first is a physico-chemical process which uses cells for chelation, adsorption, or possibly precipitation of metals. The advantages of this system include precise control of the metals-removal process in reactors or as biofilms and the use of cheap biomass produced from other industrial activities. The second system, which utilizes the activities of living cells, can be used to absorb or adsorb metals. Blooms of algae which accumulate metals can be promoted by nutrient enrichment of impoundments, and the algal mass, once decayed, can enhance the activity of sulfate-reducing bacteria in reducing environments. The biogenically-generated sulfide can contribute to the removal of metals from solution by formation of insoluble metallic sulfide compounds. The use of living cells may be a viable treatment process when large volumes of contaminated waters must be treated in holding ponds.

REFERENCES

Brierley, C.L.: 1978, Crit. Revs. Microbiol. 6, pp. 207-262.
Brierley, C.L.: 1980, Dev. Ind. Microbiol. 21, pp. 435-444.
Brierley, C.L., and Brierley, J.A.: 1980, *Biogeochemistry of Ancient and Modern environments* (Trudinger, P.A., Walter, M.R., and Ralph, B.J., eds.), Griffin Press Ltd., pp. 661-667.
Brierley, C.L., and Brierley, J.A.: 1981, *Contamination of Ground and Surface Waters by Uranium Mining and Milling. Vol. 1 Biological Processes for Concentrating Trace Elements from Uranium Mine Waters,* Final Report J0295033, U.S. Bureau of Mines, Washington, D.C., pp. 1-102.
Brierley, J.A., Brierley, C.L., and Dreher, K.T.: 1980, *First International Conference on Uranium Mine Waste Disposal* (Brawner, C.O., ed.), Society of Mining Engineers of AIME, pp. 365-376.
Charley, R.C., and Bull, A.T.: 1979, Arch. Microbiol. 123, pp. 239-244.
Crawford, S.A.: 1979, Hydrob. 62, pp. 17-31.
Eisenstadt, E.S., Fisher, S., Der, C.L., and Silver, S.: 1973, J. Bacteriol. 113, pp. 1363-1372.
Ferguson, J., and Bubela, B.: 1974, Chem. Geol. 13, pp. 163-186.
Filip, D.S., Peters, T., Adams, V.D., and Middlebrooks, E.J.: 1979, Water Res. 13, pp. 305-313.
Gadd, G.M., and Griffiths, A.J.: 1978, Micro. Ecol. 4, pp. 303-317.
Gale, N.L., and Wixson, B.G.: 1979, Dev. Ind. Microbiol. 20, pp. 259-273.
Guthrie, R.K., and Cherry, D.S.: 1979, Water Res. Bull. 15, pp. 244-248.
Gutknecht, J.: 1965, Limnol. Oceanog. 10, pp. 58-66.
Hatch, R.T., and Menawat, A.: 1979, Biotechnol. Bioeng. Symp. No. 8, pp. 191-203.
Jackson, T.A.: 1978, Environ. Geol. 2, pp. 173-189.
Jennett, J.C., Hassett, J.M. and Smith, J.E.: 1979, *International Conference on Management and Control of Heavy Metals in the Environment,* CEP Consultant, Ltd., pp. 210-217.
Kelly, D.P., Norris, P.R., and Brierley, C.L.: 1979, *Microbial Technology: Current State, Future Prospects* (Bull, A.T., Ellwood, D. C., and Ratledge, C., eds.), Cambridge Univ. Press, pp. 263-308.
Laube, V., Ramamoorthy, S., and Kushner, D.J.: 1979, Bull. Environ. Contam. Toxicol. 21, pp. 763-770.
Norberg, A.: 1981, Abstracts 28th Cong. IUPAC, p. B123.
Norris, P.R., Brierley, J.A., and Kelly, D.P.: 1980, FEMS Microbiol. Letts. 7, pp. 119-122.
Norris, P.R., and Kelly, D.P.: 1978, *Metallurgical Applications of Bacterial Leaching and Related Microbiological Phenomena* (Murr, L.E., Torma, A.E., and Brierley, J.A., eds.), Academic Press, pp. 83-102.
Norris, P.R., and Kelly, D.P.: 1979, Dev. Ind. Microbiol. 20, pp. 299-308.
Pollack, J.R., Ames, B.N., and Neilands, J.B.: 1970, J. Bacteriol. 104, pp. 635-639.
Pooley, F.D.: 1982, Nature 296, pp. 642-643.

Shumate II, S.E., Strandberg, G.W., and Parrot, J.R.: 1978, Biotechnol. Bioeng. Symp. No. 8, pp. 13-20.
Shumate II, S.E., Strandberg, G.W., McWhirter, D.A., Parrot, J.R., Bogacki, G.M., and Locke, B.R.: 1980, Biotechnol. Bioeng. Symp. No. 10, pp. 27-34.
Sterritt, R.M., and Lester, J.N.: 1979, Minerals Environ. 1, pp. 45-47.
Strandberg, G.W., Shumate II, S.E., Parrot, J.R., and North, S.E.: 1981, *Environmental Speciation and Monitoring Needs for Trace Metal-Containing Substances from Energy-Related Processes,* NBS Special Publication No. 618, U.S. Dept. of Commerce, Washington, D.C., pp. 274-285.
Tsezos, M., and Volesky, B.: 1981, Biotechnol. Bioeng. 23, pp. 583-604.
Tsezos, M., and Volesky, B.: 1982, Biotechnol. Bioeng. 24, pp. 385-401.
Wetzel, R.G.: 1966, Invest. Ind. Lakes Streams 7, pp. 147-184.
Woolfolk, C.A., and Whiteley, H.R.: 1962, J. Bacteriol. 84, pp. 647-658.

CLOSING REMARKS

C.J.F. Böttcher
Lange Voorhout 32
2514 EE The Hague, The Netherlands

I shall not try to summarize the wealth of information presented at this conference. This task will be performed in due course by some of the participants, who intend to prepare summaries.

Instead of giving a summary, I would like to mention a few observations and conclusions. Some of my observations are of a general nature. I hope they will express the feelings of all the participants. But I shall also mention some more personal observations, which are the result of the state of over-saturation I have reached by taking up such a huge amount of new knowledge. As you know, an over-saturated mixture has a tendency to form crystals or amorphous compounds, and the same process led to the following remarks.

In the announcement of this Fourth Symposium on Biomineralization, Dr. de Jong and Dr. Westbroek made several points clear:

1. that the term biomineralization is to be understood as the accumulation of metal compounds (especially salts) by living systems;
2. that in the three preceeding symposia the main emphasis was placed on the biosynthesis of calcium carbonate skeletons in invertebrates and plants; and
3. that during the third symposium, held in Japan in 1977, there was a general feeling that a new approach was needed, i.e., that biomineralization should be placed in a much wider context. In this new approach the global geochemical significance of biomineralization would be discussed; more attention should be given to the accumulation of other metals besides calcium, to deal with bio-inorganic chemistry in general; and some attention should be paid to what is at present called biotechnology, for instance, applications of biomineralization to mining and to the removal of heavy metals from waste.

Thus, the organizers of the conference defined their objectives clearly, which gives us the right to ask whether they achieved what they had in mind. I think you will all agree that the answer to this question is a wholehearted yes.

P. Westbroek and E. W. de Jong (eds.), Biomineralization and Biological Metal Accumulation, 511–514.

Another general observation has to do with the fact that this symposium has been multidisciplinary to an unusual degree. Biologists of all types, ranging from microbiologists to botanists and zoologists, took part in it, as did geologists of all varieties. Even all of the five kingdoms of chemistry (inorganic, organic, physical, analytical, and biochemistry) were represented. Remarkably enough, this multidisciplinary participation did not lead to confusion and actually raised the level of the discussion. It may therefore be said without exaggeration that this has been the sort of interdisciplinary scientific gathering strongly recommended at international ministerial conferences on science policy. Unfortunately, ministers usually do not realize how difficult it is to obtain funds for an interdisciplinary meeting that does not fit into any of the conventional categories of well-established organizations.

The lectures and discussions during this symposium have shown how geology and biology have come to an extensive interaction. Not long ago, paleontology was their main field of common interest. But modern geologists have realized that the interaction between life and its inorganic substrate - the earthcrust, water, atmosphere - has been and still is much stronger than they believed only a few decades ago.

Since the beginning of life on earth the development of both inanimate nature and living organisms has involved an extremely complicated process of interaction, which has been called co-evolution by the biologist René Dubos of Rockefeller University. For billions of years this interaction has been so strong that the biosphere, atmosphere, and earthcrust together may be considered one system in the cybernetic sense of the word, a system governed by numerous - mainly negative - feedback loops. An obvious example is provided by the influence of living organisms on the chemical composition of the atmosphere and the oceans, on the weathering of rock, on soil formation and erosion, on the water cycle, the oxygen cycle, the carbon cycle, the nitrogen cycle, and most of the mineral cycles. The resulting system could be called the geobiochemical system. But I prefer to call it the Gaia system, because a short name is always attractive. And even the opponents of the Gaia hypothesis cannot deny the reality of this geobiochemical Gaia system.

You may have noticed that I have gradually entered the domain of observations of a more personal nature. Now that I am there, let me make another personal remark, which is that if there had been more time available, one more speaker could have been added for this conference. I mean an expert in the field of biological evolution, who could have dealt with the degree to which the well-known mechanisms of evolution (natural selection and mutations, followed by much trial and error and in very rare cases trial and success) stabilize the Gaia system. His - or her - lecture would have given attention to the role of eco-systems. These sub-systems of the world-wide Gaia system contribute to its stabilization. I believe we need an analysis of this kind to find out whether the Gaia hypothesis is needed to explain facts that cannot be understood by the well-known evolutionary mechanisms.

My favourite example of an interaction developed by such mechanisms is provided by the trees of the world, which in mutual competition and cooperation, play a large role in the stabilizing influence of the water cycle on the world climate. Natural selection and mutations have led to an enormous variety among tree species, adapted to different patterns of water management. Some trees pump huge amounts of water out of the soil. But in almost rainless (but humid) regions, pine trees with hanging needles have developed. These needles collect dew, which drops to the soil, and the new result is that these trees add more water to the soil than they pump out of it. Another example is the ability of some tree species to flourish in a hot climate, whereas others can endure extremely low temperature. All these adaptations contribute to the over-all influence of trees on the water cycle of our planet.

I hope all of you will agree with me that it has been useful that during this symposium, biomineralization has now and then been placed in the context of global problems. As a physical chemist with a special interest in not only chemical thermodynamics but also the biosciences, I believe biomineralization to be an extremely important subject for anyone who wants to see the frontiers of science shifted with respect to our knowledge about the origin and early development of life. Crystals have in common with living organisms and some abiotic entities - for instance eddies and tornados - that they are examples of what is at present called the self-organization of matter. Thermodynamically, this process of organization means a decrease of entropy locally at the cost of a greater entropy increase in the environment. For instance, in the formation of crystals the environment has to take up heat from the open or closed system in which crystallization takes place.

The study of self-organizing systems had become so fashionable that a special name has been proposed for self-organization: autopoiesis. Expressed in terms of autopoiesis, one of the important aspects of biomineralization is, in my opinion, that it offers some kind of meeting-place for two at first sight totally different types of self-organization: the formation of inorganic crystals and self-organization in living organisms. Fine examples of this sort of interaction were discussed during the present symposium.

I believe that the study of biomineralization can contribute much to the knowledge of the principles and mechanisms of self-organization. And with this statement I have come to the last part of my remarks, which concerns perspectives. Dr. Westbroek mentioned in his introductory speech that the subject of biomineralization is not of purely academic interest. It has important societal aspects, not only because of special applications such as bacterial leaching of ores, but particularly because a better understanding of biomineralization could lead, for instance, to a better control of the unwanted side-effects of the exploitation of heavy-metal ores. In the future, micro-organisms could be used for the recycling of heavy metals in what we humans call waste. Biotechnology and bio-engineering are becoming fashionable subjects. There is every reason to include biomineralization among these fields of applied science. We might start in The Netherlands with the production of heavy metals from Rhine water and from poisonous river mud.

During this meeting the planet on which we live has been dealt with occasionally, with special attention to changes in the earth-crust and the biosphere. This has put us in the position of subject with Mother Earth as our object. It is the anthropocentric position. But since there are environmentalists who worry about what they call the stress on the biosphere caused by man's activities, I thought it worthwhile to ask Mother Earth what she herself thinks about such statements. Her reply was simple. She said: "Do you really believe that I feel stressed? For billions of years I have been accustomed to the tectonic cycle, to earthquakes and volcanic eruptions, to the drift of the continents, to the coming and going of ice ages. So I am both flexible and stable enough to deal with man's tiny activities. I hardly notice them. It is their stress, not mine, that worries them".

This brings me to a final remark. If the Gaia hypothesis is valid - which remains to be proven - this would be good news for Mother Earth, but not necessarity for us. It would not be pleasant for us, if, for instance, due to our activities, conditions were improved for the lower organisms, say bacteria and insects, but worsened for our species. Mother Earth probably has no special sympathy for human beings, and neither do our fellow creatures.

Thus, every addition we make to our knowledge concerning the environment in which we live can help in our struggle for survival. This means that we badly need a Fifth Symposium on Biomineralization.

LIST OF PARTICIPANTS AND CONTRIBUTORS

Arends, J.
Lab. Materia Technica
Antonius Beusinglaan 1
9713 AV Groningen
The Netherlands

Beekman, D.W.
Marine Sci. Program
Univ. of South Carolina
Columbia, SC 29208
U.S.A.

Berner, R.A.
Dept. of Geology and Geophysics
Yale University
New Haven, Connecticut
U.S.A.

Bijvoet, O.L.M.
Dept. of Clinical Endro-crinology, Building 19
University Hospital
2333 AA Leiden
The Netherlands

Blomen, L.J.M.J.
Dept. of Clinical Endro-crinology, Building 19
University Hospital
2333 AA Leiden
The Netherlands

Boddé, H.E.
Lab. Materia Technica
Antonius Beusinglaan 1
9713 AV Groningen
The Netherlands

Bontrop, R.
Dept. of Biochemistry
Univ. of Leiden
Wassenaarseweg 64
2333 AL Leiden
The Netherlands

Boon, J.
Orgast, TH Delft
de Vries van Heijstpl. 2
2628 RZ Delft
The Netherlands

Borman, A.H.
Dept. of Biochemistry
Univ. of Leiden
Wassenaarseweg 64
2333 AL Leiden
The Netherlands

Bosch, L.
Dept. of Biochemistry
Univ. of Leiden
Wassenaarseweg 64
2333 AL Leiden
The Netherlands

Böttcher, C.J.F.
Professor Emeritus of Physical Chemistry
Univ. of Leiden, Leiden
The Netherlands

Brierley, C.L.
Advanced Mineral Tech-nologies Inc.
Socorro, New Mexico
U.S.A.

Brierley, J.
Dept. of Biology
New Mexico Institute of
Mining and Technology
Socorro, NM 87801
U.S.A.

Brouwer, A.
Dept. of Geology
Univ. of Leiden
Garenmarkt 1
Leiden
The Netherlands

Burdige, D.
Scripps Institution
of Oceanography
Univ. of California
La Jolla, CA 92093
U.S.A.

Campbell, S.
Boston University
2, Cummington Street
Boston, MA 02212
U.S.A.

Cowen, J.
Marine Studies
Applied Science Bldg.
Univ. of California
Santa Cruz, CA 95064
U.S.A.

Dams, M.
Dept. of Biochemistry
Univ. of Leiden
Wassenaarseweg 64
2333 AL Leiden
The Netherlands

Dauphin, Y.
Lab. Petrologie sédimen-
taire et Palaeontologique
Université d'Orsay
91405 Orsay cedex
France

Dean, J.M.
Marine Sci. Program
Univ. of South Carolina
Columbia, SC 29208
U.S.A.

Doderer, A.
Dept. of Zoology
Free University
Amsterdam
The Netherlands

Drolshagen, H.
Inst. für Anatomie und
Physiol. der Haustiere
Universität Bonn
Katzenburgweg 7
5300 Bonn 1
West Germany

Dryer-Grossovski, B.
Biological Sci. Center
Boston University
2, Cummington Street
Boston, MA 02215
U.S.A.

Erez, J.
Marine Biological Lab.
Hebrew Univ. Elat
P.O. Box 469
Elat
Israel

Garrels, R.M.
Dept. Marine Science
Univ. of South Florida
830 First Str. South
St. Petersburg
Florida 33701,
U.S.A.

Golubic, S.
Boston University
Biological Sci. Center
2, Cummington Street
Boston, MA 02215
U.S.A.

Greaves, G.N.
SERC
Daresbury Laboratory
Warrington WA4 4AD
England

Haake, P.W.
Marine Sci. Program
Univ. of South Carolina
Columbia, SC 29208
U.S.A.

Häusle, J.
Inst. für Anatomie und Physiol. der Haustiere
Universität Bonn
Katzenburgweg 7
5300 Bonn 1
West Germany

Heijnen, O.
Inst. of Earth Sci.
Dept. Crystallography
Budapestlaan 4
Utrecht
The Netherlands

Hemleben, C.
Geologisches Institut
Sigwartstrasse 10
7400 Tübingen
West Germany

Hof-Irmsher, K.
Inst. für Anatomie und Physiol. der Haustiere
Universität Bonn
Katzenburgweg 7
5300 Bonn 1
West Germany

Huizinga, M.
Dept. of Biochemistry
Univ. of Leiden
Wassenaarseweg 64
2333 AL Leiden
The Netherlands

Isogai, F.
Inst. of School Education
Univ. of Tsukuba
Tsukuba
Japan

Jong, E.W. de
Dept. of Biochemistry
Univ. of Leiden
Wassenaarseweg 64
2333 AL Leiden
The Netherlands

Joosse, J.
Dept. of Zoology
Free University
Amsterdam
The Netherlands

Kägi, J.
Dept. of Biochemistry
Univ. of Zürich
Zürichbergstrasse 4
CH-8028 Zürich
Switzerland

Keller, J.P.
Lab. Petrologie sédimentaire et Palaeontologique
Université d'Orsay
91405 Orsay cedex
France

Kepkay, P.E.
Marine Biology Res. Div.
Scripps Inst. of Oceanography
Univ. of California
La Jolla, CA
U.S.A.

Kitano, Y.
Water Research Inst.
Nagoya University
Chikusa-ku
Nagoya 464
Japan

Kobayashi, I.
Geol. Mineral. Inst.
Niigata Univ.
Igarashi
Niigata 950-21
Japan

Kok, D.
Dept. of Biochemistry
Univ. of Leiden
Wassenaarseweg 64
2333 AL Leiden
The Netherlands

Kozawa, Y.
Dept. Oral Histology
Nihon Univ. at Matsudo
Matsudo, Chiba 271
Japan

Krampitz, G.
Inst. für Anatomie und Physiol. der Haustiere
Universität Bonn
Katzenburgweg 7
5300 Bonn 1
West Germany

Kretsinger, R.H.
Dept. of Biology
Univ. of Virginia
Gilmer Hall
Charlottesville, VI 22901
U.S.A.

Kuenen, G.
Dept. of Microbiology
TH Delft
Julianalaan 67a
2628 BC Delft
The Netherlands

Leeuw, J. de
Orgast
TH Delft
de Vries van Heijstpl. 2
2628 RZ Delft
The Netherlands

Lehninger, A.L.
Dept. of Physiological
Chemistry
John Hopkins Univ.
725 N. Wolfe Street
Baltimore, Maryl. 21205
U.S.A.

Lovelock, J.E.
Coombe Mill
St. Giles on the Heath
Launceston
Cornwall
England

Lowenstam, H.A.
Div. of Geology and
Planetary Sciences
Cal. Inst. Technology
Pasadena, CA 91109
U.S.A.

Macey, D.J.
School of Environmental
and Life Sciences
Murdoch University
Murdoch, W. Australia
Australia 6150

Mann, S.
Dept. of Inorganic Chem.
South Parks Road
Oxford OX1 3QR
England

Mano, K.
Inst. of School
Education
Univ. of Tsukuba
Tsukuba
Japan

Margulis, L.
Biological Sci. Center
Boston University
2, Cummington Street
Boston, MA 02215
U.S.A.

Mason, A.Z.
Dept. of Zoology
Univ. of Reading
Whiteknights
Reading RG6 2AJ
England

Medema, D.
Sittingbourne Res.
Centre
Shell Biosciences Lab.
Sittingbourne
Kent ME9 8AG
England

Monty, C.
Dept. of Geology
Univ. of Liège
Liège
Belgium

Mutvei, H.
Swedish Museum of
Natural History
Dept. of Paleozoology
Stockholm
Sweden

Nakahara, H.
Dept. of Anatomy
Josai Dental College
Sakado
Saitama 350-02
Japan

Nancollas, G.H.
Dept. of Chemistry
Univ. of New York
Acheson Hall
Buffalo, NY 14214
U.S.A.

Nealson, K.H.
*Scripps Institution of Oceanography
Univ. of California
La Jolla, CA 92093
U.S.A.*

Nip, M.
*Dept. of Biochemistry
Univ. of Leiden
Wassenaarseweg 64
2333 AL Leiden
The Netherlands*

Omori, M.
*Azabu Veterinary College
1-17-71 Fuchinobe
Sagamihara
Kanagawa 229
Japan*

Parker, S.B.
*Dept. of Inorganic Chem.
South Parks Road
Oxford OX1 3QR
England*

Perry, C.C.
*Dept. of Inorganic Chem.
South Parks Road
Oxford OX1 3QR
England*

Pol, W. van de
*Dept. of Biochemistry
Univ. of Leiden
Wassenaarseweg 64
2333 AL Leiden
The Netherlands*

Pouw Kraan, I. van der
*Dept. of Biochemistry
Univ. of Leiden
Wassenaarseweg 64
2333 AL Leiden
The Netherlands*

Ross, M.D.
*Dept. of Anatomy and Cell Biology
Univ. of Michigan
Ann Arbor, Michigan
U.S.A.*

Sabsay, B.
*Northwestern Univ.
Dental School
Chicago, Ill. 60611
U.S.A.*

Sawada, K.
*Dept. of Chemistry
Univ. of New York
Acheson Hall
Buffalo, NY 14214
U.S.A.*

Schors, R.C. van de
*Inst. of Earth Sci.
Free University
De Boelelaan 1085
2081 HV Amsterdam
The Netherlands*

Schram,
*Sittingbourne Res. Centre
Shell Biosciences Lab.
Sittingbourne
Kent ME9 8AG
England*

Schuttringer,
*Dept. of Chemistry
Univ. of New York
Acheson Hall
Buffalo, NY 14214
U.S.A.*

Sikes, C.S.
*Univ. of South Carolina
Mobile, Alabama 36688
U.S.A.*

Silver, S.
*Dept. of Biology
Washington Univ.
Campus Box 1137
St. Louis, MO 63130
U.S.A.*

Simkiss, K.
*Dept. of Zoology
Univ. of Reading
Whiteknights
Reading RG6 2AJ
England*

Skarnulis, A.J.
Dept. of Chemical Crystallography
Univ. of Oxford
Oxford OX1 3QR
England

Spuy, R.
Dept. of Biochemistry
Univ. of Leiden
Wassenaarseweg 64
2333 AL Leiden
The Netherlands

Stoltz, J.
Dept. of Biology
Boston University
2, Cummington Street
Boston, MA 02215
U.S.A.

Storms, J.J.H.
BION
Koningin Sophiestr. 124
2595 TM Den Haag
The Netherlands

Tanke-Visser, J.
Dept. of Biochemistry
Univ. of Leiden
Wassenaarseweg 64
2333 AL Leiden
The Netherlands

Taylor, M.
Dept. of Zoology
Univ. of Reading
Whiteknights
Reading RG6 2AJ
England

Traub, W.
Structural Chem. Dept.
Weizmann Inst. of Sci.
Rehovot
Israel

Veis, A.
Dept. of Biochemistry
Northwestern Univ.
Medical School
303 E. Chicago Ave
Chicago, Ill. 60611
U.S.A.

Volcani, B.E.
Scripps Institution of Oceanography
Univ. of California
La Jolla, CA 92093
U.S.A.

Vos, P.
Inst. of Earth Sci.
Budapestlaan 4
Utrecht
The Netherlands

Vrind, H. de
Dept. of Biochemistry
Univ. of Leiden
Wassenaarseweg 64
2333 AL Leiden
The Netherlands

Wal, P. van der
Dept. of Biochemistry
Univ. of Leiden
Wassenaarseweg 64
2333 AL Leiden
The Netherlands

Webb, J.N.
School of Mathematical and Physical Sciences
Murdoch University
Murdoch 6150
Western Australia

Watson, A.
Marine Biological Ass.
The Laboratory
Citadell Hill
Plymouth PL1 2PB
England

Weiner, S.
Dept. of Isotopes
Weizmann Inst. of Sci.
Rehovot
Israel

Westbroek, P.
Dept. of Biochemistry
Univ. of Leiden
Wassenaarseweg 64
2333 AL Leiden
The Netherlands

Williams, R.J.P.
Dept. of Inorganic Chem.
South Parks Road
Oxford OX1 3QR
England

Wheeler, A.P.
Clemson University
Clemson, SC 29631
U.S.A.

Whitfield, M.
Marine Biological Ass.
The Laboratory
Citadell Hill
Plymouth PL1 2PB
England

Wilson, C.A.
Marine Sci. Program
Univ. of South Carolina
Columbia, SC 29208
U.S.A.

Wilt, G.J. van der
Dept. of Zoology
Free University
de Boelelaan 1085
1081 HV Amsterdam
The Netherlands

With, N.D. de
Dept. of Zoology
de Boelelaan 1085
1081 HV Amsterdam
The Netherlands

SUBJECT INDEX